Bacterial Resistance to Antimicrobials

Bacterial Resistance to Antimicrobials

edited by

Kim Lewis
Northeastern University
Boston, Massachusetts

Abigail A. Salyers
University of Illinois at Urbana-Champaign
Urbana, Illinois

Harry W. Taber
New York State Department of Health and
State University of New York at Albany
Albany, New York

Richard G. Wax
Pfizer Global Research
Groton, Connecticut

MARCEL DEKKER, INC. NEW YORK • BASEL

ISBN: 0-8247-0635-8

This book is printed on acid-free paper.

Headquarters
Marcel Dekker, Inc.
270 Madison Avenue, New York, NY 10016
tel: 212-696-9000; fax: 212-685-4540

Eastern Hemisphere Distribution
Marcel Dekker AG
Hutgasse 4, Postfach 812, CH-4001 Basel, Switzerland
tel: 41-61-261-8482; fax: 41-61-261-8896

World Wide Web
http://www.dekker.com

The publisher offers discounts on this book when ordered in bulk quantities. For more information, write to Special Sales/Professional Marketing at the headquarters address above.

Current printing (last digit):
10 9 8 7 6 5 4 3 2 1

PRINTED IN THE UNITED STATES OF AMERICA

Foreword

In the preface to this volume, the editors point out that antibiotic resistance has substantial international aspects, chiefly because resistance so often results from everyday clinical use. Historically, however, there is another area related to resistance in which governments—individually or jointly—have had a mutual interest, and in which the need for effective policies is likely to increase. This is the planned mass attack on specific diseases, organized in campaigns to eradicate them.

As the word implies, eradication means tearing out by the roots, as was done successfully with smallpox and is in prospect for poliomyelitis. Both attacks, to be sure, involved an immunological, rather than therapeutic approach. Yet, antibiotic treatment was the basis for a mid-twentieth-century attack on yaws, a nonvenereal spirochetal disease spread through interpersonal contact. A full discussion of eradication is beyond the scope of this foreword, but the lessons learned have direct implications for confronting antimicrobial resistance.

In 1954, the governing body of the Pan American Sanitary Bureau, now known as the Pan American Health Organization (PAHO), Regional Office of the World Health Organization (WHO), adopted a resolution calling for the eradication of four diseases: smallpox, urban yellow fever, malaria, and yaws. At the time, the organization's director was that great apostle of disease eradication, Fred L. Soper, who, as a Rockefeller Foundation staff member, had led the successful campaign to eradicate *Anopheles gambiae*, a particularly dangerous malaria vector, from Brazil. Soper was personally persuaded that all four target diseases could be eradicated, with sufficient determination and resources.

Events turned out otherwise, although the attack on smallpox was a great success worldwide, with the last naturally occurring case in 1977. For the other three, however, the outcome has not been so salutary. For urban fever and malaria, resistance to insecticides soon became an insurmountable obstacle to the elimination of the vector, in many ways paralleling the experience with antibiotics.

Yaws was different, in that the primary problem was more political than technical. The instrument was to be mass treatment of the population with penicillin, known to be highly effective against the disease. At the time, it was thought that the development of strains of the spirochete resistant to penicillin was either unlikely or an acceptable risk. Considerable progress was made, but not without incident.

One such political situation involved the Dominican Republic, which like its neighbor, Haiti, had a significant prevalence of yaws. At the time, the Dominican Republic was under the dictatorship of Rafael Leonidas Trujillo, who signed the penicillin-treatment decrees with some pomp and circumstance. Nevertheless, and without consulting health professionals, at the end of the first year he announced that the campaign had been successful and terminated it! The result, of course, was that yaws returned. Other difficulties subsequently developed with the effort, and there was never the will and drive that are prerequisite for the 100% performance essential for the success of an eradication effort.

In a similar episode, an active malaria-eradication program in Sri Lanka, before the development of resistance to insecticides, had made excellent progress, with reported cases of malaria down to a few hundred. Against the advice of professionals, the program was cut drastically, and malaria soon returned many thousandfold to its original destructive level.

It might be supposed that problems associated with major infectious diseases relate largely to the Third World. Yet, due to globalization of travel, economics, and politics, these diseases can readily move from Third- to First-World countries and establish—or re-establish—themselves. As an example, the London East End borough of Newham currently has a tuberculosis prevalence greater than that of India; the disease enters the First-World country as latent TB in those fleeing conflict and poverty elsewhere, and is then amplified by continued poverty, homelessness, and poor nutrition.

In seeking to minimize the spread of resistance, whether to antimicrobials or to insecticides/pesticides, on an international scale, one must recognize that many individual nations are involved, not a single worldwide heath authority. Of special significance are the role, capability, and limitations of WHO.

WHO had its origins in an international conference held in New York in 1946 that created an interim commission leading to the establishment two years later of an independent agency, having its own budget and staff rules, within the United Nations family. WHO is governed by the World Health Assembly, in which each of the approximately 190 member nations has a vote. The WHO secretariat is charged by the countries that brought it into being, and supervise and finance it, to set international standards for materials and programs and then, as requested, to provide assistance and guidance—worldwide, regionally, and within individual countries—to help meet those standards.

But WHO had neither the capacity nor the legal authority to give orders or enforce rules. It does offer advice and various forms of technical cooperation, aimed at influencing medical and public health practice. Obviously, medical education is directly relevant to dealing with resistance and WHO has been active in medical education since its beginning. Physicians are taught never to use an agent which is known to be ineffective, and to use it with great caution when its effect on a particular pathogen has not been clearly demonstrated. How to develop standards and methods of instruction that would surmount differences in practice, culture, and educational systems is a problem of huge proportions that could be profitably explored by a WHO Expert Committee. In essence, WHO works by education and demonstration, not by regulation.

Tantalizing possibilities exist for the treatment and prevention of infectious disease as new therapeutic and preventive techniques are developed. Such proposals will need to be evaluated against a series of criteria such as those summarized by Aylward and colleagues,* as scientific and technological possibilities are assessed against humanitarian, financial, and political considerations. The Trujillo story is not atypical and anyone who has worked in public health, within a country or internationally, can

---

*Aylward B, Hennessey KA, Zagaria N, Olivé J-M, and Cochi S. When is a disease eradicable? 100 years of lessons learned. Am J Pub Health 2000; 90:1515–1520.

tell stories of actions or decisions based on similar budgetary/political factors. Combating antibiotic resistance will require the combined efforts of many sectors of our society. The above examples of failed efforts highlight the fact that the success of laboratory research, treatment, and policy requires collective international attention to all the links of this chain.

Myron E. Wegman, M.D., M.P.H.
Dean Emeritus, School of Public Health, University of Michigan
Professor Emeritus, University of Michigan Medical School
Former Secretary-General, Pan American Sanitary Bureau, World Health Organization

Preface

Antibiotics stand apart from other components of the physician's armamentarium in that their activities vary inversely with time. Whereas the beta-blocker that was active yesterday can be expected to lower blood pressure to the same degree tomorrow, there is a virtual certainty that the effectiveness of today's antibiotic will, with the passage of time, have been reduced or lost entirely. Microbes enlist a variety of means to defend themselves, but perhaps most significant is the fact that mechanisms evolved by one species can, through genetic exchange, be shared throughout the bacterial community. As a result, it is now clear that to successfully deal with infectious disease we must view ourselves in the way pathogens see us. As hosts for microorganisms we are not distinguishable by the continent or nation in which we live, but rather we serve as a global organism. An afflicted individual in some far-off land must be of as much concern to us as the person next door. Policies governing the proper use of antibiotics in our own nation will ultimately be only as effective as the systems in place elsewhere on the planet.

This book describes our present understanding of the nature of antimicrobial resistance and the methods by which bacteria gain these defenses. The work is structured so as to provide various perspectives, and thus discussion of a subject may be found both in chapters dealing with the bacterium, and in other chapters focusing on the mechanisms of resistance. The ultimate goal of this information is to provide guidelines for creating antibiotics effective against refractory strains, and to offer strategies to minimize the emergence and spread of resistance. Editors and authors were drawn from academia, public health, and industry, since contributions from all these sectors will be needed to provide the measures by which infectious disease can be kept under control. Each chapter begins with a summary of the concepts, so that those not actively working in the field can readily gain an overall picture of what follows.

Kim Lewis
Abigail A. Salyers
Harry W. Taber
Richard G. Wax

Contents

Contributors

Richard E. Besser, M.D. Medical Epidemiologist, Respiratory Diseases Branch, Division of Bacterial and Mycotic Diseases, National Center for Infectious Diseases, Centers for Disease Control and Prevention, Atlanta, Georgia

George T. Bonheyo, Ph.D. Postdoctoral Research Fellow, Department of Geology, University of Illinois at Urbana-Champaign, Urbana, Illinois

Stewart T. Cole, Ph.D. Professor, Unité Génétique Moléculaire Bactérienne, Institut Pasteur, Paris, France

Julian Davies, Ph.D. Emeritus Professor, Department of Microbiology and Immunology, University of British Columbia, Vancouver, British Columbia, Canada

Fernando de la Cruz, Ph.D. Professor of Genetics, Department of Molecular Biology, Medical School, University of Cantabria, Santander, Spain

Christopher Gerard Dowson, Ph.D. Professor, Department of Biological Sciences, University of Warwick, Coventry, England

George M. Eliopoulos, M.D. Professor, Department of Medicine, Harvard Medical School, and Beth Israel Deaconess Medical Center, Boston, Massachusetts

Scott K. Fridkin, M.D. Medical Epidemiologist, Divison of Healthcare Quality Promotion, Centers for Disease Control and Prevention, Atlanta, Georgia

Juan M. García-Lobo, Ph.D. Associate Professor of Microbiology, Department of Molecular Biology, Medical School, University of Cantabria, Santander, Spain

Nafsika H. Georgopapadakou, Ph.D. Department of Microbiology, Newbiotics, Inc., San Diego, California

Keeta S. Gilmore, Ph.D. Associate Professor, Department of Pathology, University of Oklahoma Health Sciences Center, Oklahoma City, Oklahoma

Michael S. Gilmore, Ph.D. Vice President for Research and GL Cross/MG McCool Professor, Departments of Ophthalmology, Microbiology, and Immunology, University of Oklahoma Health Sciences Center, Oklahoma City, Oklahoma

David C. Hooper, M.D. Associate Professor, Department of Medicine, Harvard Medical School, and Division of Infectious Diseases, Mass-General Hospital, Boston, Massachusetts

Lakshmi P. Kotra, Ph.D. Assistant Professor, Department of Chemistry, Faculty of Pharmacy, University of Toronto, Toronto, Ontario, Canada

Kim Lewis, Ph.D. Professor, Department of Biology, Northeastern University, Boston, Massachusetts

Olga Lomovskaya, Ph.D. Associate Director, Discovery Biology, Microcide Pharmaceuticals, Inc., Mountain View, California

Paul F. Miller, Ph.D. Assistant Director, Antibacterials Discovery, Pfizer Global Research and Development, Groton, Connecticut

Shahriar Mobashery, Ph.D. Professor, Department of Chemistry, Wayne State University, Detroit, Michigan

Julia Y. Morita, M.D. Medical Epidemiologist, Respiratory Diseases Branch, Division of Bacterial and Mycotic Diseases, National Center for Infectious Diseases, Centers for Disease Control and Prevention, Atlanta, Georgia

Rodger Novak, M.D. Assistant Professor, Institute of Microbiology and Genetics, Vienna Biocenter, Vienna, Austria

Alexander S. Pym, M.B.B.Chir. Research Fellow, School of Tropical Medicine, Liverpool University, Liverpool, England, and Unité de Génétique Moléculaire Bactérienne, Institut Pasteur, Paris, France

Philip Rather, Ph.D. Associate Professor, Department of Medicine, Case Western Reserve University, Cleveland, Ohio

Daniel F. Sahm, Ph.D. Chief Scientific Officer, Anti-Infective Services, Focus Technologies, Inc., Herndon, Virginia

Abigail A. Salyers, Ph.D. Professor, Department of Microbiology, University of Illinois at Urbana-Champaign, Urbana, Illinois

Jean-Pierre Samama, Ph.D. Research Director, Biological Crystallography Group, Institute of Pharmacology and Structural Biology, Centre National de la Recherche Scientifique, Toulouse, France

Nadja B. Shoemaker, M.S. Research Specialist, Department of Microbiology, University of Illinois at Urbana-Champaign, Urbana, Illinois

Harry W. Taber, Ph.D. Director, Division of Infectious Disease, Wadsworth Center, New York State Department of Health, and Professor, Department of Biomedical Sciences, School of Public Health, State University of New York at Albany, Albany, New York

Amalio Telenti, M.D., Ph.D. Chief, HIV Unit and Laboratory, Division of Infectious Diseases, and Institute of Microbiology, Centre Hospitalier Universitaire Vaudois, University of Lausanne, Lausanne, Switzerland

Fred C. Tenover, Ph.D. Associate Director for Laboratory Science, Division of Healthcare Quality Promotion, Centers for Disease Control and Prevention, Atlanta, Georgia

Krzysztof Trzciński, Ph.D. Research Fellow, Sera and Vaccines Evaluation, National Institute of Hygiene, Warsaw, Poland

Elaine I. Tuomanen, M.D. Chair and Full Member, Department of Infectious Diseases, and Director, Children's Infection Defense Center, St. Jude Children's Research Hospital, Memphis, Tennessee

Rosamund Williams, Ph.D., F.R.C.Path. Scientific Officer, Communicable Disease Surveillance and Response, World Health Organization, Geneva, Switzerland

Gerard D. Wright, Ph.D. Associate Professor, Department of Biochemistry, and Director, Antimicrobial Research Centre, McMaster University, Hamilton, Ontario, Canada

Philip J. Youngman, Ph.D.* Antibacterial Research, Millennium Pharmaceuticals, Cambridge, Massachusetts

---

Current affiliation: Elitra Pharmaceuticals, San Diego, California.

A Historical Introduction

Almost as soon as it was known that microorganisms could be killed by certain substances, it was recognized that some microbes could survive normally lethal doses and they were described as *drug-fast* (probably deriving from the fact that many of the early drugs were dyestuff derivatives and *fastness* is a textile dying term). These early studies (1–3) conceived of microbial resistance in terms of "adaptation" to the toxic agents. By 1907, Ehrlich (4) focused more clearly on the concept of resistant organisms in his discussion of the development of resistance of *Trypanosoma brucei* to *p*-roseaniline, and in 1911 Morgenroth and Kaufmann (5) reported that pneumococci could develop resistance to ethylhydrocupreine. For every new agent that killed or inhibited microorganisms, resistance became an interest as well.

Drug-fastness became a topic of importance as microbiologists sought to understand the growth, metabolism, and pathogenicity of bacteria, protozoa, and fungi. Often this research on antimicrobial agents was directed to problems of "disinfection" and related matters of public health, and the origins and properties of resistant organisms became of concern in the "fight against germs" (6). Protocols for inducing drug resistance in vivo were elaborated, and the relevance of in vitro resistance to "natural" in vivo resistance was debated in the literature of the 1930s and 1940s. One interesting aspect, now generally forgotten, was the widespread belief in bacterial life cycles as an explanation for the changing properties of bacterial cultures under what we would now call "selection." This theory of bacterial life cycles (7,8) called *cyclogeny*, held that bacteria had definite phases of growth, and that properties of bacteria such as shape, nutritional requirements, pathogenicity, antigenic reactivities, and chemical resistances were variable properties of the organism that simply reflected the growth phase of the culture. This cyclogenic variation revived an old 19th century controversy in bacteriology, namely that of Koch's monomorphism versus Cohn's polymorphism. Ferdinand Cohn believed that bacterial forms were highly variable, so that one "specie" of bacteria could exist in many shapes and with many different properties, whereas Robert Koch held that specific bacterial "species" had unique morphologies and properties that were unchanging. This debate, of course, had far-reaching implications both for problems of bacterial classification and for understanding variation and mutation of bacterial characteristics.

In the 1930s and 1940s there was great interest in the processes of microbial metabolism, from industrial as well as basic biological viewpoints. Margery Stephenson, Paul Fildes, and B. C. J. G. Knight in England developed a school of microbial biochemistry that emphasized the variability of microbial metabolism and its adaptability to specific culture conditions (9). From this research came a strong belief that bacteria could be "trained" or "adapted" to specific new culture conditions, including those of resistance to antimicrobial agents. A more extreme view of cellular metabolism was proposed by Cyril Hinshelwood, a Nobel Prize winner, who argued that all variations in cellular functions, such as enzyme inductions, changes in nutritional requirements, and drug resistances, were but readjustments of complex multiple equilibria of chemical reactions already active in the cell (10).

The basic issue, as we would see it today, that faced microbiologists in the early days of antimicrobial research is one of "adaptation versus mutation." It was passionately debated and contested by leading microbiologists from the mid 1930s until the early 1960s. Even those who viewed most microbial resistance as some sort of heritable change, or mutation, were divided on the basic problem of whether the mutations arose in response to the agent or occurred spontaneously and were simply observed after selection against the sensitive organisms. This problem was unresolved until the 1940s and 1950s, but has returned in a new form recently, as will be discussed here.

With the discovery and development of antibiotics and their medical applications, drug resistance took on new relevance, and new approaches became possible. No sooner were new antibiotics announced than reports of drug resistance appeared: sulfonamide resistance in 1939 (11), penicillin resistance in 1941 (12), and streptomycin resistance in 1946 (13), to cite a few early reports in the widely read literature. Research on resistance focused on three major problems: (1) cross-resistance to other agents; i.e., was resistance to one agent accompanied by resistance to another agent?, (2) distribution of resistance in nature; i.e., what was the prevalence of resistance in naturally occurring strains of the same organism from different sources?, and (3) induction of resistance; i.e., what regimens of drug exposure led to the induction or selection of resistant organisms?

While many practically useful results came from such research, two lines of investigation emerged that were to later prove scientifically interesting. Drug resistance provided a potent experimental tool for microbiologists who were studying bacterial genes and mutations. Rare nutritional markers were somewhat limited and such mutations often resulted in loss of function, usually recessive traits that were difficult to manipulate experimentally. For example, in 1936, I. M. Lewis (14) tested for pre-existing, spontaneous mutations to lactose utilization in a previously lactose-negative strain of *Escherichia coli*, but his results gave only indirect evidence for the random, spontaneous nature of bacterial mutation (as did the statistical approach of Luria and Delbrück in 1943). However, Lederberg and Lederberg (15) were able to use both streptomycin resistance and their newly devised replica plating technique to provide direct and convincing evidence to support the belief that mutations to drug resistance occurred even in the absence of the selective agent. Not only did such work on drug resistance clarify the nature of microbe–drug interactions, but it provided a much-needed tool for the nascent field of microbial genetics (16).

A very important clinical correlate of this new understanding of the nature of bacterial drug resistance was its application to combination chemotherapy. Since it became clear that mutations to resistance to different agents were independent events, the concept of multiple-drug therapy was developed and refined. The necessity for adequate dosages and lengths of treatment was obvious if the emergence of resistant organisms was to be avoided (17).

The second observation of basic significance was the odd phenomenon of drug dependence which was first noted for streptomycin in 1947 by Miller and Bohnhoff (18). This finding seemed to be restricted to streptomycin, but was extensively investigated at the time, and was thought to offer clues to the problems of antibiotic resistance in general. Later, however, this puzzling finding would be fundamental to understanding the functioning of the ribosome, and rather specific to the mode of action of aminoglycosides such as streptomycin. The history of this aspect of drug resistance emphasizes our inability to predict the future course of research and our failure to identify beforehand just where the likely advances will take us.

In the 1950s, in the era of many new antibiotics and the emphasis on surveys of both cross-resistance and distributions of resistance in natural microbial populations, especially in Japan, it was recognized that many strains with multiple-drug resistance were emerging. The appearance of such multiple-drug resistance could not be adequately explained on the basis of random, independent mutational events. Also, the patterns of resistance were complex and did not fit a simple mutational model. For example, resistance to chloramphenicol was rarely, if ever, observed alone, but it was common in multiply-resistant strains. Careful epidemiological and bacteriological studies of drug-resistant strains in Japan led Akiba et al. (19) and Ochiai et al. (20) to suggest that multiple-drug resistance may be transmissible both in vivo and in vitro between bacterial strains. Investigation of the possible modes of gene transfer led to the establishment of conjugal transfer of what was called the *R-factor* as the basic mechanism for transfer of multiple-drug resistance between bacteria (reviewed in Ref. 21). The R-factor was found to be one of a class of extrachromosomal genetic elements now termed *plasmids*. Plasmid biology has been greatly advanced, of course, by the subsequent study of transmissible drug resistance.

With the better understanding of the genetics of drug resistance and the classification of the types of resistance, the biochemical bases for resistance were elucidated. Knowledge of the mechanism of action of an agent led to an understanding of possible mechanisms of resistance. The specific role of penicillin in blocking cell wall biosynthesis, coupled with knowledge of the structures of bacterial cell envelopes, could explain the sensitivity of gram-positive organisms and the resistance of gram-negative organisms to this antibiotic. Likewise, understanding of its metabolic fate led to the finding that penicillin was often inactivated by degradation by β-lactamases, which provides one mechanism of bacterial drug resistance. The detailed biochemical study of the actions of antimicrobials have contributed to an understanding of the many ways in which microbes evolve to become resistant to such agents.

Not all voices for the adaptation hypothesis of drug resistance were drowned by the din of the genetic and conjugal mechanists, however. In the 1970s, mainly through the work of Samson and Cairns (22) and their colleagues, a variant of the adaptative model was revived and new mechanisms for bacterial drug resistance were discovered. Cairns and his colleagues observed that, in accord with some of the older work, bacteria could indeed be "trained" to resist certain agents by prior exposure to small, sublethal concentrations of the agent. They found that alkylating agents could induce the expression of specific genes whose products react with the alkylators, thus acting as a sink for further alkylating damage and rendering the cell hyper-resistant. While this phenomenon seems to represent a specialized pathway for dealing with alkylation damages, other examples of inducible resistance have since been discovered and, thus, a century after its first observation, microbial drug resistance is still a fruitful and surprising area of research.

In the 1970s, research on microbial drug resistance focused on detailed mechanisms of resistance in specific cases. The entry of agents into bacteria was studied and the roles of cell wall structures, outer membrane proteins, and capsular materials were found to be important. Mutations in genes encoding the outer membrane proteins were associated with altered sensitivity to some antibiotics. These proteins, called *porins*, and encoded by *Omp* genes, are involved in transport of normal physiological compounds, such as amino acids, as well as antimicrobial agents. Later studies demonstrated that a factor of major importance in resisting the action of antimicrobial agents is the synergy between the outer membrane permeability barrier and multidrug efflux pumps (see Chap. 4). These mechanisms of drug entry, and the concomitant pathways of drug resistance, explained the nature of broad-spectrum antibiotics such as tetracycline. Mutations in genes controlling drug entry also explained some of the puzzling observations on the simultaneous appearance of multiple drug resistance. Changes in the lipopolysaccharides on the bacterial cell surface were also found to alter sensitivity to antibiotics. With such studies, it became possible to begin to give a detailed chemical explanation for the very old observations relating virulence, sensitivity, and "rough/smooth" colony dimorphisms.

The new approaches and tools of molecular biology and genetics that were applied to bacterial drug resistance in the 1970s have continued to yield detailed, specific, and useful information. In the 1980s the origin of resistance to new antibiotics and the phenomenon of transfer of drug resistance between organisms yielded to investigations that brought epidemiology and laboratory work into close collaboration. Starting in the 1970s and continuing well into the 1980s there was a deepening understanding of the biology of bacterial plasmids. Driven by their importance as vectors of drug resistance, as markers of specific biotypes of bacteria, and as tools for genetic engineering in the biotechnology industry, plasmids were intensively studied at the molecular as well as the epidemiological level. It was observed that interspecific transfer of genes occurred widely in nature, often plasmid-mediated. Some of the chromosomal determinants of drug resistance could be spread by interspecific recombination as well. An important (and in hindsight, belated) realization in the 1980s was the key role played by the transfer of drug-resistance plasmids from the bacteria of lower animals to those bacteria that colonize humans.

The development of antibiotic resistance in veterinary bacteria was shown to result from selection secondary to the common practice of adding antibiotics or antibiotic production byproducts to animal feed. The potential for transfer of drug resistance to human bacteria by the widespread mechanisms of gene transfer both within and between species became appreciated as multiple-drug resistance became a worldwide concern. The public recognition and discussion of drug resistance has been a major theme in the bacteriology of the 1990s. While for many years there have remained very few agents against which widespread resistance has not developed, recent reports suggest that certain bacteria have now devised pathways of resistance even to agents such as vancomycin. From what is known about the spread of resistance, it will not be surprising if even these antibacterial bastions soon fall (23,24).

William C. Summers
Yale University School of Medicine
New Haven, Connecticut

References

1. Kossiakoff MG. De la propriété que possédent les microbes de s'accomoder aux milieux antiseptiques. Ann Inst Pasteur 1887; 1:465–476.
2. Effront J. Koch's Jahresber Gärungorganisimen 1891; 2:154 (quoted in Schnitzer RJ, Grunberg E. Drug Resistance of Microorganisms. New York: Academic Press, 1957, p 1).
3. Davenport CB, Neal HV. On the acclimatization of organisms to poisonous chemical substances. Arch Entwicklungsmech Organ 1895–1896; 2:564–583.
4. Ehrlich P. Chemotherapie trypanosomen-studien. Berl Klin Wochenschr 1907; 44:233–238.
5. Morgenroth J, Levy R. Chemotherapie der pneumokokkeninfektion. Berl Klin Wochenschr 1911; 48:1560.
6. Tomes N. The Gospel of Germs: Men, Women and Microbes in American Life. Cambridge: Harvard University Press, 1998.
7. Enderlein G. Bakterien Cyclogenie: Prologomena zu Untersuchungen über Bau, geschlechtliche und ungeschlechtliche Fortpflanzung und Entwicklung der Bakterien. Berlin: W. de Guyter, 1925.
8. Hadley P. Microbic dissociation. J Infect Dis 1927; 40:1–312.
9. Stephenson M. Bacterial Metabolism. London: Longmans, Green, 1930.
10. Hinshelwood C. The Chemical Kinetics of the Bacterial Cell. Oxford: Oxford University Press, 1946.

11. Maclean IH, Rogers KB, Fleming A. M. & B. 693 and pneumococci. Lancet 1939; i:562–568.
12. Abraham EP, Chain E, Fletcher CM, Gardner AD, Heatley NG, Jennings MA, Florey HW. Further observations on penicillin. Lancet 1941; ii:177–189.
13. Murray R, Kilham L, Wilcox C, Finland M. Development of streptomycin resistance of Gram-negative bacilli in vitro and during treatment. Proc Soc Exp Biol Med 1946; 63:470–474.
14. Lewis IM. Bacterial variation with special reference to behavior of some mutable strains of colon bacteria in synthetic media. J Bacteriol 1934; 28: 619–639.
15. Lederberg J, Lederberg EM. Replica plating and indirect selection of bacterial mutants. J Bacteriol 1952; 63:399–406.
16. Brock TD. The Emergence of Bacterial Genetics. Cold Spring Harbor, NY: Cold Spring Harbor Laboratory Press, 1991.
17. Szybalski W. Theoretical basis of multiple chemotherapy. Tuberculology 1956; 15: 82–85.
18. Miller CP, Bohnhoff M. Development of streptomycin-resistant variants of meningococcus. Science 1947; 105:620-621.
19. Akiba T, Koyama K, Ishiki Y, Kimura S, Fukushima T. On the mechanism of the development of multiple-drug-resistant clones of *Shigella*. Jap J Microbiol 1960; 4:219–227.
20. Ochiai K, Yamanaka T, Kimura K, Sawada O. Inheritance of drug resistance (and its transfer) between *Shigella* strains and between *Shigella* and *E. coli* strains [in Japanese]. Nihon Iji Shimpo 1959; 1861:34–46.
21. Watanabe T. Infective heredity of multiple drug resistance in bacteria. Bact Rev 1963; 27:87–115.
22. Samson L, Cairns J. A new pathway for DNA repair in *Escherichia coli*. Nature 1977; 267:281–283.
23. Bryan LE., ed. Antimicrobial Drug Resistance. New York: Academic Press, 1984.
24. Levy SB. The Antibiotic Paradox. New York: Plenum Press, 1992.

Bacterial Resistance to Antimicrobials

1

The Ecology of Antibiotic Resistance Genes

Abigail A. Salyers, Nadja B. Shoemaker, and George T. Bonheyo
University of Illinois at Urbana-Champaign, Urbana, Illinois

Early studies of antibiotic resistance genes focused on mechanisms of resistance and on transfer and regulation of resistance genes. In recent years, a new area of concern has surfaced: the ecology of resistance genes. The importance of understanding how resistance genes arise and spread has been underscored recently by a number of public policy concerns involving antibiotic resistance. Examples are the debate over the safety of genetically engineered plants as food for humans and animals. And more recently the debate over agricultural use of antibiotics as growth promoters in the swine and poultry industries. Also at issue is the increasing number of miniepidemics of multi–drug-resistant strains in hospitals and nursing homes. All of these cases have in common that the questions they raise are ecological in nature. Increasingly, it is becoming clear that antibiotic resistance needs to be viewed as an ecological problem. To make matters more complicated, it is important to consider not only the ecology of resistant bacterial strains but also the ecology of resistance genes, because resistance genes can move freely between different bacteria. This chapter addresses some of the questions and problems that arise from considering antibiotic resistance from an ecological perspective.

1 INTRODUCTION

Every year, hundreds of thousands of tons of antibiotics enter the environment. Although precise numbers are still not available for many countries, it appears that about half of these antibiotics are used to treat or prevent human infections and about half are used in agriculture. Moreover, the lifetime of an antibiotic does not end with the person or animal treated with the antibiotic. An unknown amount of active antibiotic is released into the environment by sewage treatment plants and runoff from manure. In the body of an animal or a human, most of the antibiotics that pass through the liver will have been inactivated by addition of a sulfate or glucuronide group. Yet, bacteria in the colon and in soil are capable of removing these groups from antibiotics (1). Moreover, although some resistant bacteria inactivate antibiotics, many use resistance mechanisms such as efflux or changing the antibiotic target that leave the antibiotic fully active. Scientists are finding easily detectable antibiotic residues in water from waste water treatment plants (2). More indirect evidence that antibiotics may move widely in the environment comes from studies showing that antibiotic-resistant bacteria can be isolated from groundwater, sludge, and soil (3–7). To date, only a small number of studies of this type have been published, but what results have been obtained raise the ecological question of how widespread the effects of antibiotics dumped into the external environment actually are.

Another ecological question that has only recently started to receive much attention is the question of whether resistant strains persist in an environment in the absence of antibiotic selection. Some studies have indicated that at least some resistant bacteria may be fit enough to persist for long periods of time. For example, a survey of *Streptococcus pyogenes* strains in Finland before and after a sizable drop in use of erythromycin shows that the incidence of erythromycin-resistant isolates declined significantly once use levels dropped (8). However, the incidence of resistant strains did not drop to zero but rather appeared to be leveling off at around 5–10%. Is it possible that there is a hard core of resistant strains that are also able to persist in the absence of selection and, if so, how do these strains differ from the ones that were outcompeted by sensitive strains? Other studies have found strains resistant to streptomycin or chloramphenicol long after use of these antibiotics declined (9). The persistence of resistant strains raises the question of whether antibiotics are the only selection pressures involved in maintaining resistant strains in the environment.

In subsequent sections of this chapter, a more detailed analysis of these questions is presented. However, since most of the ecological questions arise in connection with real-world political and economic issues, it is instructive first to examine some real-world examples of the sort of ecological questions that have arisen in recent years. Such examples make it easier to understand what types of information are needed and what types of analyses are likely to be appropriate.

2 THE ANTIBIOTIC-RESISTANT MARKER GENE CONTROVERSY

In 1995, antibiotechnology activists concerned about the safety of genetically engineered plants raised the question of whether antibiotic resistance genes present in the plant might be released from the plant cells during digestion and then be taken up by intestinal bacteria, making them resistant to antibiotics (10). The resistance genes had been used as selectable markers in early stages of cloning and had not been removed when the cloned DNA was introduced into the plant. Two types of marker genes were at issue, the ampicillin resistance (*bla*) gene found on many widely used cloning vectors and an aminoglycoside resistance gene (*nptII*). Although at first this might seem to be a plausible scenario and a frightening one, scientists who were asked to evaluate the risk rapidly realized how unlikely this series of events would be (10). First, DNA released from plant cells in the stomach and small intestine would be extensively degraded by nucleases. In an experiment in which mice were fed DNA, very small fragments (300 bp or less) were the main product recovered in the intestine (11). The second step, uptake of DNA by intestinal bacteria, also seemed unlikely. Although some strains of bacteria that might pass through the intestine, such as *Bacillus subtilis*, *Haemophilus influenzae*, and *Streptococcus pneumoniae*, are known to be capable of natural transformation, the bacteria that normally reside in the colons of humans and animals have so far been found not to be naturally transformable (10). The possibility remains, however, that natural transformation systems in intestinal bacteria have been missed owing to failure to use the right conditions. Generally, the genes necessary for natural transformation are regulated, and the time window for active uptake can be quite short. Failure to guess the right conditions would lead investigators to conclude that a species was not naturally transformable.

It is interesting to note that although naturally transformable bacteria seem to be uncommon in the microflora of the human body, many soil bacteria are naturally transformable (12). A speculation is that in nutrient-poor environments, bacteria use natural transformation to obtain DNA for use as a carbon and energy source. Whatever the explanation, the flow of resistance genes might show a different pattern in the case of soil bacteria than in the case of intestinal bacteria.

Once the DNA has entered the bacterium, it must be fixed in the chromosome. Unless an identical gene was already present, this would have to occur by illegitimate recombination, a poorly understood process that seems to be less efficient than homologous recombination. Still, since the process is poorly understood, it is possible that there is considerable species to species variation in efficiency of this type of recombination. The one case in which natural transformation appears to have been responsible for the acquisition of antibiotic resistance genes is penicillin-resistant *S. pneumonia*, which may have acquired portions of a penicillin-binding protein from another *Streptococcus* species (13), but here homologous recombination between DNA segments from very closely related organisms was responsible.

Finally, the gene has to be expressed, which might require promoter mutations or insertion of an IS element in the promoter region. Such events can be demonstrated in the laboratory and have occurred in clinical isolates, but they seem to require a strong and sustained selective pressure, such as that encountered in hospitals. Whether this type of selection pressure is exerted outside of hospitals is uncertain. A strong and constant selective pressure is present in agriculture due to use of antibiotics for prophylaxis and growth promotion, but ampicillin and aminoglycosides are not antibiotics commonly used for these purposes.

In the end, since scientists could only say that the probability of the marker genes creating new antibiotic-resistant strains was very low but not zero, the argument came down to the consequences of such an event if it did occur. The resistance genes used in cloning vectors were isolated in the 1970s, which was before the advent of the newer types of resistance genes that plague us today. The TEM-1 β-lactamase encoded by the *bla* gene is easily handled by modern antibiotics and by combinations of ampicillin and a β-lactamase inhibitor that have become popular in recent years. Thus, it seemed highly unlikely that marker genes from transgenic plants would, if they actually made the arduous journey from the plant genome to a bacterial genome, have any significant impact in today's clinical environment (10).

An examination of the arguments made during the marker gene debate reveals some of the holes in the research database. It also illustrates the difficulty of fitting the ecology of antibiotic resistance genes into the reductionist mold of traditional antibiotic resistance research. First, there is the problem of whether phenomena that are demonstrable under laboratory conditions actually occur at a significant level in natural settings such as the human or animal intestine. The converse is also possible; namely, phenomena that are not demonstrable under laboratory conditions (e.g., natural transformation of *Escherichia coli* or *Bacteroides* species) might occur in natural settings because of differences in conditions. Also at issue is the relative importance, in natural settings, of the various types of gene transfer such as transformation, phage transduction, and conjugation. How does one conduct experiments to test gene transfer in vivo? Before considering this question further, it is worth looking at another example of the sort of questions that arise in connection with the ecology of resistance genes.

3 ANTIBIOTIC USE IN AGRICULTURE

An emerging political debate in which the ecological side of antibiotic resistance will play a key role is the debate over use of antibiotics in agriculture. Initially, attention was focused primarily on multi–drug-resistant *Salmonella* and *Campylobacter* species; that is, bacteria that cause immediate disease. Since antibiotics are not appropriate for the treatment of ordinary diarrheal disease, which is usually a self-limiting infection, the main significance of antibiotic-resistant diarrheal pathogens is the small number of cases in which infected people developed systemic infections, which must be treated by antibiotics. This was demonstrated in a recent Danish outbreak of salmonellosis in which two people died of systemic infections (14). In both cases, treatment failure was attributed to the fact that the strain involved, *Salmonella typhimurium* DT104, was resistant to several antibiotics; resistance may have arisen as a result of antibiotic use.

In recent years, attention has shifted to a much more subtle but potentially much more dangerous effect of agricultural antibiotic use. Intestinal bacteria such as *Enterococcus* species do not cause gastroenteritis, but they can cause serious postsurgical infections (15). A person's own microflora can be the source of infection in such cases, especially when surgery involves the body cavity where perforation of the colon may occur. *Enterococcus* species are also found in the intestinal tracts of farm animals, and the use of antibiotics as growth promoters or for prophylaxis selects for resistant strains of these species. These resistant strains have been isolated from food for sale in supermarkets, and could thus be entering the human intestine on a near-daily basis (see, e.g., Refs. 16 and 17). There is some uncertainty as to whether the enterococci from animals can colonize the human intestinal tract or would be rapidly displaced by the human enterococcal strains. Moreover, since enterococci and other ingested bacteria must pass through the stomach before reaching the intestine, they would be stressed in addition to being in an alien environment, thus further reducing the likelihood that they could compete effec-

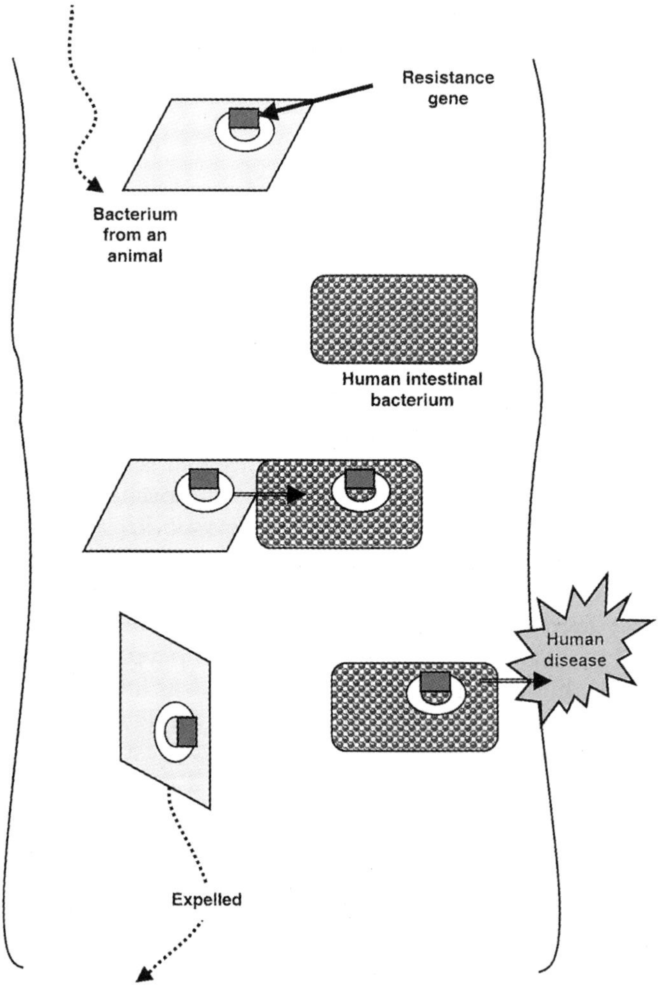

Figure 1 Bacteria that pass through the intestinal tract can contribute to the resistance of the intestinal microflora, even if they do not remain in the site, by donating resistance genes (in the case illustrated here, on a plasmid) to members of the microflora. Later, if the colon is perforated or if colon contents contaminate hands or the environment, members of the human microflora can cause disease.

tively with the resident strains. There is evidence that strains of enterococci isolated from animals and the food supply are not the same as those isolated from animals (18).

Even if multi–drug-resistant enterococci ingested in food are rapidly eliminated from the human colon, however, it is possible that they could transfer resistance genes to human strains of intestinal bacteria (19) (Fig. 1). The more drug-resistant a person's microflora, the greater the risk of a later untreatable postsurgical infection. Given this proposed scenario, the question arose as to how likely such gene transfers were to occur during the few days the ingested bacteria would reside in the human colon before being eliminated. At first glance, such transfers might seem highly unlikely. Granted, conjugal transfer of DNA can occur within a few hours or less, so there would be plenty of time for transfer to occur. Also, there are many surfaces, such as food particles, for the bacteria to colonize. Many mating systems of naturally occurring bacteria require that the donor and recipient be on a solid surface rather than suspended in solution. Yet, under optimal laboratory conditions, transfer frequencies of most naturally occurring conjugal elements tend to be low; on the order of 10^{-5} transconjugants per recipient or less. Could such infrequent events occur often enough to produce new resistant bacteria? The evidence is still spotty, but it supports the hypothesis that such horizontal gene transfer events do occur in nature (18–20).

The same questions arise here as in the case of marker genes from genetically engineered plants. To what extent can one extrapolate from the results of laboratory experiments to deduce what would happen in the colon? How could one evaluate experimentally the frequency of gene transfer events between bacteria in the human colon? How important is conjugation or transformation as a vehicle for the spread of antibiotic resistance genes in an environment like the colon?

4 HOSPITAL OUTBREAKS

A third example of a resistance issue that turns out to revolve around ecological questions is the increasing incidence in hospitals and nursing homes of miniepidemics of infections caused by antibiotic-resistant bacteria. It is often assumed that such outbreaks are caused by a single resistant strain that is transmitted from patient to patient via the hands of staff members. In some cases, this is undoubtably what happens, but is this true for all cases? Yet, it is also possible that self-infection with the patient's microflora occurs more often than we think. A recent study of a program in which incoming patients were screened for carriers of vancomycin-resistant enterococci, where carriers were isolated and treated, concluded that this preventive measure was cost effective. The fact that the program was effective suggests that patients can become colonized in the community and bring their resistant bacteria into the hospital. Other recent evidence supports the contention that methicillin-resistant *Staphylococcus aureus* (MRSA) strains are much more commonly found in people in the community than was previously thought (22). If the community is the source of most MRSA or other multi–drug-resistant strains, the appropriate intervention is to discover and control prescribing practices in the community that select for resistance and to prevent carriers from being in contact with other patients if they enter the hospital or nursing home. Assuming that MRSA and other resistance problems arise in the hospital would lead to a different set of intervention strategies, which could well prove to be ineffective if the assumption is incorrect.

Another question that needs to be addressed is the extent to which gene transfer contributes to outbreaks of resistant strains in hospitals and nursing homes. If gene transfer occurs readily in the human intestine or elsewhere on the human body, it is possible that newly resistant pathogens could arise owing to horizontal gene transfers from bacteria ingested in food to pathogens passing through the colon (18,19,23). At present, there is little evidence that such transfer events make a significant contribution to hospital outbreaks caused by antibiotic-resistant bacteria, but this is not a convincing argument that such transfer events are not important, because there have been so few attempts to investigate this hypothesis. Absence of evidence is not evidence of absence.

5 ASSESSING IN VIVO GENE TRANSFER

Any ecological approach to resistance gene transfer will have to confront the question of how to tie results of laboratory experiments to what is actually happening in vivo. In the area of environmental microbiology, some scientists have gone so far as to categorize pure culture laboratory studies as useless for understanding how microbes function in the environment. This rather discouraging point of view may sound reasonable at first, but it has a fundamental flaw. It overlooks the fact that pure culture studies cannot only generate hypotheses for testing in the real-world setting but also provide probes, such as DNA probes or antibodies, for testing the hypothesis. Scientists interested in antibiotic resistance have taken the more optimistic approach of assuming that results of laboratory studies based on pure cultures can be a starting point for learning more about resistance ecology in a real-world context.

Two approaches have been taken to bridging the in laboratorio to in vivo gap. One is to use animal models to determine how readily gene transfer events occur in the intestine. These few studies that have taken this approach so far have focused on conjugation rather than natural transformation or phage transduction. In one case, the transfers being monitored occurred between different strains of the same *Lactobacillus* species (24). A transmissible plasmid was followed in the intestines of mice that had a normal microflora from which the *Lactobacillus* species had been selectively "deleted" with antibiotics so that isogenic donor and recipient *Lactobacillus* strains could be introduced. In this case, transfer was detected within weeks.

In another experiment, germ-free mice were colonized with a donor and recipient from different genera (25). Again, a conjugal element was used. Transfer events were detected after long periods but the frequency was very low. There are a number of problems with this type of study. First, the experimental animal is not a human and, in the best-controlled experiments, those involving germ-free animals, lacks a normal intestinal microflora. Second, the transmissible elements being tested may or may not be typical of what is found in natural isolates. Since little information is available about the prevalence in nature of the plasmids and other transmissible elements that are currently being studied in the laboratory, the choice of which element to follow is problematic. The results obtained with one type of element and one type of gene transfer mechanism may not hold for other types of element and gene transfer mechanism. Those who have worked with natural isolates have long suspected that conjugation was the predominant mode of horizontal gene transfer in nature (20). This hypothesis, however, has not been proven.

Another approach to assessing resistance gene transfer in vivo is to survey bacterial isolates for resistance genes that have also been found in other species or genera. This approach is based on the assumption the presence of resistance genes with virtually identical DNA sequences in bacteria from different species or genera is proof that horizontal gene transfer of some kind has occurred. The requirement for high sequence identity (>95%, although this is somewhat arbitrary) eliminates the possibility of convergent evolution. Convergent evolution is the independent selection in different bacterial strains of proteins with virtually identical amino acid sequences. This can occur if changes in the amino acid sequence of a protein lead to big differences in its activity. Since the amino acid sequence of a protein can be maintained with differences at the DNA sequence level as great as 20%, and since the selection pressure is on the amino acid sequence and not the DNA sequence, the requirement that DNA sequences be more than 95% identical virtually assures that horizontal gene transfer and not convergent evolution is responsible for the phenomenon.

Although there have been thousands of papers surveying phenotypic resistance in bacterial isolates, there are still relatively few reports that characterize the resistance at the DNA sequence level. In recent years, DNA hybridization has been used to determine the genotype of resistant strains. This approach has the drawback that scientists rarely report the range in DNA sequences detected by their hybridization methods. Since even under stringent hybridization conditions, a probe can react with genes that differ by as much as 20%, the hybridization approach does not rule out convergent evolution. With the advent of polymerase chain reaction (PCR) and rapid sequencing, obtaining and comparing the sequences of resistance genes in different strains has become feasible.

Some work done in our laboratory illustrates the type of information that emerges from the DNA sequence approach to the ecology of resistance genes. We found that alleles of a tetracycline resistance gene, *tetQ*, could be detected in human colonic *Bacteroides* species and in strains of *Prevotella ruminicola* found in the rumen and intestine of livestock animals (26). The sequences of the *tetQ* alleles were 95% identical or higher; indicating that horizontal gene transfer was responsible for the presence of *tetQ* in these very diverse genera. In another study, we found a *Bacteroides* erythromycin resistance gene, *ermG*, which was 99.5% identical at the DNA sequence level to an *ermG* gene found previously in *Bacillus sphaericus* (27). Both of these findings indicate that horizontal gene transfer between very diverse organisms and organisms normally found in different sites is occurring in nature. What this sequence information does not tell us is the direction of transfer, nor does it tell us whether a third, or fourth, or fifth party was involved. All the sequence data proves is that there is some genetic conduit open between the two organisms in which the same gene was found.

The next challenge is to determine the direction and mechanism of transfer. This may require examination of DNA sequences around the resistance genes and an assessment of whether the gene is still transmissible. In the case of the *tetQ* genes found in isolates of *Prevotella ruminicola*, there was a fragment of a downstream gene, *rteA*, which is present on the conjugative transposons that carry *tetQ* in *Bacteroides* species (26). Laboratory experiments had shown that *Bacteroides* species could transfer conjugative transposons to *P. ruminicola*. The scenario suggested by these findings is that some *tetQ*-carrying conjugative element entered *P. ruminicola* from a *Bacteroides* strain and then experienced deletions and rearrangements that left only *tetQ* and a remnant of its normally linked gene in the new host. In the case of the *ermG* example, there are no such clues. What needs to be determined is how to use most effectively the information about sequences adjacent to antibiotic resistance genes. Using sequences adjacent to the resistance gene as a guide to direction of transfer has the limitation that so far only relatively few plasmids and other transmissible elements have been sequenced. Also, scientists have tended to focus on an element they have identified in one strain without asking whether it is prevalent in the environment of interest. This means that the sequences in the databases for plasmids, conjugative transposons and other transmissible elements do not necessarily reflect the abundance of these elements in natural settings. Nonetheless, as the sequence databases continue to grow, more information will become available making this approach more feasible.

In a recent survey of *Bacteroides* strains isolated from the human colon, we showed that carriage of a *tetQ* gene had risen from 20 to 30% in pre-1970 isolates to over 80% in isolates from the late 1990s (28). The sequences of the gene in different strains were more than 95% identical; indicating horizontal transfer. By using available information about the characteristics of conjugative transposons and plasmids found in *Bacteroides* species, we were able to show that this apparent epidemic of horizontal gene transfer had been due to a particular type of conjugative transposon. We also found an *ermB* in some isolates that was over 99% identical to one that had been found previously in gram-positive pathogens. In this case, the sequences of DNA adjacent to the resistance gene were similar to those found in gram-positive bacteria; raising the possibility that the direction of transfer had been from the gram-positive bacteria into *Bacteroides*. Further evidence for the direction of transfer can be obtained if it can be shown that a gene is much more common in one group of bacteria than in another.

Before leaving this subject, it is useful to raise a question that is almost never asked. There are numerous examples of the transmission of resistance genes by transformation and conjugation but no known examples of transfer of resistance genes by bacteriophages except in laboratory experiments where resistance genes were used as markers in phage transduction experiments. Do bacteriophage in nature ever carry antibiotic resistance genes? A class of conjugal elements called conjugative transposons, some of which do carry resistance genes, share a number of properties with phages. These properties include similar integrase proteins, similar mechanisms of integration, and, in some cases, a propensity for integrating into the ends of tRNA genes. Are these elements defective phages that can no longer spread by lysis and subsequent infection but have acquired the ability to transfer by conjugation (29)? Some eukaryotic viruses transfer from cell to cell by fusing the membranes of donor and recipient. Such a mode of transfer has not been described in the case of bacteriophages, but the study of bacteriophages has been limited to only a few examples. Little is known about the true diversity of bacteriophages. Bacteriophages have been implicated in the transfer of virulence genes, including genes encoding toxins that have no discernable benefit to the phage, so the possibility that phage are conduits for resistance genes should not be dismissed lightly.

6 PERSISTENCE OF RESISTANCE GENES

Most of the studies of resistance gene transfer have focused on acquisition of the resistance genes. An ecological view of the flow of resistance genes has to take into account not only acquisition of genes but their persistence once they are acquired. A dogma that has been widely accepted until recently is that antibiotic-resistant bacteria are always less fit in the absence of antibiotic selection than susceptible strains. This dogma arose from early studies of mutants resistant to protein synthesis inhibitors such as streptomycin. It is not surprising that a mutation in a ribosomal protein or ribosomal RNA gene would render a bacterium less fit than wild type, at least initially, but this type of fitness toll may not occur in all resistant strains. There are now two reasons to challenge the dogma of the intrinsic unfitness of antibiotic-resistant bacteria. First, Lewin and colleagues (30) have shown that fitness of resistant strains can change if the bacteria experience prolonged antibiotic pressure. A streptomycin-resistant mutant was initially less fit than wild type. But when the mutant was passed many times in medium containing streptomycin, it became able to compete equally with the wild type in the absence of antibiotic. Compensatory mutations had occurred that reduced the fitness toll due to the original mutation.

A second reason for questioning the unfitness dogma is that many resistance genes have regulated expression, so that the resistance protein is only made when the antibiotic is present. Moreover, a resistance gene may exact a lower fitness toll if it is present on a low copy number plasmid or is integrated into the chromosome. In our study of *tetQ* distribution, we found that over 80% of *Bacteroides* isolates obtained in the late 1990s from healthy people with no recent history of tetracycline therapy carried the tetracycline resistance gene *tetQ* (28). The *tetQ* gene was located on a conjugative transposon. This high level of carriage of both *tetQ* and the element that transfers it, in the absence of antibiotic selection, indicates not only that the conjugative transposon that carries it is readily acquired but also that it is stably maintained under natural conditions.

Another possible explanation for the maintenance of a resistance gene in the absence of the corresponding antibiotic is that the resistance gene is linked to other resistance genes. Elements such as integrons or plasmids or conjugative transposons that carry more than one type of resistance gene are likely to be selected for as a single entity. Thus, the continued maintenance of a plasmid or integron carrying a gene for resistance to antibiotics X and Y can be ensured either by selection for X or for Y. Even if the antibiotic that originally selected for resistance to X is no longer used but antibiotic Y is, there is selection for the combination of genetically linked genes. Moreover, some genes confer resistance to more than one class of antibiotic. Examples are efflux pumps that pump more than one class of antibiotic or the macrolide-lincomycin-streptogramin (MLS) type of *erm* genes. Any of the antibiotics to which they confer resistance will select for retention of such multifunctional genes.

The fact that clusters of resistance genes found in some integrons sometimes include genes that confer resistance to cadmium or mercury or resistance to quaternary ammonium disinfectants raise the still more troubling possibility that compounds that are not antibiotics could select for the maintenance of resistance genes. This possibility needs to be examined in more detail. If metal pollution or disinfectants can select for maintenance of bacteria resistant to antibiotics, this fact needs to be factored into decisions about efforts to control the spread of resistance.

7 DESIGNING APPROPRIATE EXPERIMENTAL CONTROLS

A little-discussed aspect of resistance ecology studies is controls. How to design positive and negative controls in an ecological experiment may not be as straightforward as designing positive and negative controls in a laboratory experiment. In laboratory experiments, for example, geography is not an issue. An experiment conducted in one laboratory will yield exactly the same result as the same experiment conducted in another part of the same laboratory or in a different laboratory. In ecological studies, such factors as location matter. Suppose the goal is to conduct an experiment to test whether use of antibiotics in a hospital increases the incidence of resistant strains. Currently, the design of such an experiment would likely focus exclusively on the hospital environment itself. It might take into account differences in resistance patterns in different parts of the hospital, but it might well ignore the movement of resistant strains into and out of the hospital. Similarly, few studies of the impact of agricultural use of antibiotics check the resistant pattern of bacteria from wild areas nearby or bacteria in soil. Yet, it is conceivable that antibiotic-laden runoffs from manure could select for resistant bacteria in the intestines of wild animals or soil and these bacteria might be cycled back into the intestines of farm animals. Antibiotic-resistant bacteria can sometimes be found in the intestines of animals that have not been exposed to antibiotics. In such cases, an environmental source of the bacteria should be considered.

One of the challenges for scientists interested in the ecology of antibiotic-resistant bacteria, whether at the phenotypic or genotypic level, is to define adequate controls. Unfortunately, there is probably no single rule for the definition of controls that will fit all situations. Environmental microbiologists who are preparing to conduct ecological studies spend time at the site where the study is to be done looking carefully at the geography and human or animal ecology of the site. By contrast, those who conduct ecological studies in clinical contexts are all too often content simply to acquire a set of strains isolated by someone else.

8 THE NEW PERSPECTIVE

Even though taking an ecological view of antibiotic resistance problems is still a relatively new idea, already several lessons have been learned. First, commensal bacteria are important players. The notion that commensal bacteria might serve as reservoirs for resistance genes and could transfer these genes to bacteria that do cause disease is not a new one. Yet, evidence that such transfers can and do occur is only now beginning to emerge. Every time a person is treated with an antibiotic, that person's normal microflora is treated as well as the target pathogen. Similarly, as excreted antibiotics move into the external environment, the microbial populations of soil and water are subjected to a selective pressure that favors the emergence and maintenance of resistant strains.

Second, it is important to move beyond phenotype to genotype in resistance studies. Once a new resistance gene or a new variant of an old gene appears, it may not stay within the strain in which it arose. Natural isolates of bacteria are replete with transmissible elements and appear to exchange these elements freely under natural conditions. Such exchanges can occur across species and genus lines. Evidence that such transfers actually occur and may occur rather frequently is beginning to accumulate (see, e.g., Refs. 27 and 28). Rising resistance in commensals, especially those related to important human pathogens, should serve as a warning that the resistance gene responsible could well make its way into more dangerous bacteria. Also, the commensals themselves are becoming an infection problem, especially in people with undermined defenses.

Third, the stability of resistance genes is as important as their acquisition. In some cases, resistance genes are readily lost in the absence of antibiotic selection. In others, they are stably maintained. In evaluating various types of uses of antibiotics for their impact on resistance, it will be important to consider how the amount of antibiotic and duration of use affect the stability of resistance genes. Long-term, low-dose administration may well be more dangerous in this respect than the short, high-dose regimens normally used in therapy if it gives bacteria the chance to acquire additional mutations that cause a resistance gene to be more stably maintained.

Finally, it is becoming clear there is no single ecology of resistance but rather multiple ecologies that may overlap. In one case, a single resistant clone may arise and spread widely. In another case, multiple resistant strains may have arisen on separate occasions as a result of horizontal gene transfer. In one setting, the incidence of one type of resistant strain may decline precipitously if use of an antibiotic ceases. In another, cessation of antibiotic use may have no effect on the incidence of resistant strains. Indeed, patterns of resistance emergence and sustainability are likely to differ from one bacterial species to another and from one antibiotic to another, even in the same setting. Each case must be viewed with fresh eyes and an open mind. The notion that there is one simple explanation for the rise in incidence of resistant bacteria that fits all settings may be emotionally satisfying but is ultimately counterproductive and even dangerous.

REFERENCES

1. Goldman P. Biochemical pharmacology and toxicology involving the intestinal flora. In: Human Intestinal Microflora in Health and Disease. New York: Academic Press, 1983:241–263.
2. Raloff J. Drugged waters. Science News 1998; 153:187–189.
3. McKeon DM, Calabrese JP, Bissonnette GK. Antibiotic resistant gram-negative bacteria in rural groundwater supplies. Water Res 1995; 29:1902–1908.
4. Droge M, Puhler A, Selbitschka W. Phenotypic and molecular characterization of conjugative antibiotic resistance plasmids isolated from bacterial communities of activated sludge. Mol Gen Genet 2000; 263:471–482.
5. Waters B, Davies J. Amino acid variation in the GyrA subunit of bacteria potentially associated with natural resistance to fluoroquinolone antibiotics. Antimicrob Agents Chemother 1997; 41:2766–2769.
6. Papandreou S, Pagonopoulou O, Vantarakis A, Papapetropoulou M. Multi-antibiotic resistance of gram-negative bacteria isolated from drinking water samples in southwest Greece. J Chemother 2000; 12:267–273.
7. Andersen SR, Sandaa RA. Distribution of tetracycline resistance determinants among gram-negative bacteria isolated from polluted and unpolutted marine sediments. Appl Environ Microbiol 1994; 60:908–912.
8. Seppala H, Klaukka T, Vuopio-Varikila J, Muotiala A, Helenius H, Lager K,

Huovinen P. The effects of changes in the consumption of macrolide antibiotic on erythromicin resistance in group A streptococci in Finland. N Engl J Med 1997; 337:441–446.

9. Salyers AA, Amabile-Cuevas CF. Why are antibiotic resistance genes so resistant to elimination? Antimicrob Agents Chemother 1997; 41:2321–2325.
10. Salyers AA. Genetically engineered foods: safety issues associated with antibiotic resistance genes. 1997. Posted on the website: *www.roar.antibiotic.org*.
11. Schubbert R, Lettman C, Doefler W. Ingested foreign DNA (phage M13) survives transiently in the gastrointestinal tract and enters the bloodstream of mice. Mol Gen Genet 1994; 242:495–504.
12. Lorenz MG, Wachernagel W. Bacterial gene transfer by natural transformation in the environment. Microbiol Rev 1994; 58:563–602.
13. Hakenbeck R. Transformation in *Streptococcus pneumoniae*: mosaic genes and the regulation of competence. Res Microbiol 2000; 151:453–456.
14. Molbak K, Baggesen DL, Aarestrup FM. An outbreak of multidrug-resistant quinolone-resistant *Salmonella enterica* serotype typhimurium DT104. N Engl J Med 1999; 341:1420–1425.
15. Woodford N. Glycopeptide-resistant enterococci: a decade of experience. J Med Microbiol 1998; 47:849–862.
16. Woodford N, Adebiyi A-M, Palepou MF, Cookson BD. Diversity of *vanA* glycopeptide resistance elements in enterococci. Antimicrob Agents Chemother 1998; 42:502–508.
17. Murray BE. Vancomycin-resistant enterococci from nosocomial, community and animal sources in the United States. Antimicrob Agents Chemother 1996; 102:2605–2609.
18. van den Braak N, van Belkum A, van Keulen M, et al. Molecular characterization of vancomycin-resistant enterococci from hospitalized patients and poultry products in the Netherlands. J Clin Microbiol 1998; 36:1927–1932.
19. Simonsen GS, Haaheim H, Dahl KH, Kruse H, Lovseth A, Olsvik O, Sundsfjord A. Transmission of VanA-type vancomycin-resistant enterococci and *vanA* resistance elements between chickens and humans at avoparcin-exposed farms. Microb Drug Resist 1998; 4:313–318.
20. Salyers AA, Cooper AJ, Shoemaker NB. Lateral broad host range gene transfer in nature: how and how much? In: Horizontal Gene Transfer. Syvanen M, Kado C, eds. New York: Chapman & Hall, 1998:40–52.
21. Papia G, Louie M, Tralla A, Johnson C, Collins V, Simor AE. Screening high risk patients for methicillin-resistant *Staphylococcus aureus* on admission to the hospital: is it cost effective? Infect Control Hosp Epidemiol 1999; 20:473–477.
22. Shopsin B, Mathema B, Martinez J, Ha E, Campo M L, Fierman A, Krasinski K, Kornblum J, Alcabes P, Waddington M, Riehman M, Kreiswirth BN. Prevalence of methicillin-resistant and methicillin-susceptible *Staphylococcus aureus* in the community. J Infect Dis 2000; 182:359–362.
23. Teuber M, Meile L, Schwartz Y. Acquired antibiotic resistance in lactic acid bacteria from food. Antonie van Leeuwenhoek 1999; 76:115–137.
24. McConnell MA, Mercer AA, Tannock GW. Transfer of plasmid pAMβ1 between members of the normal microflora inhabiting the murine digestive tract

and modification of the plasmid in a *Lactobacillus reuteri* host. Microbial Ecol Health Dis 1991; 4:343–355.
25. Doucet-Populaire F, Trieu-Cuot P, Andremont A, Courvalin P. Conjugal transfer of plasmid DNA from *Enterococcus faecalis* to *Escherichia coli* in the digestive tracts of gnotobiotic mice. Antimicrob Agents Chemother 1992; 36: 502–502.
26. Nikolich M, Hong G, Shoemaker N, Salyers AA. Evidence that conjugal transfer of a tetracycline resistance gene (*tetQ*) has occurred very recently in nature between the normal microflora of animals and the normal microflora of humans. Appl Environ Microbiol 1994; 60:3255–3260.
27. Cooper AJ, Shoemaker NB, Salyers AA. The erythromycin resistance gene from the *Bacteroides* conjugative transposon Tc^rEm^r 7853 is nearly identical to *ermG* from *Bacillus sphaericus*. Antimicrob Agents Chemother 1996; 40: 506–508.
28. Shoemaker NB, Vlamakis H, Hayes K, Salyers AA. Evidence for extensive resistance gene transfer among *Bacteroides spp* and between *Bacteroides* and other genera in the human colon. Appl Environ Microbiol 2001: 67:561–568.
29. Salyers AA, Shoemaker NB, Frias J. Conjugative transposons: transmissible resistance islands. In: Pathogenicity Islands and Other Mobile Virulence Elements. (Kaper JB, Hacker J, eds. Washington, DC: ASM Press, 2000:331–346.
30. Levin BR, Perrot V, Walker N. Compensatory mutations, antibiotic reistance and the population genetics of adaptive evolution in bacteria. Genetics 2000; 154:985–997.

2

Antibiotic Resistance: How Bacterial Populations Respond to a Simple Evolutionary Force

Fernando de la Cruz
and Juan M. García-Lobo
University of Cantabria, Santander, Spain

Julian Davies
University of British Columbia, Vancouver, British Columbia, Canada

Antibiotics can be thought to exert a simple selective pressure upon bacteria. However, analysis of bacterial response to antibiotics indicates that nothing is simple in nature. The very concepts of antibiotic susceptibility and resistance are far from unequivocal. We neither understand sufficiently the fate of susceptible bacteria nor the differential steps by which they become resistant. Antibiotic resistance develops in the laboratory by point mutations, as could be predicted. Antibiotic resistance in nature, however, brings into play a series of unsuspected resources that bacteria use against selective challenges. The bacterial response can be characterized by the properties of diversity, promiscuity, rapidity, persistence, and novelty. It accrues genes of obscure and ancient origin; a circumstance suggesting that antibiotics play ecological roles not related to their antimicrobial activity. There is an interplay between mutation and acquisition in the development of most observed resistance mechanisms. All the evidence taken together suggests a three-stage process for antibiotic resistance selection in natural bacterial populations: proaction, acquired response, and selection. These are complex, with the result that the mechanistic responses of bacteria to an (apparently) simple selective pressure are in most cases unpredictable.

1 INTRODUCTION

Prior to the introduction of antibiotics, some 50 years ago, natural populations of human/animal bacterial pathogens or commensal bacteria were antibiotic susceptible. This is known because analyses of collections of bacteria dating from the preantibiotic era fail to show resistance to the antibiotics commonly used in infectious disease therapy (1). In the year 2000, a high proportion of the same bacterial populations are functionally resistant to most antibiotics that have been used extensively during the period of antibiotic use. If we consider this event as a global experiment on bacterial evolution, we may draw conclusions on the mechanisms by which bacteria react to selective pressure (2,3).

The purpose of this chapter is to summarize some general concepts with respect to the routes by which antibiotic resistance (Ab^R) genes emerged, particularly in human pathogens. Awareness of these routes may help to devise new strategies to control the dissemination of Ab^R, which appears to be approaching a catastrophic situation for infectious disease treatment. Considerations of these routes has led the authors to view the genetic structure of the bacterial world in a different light. First, the very nature of Ab^R selection is problematic, since concepts such as those of life and death are far from being obvious when they refer to prokaryotic organisms. Second, the evidence suggests that, in many instances, in nature a bacterial response to the antibiotic challenge does not occur by mutation and selection of the fittest individuals but rather involves a response of complex populations involving a sharing of Ab^R genes of ancient and obscure origin. This issue either reflects our absolute ignorance of the natural role of antibiotics or hints at the immense reservoir of hidden bacterial genetic diversity.

Only general trends will be discussed in this chapter, and no comprehensive analysis of the different aspects of Ab^R will be attempted. For this, readers should consult reviews on resistance to specific antibiotics which may be found in Table 1 (see Section 3) and elsewhere (4,5), as well as in the accompanying chapters in this book.

2 THE CONCEPTS OF ANTIMICROBIAL SUSCEPTIBILITY AND RESISTANCE: WHAT IS LIFE FOR A UNICELLULAR PROKARYOTE?

Fifty years of the antibiotic-based war against bacteria have not eliminated any significant human pathogen. Rather, if they were originally susceptible, they are becoming increasingly resistant to the antibiotics most used against them. For the newer antibiotics, acquisition of resistance appears to be only a matter of time. The overwhelming majority of enterobacteria are now resistant to sulfonamides, the first antibiotic introduced in clinical practice (around 1937). A high proportion are resistant to the first wave of broad-range penicillins (introduced in 1940) and also to, for example, streptomycin, chloramphenicol, and tetracyclines. These antibiotics were substituted by new antibiotic families, or improved versions of old families, with shorter and shorter effective functional lives. In summary, although many bacterial pathogens are theoretically susceptible to antibiotics, clinical practice recommends limited use, since resistance immediately and inevitably appears; it can be said they are functionally resistant. How does the use of antibiotics affect a susceptible population?

1. The so-called *bactericidal* antibiotics irreversibly disrupt bacterial multiplication. When the antibiotic is eliminated from the culture medium the bacteria do not recover their capacity to grow. The bactericidal action of antibiotics such as the penicillins has been studied in some detail: In this case, since the structure of the cell wall is sabotaged, bacteria lyse (often by the triggered production of autolysins) and are effectively destroyed. The situation is not so clear with other antibiotics such as aminoglycosides for which cell integrity is preserved, and the bacteria continue to exist with a subset of normal metabolic reactions. Important among them for the purpose of this chapter, bacteria can still participate in the exchange of genetic information (6). It is not easy to understand why some inhibitors of protein synthesis, such as the aminoglycosides, are bactericidal whereas others, such as the tetracyclines and macrolides, have a static action against bacteria. In addition, in any population of antibiotic "killed" bacteria, a small number of survivors can recuperate. These are known as "persistors" that when regrown remain susceptible to the bactericidal action of the antibiotic (7). Similarly, "injured" bacteria can grow on nonselective media but not on certain selective media in which their noninjured predecessors can still grow (8). These phenomena have been little studied, but they suggest to us that the transit from life to death in bacteria is not as clear-cut and obvious as it is in complex multicellular organisms. It seems that 99% of the existing bacteria in natural populations (e.g., soil, marine

environments) are not culturable (9), and culturable bacteria often enter stages of nonculturability (10). A significant proportion of them may remain nonculturable because of the presence or absence of specific bioactive microbial compounds. In any case, the genetic information preserved in the nonculturable population of bacteria is still subject to mutations that can be transferred to survivors. In terms of gene flux, nonculturable bacteria may be significant recipients of foreign genetic material: We speculate that these bacteria may have restriction and modification functions depressed, and thus they may offer open gates for gene colonization.

2. *Bacteriostatic* antibiotics disrupt bacterial multiplication, but the bacteria recover after removal of the antibiotic. As mentioned above, most antibiotic inhibitors of translation (also transcription, replication, or other biosynthetic process) lead to bacterial stasis. As in the case of bactericidal antibiotics, a number of events of DNA metabolism, as well as DNA transactions, can still occur in these states of suspended growth. Therefore, although susceptible bacteria apparently are inhibited by the action of the static drug, there are underlying genetic and biochemical processes taking place that may ultimately overcome growth inhibition. When bacterial multiplication is stopped within a human organism, for example, cessation of growth gives time for the immune defenses to eliminate the pathogen. As a result, the antibiotic is considered to be effective. But when bacterial multiplication is stopped in a "neutral" environment, the bacteria may persist in a state of arrest for a very long time, thus maximizing the chances of modifying their genetic background (11) so they eventually regain the capacity for multiplication (either for themselves or for the "global organism" if the genetic information can be passed to others, as discussed above). Our contention is that the main "purpose" of antibiotics is not to destroy neighboring sensitive bacteria but to control their multiplication (as do hormones or growth modulators in multicellular organisms).

Within a bacterial population which is susceptible to a given antibiotic, development of low-level resistance is very common and very important in evolutionary terms. This phenomenon, which has not received sufficient attention in our opinion, is elegantly discussed by Baquero (12). In any antibiotic treatment, the drug reaches the bacterial population as a concentration gradient. Only in the so-called selective compartment are targeted bacteria eliminated. But there are *always* insufficiently selective compartments in which the subinhibitory concentration of antibiotic provide the appropriate scenario to test different resistance strategies so that usually at least low-level resistance variants of the targeted populations are selected. This probably constitutes the most active evolutionary environment in which a plethora of bacterial variants can thrive. In addition, a small proportion of any existing bacterial population is always fully resistant to selective concentrations of the antibiotic as a result of spontaneous mutation, or may acquire resistance genes from the extended genome (13,14).

In summary, although antibiotics effectively stop bacterial multiplication, more often than not they fail to destroy all susceptible organisms. Temporarily or permanently damaged bacteria can still engage in processes of DNA metabolism and DNA transactions which may overcome the multiplication barrier (for the strain if not for the individual). Thus, the appearance of resistance, which was predicted to be negligible from the initial in vitro studies, is omnipresent.

3 RESPONSE IN THE LABORATORY COMPARED TO RESPONSE IN NATURE: WHY ARE WE SO BAD AT MAKING PREDICTIONS?

When we take a pure bacterial strain in the laboratory and challenge it with an antibiotic, it is generally found that a proportion (10^{-7}–10^{-10} depending on the antibiotic) is resistant to that antibiotic. Analysis of the resistant variants uncover the simplest mechanism of resistance: point mutations. Mutations can occur in the target gene or its regulators and result in an insensitive target. Mutations can also occur in any of a set of membrane proteins resulting in reduced uptake or increased efflux of the antibiotic. We may consider these point mutations as a "proactive response," since they occur prior to the antibiotic challenge and constitute the most elementary mechanism of bacterial protection against an antibiotic. It should be noted that these never lead to antibiotic inactivation; the naïve bacterial population had not previously encountered the drug, so they cannot act against it in such a sophisticated biochemical manner. These are the ways by which laboratory populations of bacteria become resistant to antibiotics, and they come as no surprise to an evolutionary biologist. Since the frequency of appearance of resistance was so low, in the early days of antibiotic use, it was predicted that mutation would be unlikely to pose a serious threat to antimicrobial therapy.

Strikingly different was the outcome of the analogous "experiment" when it took place on a much larger, environmental scale. When natural populations of commensal and pathogenic bacteria were exposed to antibiotics at concentrations higher than experienced in their natural habitats, and for long periods of time in the fight against disease, in animal husbandry, and in other agricultural practices, some bacteria became resistant to certain antibiotics. However, resistance appeared in a very different form—not by mutation but usually through the acquisition of exogenous

TABLE 1 Examples of Mechanisms of Antibiotic Resistance: A Comparison Between Laboratory and Naturally Encountered Mechanisms

Antibiotic type	Producing organism	Laboratory mutations	Naturally encountered mechanisms	Homologous proteins
1. Protein synthesis				
Aminoglycosides (41)	*Streptomyces*	Target mutation (30S ribosomal subunit *rpsL*)	Antibiotic inactivation: Acetlylation Phosphorylation Nucleotidylation Membrane impermeability	Histone-acetylases Protein kinases ?
Tetracyclines (29,42)	*Streptomyces*	Antibiotic efflux (*mar*)	Antibiotic efflux Target protection	Major facilitator superfamily EF-Tu, EF-G
Chloramphenicol (19–21)	*Streptomyces*	Antibiotic efflux (*mar*)	Antibiotic inactivation (major) *cat* Antibiotic efflux (minor) *cml*	Host acetylases Major facilitator superfamily
MLS group (43)	*Streptomyces*	Target mutation (50S ribosomal subunit)	Target modification (major) Target mutation (minor) Antibiotic inactivation (minor)	

2. Cell wall synthesis				
β-lactams (44)	*Penicillium* (fungi) *Cephalosporium* (fungi) Several bacteria	Reduced uptake (*ompF*) Antibiotic inactivation (*ampC* mutation and increased expression)	Antibiotic inactivation Target modification	
Glycopeptides (45)	*Streptomyces* *Amycolatopsis*	?	Target modification (murein, *vanA, vanB*, ...)	
Fosfomycin (46)	*Streptomyces*	Reduced uptake	Antibiotic inactivation	
3. DNA and RNA synthesis				
Fluoroquinolones (47)	*Synthetic*	Target mutation (*gyrAB, parEC*)	Target mutation (major) Antibiotic efflux (minor)	Topoisomerase I and IV Major facilitator subfamily
Rifampicin (48)	*Amycolatopsis*	Target mutation (*rpoB*)	Target mutation (major) Antibiotic inactivation (minor)	RNA polymerase (β subunit)
4. Folic acid biosynthesis				
Sulfonamides (49)	*Synthetic*	Target mutation (*folP*)	Target bypass	Dihydropteroate synthase
Trimethoprim (49)	*Synthetic*	Target bypass (*thyA*) Target mutation (*dfr*)	Target mutation (*dfr*)	Dihydrofolate reductase

Ab^R genes of obscure origin that lead principally to efflux or inactivation of the antibiotic. Acquisition of resistance mechanisms by a small fraction of the population, prior to the challenge, provided defense against the antibiotics. Table 1 summarizes some of the principal mechanisms of Ab^R and compares the mechanisms found in the laboratory with those found in nature.

Apart from the case of resistance to the synthetic fluoroquinolones and to rifampicin (with a few exceptions), resistance mechanisms are usually acquired through the intervention of mobile genetic elements (plasmids or transposons) rather than being generated by mutation. The first conclusion that can be drawn from Table 1 is that it seems easier or more favorable for bacteria to pick up existing resistance genes from the extended gene pool than to submit a susceptible population to de novo mutations. For example, in Japan in the 1950s there was a significant epidemic of *Shigella* dysentery. Development of resistance occurred rapidly with the appearance of transmissible multi–drug resistance, which was amply documented in the scientific literature (reviewed in Ref. 15). A special situation appears when the acquired resistance genes become integrated in the chromosome, hampering their identification. This is well illustrated by the *mecA* determinant of methicillin resistance in *Staphylococcus aureus* (MRSA). The *mecA* DNA sequence appears only in MRSA strains and it seems to have been acquired by horizontal gene transfer (HGT) from coagulase-negative staphylococci. A similarly interesting situation occurs in the streptococci and *Neisseria* where an acquired penicillin-binding protein gene (*pbp2*), or a DNA fragment thereof, recombines with the *pbp2* gene of the host to generate a mosaic gene encoding a protein that retains normal cell wall function but no longer binds penicillin effectively (16,17).

Nonetheless, resistance due to mutational alterations is still significant for specific antibiotic classes under certain conditions. Briefly, the specific situations leading to resistance mutations are 1) When a non-natural compound is used as antibiotic. This is particularly true for the fluoroquinolone antibiotics and for rifampicin. We assume that since these structures do not exist in nature, bacteria still acted in a proactive fashion as discussed above. 2) When the disease state is characterized by large populations of an infecting organism, such as *Mycobacterium tuberculosis*, that is genetically isolated. Here there is no accessibility to external genetic input, so resistance must arise as it does in a laboratory setting.

It must be emphasized that mutation of the acquired resistance determinants is usually a significant factor; for example, in the development of extended-spectrum β-lactamases, as discussed below, or by adjustment of antibiotic efflux systems. Additionally, it is very likely that one of the early steps in the development of Ab^R in many bacterial species is likely to be a mutation leading to a low-level of resistance, as happens with *mar* mutants of *Escherichia coli* (18).

4 OVERALL CHARACTERISTICS OF BACTERIAL ACQUIRED RESPONSE AGAINST ANTIBIOTICS

The observations stated above indicate that we have been exposed to an evolutionary phenomenon by the genetic resources that allow bacteria to respond to a simple force that threatens their survival. Instead of surviving antibiotic challenge primarily as a result of low-frequency mutations, it has become obvious that bacteria bring into play a broad range of biological resources. Five properties characterize bacterial acquired response:

1. *Diversity*. Bacteria may respond through the use of several different mechanisms of resistance, usually observed in the same species but also in different genera. Chloramphenicol resistance, for example, can appear by antibiotic inactivation or by active efflux (19). Even for a given resistance mechanism, different variant proteins may be brought into play. For example, there are several clearly different chloramphenicol efflux mechanisms (20) and at least three classes of inactivating chloramphenicol acetyltransferases (21). The proteins in a resistance family are obviously related but sufficiently different so that one can conclude that they did not diverge within the time frame of antibiotic use.
2. *Promiscuity*. Bacteria respond as global organisms (22). In spite of the existence and appearance of many resistance mechanisms, many of the most successful Ab^R genes appear, in identical or almost identical formats, in widely different bacterial genera. Thus, if we submit a given population of bacteria to a challenge and they survive, one can be reasonably certain that the same mechanism of survival will appear in many other unrelated cases.
3. *Rapidity*. Bacteria can adapt very quickly depending on the strain and on the antibiotic. Acquired resistance to recently introduced antibiotics has been seen within a few years of initial introduction.
4. *Persistence*. The ecological impact of massive use of antibiotics in a given niche, for instance, the human gut, is (as far as we can ascertain) less than expected. No known organism was wiped out; none was substituted by fitter organisms. There are obviously local alterations of the flora, and even transient coloniza-

tion from new organisms, but in the long run the same or very similar microorganisms as those initially present regain their prevalent place in the ecosystem. There is a proviso here, since recent studies of the complex microbial population of the human gastrointestinal tract have shown that it is to a large part uncharacterized and also extremely variable.

5. *Novelty*. "New" genes are brought into play. They were not available in the previous history of any bacterial species. They seem to appear spontaneously from the apparently immense gene reservoirs available in bacteria present in, for example, soil and marine sediments (see below).

5 ACQUIRED ANTIBIOTIC RESISTANCE BRINGS INTO PLAY A VARIETY OF ANCIENT GENES

As mentioned previously, the majority of resistance genes bear little or no relationship to bacterial genes not related to Ab^R. In the absence of such relationships, they must have been delivered to the bacterial pathogen by HGT; but to date no clear ancestors (metabolic counterparts) have been found for many of the Ab^R genes. They code for proteins that belong to families distinct from known protein families (however, in many cases, they belong to protein superfamilies, as will be discussed below). Thus, they are likely to be very ancient genes, and their evolutionary roots may be difficult or impossible to trace (it is like asking what is the origin of glucose-6-P-dehydrogenase!) The origin of Ab^R genes becomes a problem of the origin of protein families.

Given the immense numbers and diversity of the bacterial pool (23), it is always possible that there is *one* organism for which the prevalent resistance mechanism is active in nature (in the sense described above). In these cases, Ab^R genes may result from rapid gene evolution, causing significant divergence from a parent gene of unknown bacterial origin (the clearest example may be that of resistance to the synthetic antibiotics trimethoprim and sulfonamides). The evolved resistance genes are then disseminated by HGT and are finally identified as mobile genes in many bacterial genera.

Unambiguously finding the ancestor may be an impossible task in most cases. However, examples of identifiable precursors are still recognizable in the origin of class A β-lactamase genes. Indigenous chromosomal genes for class A β-lactamases are present in a series of gram-positive and gram-negative organisms (24) and may have served as precursors for the lactamase genes found on mobile elements.

A similar example is that of a gene encoding streptogramin-A acetyl transferase activity, which has been found in the chromosome of all strains of *Yersinia enterocolitica*, a species already highly resistant to streptogramins owing to impermeability (25). The corresponding protein is 60% identical to those conferring streptogramin resistance in *Streptococcus* and *Staphylococcus* plasmids. Could this be the ancestor of their now mobile prevalent successors? It should be emphasized that a functional correlation is not enough to establish lineage between a modern Ab^R gene and its precursor. A close evolutive distance, as in the examples above, is also required to establish this relationship unambiguously. This closeness does not exist in the case of the gentamicin acetylating enzyme (AAC2) found in the chromosomes of *Providencia* (26) and *Mycobacterium fortuitum* (27). Even when these genes become Ab^R determinants upon overexpression, as can be shown, they have never been used by nature in the process of evolution of resistance.

From a biochemical standpoint, a wide range of different mechanisms of resistance can be acquired by various genetic processes, including uptake and assimilation of DNA not carried by mobile genetic elements. In the majority of cases, these resistance determinants could not have been derived by mutation, since they require biochemical reactions that are clearly foreign to the host bacterium (but not necessarily foreign to other bacteria, such as antibiotic-producing streptomycetes). Exceptions to this are the cases when an enhanced system of antibiotic efflux is acquired by HGT (28,29).

Resistance to sulfonamides and trimethoprim is the result of bypass mechanisms, since the resistant strain carries two enzymes that perform the same biochemical function; one sensitive (the native enzyme) and one resistant (acquired) to inhibition. The resistance enzymes have altered kinetic constants for their substrates.

6 ECOLOGICAL ROLES FOR "NATURAL" ANTIBIOTICS MAY EXPLAIN THE ANCIENT NATURE OF SOME RESISTANCE GENES

Although the definition of an antibiotic originally referred to a microbial product, this now encompasses any low molecular weight compound that inhibits the growth of microbes (the term *antimicrobial* is correct and more general and is also in common use: antimicrobial agent = antibiotic). Examination of the mechanisms of resistance to "natural" antibiotics compared to chemically synthesized compounds reveals interesting differences (see Table 1); several different types of mutation give rise to resistance to both classes but, in contrast, resistance to synthetic (nonnatural) inhibitors only rarely develops by gene acquisition. The principal exception is that of efflux. Additional exceptions are those of resistance to sulfonamides and to trimethoprim for which the source organism(s) remain a mystery. The fact that resistance to naturally occurring antibiotics (or their close analogues) is frequently acquired implies that there is a link between the production of these antibiotics and their cognate resistance determinants.

The natural biological functions of the low molecular weight microbial metabolites that are defined as antibiotics are unclear (30). These compounds are produced in relatively small quantities in nature (much lower than the yields obtained from industrially improved strains of the pharmaceutical industry) and the concentrations at which antibiotics are used for therapeutic purposes considerably exceed their concentrations in the environment. Also, most antibiotics are strongly adsorbed by soil and their efficacy as inhibitors is likely to be reduced. Finally, antibiotics are sometimes transcriptional modulators of their corresponding resistance genes (i.e., tetracycline in Tn*1721* and conjugative transposons), underscoring the process of adaptation between antibiotics and resistance mechanisms. At least partly because of these reasons, it has been proposed that so-called antibiotics rarely play this role in nature and that they play roles as intercellular and intracellular modulators of biochemical function, although adsorption by soil, for example, may also influence this activity. However, since most bacteria in the environment exist as complex multicomponent complexes (biofilms, aggregates, or consortia), this may permit the low molecular weight metabolites to exercise their effects over other microbes through the matrix. The physiological consequences of subinhibitory concentrations of antibiotics (those likely to be encountered in soil, for instance) are largely unknown. To take one example, streptomycin leads to error-prone protein synthesis; since the error frequency in normal protein synthesis is quite high (10^4–10^5 per amino acid position), increasing this error frequency means that many proteins produced in susceptible bacteria in the presence of a subinhibitory concentration of streptomycin will likely contain amino acid substitutions. Thus, if antibiotics are produced under stress conditions, they can help bacteria to "explore" the catalytic possibilities of the protein surfaces they have available. In this way, new metabolic or detoxifying activities might be found. These "protein mutations" would allow bacteria to survive for a few generations, sufficient to permit the appearance of classic gene mutations resulting in antibiotic resistance. This argument is further reinforced if we take into account the fact that streptomycin is itself mutagenic (31).

These arguments do not say that compounds such as streptomycin, erythromycin, and tetracycline, are not antibiotics in the environment; however, we emphasize that they could exercise a number of different environmental functions that remain to be identified. The homoserine lactones and butyrolactones provide good models. These secondary metabolites were identified as being critical quorum-sensing signaling agents, but some act as growth inhibitors when studied in the laboratory (32). Suffice it to say that genuine studies of the natural role of the enormous structural diversity produced by microbial secondary biosynthetic pathways are few and far between. As scientists, we find this an intellectual void that needs to be filled. It is encouraging that more attention is now being focused on interactions among communities of bacteria (e.g., "biofilm" studies) and to the molecules that are secreted by some members of a community that affect the physiology of neighboring bacteria. Antibiotics and other microbial metabolites could be somehow considered as being hormones or growth modulators, that is, microbial signals, the effects of which are essential if we want to understand the nature of bacterial communities.

7 INTERPLAY BETWEEN MUTATION AND ACQUISITION IN THE DEVELOPMENT OF MOST OBSERVED ANTIBIOTIC RESISTANCE MECHANISMS

When large amounts of antibiotics were introduced into the environment at the start of the antibiotic era, they were inflicted on a largely unprepared population of microbes. Studies of the genetics of Ab^R in *E. coli* and other laboratory strains showed that mutants could be isolated at various frequencies depending on the selecting concentration. Quite high frequencies could be obtained with low concentrations of antibiotics—as were likely to be encountered in nature and even in patients. It can be assumed, therefore, that the initial response to clinical use of antibiotics was the appearance of populations of bacteria resistant owing to mutation. The biochemical nature of the mutations has not been well studied, although there is evidence that enhanced efflux was a primary response (28). Even if such resistance is at a low level and somewhat nonspecific, it would suffice to provide the microbes with some measure of protection against a variety of antibiotics (12). It must be assumed that increased use of higher concentrations of antibiotics not only led to increased levels of resistance (additional mutations) but the development of resistance in a wider spectrum of bacterial species.

The case of the extended-spectrum β-lactamases (ESBLs) which have evolved through *natural* protein engineering under the selective pressure of a series of increasingly potent β-lactam and related antibiotics is perhaps the best studied example of the interplay between acquisition and mutation in resistance development. The alterations in enzyme specificity and kinetic efficacy resulting from a series of single base (single amino acid) substitutions has been recorded in detail (http://www.lahey.org/lcinternet/studies/webt.htm) and provides a spectacular example of the enzyme evolution in response to drug selection. The process is ongoing! (33). The ESBLs are not the only example of mutational change of Ab^R, which has also been characterized in the cases of trimethoprim (34) and quinolone resistance (35).

The role of compensatory mutations in the evolution of Ab^R has been revealed recently, but there are undoubtedly many other instances to be identified. Simply described, the acquisition of a resistant phenotype by mutation or gene acquisition has been shown to reduce the "fitness" of bacteria. This has been demonstrated by the studies of Lenski (36,37) and Levin (38). However, second site mutations rapidly compensate for this reduced fitness (39). It is evident from studies of Sköld (40) that compensatory mutations occur in other resistance genes such as *sul* (sulfonamide resistance). The mechanism by which compensatory mutations operate can only be surmised, since there are limited biochemical studies of the phenomenon. The assumption is that these mutations reverse the negative effect on the function (structure?) of the altered protein or macromolecule. What happens in the case of plasmid compensation? This is particularly interesting, since reversion of the original resistance allele or loss of the plasmid (leaving the compensatory mutation with nothing to compensate!) reduces fitness; thus the presence of the two mutations is necessary for normal growth. In essence, fitness now becomes dependent on the resistance mutation or resistance plasmid in the absence of the antibiotic, making it unlikely that Ab^R can be completely eliminated in a population by discontinuing antibiotic use. This "catch 22" situation might explain why plasmids carrying streptomycin and chloramphenicol resistance are still frequent in the bacterial population even though these antibiotics are very rarely used. There is now a selective advantage to maintaining such resistance determinants in the population when compensation has been genetically fixed. Given the prevalence of multi–drug-resistant *M. tuberculosis* carrying as many as five resistance mutations, the role of genetic compensation in such strains must be significant.

8 GENETIC ARCHEOLOGY—WHAT HAPPENED WHEN ANTIBIOTICS WERE FIRST USED?

The genetic archeology of Ab^R is a complex subject—like any archeological investigation. One is studying past events by the examination of "fossils" and laboratory reconstructions. In the case of Ab^R, there are many factors to be considered: the sources of the genes, acquisition of resistance genes and their expression (which is necessary for the new host to gain a selective advantage), adaptation of the gene to the new host and of the host to the acquired genetic and physiological burden, and dissemination of the resistance genes for the benefit of the population. As with classic archeology, not all of the components contributing to an event can be identified and the event must be described in the simplest terms. In the case of Ab^R, the populations are large, complex, and ill defined; resistance developed as a result of a bacterial community activity. It is now known that some 99% of bacteria in a given environment cannot be grown in the laboratory; thus, as with all bacterial population studies, the organisms that can be grown represent the tip of the iceberg—the same must be true for resistance genes and organisms. All of the gene-tracking studies done in epidemiological studies of resistance have been carried out with pathogens and one day we must ask what has happened in the nonpathogenic majority.

In summary, we would like to propose a scenario with three stages to explain most processes of Ab^R selection:

1. *The proactive stage.* There are preexisting genes in the chromosomes of one or a number of bacterial genera which can give rise to Ab^R genes by mutation. These can be either ancient resistance genes of antibiotic-producing organisms or genes of unknown function in any of the dozens of thousands of bacterial genera, most of them unculturable (and thus uncountable). Bacterial diversity is thus at the root of any significant resistance mechanism.
2. *The acquired response stage.* A series of selection gradients act on complex populations. Some bacteria had received resistance genes, especially as mobile genetic elements recruited and spread through the population. These newly integrated genes interact with the "stable" fraction of bacterial genomes and adapt to them. Following the waves of gene shuffling and adaptation, bacteria can explore and ascend the selection gradients. Among the genes mobilized by selection in the different population compartments described above there is competition, as well as sharing.
3. *The selection stage.* When we look at the selected bacterial population we only see the surviving genes which have already spread among the global bacterial population. Therefore, we see only a few of the resistance genes; the last chapter of a long series that probably involved thousands of genes and hundreds of mechanisms in many diverse species. We have lost track of the quantum leaps that make the appearance of a given resistance mechanism predictable. Thus, bacterial response to a simple evolutionary force is complex and, in the specific mechanisms selected, unpredictable.

9 CONCLUSIONS

The mechanisms used by bacteria to respond to a simple evolutionary challenge were strikingly different to what was expected and analysis of these mechanisms has uncovered a whole new area of microbiological research (mobile genetic elements). Antibiotics themselves may be but the tip of the iceberg covering just another hidden aspect of microbiological nature: the social life of microbes. This is just another example of biological serendipity by which an unrelated observation uncovers new subjects of enquiry: What is the role of antibiotics in nature? Are bacteria really organized in networks of communicating organisms? Are many of the nonculturable bacteria of soil or sea in intermediate stages between life and death? How is death defined in bacteria?

REFERENCES

1. Hughes VM, Datta N. Conjugative plasmids in bacteria of the 'pre-antibiotic' era. Nature 1983; 302:725–726.
2. Levy SB. The antibiotic paradox: how miracle drugs are destroying the miracle. New York: Plenum Press, 1992:xiv, 279.
3. de la Cruz F, Davies J. Horizontal gene transfer and the origin of species: lessons from bacteria. Trends Microbiol 2000; 8:128–133.
4. Davies J, Webb V. Antibiotic resistance in bacteria. In: Krause RM, ed. Emerging Infections. San Diego: Academic Press, 1998.
5. Chadwick D, Goode J. Antibiotic resistance: origins, evolution, selection, and spread. Ciba Foundation symposium 207. Chichester: Wiley, 1997:ix, 250.
6. Heinemann JA. How antibiotics cause antibiotic resistance. Drug Discovery Today 1999; 4:72–79.
7. Bryan LE, Godfrey AJ, Schollardt T. Virulence of *Pseudomonas aeruginosa* strains with mechanisms of microbial persistence for beta-lactam and aminoglycoside antibiotics in a mouse infection model. Can J Microbiol 1985; 31: 377–380.
8. Hurst A. Bacterial injury: a review. Can J Microbiol 1977; 23:935–944.
9. Amann RI, Ludwig W, Schleifer KH. Phylogenetic identification and in situ detection of individual microbial cells without cultivation. Microbiol Rev 1995; 59:143–169.
10. Bloomfield SF, Stewart GS, Dodd CE, Booth IR, Power EG. The viable but non-culturable phenomenon explained? Microbiology 1998; 144:1–3.
11. Bridges BA. The role of DNA damage in stationary phase ('adaptive') mutation. Mutat Res 1998; 408:1–9.
12. Baquero F, Negri MC, Morosini MI, Blazquez J. The antibiotic selective process: concentration-specific amplification of low-level resistant populations. Ciba Found Symp 1997; 207:93–105.
13. Davison J. Genetic exchange between bacteria in the environment. Plasmid 1999; 42:73–91.

14. Arber W. Genetic variation: molecular mechanisms and impact on microbial evolution. FEMS Microbiol Rev 2000; 24:1–7.
15. Davies J. Vicious circles: looking back on resistance plasmids. Genetics 1995; 139:1465–1468.
16. Spratt BG. Hybrid penicillin-binding proteins in penicillin-resistant strains of *Neisseria gonorrhoeae*. Nature 1988; 332:173–176.
17. Spratt BG, Bowler LD, Zhang QY, Zhou J, Smith JM. Role of interspecies transfer of chromosomal genes in the evolution of penicillin resistance in pathogenic and commensal *Neisseria* species. J Mol Evol 1992; 34:115–125.
18. Alekshun MN, Levy SB. The mar regulon: multiple resistance to antibiotics and other toxic chemicals. Trends Microbiol 1999; 7:410–413.
19. Nilsen IW, Bakke I, Vader A, Olsvik O, El-Gewely MR. Isolation of cmr, a novel *Escherichia coli* chloramphenicol resistance gene encoding a putative efflux pump. J Bacteriol 1996; 178:3188–3193.
20. Bissonnette L, Champetier S, Buisson JP, Roy PH. Characterization of the nonenzymatic chloramphenicol resistance (cmlA) gene of the In4 integron of Tn1696: similarity of the product to transmembrane transport proteins. J Bacteriol 1991; 173:4493–4502.
21. Murray IA, Shaw WV. O-Acetyltransferases for chloramphenicol and other natural products. Antimicrob Agents Chemother 1997; 41:1–6.
22. Doolittle WF. Lateral genomics. Trends Cell Biol 1999; 9:M5–8.
23. Whitman WB, Coleman DC, Wiebe WJ. Prokaryotes: the unseen majority. Proc Natl Acad Sci USA 1998; 95:6578–6583.
24. Campbell JI, Scahill S, Gibson T, Ambler RP. The phototrophic bacterium *Rhodopseudomonas capsulata* sp108 encodes an indigenous class A beta-lactamase. Biochem J 1989; 260:803–812.
25. Seoane A, García-Lobo JM. Identification of a streptogramin acetyltransferase gene in the chromosome of *Yersinia enterocolitica*. Antimicrobial Agents Chemother 2000; 44:905–909.
26. Rather PN, Orosz E, Shaw KJ, Hare R, Miller G. Characterization and transcriptional regulation of the 2′-N-acetyltransferase gene from *Providencia stuartii*. J Bacteriol 1993; 175:6492–6498.
27. Ainsa JA, Martin C, Gicquel B, Gomez-Lus R. Characterization of the chromosomal aminoglycoside 2′-N-acetyltransferase gene from *Mycobacterium fortuitum*. Antimicrob Agents Chemother 196; 40:2350–2355.
28. Saier MH, Jr, Paulsen IT, Sliwinski MK, Pao SS, Skurray RA, Nikaido H. Evolutionary origins of multidrug and drug-specific efflux pumps in bacteria. Faseb J 1998; 12:265–274.
29. Roberts MC. Tetracycline resistance determinants: mechanisms of action, regulation of expression, genetic mobility, and distribution. FEMS Microbiol Rev 1996; 19:1–24.
30. Davies JE. Origins, acquisition and dissemination of antibiotic resistance determinants. Ciba Found Symp 1997; 207:15–27.
31. Ren L, Rahman MS, Humayun MZ. *Escherichia coli* cells exposed to streptomycin display a mutator phenotype. J Bacteriol 1999; 181:1043–1044.
32. Schripsema J, de Rudder KE, van Vliet TB, et al. Bacteriocin small of *Rhizobium*

leguminosarum belongs to the class of N-acyl-L-homoserine lactone molecules, known as autoinducers and as quorum sensing co-transcription factors. J Bacteriol 1996; 178:366–371.
33. Petrosino J, Cantu C, 3rd, Palzkill T. beta-Lactamases: protein evolution in real time. Trends Microbiol 1998; 6:323–327.
34. Adrian PV, du Plessis M, Klugman KP, Amyes SGB. New trimethoprim-resistant dihydrofolate reductase cassette, dfrXV, inserted in a class 1 integron. Antimicrob Agents Chemother 1998; 42:2221–2224.
35. Rattan A. Mechanisms of resistance to fluoroquinolones. Natl Med J India 1999; 12:162–164.
36. Papadopoulos D, Schneider D, Meier-Eiss J, Arber W, Lenski RE, Blot M. Genomic evolution during a 10,000-generation experiment with bacteria. Proc Natl Acad Sci USA 1999; 96:3807–3812.
37. Elena SF, Ekunwe L, Hajela N, Oden SA, Lenski RE. Distribution of fitness effects caused by random insertion mutations in *Escherichia coli*. Genetica 1998; 103:349–358.
38. Andersson DI, Levin BR. The biological cost of antibiotic resistance. Curr Opin Microbiol 1999; 2:489–493.
39. Andersson DI, Bjorkman J, Hughes D. [Antibiotic resistance here to stay? Compensatory mutations restore virulence of resistant bacteria]. Lakartidningen 1998; 95:3940, 3943–3944.
40. Swedberg G, Fermer C, Skold O. Point mutations in the dihydropteroate synthase gene causing sulfonamide resistance. Adv Exp Med Biol 1993; 338: 555–558.
41. Davies J, Wright GD. Bacterial resistance to aminoglycoside antibiotics. Trends Microbial 1997; 5:234–240.
42. Schnappinger D, Hillen W. Tetracyclines: antibiotic action, uptake, and resistance mechanisms. Arch Microbiol 1996; 165:359–369.
43. Leclercq R, Courvalin P. Intrinsic and unusual resistance to macrolide, lincosamide, and streptogramin antibiotics in bacteria. Antimicrob Agents Chemother 1991; 35:1273–1276.
44. Pitout JD, Sanders CC, Sanders WE, Jr. Antimicrobial resistance with focus on beta-lactam resistance in gram-negative bacilli. Am J Med 1997; 103:51–59.
45. Arthur M, Reynolds P, Courvalin P. Glycopeptide resistance in enterococci. Trends Microbiol 1996; 4:401–407.
46. Leon J, Garcia-Lobo JM, Navas J, Ortiz JM. Fosfomycin-resistance plasmids determine an intracellular modification of fosfomycin. J Gen Microbiol 1985; 131:1649–1655.
47. Piddock LJ. Mechanisms of fluoroquinolone resistance: an update 1994–1998. Drugs 1999; 58:11–18.
48. Enright M, Zawadski P, Pickerill P, Dowson CG. Molecular evolution of rifampicin resistance in *Streptococcus pneumoniae*. Microb Drug Resist 1998; 4:65–70.
49. Huovinen P, Sundstrom L, Swedberg G, Skold O. Trimethoprim and sulfonamide resistance. Antimicrob Agents Chemother 1995; 39:279–289.

3

Global Response Systems That Cause Resistance

Paul F. Miller
Pfizer Global Research and Development, Groton, Connecticut

Philip Rather
Case Western Reserve University, Cleveland, Ohio

The majority of attention to antibiotic resistance mechanisms has been justifiably focused on those factors that are highly transmissible among species and that lead to high levels of resistance to a specific class of antibiotics. Less is known about the ability of bacteria to alter their susceptibility to noxious agents by modulating their own intrinsic physiological systems to affect resistance. We will describe two of the better-studied examples of this latter situation, both of which occur in gram-negative species.

In the first example, the *mar/sox* regulatory network found in *Escherichia coli* was described. This system acts to modulate factors that limit the accumulation of a wide range of noxious agents, including several clinically important antibiotics. As such, a network of sensory and regulatory factors will be discussed that operate to control the expression of genes whose products either actively extrude antibiotics or enhance the effectiveness of external permeability barriers. As Lewis and Lomovskaya will specifically address efflux pumps in Chapter 4, our discussion will focus on the structure and function of the *marRAB* and *soxRS* regulatory loci. Evidence will be reviewed describing the high degree of molecular redundancy shared by these two regulatory systems, leading toward the concept that these are two semi-independent sensory systems that control a nearly identical set of target genes, although in quantitatively different ways. These differences may reflect the distinct types of signals that are sensed by the two systems, such that a protective response to inducers of one (e.g., superoxide-generating compounds for *soxRS*) may require a slightly different gene expression pattern than would the response to inducers of the second (phenolic agents and antibiotics for *marRAB*).

In the second example, we described the regulatory mechanisms controlling the *aac(2′)-Ia* gene in *Providencia stuartii*. The *aac(2′)-Ia* gene is a member of a growing family of chromosomally encoded aminoglycoside acetyltransferases that are intrinsic to certain bacterial species. Although the role of these acetyltransferases is largely unknown, the AAC(2′)-Ia enzyme in *P. stuartii* functions as a peptidoglycan *O*-acetyltransferase. Given the possibility of diverse functions for these enzymes, we anticipate that the regulation of these genes will involve distinct mechanisms. However, the information on *aac(2′)-Ia* expression that has been compiled to date may serve as a useful preliminary model for other systems.

1 INTRODUCTION

Microorganisms live in intimate proximity to their environment. For free-living species, this situation equates to the constant threat of exposure to a wide variety of potentially toxic agents produced either deliberately (e.g., by other organisms for defense against microbial attack) or as a consequence of normal organic turnover. Similarly, commensal and pathogenic organisms must protect themselves from both specific and nonspecific agents elicited by the host. Not surprisingly then, unicellular species have evolved an elaborate array of defenses designed to reduce or prevent the accumulation of unwanted toxic substances. There is, for example, a remarkable inventory of efflux systems that can be identified in the genomes of specific bacteria. The mechanisms by which efflux pumps operate are discussed elsewhere in this volume (see Chap. 4).

With such a genetic investment in defense systems, it also makes sense that these organisms would possess similarly intricate regulatory mechanisms which allow them to control the deployment of these systems. In this chapter, we will highlight our understanding of a few of the better-characterized regulatory systems, including global resistance systems and intrinsic modifying enzymes. Although the systems described in this chapter have been studied primarily in *Escherichia coli* and *Providencia stuartii*, it is reasonable to expect that these systems will serve as formal paradigms for as yet undiscovered control networks in other bacterial species.

2 GLOBAL REGULATORS OF ANTIBIOTIC RESISTANCE IN *E. COLI*

2.1 The *mar* Regulatory Locus

Undoubtedly the best-characterized global antibiotic resistance regulatory system is the *mar* system in *E. coli*. An excellent review of the molecular genetics of this system has been published recently (1). Much of the detailed work described in that chapter will be only summarized here, and the reader is encouraged to look to that source for additional detailed information. The *mar* locus was first described in 1983 in the pioneering studies of George and Levy. As a component of an ongoing effort to understand the mechanisms contributing to tetracycline resistance, these investigators identified a locus on the *E. coli* chromosome that was associated with the frequent emergence of low-level resistant strains (2). Moreover, it was shown that these tetracycline-resistant (tet^r) strains had also acquired a simultaneous resistance to other structurally unrelated antibiotics including chloramphenicol, rifampicin, and fluoroquinolones (2); mechanistically this phenotype was associated with reduced accumulation and efflux of the affected agents (2–4). The substrate spectrum for this system was later expanded to include certain organic solvents and disinfectants (5,6). A Tn*5* insertion at the 34-min region of the chromosome reversed the resistance phenotype for all of these agents, and identified the genetic locus designated as *mar* (for *m*ultiple *a*ntibiotic *r*esistance) (7). DNA sequence analysis of cloned genetic segments that could complement the Mar^- phenotype associated with either the Tn*5* insertion or a larger chromosomal deletion encompassing this region revealed a three-gene regulatory operon, designated *marRAB* (8–11). The Tn*5* insertion originally isolated by George and Levy was located in the second gene, called *marA*. Overexpression of this gene by itself was shown to be sufficient to confer the Mar phenotype in all cell types, including strains deleted for this region of the chromosome (12). The deduced protein product of this gene, MarA, is related by amino acid sequence similarity to a family of transcriptional activators, the prototype for which is the AraC regulator that controls genes involved in the metabolism of arabinose (13). This observation suggested that the Mar phenotype resulting from a mutation at the *mar* locus was likely due to an indirect mechanism, with MarA serving to control the expression of genes located elsewhere on the chromosome. It is presumably these target genes that are the more direct effectors of antibiotic resistance.

If overexpression of *marA* is sufficient to confer a Mar phenotype, then the *mar* locus must be capable of controlling the expression of *marA*. This proved to be the case, and the first gene in the operon, *marR*, was determined to play a critical role in this process (9). Unlike MarA, the MarR protein, at the time of its sequencing, bore little similarity to any known genes. However, analysis of selected Mar isolates showed that the majority of these bore mutations in *marR* and concomitantly exhibited elevated levels of the *marRAB* transcript (9,11,14). Introduction of a wild-type copy of *marR* in *trans* on a plasmid reversed the Mar phenotype, indicating that the *marR* mutations were recessive and that this gene encoded a repressor of *marRAB* operon expression. Results of genetic experiments suggested that the target for MarR repression is the operator/promoter region of the *marRAB* operon, *marOP*, as one could titrate the repressing activity of MarR simply by introducing additional copies of *marOP* on a plasmid (9,14). This finding was confirmed biochemically by showing that purified MarR protein bound specifically to *marOP* DNA sequences (15).

At roughly the same time as the original George and Levy experiments, it was noted that exposing *E. coli* cells to the weak aromatic acid salicylate (SAL) induced a condition of *p*henotypic *a*ntibiotic *r*esistance subsequently referred to as Par (16). Notably, SAL treatment conferred resistance to the same diverse group of antibiotics as was observed for the *mar* mutants. These findings converged mechanistically when it was found, through the use of a *mar-lacZ* fusion, that SAL treatment led to an induction of *marRAB* expression (17). Importantly, this was the first observation that connected the *mar* regulatory locus with what might be going on outside of the cell. Deletion of the *marRAB* operon led to a greatly reduced responsiveness to SAL as an inducer of antibiotic resistance, and to a hypersensitivity to many of the same agents that were found affected by the original *mar* mutants (11,12,17). The extent to which this hypersensitivity was observed depended on the specific *E. coli* strain background in use (8,11,18).

These studies suggested that the following hierarchy could explain inducible antibiotic resistance mediated by the *marRAB* system. The *mar* locus is normally maintained in a quiescent state owing to the autorepressor activity of the *marR* gene product. Exposure to a specific inducer such as SAL leads to the binding of the inducer by MarR, antagonizing its ability to mediate transcriptional repression of the *marRAB* operon. This results in an increase in transcription of the *marRAB* genes, leading to an increase in the abundance of the products of these genes in the cell. MarA, the proximal activator of target genes involved in the antibiotic resistance response, thus becomes available in sufficient quantities to diffuse to other sites on the chromosome and activate its target genes. A more detailed discussion of the targets and inducers in the *mar* regulatory network will be discussed below.

2.2 The *soxRS* System

Exposure of *E. coli* cells to various redox-cycling agents such as paraquat leads to the induction of a number of genes which collectively constitute the superoxide stress response (19). Constitutive mutants have been selected in which the expression of these target genes is elevated in the absence of any inducing agent. Such regulatory mutants typically map to the *soxR* locus; located at 92 min on the *E. coli* chromosome (20). Notably, these constitutive regulatory mutants also exhibit a concomitant antibiotic resistance phenotype, which is remarkably similar to that associated with *mar* mutants. In addition, one such regulatory mutant with a very similar phenotype, known as *soxQ1*, mapped to the *marA* locus (21).

Molecular dissection of the *soxR* locus revealed two divergently transcribed regulatory genes, *soxR* and *soxS*. The constitutive *sox* mutants mapped to *soxR*, and have been referred to as *soxR*(Con) alleles to distinguish them from nonfunctional mutants. Gene-expression studies showed that the expression of *soxR* is unaffected by either superoxide-generating agents or the constitutively activating mutations (22,23). In contrast, expression of *soxS* is induced by redox-cycling agents as well as by *soxR*(Con) mutants, and an intact *soxR* gene is required for induction of *soxS* expression as well as that of superoxide stress response target genes (22,23). Similar to findings described above for *marA*, overexpression of *soxS* was shown to be sufficient to both activate the expression of superoxide stress response target genes as well to confer the antibiotic resistance phenotype (22,23). These findings, combined with the recognition that the SoxR protein contains iron-sulfur clusters in its C-terminal region that are characteristic of those involved with redox reactions, suggested that SoxR activity (and not expression) may be modulated in response to superoxide radicals, and led to a better molecular understanding of the two-stage model for control of this regulon (24,25). In this model, exposure to agents or conditions leading to an accumulation of superoxide radicals results in the conversion of inactive SoxR to an activated form. Activated SoxR then induces the transcription of the adjacent *soxS* gene, whose product stimulates the expression of the unlinked regulon genes, the products of which presumably engender resistance to superoxide radical–generating agents and Mar-type antibiotics. This would then explain why *soxR*(Con) mutants are active in the absence of a small molecule activator.

Additional observations tied the *soxRS* regulon to the *mar* system. Along with the observations that *soxR*(Con) mutants have a Mar phenotype, and that the *soxQ1* mutant mapped near *marA*, another mutant that was initially selected based on its strong Mar phenotype was found to map to the *soxR* locus (26). Reconciliation of these genetic observations began when it was recognized that MarA and SoxS, the proximal activators in these regulatory systems, are closely related members of the AraC family of transcription factors (13). Thus, overexpression of either *soxS* or *marA* leads to both a Mar phenotype as well as induction of the superoxide stress response target genes. However, these regulators do not behave in completely redundant ways, as there appear to be quantitative differences in the effects of these activators on the different target genes that have been studied to date. For example, *marA* overexpression tends to produce a greater level of antibiotic resistance and a smaller induction of superoxide stress response target genes, such as *nfo*, than does *soxS* (21,26).

2.3 Rob—A Third Regulator?

E. coli contains another gene whose product exhibits significant amino acid sequence similarity to MarA and SoxS. This protein, known as Rob, was first identified as a factor that binds to the chromosomal origin of replication (27). It is larger than either MarA or SoxS, and it appears to contain an additional domain not found in the other two proteins. It is also different in that it is constitutively expressed at high levels, increasing in concentration as cells transition from logarithmic to stationary phase. Although higher level induction of Rob accumulation has been shown to confer a Mar phenotype, and purified Rob protein has been shown to bind to MarA/SoxS target promoters in vitro (28,29), a physiological role for this protein in antibiotic resistance has yet to be observed. In addition, mutants affecting intrinsic antibiotic resistance have yet to be linked to the *rob* gene. For these reasons, this interesting and mysterious protein will not be described further here.

2.4 A Single Regulon with Two Activators

As has been proposed recently, it now seems reasonable to consider the existence of a single stress response regulon that is controlled by multiple related regulators (30). This could be called the *mar* regulon, as has been proposed, or be referred to by a more general descriptor to reflect the distinct stresses that lead to its activation. Regardless, the important consequence from the perspective of this chapter is that intrinsic antibiotic resistance is affected. We shall now consider more distal and proximal components of this pathway.

2.5 Regulon Targets and Antibiotic Resistance

Recent work has led to a greater understanding of the target-binding site in MarA and SoxS responsive promoters (30–32). Work with MarA has suggested that this activator interacts with target promoters as a monomeric protein, and that it can bind in either of two orientations to effect transcription. However, the orientation of the binding site in a given promoter must be as it originally exists in that element; inverting it leads to a loss of MarA responsiveness. In addition, distinct spacing rules appear to exist regarding the distance between the "marbox" and the binding sites for RNA polymerase (RNP), depending on whether the marbox is present in the + or − orientation. Marboxes that are located on the opposite strand from that of the RNP–binding sites (−35 and −10 sequences) are positioned further upstream than are those that are found on the same strand as the RNP-binding site (30,33,34). It has been proposed that these positions and orientations allow MarA to interact productively with RNP in either orientation.

Marboxes that have been found upstream from a number of target promoters in *E. coli* have been aligned to generate a consensus-binding site (30). Despite significant experimental work, this consensus remains quite degenerate. From the x-ray crystal structure studies of MarA, it has been proposed that MarA interacts with specific promoter elements by way of an interaction of complementary shapes that are held together by Van der Waals forces (35). By inference, it seems reasonable to expect that many of the mechanistic observations made for MarA will also be applicable to SoxS. This is supported by the biochemical studies that have been conducted with this latter protein, and its interaction with known target genes (31,32,34,36). Thus, several of the genes containing marbox elements in their promoters have been implicated by both genetic and biochemical methods as specific targets for MarA and/or SoxS control. Because of the focus of this volume, those key target genes implicated in antibiotic resistance will be discussed in further detail here.

2.5.1 *micF*

One of the earliest physiological observations associated with the Mar phenotype was a downregulation of the major outer membrane porin OmpF (37); this has also been observed following SAL treatment (38). This porin forms a large outer membrane channel through which low molecular weight, water-soluble compounds can diffuse. Thus, a reduction in the abundance of this channel in *mar* mutants fits well with the reduced antibiotic accumulation phenotype observed with these strains. Studies of OmpF regulation revealed that one form of negative control involved a posttranscriptional repression mechanism mediated by the antisense RNA *micF* (39,40). Experiments with *micF-lacZ* fusions as well as *micF-lacZ* deletions demonstrated that *mar* mutants have elevated levels of *micF* expression, and that *mar*-mediated downregulation of OmpF requires an intact *micF* gene (12,41). However, using strains deleted for the *ompF* gene, it was also shown that a simple loss of OmpF from the outer membrane was not sufficient to confer a Mar phenotype (12). Thus, additional *marA* targets appeared to be required for a full Mar phenotype.

2.5.2 *acrAB* and *tolC*

Accumulating experimental evidence on the structure and function of efflux pumps in gram-negative organisms (42) suggested that one of these export systems might play a role in *mar*-mediated antibiotic resistance. Subsequent genetic studies then showed that the multi–drug efflux pump encoded by the *acrAB* genes is required for the Mar phenotype, as a deletion of *acrAB* completely eliminated the Mar phenotype associated with *mar* mutants (43). Subsequently, it was noted that the promoter for the *acrAB* operon, as well as that of the *tolC* gene, whose product forms the outer membrane channel component of the AcrAB pump, contains a marbox element (30,44), which is bound by both MarA and SoxS in vitro.

2.5.3 *marRAB*

The promoter for the *marRAB* operon also contains a marbox element and is subject to autoactivation (45). This observation helped to rationalize earlier studies which showed that high-level expression of ether *soxS* or *marA* led to increased *marRAB* operon expression. The marbox in the *marRAB* promoter region is one of the most MarA-responsive elements studied to date (30).

As mentioned above, the SoxS protein is expected to bind to virtually the same set of target gene promoters as has been identified for MarA. This has been largely substantiated experimentally, and in many cases a SoxS interaction was demonstrated first (36). If this were true, then the explanation for the different affects of *marA* versus *soxS* induction on multiple antibiotic resistance, or the superoxide stress response, must lie in the quantitative ways in which these two regulators interact with their target promoters. This hypothesis is supported by recent evidence (46). The marbox elements in different regulon promoters respond differently to MarA or SoxS induction. This difference was shown to be due to specific nucleotide sequence differences among the various marbox elements, and it was possible to vary the responsiveness of a promoter to MarA compared with SoxS by changing the sequence of a specific marbox (46). These findings may also provide an explanation for a perplexing observation associated with certain bases in the proposed consensus sequences. Some of the invariant positions in the consensus have nonetheless been shown to be dispensable for MarA responsiveness. Although one can consider it to be reasonable to propose that MarA and SoxS control an almost identical set of target genes (although in quantitatively different ways), it seems possible that these positions may be more important for SoxS binding than they are for MarA.

2.6 Mechanisms of Regulon Induction and Physiological Roles

Although much recent work has focused on the mechanisms by which MarA and SoxS interact with regulon target promoters, early studies were actually driven by observations that gave insights into regulon induction. For the *mar* system, this work centered on the phenolic compound salicylate (SAL) and its ability to stimulate *marRAB* expression (17). As mentioned earlier, *marRAB* induction involves antagonism of the MarR repressor; apparently by a direct interaction with SAL (15). The poor solubility of MarR in a purified form along with the relatively weak affinity of SAL for MarR has made further biochemical characterization of this interaction difficult. An understanding of the molecular details of the induction mechanism will require the determination of the MarR crystalline structure and, perhaps, the identification of more potent MarR-binding molecules.

In contrast, *soxRS* induction by superoxide-inducing agents is somewhat better understood. Genetic and biochemical experiments demonstrated that superoxide radicals, or some related species, indirectly activate *soxS* transcription via their effects on SoxR (24,25). As stated earlier, SoxR activation involves a cluster of iron-sulfur centers near the 3′ end of the protein.

In a potentially intriguing connection, the *mar* and *sox* regulatory system may be linked at the sensory level, much in the same way that they share target genes. In a series of preliminary studies, a collection of naturally occurring, plant-derived phenolic compounds was tested for their ability to induce either *marA* or *soxS* expression (48). It was noted that certain naphthoquinones, which were known to induce *soxS* expression, were also effective inducers of *marA* transcription (17,47). These observations led to the proposal that compounds of this sort may be the true inducers (and substrates?) (49), or may be related to the inducers of a progenitor stress response system that has subsequently duplicated and diverged into the present-day *mar/sox* system. This proposal is supported by the finding that MarR is related at the amino acid sequence level to regulatory proteins found in other bacterial species that are known to respond to phenolic compounds (50). Again, structural studies describing the specific interactions between MarR and the inducers that it binds will lead to a better understanding of the kinds of compounds that induce the *mar* system and will help shed light on this question.

In summary, the highly overlapping *mar* and *sox* systems represent intrinsic, inducible stress response networks in *E. coli* and other enteric species (1,51). The complexity of this regulatory network suggests that more significantly diverged microbes may have also evolved their own strategies to counter these same environmental challenges even if they lack recognizable *mar* and *sox* homologues. *mar* Mutants have been identified among clinical antibiotic-resistant isolates of *E. coli* (52), and a Mar phenotype has been observed among several other gram-negative quinolone-resistant strains (53). Observations of this sort raise the possibility that the role of *mar*-type mechanisms is substantially underappreciated in considerations of the factors affecting both intrinsic as well as acquired antibiotic resistance. Thus, the identification and characterization of these systems can only help us in our efforts to predict, avoid, and counteract antibiotic resistance.

3 INTRINSIC ACETYLTRANSFERASES IN BACTERIA

Although the above section describes a mechanism by which certain gram-negative bacteria can alter their permeability barriers to afford antibiotic resistance, a different approach involving antibiotic inactivation will now be presented. In this case, antibiotic resistance is restricted to a particular chemical class of agents, the aminoglycosides, but with apparent broad specificity among constituent components of this group. Because of the regulatory nature of many of the mutations described below, it is appropriate to consider this as an additional example of intrinsic global resistance.

Aminoglycoside resistance in bacteria is primarily mediated by the presence of plasmid-encoded modifying enzymes (54). These enzymes modify the aminoglycosides by acetylation, phosphorylation, or adenylylation (54). In addition to these plasmid-encoded enzymes, an expanding list of chromosomally encoded aminoglycoside acetyltransferases has been identified. For each of these enzymes, the corresponding gene appears to be intrinsic to the bacterial species in which it is found (55,56).

Therefore, it is possible that these intrinsic acetyltransferases act as housekeeping enzymes involved in the acetylation of cellular substrates.

Since these enzymes also acetylate aminoglycosides, there may be structural similarities between aminoglycosides and the cellular substrates for these enzymes. The AAC(2′)-Ia enzyme in *Providencia stuartii* has a role in peptidoglycan acetylation (see below) and the AAC(2′)-Id enzyme in *Mycobacterium smegmatis* has a role in lysozyme resistance, indicating a possible function related to the cell wall (57). Recently, the AAC(6′)-Iy enzyme has been identified in *Salmonella enterica* subsp. *enterica* serotype *Enteritidis* (58). The potential function of this enzyme may be related to sugar metabolism.

3.1 AAC(2′)-Ia in *Providencia stuartii*

3.1.1 Physiological Functions

Early studies on mutants which overexpressed the AAC(2′)-Ia enzyme indicated that they possessed altered cell morphology, forming small rounded cells. To address further the role of AAC(2′)-Ia, a null allele was created by introducing a frameshift mutation into the *aac(2′)-Ia* coding region by allelic replacement (59). The loss of *aac(2)-Ia* resulted in cells with a slightly elongated phenotype (59). Furthermore, the staining properties of *aac(2′)-Ia* mutant cells with uranyl acetate was altered relative to wild-type cells. The basis for this phenotype is unknown; however, it suggests changes in the surface properties of cells. These data suggested a possible role for AAC(2′)-Ia that is related to the cell envelope. Work done by Payie and Clarke has revealed that AAC(2′)-Ia functions as a peptidoglycan *O*-acetyltransferase (60). The *O*-acetylation of peptidoglycan is a modification that regulates the activity of autolytic enzymes involved in peptidoglycan breakdown and turnover (61,62). The altered cell morphology seen in cells with changes in *aac(2′)-Ia* expression may be due to the changes in the activity of autolytic enzymes. The AAC(2′)-Ia enzyme is capable of obtaining acetate from peptidoglycan, *N*-acetylglucosamine and from acetyl-coenzyme A (acetyl-CoA) (60). Interestingly, the AAC(2′)-Ia enzyme is released by osmotic shock and may be located in the periplasm. Since acetyl-CoA is located within the periplasm, the use of this substrate as a source of acetate would require a mechanism for transfer into the periplasm. The mechanism for such a transfer is unknown in *P. stuartii*.

3.1.2 Genetic Regulation

Studies on the regulation of *aac(2′)-Ia* have been conducted using *lacZ* reporter gene fusions to the *aac(2′)-Ia* promoter region. Early studies demonstrated that *aac(2′)-Ia* transcription was not inducible by subinhibitory amounts of aminoglycoside antibiotics (63). Using these fusions, two approaches have been used to identify gene products that act in *trans* to regulate *aac(2')-Ia*. The first approach involved selecting spontaneous gentamicin-resistant mutants of a *P. stuartii* strain harboring an *aac(2')-lacZ* fusion on a low-copy plasmid. One mechanism for the increased gentamicin resistance of these mutants would be increased expression of the chromosomal *aac(2')-Ia* gene. During these isolations, the predominant class of mutants were darker blue in the presence of X-gal, indicating increased transcription from the *aac(2')-Ia* promoter region on the plasmid. A second approach to identify regulatory mutants involved identifying transposon insertions (mini-Tn5Cm) that activated the *aac(2')-lacZ* fusion. Insertions that resulted in *aac(2')-lacZ* activation were then tested for increased expression of the chromosomal *aac(2')-Ia* gene. Using both of these strategies, genes designated *aar* (*a*minoglycoside *a*cetyltransferase *r*egulator) have been identified. The surprising number of regulatory genes that have been identified suggests the importance of modifying *aac(2')-Ia* expression in response to various environmental conditions. This would allow cells to fine tune the levels of peptidoglycan acetylation and regulate autolysis.

The *aar* genes are grouped into two classes. The first class of genes act phenotypically as negative regulatory genes, since loss of function mutations increase *aac(2')-Ia* expression. The second class of regulatory genes are those which act in a positive manner and are required for normal levels of *aac(2')-Ia* expression.

3.1.3 Negative Regulators

aarA. The *aarA* gene encodes a very hydrophobic polypeptide of 31.1 kD in size (64). The AarA protein contains at least two possible transmembrane domains, suggesting that it is an integral membrane protein. Homology searches of the databases with AarA resulted in no significant matches to other proteins. The AarA protein is required for the production or activity of an extracellular pheromone signal, AR-factor, that acts to reduce *aac(2')-Ia* expression. The *aarA* gene was identified as a mini-Tn5Cm insertion that increased gentamicin resistance levels eightfold above wild type. The *aarA* mutants increase *aac(2')-Ia* transcription 3- to 10-fold depending on the growth phase of cells. Null mutations in *aarA* are highly pleiotropic, and additional phenotypes include loss of production of a diffusible yellow pigment and a cell-chaining phenotype that is most prominent in cells at mid-log phase.

aarB. The *aarB3* mutation originally designated *aar3* (63) results in a 10- to 12-fold increase in *aac(2')-Ia* transcription. In the *aarB3* background, the levels of aminoglycoside resistance are increased 128-fold above wild type, suggesting that this mutation further increases aminoglycoside resistance in a manner independent of *aac(2')-Ia* expression. The *aarB3* mutation also results in altered cell morphology and a slow growth phenotype. The identity of the *aarB* gene has not been determined.

aarC. The *aarC* gene encodes a homologue of *gcpE*, a protein widely distributed in bacteria. A missense allele, *aarC1*, resulted in a number of pleiotropic phenotypes including slow growth, altered cell morphology, and increased *aac(2')-Ia* expression at high cell density (65). The biochemical function of AarC remains to be determined.

aarD. The *aarD* gene was identified by a mini-Tn5Cm insertion that resulted in a fivefold activation of an *aac(2')-lacZ* fusion and a threefold increase in the levels of *aac(2')-Ia* mRNA accumulation (66). In addition, a 32-fold increase in aminoglycoside resistance was observed in *aarD* mutants, relative to wild-type *P. stuartii*. The *aarD* locus encodes two polypeptides which are homologues of the *E. coli* CydD and CydC proteins (66–68). The CydD and CydC proteins act in a heterodimeric ABC transporter complex required for formation of a functional cytochrome d oxidase complex (69–72). *P. stuartii aarD* mutants exhibit phenotypic characteristics consistent with a defect in the cytochrome d oxidase, including hypersusceptibility to the respiratory inhibitors Zn2+ and toluidine blue (66).

The increased *aac(2')-Ia* expression observed in the *aarD1* background contributes minimally to the overall increase in gentamicin resistance, since introduction of the *aarD1* mutation into an *aac(2')-Ia* mutant strain also results in a 32-fold increase in gentamicin resistance. Previous studies have demonstrated that uptake of aminoglycosides is dependent on the presence of a functional electron transport system (73–75). Since electron transport is defective in the *aarD1* background (66), it is probable that a decrease in aminoglycoside uptake accounts for the high level of resistance observed in *aarD* mutants. However, the mechanism that contributes to increased *aac(2')-Ia* transcription is unknown. A direct role for *aarD* in the regulation of *aac(2')-Ia* is unlikely, since ABC transporters are not known to function as transcriptional regulators (76). A regulatory protein may couple changes in the redox state of the membrane to *aac(2')-Ia* expression (see below) (77). Mutations in *aarD* are predicted to alter the redox state of the membrane and thus indirectly affect *aac(2')-Ia* expression.

aarG. The *aarG* gene encodes a protein with similarity to sensor kinases of the two-component family with the strongest identity to PhoQ (57%). Immediately upstream of *aarG* is an open reading frame designated *aarR* which encoded a protein with 75% amino acid identity to PhoP, a response regulator (78,79). The regulatory phenotypes associated with the *aarG1* mutation may result from a failure to phosphorylate the putative response regulator AarR, which functions as a repressor of *aarP* and possibly *aac(2')-Ia*

A recessive mutation (*aarG1*) results in an 18-fold increase in the expression of β-galactosidase from an *aac(2')-lacZ* fusion (78). Direct measurements of RNA from the chromosomal copy of *aac(2')-Ia* have confirmed this increase occurs at the level of RNA accumulation. Taken together, these results demonstrate that loss of *aarG* results in increased *aac(2')-Ia* transcription. The *aarG1* allele also results in enhanced expression of *aarP*, encoding a transcriptional activator of *aac(2')-Ia* (see below) (80). Genetic experiments have shown that in an *aarG1, aarP* double mutant, the expression of *aac(2')-Ia* is significantly reduced over that seen in the *aarG1* background. However, the levels of *aac(2')-Ia* in this double mutant are still significantly higher than in a strain with only an *aarP* mutation. Therefore, the *aarG1* mutation increases *aac(2')-Ia* expression by both *aarp*-dependent and *aarP*-independent mechanisms.

The *aarG1* allele confers a multiple antibiotic resistance phenotype (Mar) to *P. stuartii* resulting in increased resistance to tetracycline, chloramphenicol, and fluoroquinolones. This Mar phenotype in the *aarG1* background is partially due to overexpression of *aarP*, which is known to confer a Mar phenotype in both *P. stuartii* and *E. coli* (see below). However, an *aarP*-independent mechanism also accounts for increased levels of intrinsic resistance in the *aarG1* background. This mechanism could involve increased expression of a second activator with a target specificity similar to that of AarP.

3.1.4 Positive Regulators of *aac(2')-Ia*

aarE. The *aarE* gene is *ubiA*, which encodes an octaprenyltransferase required for the second step of ubiquinone biosynthesis (81). Although the *aarE* mutations increase aminoglycoside resistance, the accumulation of *aac(2')-Ia* mRNA is significantly reduced in the *aarE1* background. The loss of ubiquinone function is predicted to decrease the uptake of aminoglycosides, which accounts for the high-level aminoglycoside resistance. The decreased *aac(2')-Ia* mRNA accumation may reflect a requirement for ubiquinone either directly or indirectly in a regulatory process involved in *aac(2')-Ia* mRNA expression.

aarF. The *aarF* locus of *P. stuartii* acts as a positive regulator of *aac(2')-Ia* expression with the level of *aac(2')-Ia* mRNA decreased in an *aarF* null mutant (77). Despite the lack of *aac(2')-Ia* expression, *aarF* null mutants exhibit a 256-fold increase in gentamicin resistance over the wild-type strain. *P. stuartii aarF* null mutants also exhibit severe growth defects under aerobic growth conditions and have been found to lack detectable quantities of the respiratory cofactor ubiquinone. The wild-type *aarF* gene has been cloned and encodes a 62.5-kD polypeptide which exhibits extensive amino acid identity to YigR from *E. coli* (82). Disruption of the *yigR* gene has confirmed that this locus is required for ubiquinone production in *E. coli*. Heterologous complementation studies demonstrate that *aarF* and the *E. coli yigR* genes are functionally equivalent.

The high-level gentamicin resistance observed in the *aarF* and *yigR* mutants is likely associated with decreased accumulation of the drug resulting from the absence of aerobic electron transport. It seems unlikely that *aarF* is directly involved in the regulation of *aac(2')-Ia*. It has been proposed that a reduced form of ubiquinone acts as an effector molecule in an uncharacterized regulatory pathway that activates the expression of *aac(2')-Ia* (77). In ubiquinone-deficient *aarF* mutant strains, this regulatory cascade would be disrupted resulting in decreased *aac(2')-Ia* expression (see below).

aarP. The *aarP* gene was originally isolated from a multicopy library of *P. stuartii* chromosomal DNA based on the ability to activate *aac(2')-Ia* expression in *trans* (80). The presence of *aarP* in multiple copies led to an eightfold increase in *aac(2')-Ia* mRNA accumulation. Studies utilizing an *aac(2')-lacZ* transcriptional fusion demonstrate that this increase results from an activation of *aac(2')-Ia* transcription. Chromosomal disruption of the *aarP* locus resulted in a fivefold reduction in *aac(2')-Ia* mRNA levels and eliminated the induction of *aac(2')-Ia* expression normally observed during logarithmic growth (80). Expression of *aarP* has been shown to be increased in the *aarB*, *aarC*, and *aarG* mutants, demonstrating that *aarP* contributes to the overexpression of *aac(2')-Ia* in these mutant backgrounds (63,65,78).

The *aarP* gene encodes a 16-kD protein which contains a putative DNA binding helix-turn-helix motif and belongs to the AraC/XylS family of transcriptional activators (80,83). The AarP protein exhibits extensive homology with the *E. coli* MarA and SoxS proteins that were discussed above. AarP exhibits high homology to MarA and SoxS in the helix-turn-helix domain and was found to activate targets of both MarA and SoxS in vivo (80). The purified AarP protein binds to a wild-type *aac(2')-Ia* promoter fragment in electrophoretic mobility shift assays (84).

Expression of *aarP* appears to be governed by a mechanism that differs from those controlling MarA and SoxS expression. Unlike the MarA and SoxS proteins, which are located in operons containing a gene which regulates their expression, the *aarP* message appears to be monocistronic. Expression of *aarP* is not elevated in the presence of salicylate, a potent inducer of MarA. Recent studies of *aarP* expression have revealed that the AarP message accumulates as cell density increases (85). At least three *aar* genes (*aarB*, *aarC*, and *aarG*) are involved in *aarP* regulation (63,65,78). In addition, we have recently identified a role for the stationary phase starvation protein SspA as an activator of *aarP* (85). The SspA protein is a global regulator that is proposed to interact with RNA polymerase during starvation and redirect new gene expression (86,87).

3.1.5 Role of Quorum Sensing in *aac(2′)-Ia* Regulation

The regulation of *aac(2′)-Ia* expression is mediated by cell to cell signaling (88). The accumulation of *aac(2′)-Ia* mRNA exhibits two levels of growth phase–dependent expression. First, as cells approach mid-log phase, a significant increase is observed relative to cells at early-log phase. This increase at mid-log phase is the result of increased *aarP* expression. Second, as cells approach stationary phase, the levels of *aac(2′)-Ia* mRNA are decreased to levels that are at least 20-fold lower than those at mid-log phase. This decrease at high density is mediated by the accumulation of an extracellular factor (AR factor) (88). The growth of *P. stuartii* cells in spent (conditioned) media from stationary phase cultures resulted in the premature repression of *aac(2′)-Ia* in cells at mid-log phase. The ability to produce AR factor is dependent on the AarA protein described previously.

In summary, the large number of genes that influence *aac(2′)-Ia* regulation suggest that the expression of *aac(2′)-Ia* and the subsequent *O*-acetylation of peptidoglycan must be tightly controlled in *P. stuartii*. The AAC(2′)-Ia enzyme represents a minor *O*-acetyltransferase in *P. stuartii* (59). The physiological function of AAC(2′)-Ia may be to "fine tune" the levels of peptidoglycan *O*-acetylation in response to different environmental conditions of phases of growth. For example, in cells at mid-log phase, there is a burst of *aac(2′)-Ia* expression that may be required for peptidoglycan turnover in rapidly growing cells. As cells increase in density and approach stationary phase, the accumulation of AR factor leads to decreased *aac(2′)-Ia* expression at stationary phase. This may reflect a requirement for lower peptidoglycan turnover at stationary phase. The additional levels of *aac(2′)-Ia* regulation, namely, the role of ubiquinone and/or electron transport, are understood in less detail. The simplest model, proposed earlier, is that *aac(2′)-Ia* expression is also coupled to electron transport via regulatory protein(s) that sense the redox status of the cell. The AarG/AarR two-component system may have a role in this process. At the present time, interplay among the *aar* genes, electron transport, and quorum sensing in controlling *aac(2′)-Ia* expression is being investigated. The mechanisms identified may serve as a model for the regulation of other chromosomally encoded acetyltransferases. In addition, the identification of physiological roles for the other intrinsic acetyltransferases will allow us to better predict how the modification of intrinsic genes can lead to antibiotic resistance.

4 CONCLUSIONS

Although the *mar/sox* and *aac(2′)-Ia* systems differ significantly at both the genetic and physiological levels, there are important similarities worth noting. As global, intrinsic resistance systems, they both contain regulatory components as key factors controlling the expression of resistance determinants. In the case of MarA/SoxS and AarP, the products of key regulatory genes are remarkably conserved. In addition, there is a common element of environmental sensing that is shared by these two systems. These observations support the notion that the resistance phenotypes observed for specific regulatory mutants are physiologically relevant, because they result from changes in the activity of factors that control the expression of otherwise normal effector genes. As the best-studied examples of global resistance systems, the *mar/sox* and *aac(2′)-Ia* networks provide models for how subtle yet effective pathways affecting antibiotic resistance may lie buried with in the complex genomes of many organisms. It is interesting to note that the kinds of antibiotics affected by the two systems are almost entirely complementary: Aminoglycosides are among the only kinds of agents that are not impacted by the *mar/sox* pathway. Perhaps the *aac(2′)-Ia* system evolved divergently to address this gap? Regardless, these pathways provide paradigms that should assist future investigators in the characterization of other global systems that affect antibiotic susceptibility.

REFERENCES

1. Alexshun MN, Levy SB. Regulation of chromosomally mediated multiple antibiotic resistance: the *mar* regulon. Antimicrob Agents Chemother 1997; 41:2067–2075.
2. George AM, Levy SB. Amplifiable resistance to tetracycline, chloramphenicol and other antibiotics in *Escherichia coli*: involvement of a non-plasmid-determined efflux of tetracycline. J Bacteriol 1983; 155:531–540.
3. Cohen SP, McMurry LM, Hooper DC, Wolfson JS, Levy SB. Cross-resistance to fluoroquinolones in multiple-antibiotic-resistant (Mar) *Escherichia coli* selected by tetracycline or chloramphenicol: decreased drug accumulation asso-

ciated with membrane changes in addition to OmpF reduction. Antimicrob Agents Chemother 1989; 33:1318–1325.
4. McMurry LM, George AM, Levy SB. Active efflux of chloramphenicol in susceptible *Escherichia coli* strains and in multiple–antibiotic-resistant (Mar) mutants. Antimicrob Agents Chemother 1994; 38:542–546.
5. White DG, Goldman JD, Demple B, Levy SB. Role of the *acrAB* locus in organic solvent tolerance mediated by expression of *marA*, *soxS*, or *robA* in *Escherichia coli*. J Bacteriol 1997; 179:6122–6126.
6. Moken MC, McMurry LM, Levy SB. Selection of multiple–antibiotic-resistant (Mar) mutants of *Escherichia coli* by using the disinfectant pine oil: roles of the *mar* and *acrAB* loci. Antimicrob Agents Chemother 1997; 41:2770–2772.
7. George AM, Levy SB. Gene in the major cotransduction gap of the *Escherichia coli* K-12 linkage map required for the expression of chromosomal resistance to tetracycline and other antibiotics. J Bacteriol 1983; 155:541–548.
8. Hächler H, Cohen SP, Levy SB. *marA*, Regulated locus which controls expression of chromosomal multiple antibiotic resistance in *Escherichia coli*. J Bacteriol 1991; 173:5532–5538.
9. Cohen SP, Hächler H, Levy SB. Genetic and functional analysis of the multiple antibiotic resistance (*mar* locus in *Escherichia coli*. J Bacteriol 1993; 175:1484–1492.
10. Sulavik MC, Gambino LF, Miller PF. Analysis of the genetic requirements for inducible multiple-antibiotic resistance associated with the *mar* locus in *Escherichia coli*. J Bacteriol 1994; 176:7754–7756.
11. Martin RG, Nyantakyi PS, Rosner JL. Regulation of the multiple antibiotic resistance (*mar*) regulon by *marORA* operator sequences in *Escherichia coli*. J Bacteriol 1995; 177:4176–4178.
12. Gambino L, Gracheck SJ, Miller PF. Overexpression of the MarA positive regulator is sufficient to confer multiple antibiotic resistance in *Escherichia coli*. J Bacteriol 1993; 175:2888–2894.
13. Gallegos MT, Schleif R, Bairoch A, Hofmann K, Ramos JL. AraC/XylS family of transcriptional regulators. Mol Microbiol 1997; 61:393–410.
14. Ariza RR, Cohen SP, Bachhawat N, Levy SB, Demple B. Repressor mutations in the *marRAB* operon that activate oxidative stress genes and multiple antibiotic resistance in *Escherichia coli*. J Bacteriol 1994; 176:143–148.
15. Martin R, Rosner JL. Binding of purified multiple antibiotic-resistance repressor protein (MarR) to *mar* operator sequences. Proc Natl Acad Sci USA 1995; 92:5456–5460.
16. Rosner JL. Nonheritable resistance to chloramphenicol and other antibiotics induced by salicylates and other chemotactic repellants in *Escherichia coli*. Proc Natl Acad Sci USA 1985; 82:8771–8774.
17. Cohen SP, Levy SB, Foulds J, Rosner JL. Salicylate induction of antibiotic resistance in *Escherichia coli*: activation of the *mar* operon and a *mar*-independent pathway. J Bacteriol 1993; 175:7856–7862.
18. Sulavik MC, Gambino LF, Miller PF. Analysis of the genetic requirements for inducible multiple-antibiotic resistance associated with the *mar* locus in *Escherichia coli*. J Bacteriol 1994; 176:7754–7756.

19. Demple B. Regulation of bacterial oxidative stress genes. Ann Rev Genet 1991; 25:315–337.
20. Wu J, Weiss B. Two divergently transcribed genes, *soxR* and *soxS*, control a superoxide response regulon of *Escherichia coli*. J Bacteriol 1991; 173:2864–2871.
21. Greenberg JT, Monach P, Chou JH, Josephy PD, Demple B. Positive control of a global antioxidant defense regulon activated by superoxide-generating agents in *Escherichia coli*. Proc Natl Acad Sci USA 1990; 87:6181–6185.
22. Nunoshiba T, Hidalgo E, Amabile-Cuevas CF, Demple B. Two-stage control of an oxidative stress regulon: the *Escherichia coli* SoxR protein triggers redox-inducible expression of the *soxS* gene. J Bacteriol 1992; 174:6054–6060.
23. Wu J, Weiss B. Two-stage induction of the *soxRS* (superoxide response) regulon of *Escherichia coli*. J Bacteriol 1992; 174:3915–3920.
24. Hidalgo E, Demple B. An iron-sulphur center essential for transcriptional activation by the redox-sensing SoxR protein. EMBO J 1994; 13:138–146.
25. Wu J, Dunham WR, Weiss B. Overproduction and physical characterization of SoxR, a [2Fe-2S] protein that governs an oxidative response regulon in *Escherichia coli*. J Biol Chem 1995; 270:10323–10327.
26. Miller PF, Gambino LF, Sulavik MC, Gracheck SJ. Genetic relationship between *soxRS* and *mar* loci in promoting multiple antibiotic resistance in *Escherichia coli*. Antimicrob Agents Chemother 1994; 38:1773–1779.
27. Skarstad K, Thony B, Hwang DS, Kornberg A. A novel binding protein of the origin of the *Escherichia coli* chromosome. J Biol Chem 1993; 268:5365–5370.
28. Ariza RR, Li Z, Ringstad N, Demple B. Activation of multiple antibiotic resistance and binding of stress-inducible promoters by *Escherichia coli* Rob protein. J Bacteriol 1995; 177:1655–1661.
29. Nakajima H, Kobayashi K, Kobayashi M, Asako H, Aono R. Overexpression of the *robA* gene increases organic solvent tolerance and multiple antibiotic and heavy metal ion resistance in *Escherichia coli*. Appl Environ Microbiol 1995; 61:2303–2307.
30. Martin RG, Gillette WK, Rhee S, Rosner JL. Structural requirements for marbox function in transcriptional activation of *mar/sox/rob* regulon promoters in *Escherichia coli*: sequence, orientation and spatial relationship to the core promoter. Mol Microbiol 1999; 34:431–441.
31. Li Z, Demple B. Sequence specificity for DNA binding by *Escherichia coli* SoxS and Rob proteins. Mol Microbiol 1996; 20:937–945.
32. Fawcett WP, Wolf RE Jr. Purification of a MalE-SoxS fusion protein and identification of the control sites of *Escherichia coli* superoxide-inducible genes. Mol Microbiol 1994; 14:669–679.
33. Jair K-W, Martin RG, Rosner JL, Fujita N, Ishihama A, Wolf RE Jr. Purification and regulatory properties of MarA protein, a transcriptional activator of *Escherichia coli* multiple antibiotic and superoxide resistance promoters. J Bacteriol 1995; 177:7100–7104.
34. Jair K-W, Fawcett WP, Fujita N, Ishihama A, Wolf RE Jr. Ambidextrous transcriptional activation by SoxS: requirement for the C-terminal domain of the RNA polymerase alpha subunit in a subset of *Escherichia coli* superoxide-inducible genes. Mol Microbiol 1996; 19:307–317.

35. Rhee SR, Martin RG, Rosner JL, Davies DR. A novel DNA-binding motif in MarA: the first structure for an AraC family transcriptional activator. Proc Natl Acad Sci USA 1998; 95:10413–10418.
36. Li Z, Demple B. SoxS, an activator of superoxide stress genes in *Escherichia coli*. Purification and interaction with DNA. J Biol Chem 1994; 269:18371–18377.
37. Cohen SP, McMurry LM, Levy SB. *marA* locus causes decreased expression of OmpF porin in multiple–antibiotic-resistant (Mar) mutants of *Escherichia coli*. J Bacteriol 1988; 170:5416–5422.
38. Rosner JL, Chai T-J, Foulds J. Regulation of OmpF porin expression by salicylate in *Escherichia coli*. J Bacteriol 1991; 173:5631–5638.
39. Mizuno T, Chou M-Y, Inouye M. A unique mechanism regulating gene expression: translational inhibition by a complementary RNA transcript (micRNA). Proc Natl Acad Sci USA 1984; 81:1966–1970.
40. Anderson J, Delihas N. *micF* RNA binds to the 5′ end of *ompF* mRNA and to a protein from *Escherichia coli*. Biochemistry 1990; 29:9249–9256.
41. Chou JH, Greenberg JT, Demple B. Posttranscriptional repression of *Escherichia coli* OmpF protein in response to redox stress: positive control of the *micF* antisense RNA by the *soxRS* locus. J Bacteriol 1993; 175:1026–1031.
42. Nikaido H. Prevention of drug access to bacterial targets: permeability barriers and active efflux. Science 1994; 264:382–388.
43. Ma D, Cook DN, Alberti M, Pon NG, Nikaido H, Hearst JE. Genes of *acrA* and *acrB* encode a stress-induced efflux system of *Escherichia coli*. Mol Microbiol 1995; 16:45–55.
44. Aono R, Tsukagoshi N, Yamamoto M. Involvement of outer membrane protein TolC, a possible member of the *mar-sox* regulon, in maintenance and improvement of organic solvent tolerance of *Escherichia coli* K-12. J Bacteriol 1998; 180:938–944.
45. Martin RG, Jair K-W, Wolf RE Jr, Rosner JL. Autoactivation of the *marRAB* multiple antibiotic resistance operon by the MarA transcriptional activator in *Escherichia coli*. J Bacteriol 1996; 178:2216–2223.
46. Martin RG, Gillette WK, Rosner JL. Promoter discrimination by the related transcriptional activaters MarA and SoxS: differntial regulation by differential binding. Mol Microbiol 2000; 35:623–634.
47. Seoane A, Levy SB. Characterization of MarR, the repressor of the multiple antibiotic resistance (*mar*) operon in *Escherichia coli*. J Bacteriol 1995; 177:3414–3419.
48. Miller PF, Sulavik M, Gambino L, Dazer M. Roles of the *marRAB* and *soxRS* regulators in protecting *Escherichia coli* from plant-derived phenolic agents. 96th General Meeting of the American Society for Microbiology, New Orleans, May 19–23, 1996.
49. Miller PF, Sulavik MC. Overlaps and parallels in the regulation of intrinsic multiple-antibiotic resistance in *Escherichia coli*. Mol Microbiol 1996; 21:441–448.
50. Sulavik MC, Gambino LF, Miller PF. The MarR repressor of the multiple antibiotic resistance (*mar*) operon in *Escherichia coli*: prototypic member of a family of bacterial regulatory proteins involved in sensing phenolic compounds. Mol Medicine 1995; 1:436–446.

51. Cohen SP, Yan W, Levy SB. A multidrug resistance regulatory chromosomal locus is widespread among enteric bacteria. J Infect Dis 1993; 168:484–488.
52. Maneewannakul K, Levy SB. Identification of *mar* mutants among quinolone-resistant clinical isolates of *Escherichia coli*. Antimicrob Agents Chemother 1996; 40:1695–1698.
53. Hooper DC, Wolfson JS. Bacterial resistance to the quinolone antimicrobial agents. Am J Med 1989; 87:17S–23S.
54. Shaw KJ, Rather PN, Hare RS, Miller GH. Molecular genetics of aminoglycoside resistance genes and familial relationships of the aminoglycoside-modifying enzymes. Microbiol Rev 1993; 57:138–163.
55. Rather PN. Origins of the aminoglycoside modifying enzymes. Drug Res Updates 1998; 5:285–291.
56. Rather PN, Macinga DR. The chromosomal 2′-N-acetyltransferase of *Providencia stuartii*: physiological functions and genetic regulation. Front Biosci 1999; 4:132–140.
57. Ånsa JA, PÇrez E, Pelicic V, Berthet F-X, Gicquel B, Martin C. Aminoglycoside 2′-N-acetyltransferase genes are universally present in mycobacteria: characterization of the aac(2′)-Ic from *Mycobacterium tuberculosis* and the aac(2′)-Id gene from *Mycobacterium smegmatis*. Mol Microbiol 1997; 24:431–441.
58. Magnet S, Courvalin P, Lambert T. Activation of the cryptic aac(6′)-Iy aminoglycoside resistance gene of Salmonella by a chromosomal deletion generating a transcriptional fusion. J Bacteriol 1999; 181:6650–6655.
59. Payie KG, Rather PN, Clarke AJ. Contribution of gentamicin 2′-N-acetyltransferase to the O-acetylation of peptidoglycan in *Providencia stuartii*. J Bacteriol 1995; 177:4303–4310.
60. Payie KG, Clarke AJ. Characterization of gentamicin 2′-N-acetyltransferase from *Providencia stuartii*: its use of peptidoglycan metabolites for acetylation of both aminoglycosides and peptidoglycan. J Bacteriol 1997; 179:4106–4114.
61. Clarke AJ, Dupont C. O-acetylated peptidoglycan: its occurrence, pathobiological significance, and biosynthesis. Can J Microbiol 1992; 38:85–91.
62. Clarke AJ. Extent of peptidoglycan O-acetylation in the tribe Proteae. J Bacteriol 1993; 175:4550–4553.
63. Rather PN, Orosz E, Shaw KJ, Hare R, Miller GH. Characterization and transcriptional regulation of the 2′-N-acetyltransferase gene from *Providencia stuartii*. J Bacteriol 1993; 175:6492–6498.
64. Rather PN, Orosz E. Characterization of *aarA*, a pleiotrophic negative regulator of the 2′-N-acetyltransferase in *Providencia stuartii*. J Bacteriol 1994; 176: 5140–5144.
65. Rather PN, Solinsky K, Paradise MR, Parojcic MM. *aarC*, an essential gene involved in density dependent regulation of the 2′-N-acetyltransferase in *Providencia stuartii*. J Bacteriol 1996; 179:2267–2273.
66. Macinga DR, Rather PN. *aarD*, a *Providencia stuartii* homologue of *cydD*: role in 2′-N-acetyltransferase expression, cell morphology and growth in the presence of an extracellular factor. Mol Microbiol 1996; 19:511–520.
67. Delaney JM, Wall D, Georgopoulos C. Molecular characterization of the *Esche-*

richia coli htrD gene: cloning, sequence, regulation and involvement with cytochrome d oxidase. J Bacteriol 1993; 175:166–175.

68. Poole RK, Hatch L, Cleeter MW, Gibson JF, Cox GB, Wu G. Cytochrome bd biosynthesis in *Escherichia coli*: the sequences of the *cydC* and *cydD* genes suggest that they encode the components of an ABC membrane transporter. Mol Microbiol 1993; 10:421–430.
69. Bebbington KJ, Williams HD. Investigation of the role of the *cydD* gene product in the production of a functional cytochrome d oxidase. FEMS Microbiol Lett 1993; 112:19–24.
70. Poole RK, Gibson F, Wu G. The *cydD* gene product, component of a heterodimeric ABC transporter, is required for assembly of periplasmic cytochrome c and of cytochrome bd in *Escherichia coli*. FEMS Microbiol Lett 1994; 117:217–224.
71. Poole RK, Williams HD, Downie JA, Gibson F. Mutations affecting the cytochrome d–containing oxidase complex of *Escherichia coli* K-12: identification and mapping of a fourth locus, *cydD*. J Gen Microbiol 1989; 135:1865–1874.
72. Georgiou CD, Fang H, Gennis RB. Identification of the *cydC* locus required for expression of the functional form of the cytochrome d terminal oxidase complex in *Escherichia coli*. J Bacteriol 1987; 169:2107–2112.
73. Taber HW, Mueller JP, Miller PF, Arrow AS. Bacterial uptake of aminoglycoside antibiotics. Microbiol Rev 1987; 51:439–457.
74. Bryan LE, Kwan S. Roles of ribosomal binding, membrane potential, and electron transport in bacterial uptake of streptomycin and gentamicin. Antimicrob Agents Chemother 1983; 23:835–845.
75. Bryan LE, van den Elzen HM. Effects of membrane energy mutations and cations on streptomycin and gentamicin accumulation by bacteria: a model for entry of streptomycin and gentamicin in susceptible and resistant bacteria. Antimicrob Agents Chemother (1977, 12:163–177.
76. Faith MJ, Kolter R. ABC transporters: bacterial exporters. Microbiol Rev 1993; 57:995–1017.
77. Macinga DR, Cook GM, Poole RK, Rather PN. Identification and characterization of *aarF*, a locus required for production of ubiquinone in *Providencia stuartii* and *Escherichia coli* and for expression of 2′-N-acetyltransferase in *P. stuartii*. J Bacteriol 1998; 180:128–135.
78. Rather PN, Paradise MR, Parojcic MM, Patel S. A regulatory cascade involving *aarG*, a putative sensor kinase, controls the expression of the 2′-N-acetyltransferase and an intrinsic multiple antibiotic resistance (Mar) response in *Providencia stuartii*. Mol Microbiol 1998; 28:1345–1353.
79. Groisman EA, Heffron F, Solomon F. Molecular genetic analysis of the *Escherichia coli phoP* locus. J Bacteriol 1992; 174:486–491.
80. Macinga DR, Parojcic MM, Rather PN. Identification and analysis of *aarP*, a transcriptional activator of the 2′-N-acetyltransferase in *Providencia stuartii*. J Bacteriol 1995; 177:3407–3413.
81. Paradise MR, Cook G, Poole RK, Rather PN. *aarE*, a homolog of *ubiA*, is required for transcription of the *aac(2′)-Ia* gene in *Providencia stuartii*. Antimicrob Agents Chemother 1998; 42:959–962.

82. Daniels DL, Plunkett G, Burland V, Blattner FR. Analysis of the *Escherichia coli* genome: DNA sequence of the region from 84.5 to 86.5 minutes. Science 1992; 257:771–777.
83. Ramos JL, Rojo F, Zhou L, Timmis KN. A family of positive regulators related to the *Pseudomonas putida* TOL plasmid XylS and the *Escherichia coli* AraC activators. Nucl Acids Res 1990; 18:2149–2152.
84. Macinga DR, Paradise MR, Parojcic MM, Rather PN. Activation of the 2′-N-acetyltransferase gene (*aac(2′)-Ia*) in *Providencia stuartii* by an interaction of AarP with the promoter region. Antimicrob Agents Chemother 1999; 43:1769–1772.
85. Ding X, Rather PN. Unpublished data.
86. Williams MD, Ouyang TX, Flickinger MC. Starvation-induced expression of SspA and SspB: the effects of a null mutation in *sspA* on *Escherichia coli* protein synthesis and survival during growth and prolonged starvation. Mol Microbiol 1994; 11:1029–1043.
87. Williams MD, Ouyang TX, Flickinger MC. Glutathione S-transferase-*sspA* fusion binds to *E. coli* RNA polymerase and complements delta *sspA* mutation allowing phage P1 replication. Biochem Biophys Res Comm 1994; 201:123–127.
88. Rather PN, Paradise MR, Parojcic MM. An extracellular factor regulating expression of the chromosomal aminoglycoside 2′-N-acetyltransferase in *Providencia stuartii*. Antimicrob Agents Chemother 1997; 41:1749–1754.

4

Drug Efflux

Kim Lewis
Northeastern University, Boston, Massachusetts

Olga Lomovskaya
Microcide Pharmaceuticals, Inc., Mountain View, California

Resistance by efflux is widespread in the bacterial world and comes in many shapes and forms. Some pumps are specific for a particular drug. Examples include tetracycline, chloramphenicol, and macrolide-specific transporters. New derivatives of tetracyclines and macrolides have been obtained that are poorly recognized by these specific efflux mechanisms.

Other translocases, called multidrug resistance pumps (MDRs), are capable of recognizing multiple antibiotics. They represent the most prevalent type of efflux resistance. MDRs belong to five different protein families and are present in all organisms studied to date. The significance of MDRs is underscored by the fact that many of these transporters are chromosomally encoded and are expressed constitutively, which accounts for high levels of "intrinsic antibiotic resistance" in such pathogens as *Pseudomonas aeruginosa* and *B. pseudomallei*. The simpler MDRs of the small multidrug resistance (SMR) and major facilititator (MF) families almost exclusively extrude amphipathic cations like quaternary ammonium antiseptics, suggesting that these MDRs evolved to protect the cell from natural hydrophobic cationic antimicrobials. Berberine alkaloids seem to represent an example of this type of natural MDR substrate. More complex MDRs of the RND family are only found in gram-negative species, and the transporter is a multiprotein complex traversing the entire cell envelope. The RND MDR extrudes its substrates across the outer membrane, which is a good barrier for amphipathic substances. As a result, the synergistic action of the pump and the outer membrane barrier results in high levels of antibiotic resistance. RND pumps have a very broad spectrum of substrates and extrude most known antibiotics. The ability of MDRs to extrude chemically unrelated substances from the cell suggests that these transporters represent a "ready-made" resistance mechanism for present and future synthetic antibiotics. Developing therapies based on MDR inhibitors represents a rational response to this universal resistance based on MDR pumps.

1 EFFLUX TRANSPORTER FAMILIES

1.1 Major Facilitator Transporters

The major facilitator is the largest family of translocases (1) that comprises pmf-dependent uptake transporters of regular substrates, such as the LacY lactose/H^+ symporter of *Escherichia coli*, and a subfamily of drug/H^+ antiporters that includes both specific transporters and multidrug resistance pumps (MDRs) (Fig. 1). The MF transporters are likely descendants of a gene duplication event, as proposed originally for TetA where the two halves of the protein share significant homology (2). There is good evidence that TetA is in fact a dimer (3). The N end half of the MF transporters is better conserved than the C half, suggesting that the C half domains harbor the drug-binding site. MF transporters share a number of consensus sequences. A particularly interesting motif GxhyxGPhyhyGGxhy (hy = hydrophobic) is found only in the efflux transporters of this family. It was proposed that this motif forms a kink in membrane domain 5 that changes the ligand/simport coupling of intake transporters into a ligand/antiport pathway of the efflux pumps (4). This consensus is very useful in homology analysis, since its presence in a newly sequenced gene (or genome) clearly indicates that the gene codes for an efflux translocase. Both specific drug efflux transporters and MDRs are members of this this family.

1.1.1 Major Facilitator–Specific Drug Efflux Transporters

TetA-TetE, TetG, TetH and TetK, TetL, TetA(P), and OtrB proteins are tetracycline-specific efflux pumps found in the majority of pathogenic gram-negative and gram-positive bacteria, respectively (5). The most

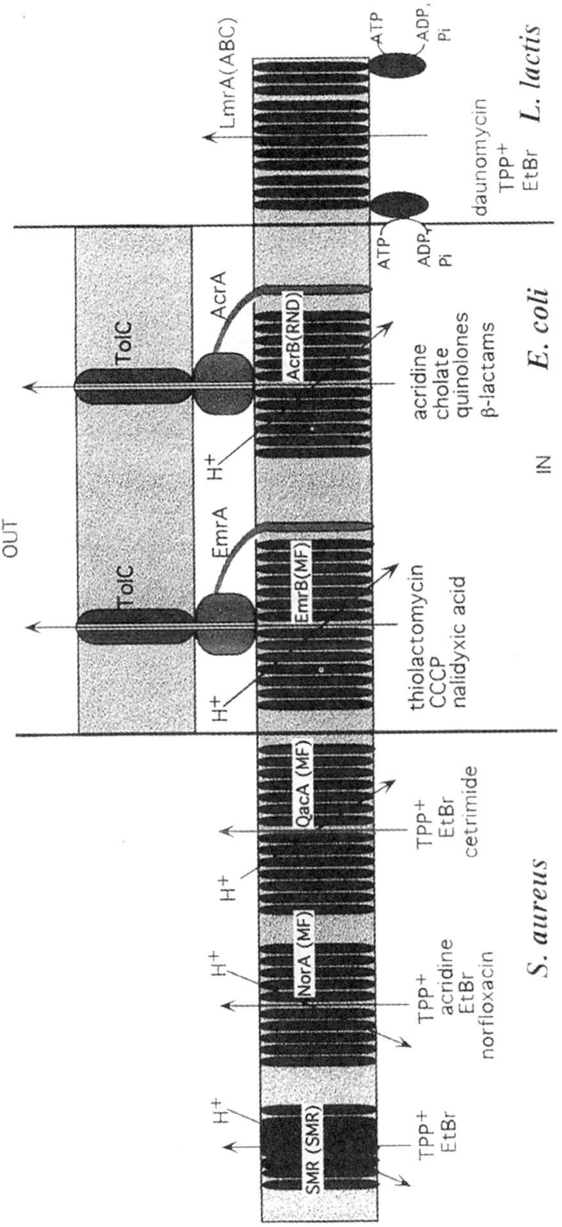

Figure 1 Bacterial multidrug pumps.

widely distributed pumps are TetA/TetB and TetK/TetL. Both types of pumps are drug/H^+ antiporters with the actual substrate being a divalent metal–tetracycline chelate $[M\text{-tet}]^+$ (6). In the case of TetA, it was shown that the protein pumps out tetracycline into the periplasm of gram-negative bacteria (7).

Chloramphenicol is extruded by the 12-TMS exporters of the MF family, coded for by the *cmlA* genes. They confer inducible resistance of 16–200 μ/mL to chloramphenicol but not to the analogue florfenicol (8,9). The genes are widespread among Gram-negative bacteria and were identified in various species of Enterobacteriaceae (10–12), in *Haemophilus influenzae* (13), and in *Pseudomonas aeruginosa* (9). They are usually located on the plasmids as part of integrons. *cmlA* genes from different species share up to 87% of amino acid identity. It is possible that the efflux capability of CmlA alone is not sufficient to provide the high-level chloramphenicol resistance that is associated with failures in therapy. Several reports have implicated these proteins in the repression of synthesis of outer membrane porins (9,11,13).

Shortly after the introduction of florfenicol, researchers in Japan identified a plasmid-located gene, *pp-flo*, which conferred resistance to florfenicol and chloramphenicol in the important fish pathogen, *Pasteurella piscicida* (14). This gene encoded a protein with approximately 50% homology to CmlA. Recently, a gene with 97% identity to *pp-flo*, designated flo_{st}, was discovered among *Salmonella* spp (15). Some of the resistant isolates included *Salmonella typhimurium* DT104, which is an emerging multi–drug-resistant strain (16,17). A Flo pump was also identified among the majority of chloramphenicol- and florfenicol-resistant pathogenic isolates of *E. coli*. The strains harboring Flo were reported to have MIC of approximately 128 μg/mL. Presently, it is not known whether the Flo pump alone is sufficient to confer this high-level resistance.

The *mef* genes (macrolide efflux) encode efflux pumps specific for 14- and 15-membered macrolides. The phenotype conferred by these pumps was designated as M-phenotype for macrolide resistance, with lincosamide and streptoGramin sensitivity (18). Mef-mediated resistance is not induced in the presence of antibiotics. Mef pumps belong to the MF family of transporters with 12-transmembrane spanning domains (19,20). The level of resistance provided by the *mef* genes is 4–16 μg/mL, which is generally lower than the *erm*-mediated resistance of 32–128 μg/mL (21).

The *mefA* gene was first identified in *Streptococcus pyogenes* (19). It was later shown that *mefA* and its close homologue, *mefE*, are widely distributed among various streptoccocci (20,22–24).

A different macrolide resistance gene designated, *mreA*, was identified in a train of *S. agalactiae* (19). The product of this gene conferred resistance to 14-, 15-, and 16-membered macrolides. MreA encodes a protein that belongs to the MF superfamily but is clearly different from the Mef proteins. Its distribution and role in clinical resistance to macrolides remain to be determined.

1.1.2 Major Facilitator MDRs

A well-studied MDR of the MF family is BMR of *Bacillus subtilis* that confers resistance to mainly hydrophobic cations, such as TPP^+ and rhodamine (25). A close homologue of BMR is NorA, a chromosomally encoded protein of *Staphylococcus aureus* whose overproduction, due to promoter mutations, causes clinically significant resistance to quinolones (26–28).

An unusually broad substrate specificity is exhibited by an *E. coli* MDR known as MdfA, or Cmr. Apart from amphipathic cations, it confers resistance to chloramphenicol and modest, but distinct, resistance to aminoglycosides (29–32).

An interesting subgroup of MF MDRs are proteins with a 14-membrane domain structure. QacA, found on plasmids of *S. aureus*, can extrude amphipathic divalent cations (33), a function that might require additional complexity. QacA confers resistance to cationic antiseptics and disinfectants—quaternary ammonium compounds (QAC) cetrimide and benzalkonium chloride, and biguanidines, such as chlorhexidine.

EmrB of *E. coli* (34) is particularly interesting in that it forms a *trans* envelope complex with a periplasmically located EmrA and outer membrane TolC. This complex structure allows the pump to "take advantage" of the outer membrane barrier that restricts passage of hydrophobic compounds that readily partition in the cytoplasmic membrane. It seems that the only way to prevent their rapid reentry into the cytoplasmic membrane is to extrude them across the outer membrane barrier. Using this strategy, the EmrAB and homologous pumps VceAB of *Vibrio cholerae* (35) and FarAB of *Neisseria gonorrhoeae* (36), and RND MDRs (see below) protect the cell from such membrane-active agents as uncouplers of oxidative phosphorylation, fatty acids, and detergents like sodium dodecylsulfate (SDS) and bile acids. Not surprisingly, MDRs capable of uncoupler extrusion have not been found in gram-positive species.

The analysis of relatedness within the MF family suggests that MDRs were not derived from some primordial MDR but rather evolved independently many times in the course of evolution (37). For example, a *S. aureus* MDR, NorA, is much closer to the tetracycline extrusion pump TetB from *E. coli* than it is to the *Lactococcus lactis* LmrP multidrug resistance pump. It seems that MDRs were derived from specific drug extrusion pumps (such as Mmr that extrudes methylenomycin from the producing *Streptomyces coelicolor*) of bacteria producing amphipathic antibiotics. Selecting for a mutation that allows an enzyme or a translocase to recognize a new substrate usually leads to broadening the spectrum of specificity rather than to switching specificity from one compound to another. We have proposed that MDRs might have been intermediates in the evolution of one specific drug extrusion pump from another (37). Some of these intermediates might have taken on a life of their own, with their unspecific amphipathic binding domain providing the cell with the capability to rid itself of a wide range of chemically unrelated amphipathic toxins.

1.2 Resistance-Nodulation-Division (RND) MDRs

The RND pump AcrAB (formerly AcrAE) is responsible for what has been known since the 1960s as the "Acr phenotype" in *E. coli* (38,39). A mutation in the pump leads to sensitivity to acridine and other hydrophobic cations, tetracycline, β-lactam, erythromycin, quinolones, and detergents like bile acids and SDS (Fig. 2). Not surprisingly, the *acrAB* mutation has been originally thought to disrupt the outer membrane. The AcrB is a large 113-kD peptide that is the translocase proper and acts as a drug/H^+ antiporter. It has a putative 12-membrane domain structure with two very large periplasmic segments. This topology was confirmed for the homologous MexB transporter from *P. aeruginosa* (40). AcrA belongs to the same membrane fusion family as EmrA. It appears that AcrAB together with TolC form a *trans* envelope structure (41). Other RND MDRs with similar composition are found only in *E. coli* and in other gram-negative bacteria such as *P. aeruginosa* and *N. gonorrhoeae* (42). Recently, structural and biochemical studies confirmed the *trans* envelope structure for the *E. coli* AcrAB-TolC pump. It was shown that AcrA is a highly asymmetrical elongated protein (43). This is compatible with the putative channel function of AcrA, connecting the inner and outer membrane and forming a continuous route for the drug molecule. It was also shown that purified AcrA is capable of association with lipid bilayers and can promote the close association or even fusion of two different liposome membranes (44). The purified AcrB protein was reconstituted in proteoliposomes containing fluorescent phospholipid analogues. AcrB extruded the fluorescent molecules from the vesicles in the presence of an artificially imposed pH gradient (44). Addition of AcrA increased the rate of transport.

Recently, two RND transporters, AmrAB-OprA in *Burkholderia cepacia* (45) and MexXY-OprM in *P. aeruginosa* (46) were reported to efflux hydrophilic and positively charged aminoglycosides. Apparently OprM serves as the outer membrane component for this pump (46,47). It was shown that MexXY is expressed in the wild-type cells of *P. aeruginosa*, since deletion of the corresponding genes rendered *P. aeruginosa* more suscep-

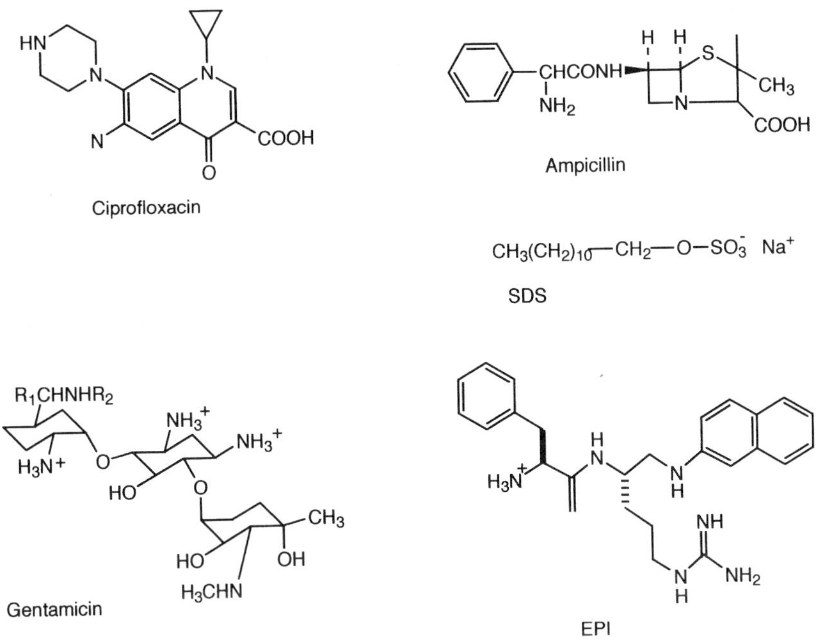

Figure 2 The structure of MDR substrates and inhibitors. Substrates of RND MDRs. Examples of several antibiotics are given, including the hydrophilic aminoglycoside gentamicin which is a substrate of MexXY MDR. Also shown is the RND MDR inhibitor EPI.

tible to various aminoglycosides (approximately eightfold). This is surprising, since MDRs do not normally extrude hydrophilic compounds. MexXY also protected the cells from hydrophobic erythromycin, showing that it is not a specific "aminoglycoside pump" but a true MDR. A gene *mexZ* homologous to repressors of active efflux systems was identified upstream of the *mexXY* locus. Preliminary data indicate that transformation of several clinical isolates of *P. aeruginosa* resistant to aminoglycosides via a "nonenzymatic" mechanism with the cloned *mexZ* gene restored their susceptibility to the wild-type level (P. Plesiat, personal communication). In another recent report it was shown that transcription of MexXY (AmrAB) was upregulated in *P. aeruginosa* clinical isolates with a permeability type of resistance compared to genotypically matched sensitive clinical isolate from the same patient (48). These data implicate MexXY in clinically relevant resistance to aminoglycosides in *P. aeruginosa*.

1.3 Small Multidrug Resistance Pumps

The peptides coding for small multidrug resistance pumps (SMRs) form only four-membrane domains, which makes them the smallest known translocases at about 110 amino acids long (reviewed in Ref. 49). The substrates are invariably hydrophobic cations, including disinfectants and antiseptics, and the spectrum is narrower than that of QacA, for example. Representative pumps that have been studied extensively are chromosomally coded EmrE of *E. coli* and plasmid-coded SMRs present in some strains of *S. aureus* (see Sec. 2 for a more detailed description). The only known non-MDR protein of this group is SugE (and its homologues) that suppresses mutations in the GroE chaperone (50). It has been suggested that the distantly related tellurite resistance transporter also belongs to this group (51).

1.4 The MATE Family

The MATE family of MDR pumps was identified quite recently (52). Similarly to MFS, they contain 12-transmembrane domains but do not share sequence similarity with any of the members of MFS family. NorM protein from *Vibrio parahaemolyticus* was the first identified member of this family (53). It mediated resistance to a rather broad range of antibacterial agents, including hydrophobic cations, aminoglycosides, and fluoroquinolones. NorM homologues were identified in *E. coli* and *Haemophilus*. The significance of the MATE proteins in clinical settings has not yet been established.

1.5 ABC MDRs

There is a large family of ATP-dependent translocases in bacteria, most commonly involved in the uptake of nutrients that requires a periplasmic binding protein. The maltose and histidine transporters are pertinent examples. There are also ABC efflux pumps, such as the hemolysin transporter of *E. coli* and ABC transporters of antibiotics. ABC transporters were found to confer specific resistance to macrolides.

A phenotype of resistance to 14- and 15-membered macrolides and streptoGramin B, but not 16-membered macrolides (MS-phenotype), was identified in several species of staphylococci. The pump-mediated antibiotic resistance is inducible and can reach a significant level of 32–128 μg/mL (54). Separate but adjacent genes primarily located on various plasmids encode the ATP-binding and the transmembrane domains of these pumps. The *msrA* and the closely related *msrB* genes encode the ATP-binding domains (54–56). Introduction of either *msrA* or *msrB* into

S. aureus is sufficient to confer MS-phenotype, indicating that unknown proteins encoded in the *S. aureus* chromosome can substitute for the original membrane components of this pump.

The ABC MDR LmrA discovered in *Lactococcus lactis* (57) remains the only reported bacterial MDR of this family. With six transmembrane domains, the pump looks like half of the duplicated mammalian P glycoprotein MDR. The pump is closely related to the P glycoprotein, and the homology is over the entire sequence range. The substrate spectrum is similar to that of P glycoprotein as well (58).

2 MECHANISM

It is tempting to simplify the analysis of a transport mechanism by separating it into two parts—the energy-dependent transport of a ligand and the selective binding. MDRs of the MF family, for example, are homologous to specific efflux antiporters and share with them a number of conserved domains thought to be involved in energy transformations and ligand movement (51). However, it appears that the path of the ligand in an MDR-driven efflux might be intimately related to the mechanism of discrimination.

It has been noted by Higgins and Gottesman (59) that drugs with intracellular targets need to be amphipathic in order to cross the cytoplasmic membrane, and by the same token intracellular compounds must be hydrophilic in order to avoid escaping into the external environment. This difference in polarity rather than in chemical structure could serve as a basis for drug/self-discrimination. Perhaps taking this logic to its limit, the same researchers proposed that it is the membrane that acted as a discriminator—any substance partitioned in the internal leaflet of the bilayer would then be picked up by the completely unspecific MDR and flipped to the outer leaflet. This would shift the equilibrium in the direction of efflux, decreasing the intracellular concentration of the drug. This specific model of substrate discrimination proposed for P glycoprotein does not appear to be correct—purified MDR clearly has binding preferences of its own (60,61), and amphipathic cations are by far better substrates than neutral molecules or amphipathic anions with similar polarity (reviewed in Refs. 61 and 62). What is truly impressive, however, is that the idea of a flippase mechanism for the P glycoprotein appears to be correct. It was found that human MDR3 that is highly homologous to P glycoprotein MDR1 is indeed a phospholipid flippase responsible for phosphatidylcholine transport into the bile (63). It was further found that P-glycoprotein can transport (flip) analogues of phospholipids with short fatty acid chains from the inner to the outer leaflet of the membrane of epithelial cells (64). This suggests that the membrane does influence the substrate specificity of the MDR through a prescreening process—only substances that partition in the membrane will be "considered" as ligands by the MDR.

Interestingly, the flippase mechanism appears to be an exception rather than a rule among MDRs. A fairly close homologue of mammalian P glycoprotein, the *L. lactis* LmrA, has been studied by Konings and coworkers (57). LmrA has an interesting structure—it is exactly half of the 12 TMS P glycoprotein and apparently forms a dimer. The substrate specificity of LmrA is similar to P glycoprotein, and LmrA was found to replace successfully P glycoprotein in mammalian cells (58). However, experiments with amphipathic dyes suggest that LmrA is not a flippase. A hydrophobic probe, TMA-DPH, that is fluorescent in the membrane inserts into the outer leaflet of the membrane and then slowly flips, like a phospholipid, to the inner leaflet. Activation of an MDR by providing an energy source caused rapid efflux of the probe into the external environment. This experiment showed that the MDRs were not acting as flippases, since merely flipping a substance from the inner to the outer leaflet would have little effect on fluorescence. Importantly, the experiments suggest that the binding site is accessible from the membrane (65). This and similar experiments indicate that the path of an LmrA ligand is from the inner leaflet of the cytoplasmic membrane all the way to the external medium. The ligand pathway was analyzed in similar experiments for *L. lactis* LmrP MDR, which belongs to the MF family of translocases. The results showed that LmrP also transports its ligands from the inner leaflet of the bilayer to the outside medium (66). LmrP belongs to the MF subfamily together with specific translocases of antibiotics. Hydrophobic antibiotics partition preferentially into the membrane, and it seems very reasonable for an antibiotic translocase to have the ability to pick its substrate from the membrane.

If microbial MDRs indeed evolved from homologous specific translocases of antibiotics, as discussed in Section 1, it would seem that MDRs had to develop a distinctly unique ability to pick up their substrates from the membrane. However, it is quite possible that binding of ligands within the membrane had actually originated in efflux translocases of hydrophobic antibiotics. The polyketide synthase of tetracenomycin is a membrane protein (67). Tetracenomycin is fairly hydrophobic and is exported by a translocase of the 14-TMS subfamily homologous to EmrB and QacA (68,69). It is reasonable to assume that the membrane synthase deposits its hydrophobic product into the membrane where it is picked up by the efflux transporter.

The RND pumps of gram-negative species apparently have the most unusual pathway for their ligands. In *P. aeruginosa*, MexAB-OprM protects the cell from β-lactams that do not penetrate into the cytoplasm (70–72). The AcrAB pump of *E. coli* protects the cell from β-lactams as well (73). It was established that in *S. typhimurium*, transport activity by the AcrAB pump increases with decreasing polarity of β-lactams (74). It was therefore proposed that the pump picks up the substrate from the outer bilayer (as well as from the inner bilayer) and transports it across the outer membrane (70) with the aid of a membrane fusion protein and an outer membrane porin TolC.

The advantage for an RND pump to pick up its substrate at the outer leaflet of the cytoplasmic membrane is evident—this allows for extrusion of fairly hydrophilic substances that will only partially partition in the membrane. The ability of the pump to transport them across the outer membrane makes this a reasonable strategy, especially for substances like β-lactams whose target is the cell wall. The interesting ability of the RND Mtr pump of *N. gonorrhoeae* to protect the cell from mammalian antimicrobial peptides such as protegrin-1 and LL-37 (75) would also be aided by an outer leaflet–outer medium transport pathway. After insertion in the membrane, antimicrobial peptides are known to form oligomeric pore structures that would no longer be subject to export.

It also seems very reasonable for the MDRs of gram-positive species like LmrA and LmrP to have an inner leaflet–outer medium ligand pathway. Many of the substrates are cations and accumulate in the cell driven by the membrane potential. The concentration of the substrate will therefore be considerably higher in the inner leaflet equilibrated with the cytoplasm rather than in the outer leaflet equilibrated with the external medium. Without being able to take an advantage of an additional permeability barrier like the outer membrane, it does seem that the optimal strategy for a pump is to take the substrate out of a site where its concentration is highest. This would allow the pump to clear the membrane from compounds present at a very low extracellular concentration—we have found that the *S. aureus* NorA pump protects the cell from nanomolar concentrations of hydrophobic cations (76).

Much less is known about the ligand binding site itself. Numerous attempts to localize it in the P glycoprotein, for example, revealed mutations that affect binding affinity and specificity to be scattered around the transmembrane segments (reviewed in Ref. 62). A similar picture emerged from a more limited mutagenesis analysis of the bacterial BmR (77–79). Interesting structural work has been done with EmrE of *E. coli*, the small MDR of the SMR family. This very hydrophobic MDR was solubilized in chloroform/methanol/water and analyzed by nuclear magnetic resonance (NMR) (80). Results indicated the presence of four α-helical bundles in good agreement with computerized predictions. A biochemical study of membrane topology with a closely homologous Smr protein from *S. aureus* using reporter protein fusions also shows a 4-TMS structure (81). Smr has a single charged residue in the membrane, a glutamate. This residue is conserved in the family and was found to be essential for drug transport (82). Even a conserved substitution with an aspartate abolished efflux. According to a proposed model, glu acts to bind a cationic drug, and a proton coming in from the external medium replaces the cation, which then continues on its way out of the cell (82). It is important to note that all substrates of Smr are cations, which agrees well with this model of translocation. However, the model does not, and hardly can, address the question of drug recognition.

An overall structure of purified P glycoprotein was recently obtained by electron micrography and Fourier projection maps of small two-dimensional crystalline arrays. From above the membrane plane the protein appears to have a diameter of about 10 nm with a large central pore of about 5 nm in diameter closed at the cytoplasmic side of the membrane. Two 3-nm lobes are exposed at the cytoplasmic face of the membrane, likely corresponding to the nucleotide binding domains (83,84). With little indication that a crystalline structure of an MDR will be obtained anytime soon, a number of laboratories have been turning to the study of multidrug sensors, small soluble cytoplasmic proteins that control the expression of MDR pumps. Three such proteins have been described so far. In *E. coli*, we identified a regulator of the EmrAB pump, EmrR, an 18-D protein that is coded by the first gene of the *emrRAB* operon (85) and belongs to the MarR family of transcriptional repressors (86). EmrR binds to such EmrAB substrates as uncouplers of oxidative phosphorylation and binding releases repression and activates transcription. Experiments with purified EmrR show that it directly binds its ligands in vitro. Scatchard analysis of equilibrium dialysis data showed a one ligand per monomer with good affinity, K_s around 1 μM for FCCP and CCCP, and Ks around 10 μM for the more hydrophilic DNP (87). The central region of the protein is fairly well conserved within the family and corresponds to a helix-turn-helix motif (86). This would indicate that the ligand-binding site would be in the divergent N- or C-terminal portion of the protein. According to our data, the C-terminal half expressed from a recombinant vector had no ligand-binding activity and the N-terminal half was insoluble. Structural studies will thus have to be done with a full-length protein. Native gel electrophoresis and molecular sieve chromatography show that the protein is a dimer (87). Efforts to localize a ligand-binding site were successful in the study of *B. subtilis* BMR multidrug sensor, and the crystalline structure of this C-terminal domain has recently been resolved (88,89). BmrR is a transcriptional activator of the MF BMR pump (90,91). BmrR binds chemically unrelated hydrophobic cations, such as TPP^+ and ethidium bromide (EtBr), substrates of the BMR pump, and activates the BMR transcription. BmrR is a 32-kD protein that forms a dimer and binds one ligand molecule per dimer. The 18.4-kD C-terminal ligand-binding domain still forms a dimer, and this peptide was crystallized both with and without the ligand TPP^+. The binding site is rather unusual—it is not obviously present in the apoprotein and is formed in the process of ligand binding. TPP^+ apparently aids unfolding and displacement of an α-helix, which exposes a hydrophobic binding pocket with a buried glutamate residue. Once past the gate, the ligand is bound by stacking and van der Waals interactions with residues of hydrophobic amino acids and by electrostatic interaction with the glutamate. This interesting and unusual structure explains the main features of selectivity. Apparently, hydrophilic molecules will not gain access to the hydrophobic site that is not even open in the apoprotein; once inside the pocket, the amphipathic ligand will be retained by hydrophobic interactions; and the presence of a strong negative charge in this hydrophobic environment will select for cationic species.

A third multidrug sensor has been recently described that regulates expression of the QacA MF MDR found on multidrug resistance plasmids of *S. aureus*. It is coded by a *qacR* gene located immediately upstream of *qacA* and is divergently transcribed from its own promoter. The QacR repressor binds hydrophobic cations (92), exclusive substrates of QacA, and belongs to the TetR family of repressors.

There is evidently intrinsic value in understanding the mechanism of drug discrimination by multidrug sensors, but will they also serve as a useful model to understand the more complex MDR pumps? The sensors share a substrate specificity with the pumps but no homology. However, it seems possible that the general principles of ligand binding/drug discrimination that emerge from the structural studies of drug sensors will be applicable to MDR pumps. At the very least, information gained from the study of the sensors will stimulate the design of experiments to test particular models of drug discrimination by MDR pumps.

3 FUNCTION AND THE SEARCH FOR NATURAL SUBSTRATES

MDR pumps can be very effective in protecting the microbial cell from toxic compounds. For example, the EmrAB pump of *E. coli* (34) confers a 60-fold increase in resistance to the antibiotic thiolactomycin (93); the AcrAB pump of *E. coli* confers a 60- to 100-fold resistance to novobiocin and erythromycin (73); and the MexAB pump of *P. aeruginosa* confers a 50-

to 100-fold resistance to some quinolones and hydrophobic antibiotics (94). These levels of resistance rival specific resistance mechanisms such as tetracycline efflux by a dedicated pump or inactivation of kanamycin by a specific acetyl transferase.

Even though MDR pumps have the potential to protect cells from antimicrobials and do this both in the laboratory and in the clinical setting, this does not necessarily mean that drug resistance is the natural function of MDRs. For example, the Blt pump of *B. subtilis* that protects cells from a panel of amphipathic cationic antimicrobials is part of an operon that codes for a putrescine acetyl transferase (95). The Blt pump extrudes putrescine from the cell, suggesting that this might be its natural function, and drug resistance is a coincidental consequence of the protein being a "sloppy translocase."

However, other data indicate that drug resistance is the natural function of MDRs. The QacA MDR pump is found on broad host range plasmids that also carry specific gentamicin and trimethoprim resistance genes (33). This context suggests QacA is a dedicated drug resistance component of the plasmid as well. The induction of MDRs by their substrates acting through MDR sensors is a very strong argument for multidrug resistance being the natural function of at least some of these translocases. The BmrR ligand-binding site discussed above seems to have evolved to accommodate a wide range of substrates that it selects largely on the basis of polarity, a salient feature of MDR pump recognition.

But if there are dedicated MDRs, what are their natural substrates? RND pumps extrude natural antibiotics, but handle artificial substrates just as well or better, and have such a broad substrate spectrum that it is unclear whether antibiotic extrusion plays a leading role outside of the clinical setting. The AcrAB pump was shown to strongly protect *E. coli* from bile acids, and it was proposed that bile acid extrusion might be the natural function of this pump (96). This is an appealing hypothesis, and bile acids might indeed be among the many natural substrates of the extremely broad-spectrum AcrAB pump. Protection from bile salts appears to be one of the functions of the Mtr RND pump of *N. gonorrhoeae* (97) and of VceAB which is a *V. cholerae* homologue of the *E. coli* EmrAB pump (35). As mentioned above, the Mtr pump of *N. gonorrhoeae* has been reported to protect the cell from small defensin peptides, and this might be part of its natural function. It does not seem, however, that RND pumps that are widely distributed among gram-negative bacteria evolved to cope with bile acids or defensins of animals. Another possible example of a natural substrate has been described recently in a study of an IfeAB RND pump from *Agrobacterium tumefaciens* (98). Mutation of the pump decreased accumulation of an isoflavonoid coumestrol that is produced by its host, the alfalfa plant. Coumestrol induced expression of the pump and the mutant was outcompeted by the wild type in colonizing the plant. However, neither the wild type nor the mutant was sensitive to coumestrol. As the researchers note, coumestrol might be acting as an inducing signal for the pump rather than its natural substrate. The issue of natural MDR substrates in most cases remains very much an open question.

Even though MDR pumps extrude structurally unrelated compounds, a general theme emerges if one considers the preferred artificial substrates of most MDRs. These substances with little exception are amphipathic cations (Fig. 3). This observation suggests that amphipathic cations represent the prototypical and existing natural substrates of MDR pumps. The simplest MDRs of the SMR family are unique in that no specific translocases are found in this family, and they might have been the first dedicated MDRs to evolve (37). This is also the only group of MDRs to have amphipathic cations as their exclusive substrates. There are many different MDR pumps in the MF family, and most of them exclusively extrude amphipathic cations. For example, the QacA pump only extrudes cations; the NorA pump of *S. aureus* extrudes cations and to a lesser extent quinolones; the BMR pump of *B. subtilis* extrudes primarily cations and neutral chloramphenicol (reviewed in Ref. 51). The BmrR regulator that activates BMR transcription has a design suggesting it specifically evolved to detect a wide range of amphipathic cations. MDRs of the RND family have a broad substrate spectrum, and all tested RND pumps extrude amphipathic cations. The preferred substrates for ABC MDRs LmrA and P glycoprotein are amphipathic cations, but some neutral compounds can be extruded as well. There are many ABC MDRs in yeast, and at least nine functional ABC MDRs are present in *S. cerevisiae* alone. The substrates of these pumps are amphipathic cations, and neutral substances such as antiyeast azoles (99,100).

The following picture emerges from this analysis. Simple MDRs like SMRs only export amphipathic cations; MF and MATE MDRs export mainly cations; RND and ABC are the most complex of the MDRs and have broad spectra of specificity that include amphipathic cations as preferred substrates. MDRs clearly prefer amphipathic cations to other substances even though they belong to five unrelated protein families. Not only are MDRs unrelated, but even the general mechanisms of drug transport are different for different MDRs, as discussed above. It appears that a similar need to protect the cell from amphipathic cations evolved in different groups of MDRs (and in different organisms) in spite of a lack of overall homology or similarity in the mechanism of action.

Quite surprisingly, one does not find amphipathic cations in a general list of natural antimicrobials. (The known cationic antibiotics of the

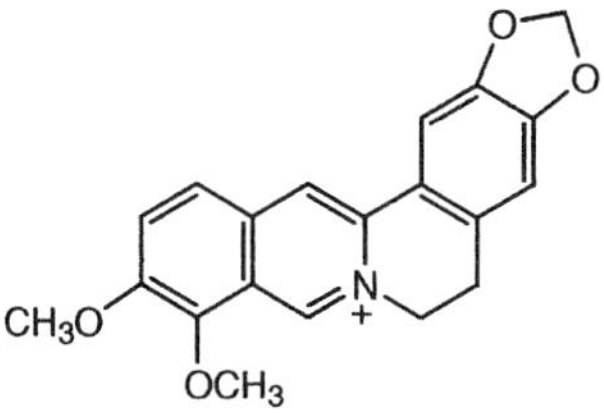

berberine

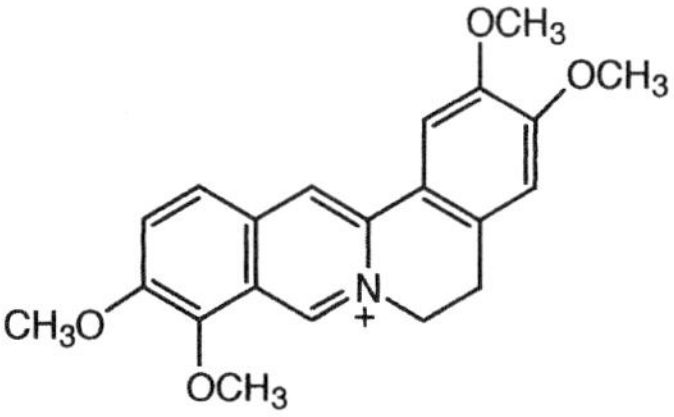

palmatine

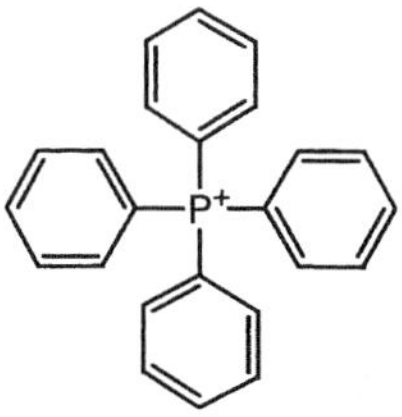

tetraphenylphosphonium

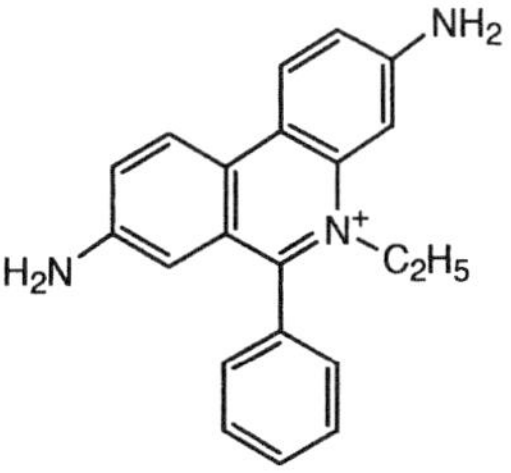

ethidium bromide

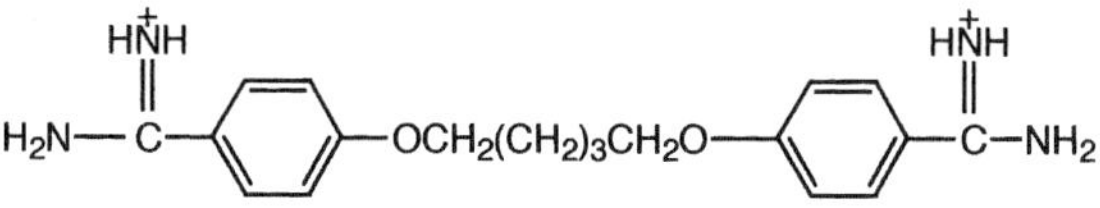

pentamidine

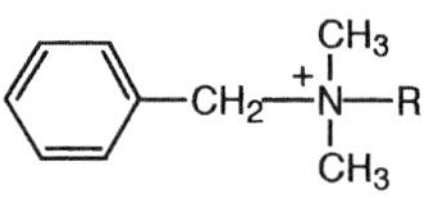

benzalkonium chloride

norfloxacin

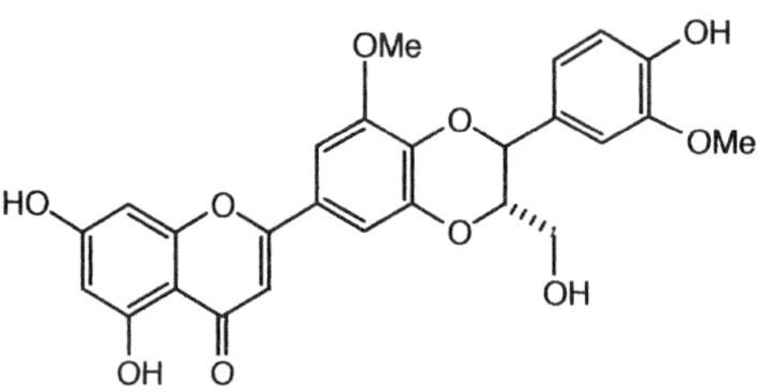

5'-methoxyhydnocarpin

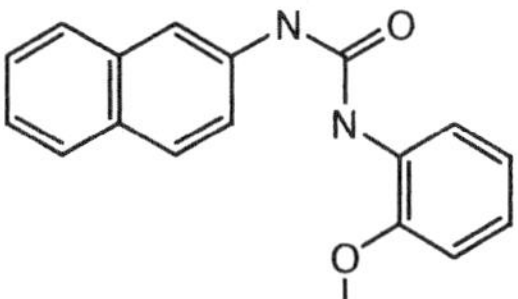

INF271

aminoglycoside group, such as streptomycin and kanamycin, are hydrophilic substances that get smuggled into the cell via specific translocases and are not generally substrates of MDR pumps.) At the same time, amphipathic cations should be among the most potent antimicrobials. A positive charge will lead to a considerable accumulation of a substance in the cell. According to the Nernst equation, there is a 10-fold accumulation of a cation (cat) for every 60 m V of the membrane potential:

$$\Delta\varphi(\text{mV}) = \frac{RT}{nF}\ln\left\{\frac{[\text{cat}]\text{in}}{[\text{cat}]\text{out}}\right\} = 60\,\text{lg}\left\{\frac{[\text{cat}]\text{in}}{[\text{cat}]\text{out}}\right\}$$

The $\Delta\varphi$ in bacteria is approximately 140 mV and approximately 180 mV in yeast plasma membrane (101), which would result in a 100- to 1000-fold accumulation of an antibiotic. Note what weak amphipathic bases (such as chlorhexidine) are also MDR substrates, but one would not expect these compounds to be among natural antibiotics, since they are extruded from the cell by the pH gradient and are therefore intrinsically less potent than neutral compounds or strong cations. The fact that strong amphipathic cations are conspicuously absent from known natural antibiotics is especially puzzling given that these substances are the preferred substrates for most MDRs. We have argued that it is precisely the existence of MDR pumps that is responsible for this apparent paradox (76). If MDRs evolved in response to natural antimicrobial amphipathic cations, then these substances would be difficult to discover in standard screens that employ cells carrying MDR pumps. In the process of drug discovery, the concentration of antimicrobials is prone to be low, and MDR substrates will be overlooked. MDR mutants can therefore be used as sensitive tools for drug discovery.

Although using MDR mutants is a reasonable (if somewhat unpredictable) way to discover possible cationic antimicrobials, another approach is to search for possible MDR substrates among known compounds. Many natural substances have been described as a result of systematic chemical analysis of organisms rather than in particular bioassay-driven purifications. One would then look for substances that are amphipathic cations of natural origin that have little or no antimicrobial activity.

Figure 3 Amphipathic cationic substrates of MDRs. Berberine and palmatine are plant isoquinoline alkaloids. Tetraphenylphosphonium has been used as a probe to measure the membrane potential. Benzalkonium chloride is an antiseptic and disinfectant. Weak bases—pentamidine is a systemic antiprotozoan and chlorhexidine is an antiseptic. 5′-Methoxyhydnocarpin is a natural MF MDR inhibitor produced by *Berberis* plants that synthesize berberine. INF is a synthetic MDR inhibitor.

Using these criteria, we identified a group of plant alkaloids whose members have little or no antimicrobial activity. These are the isoquinoline alkaloids (see Fig. 1) that are widely spread among the plant world and are found among many *Ranunculales* species, for example (102). These substances bear a resemblance to artificial MDR substrates such as ethidium bromide (see Fig. 1). They are amphipathic and have a strong positive charge which is delocalized by the conjugated ring structure, an essential feature of a good permeant cation (101).

We have chosen two representative substances of this group, berberine and palmatine, to test whether they are MDR substrates. Palmatine had very low (>200 μg/mL MIC) and berberine poor (240–120 μg/mL) activity against wild-type *S. aureus* (76). The antimicrobial activity of the alkaloids increased sharply in a *norA* mutant, with an MIC of 50 μg/mL for palmatine and 7.5 for berberine. Sensitivity to alkaloids increased further in the presence of an MDR inhibitor INF271 (provided kindly by Dr. P. N. Markham of Influx, Inc.; K. Lewis, unpublished data). It appears that the inhibitor disables both NorA and a possible additional unknown MDR(s). Thus, in the presence of the MDR inhibitor, berberine becomes an extremely potent antibiotic (MIC 0.5 μg/mL), about 10 times stronger than streptomycin. Our preliminary experiments show that berberine is also the substrate of the plasmidborne QacA pump of *S. aureus* (MIC 500 μg/mL vs 1 μg/mL in the presence of INF271). In the presence of the MDR inhibitor, yeast become very sensitive to berberine as well (MIC 120 μg/mL vs 1 μg/mL).

Screening natural compound libraries that is being performed by drug discovery companies in search of new antibiotics will produce a catalogue of natural MDR substrates, especially if strains lacking MDR pumps are used as targets. This work will might lead to the discovery of new classes of antibiotics that are hydrophobic cations.

Why should plants keep on making isoquinoline alkaloids if microorganisms have MDR pumps that can render these substances essentially ineffective? One possibility is that isoquinoline alkaloids are not antimicrobial compounds in vivo and have a different function, such as antiherbivoral. A more interesting possibility is that in response to bacterial resistance mechanisms, plants have developed MDR inhibitors that act synergistically with isoquinoline alkaloids. We have tested this hypothesis using a berberine-producing *Berberis fremontii*. An extract of the plant has at least two different MDR inhibitors that act synergistically with berberine in inhibiting the growth of *S. aureus* (103). The first one identified is 5′-methoxyhydnocarpin (5′-MHC), a flavonolignan that has been reported previously as a minor contaminant of chaulmoogra oil (a traditional antileprosy remedy [104]) from seeds of *Hydnocarpus* trees, and no biological activity for this compound has been described. 5′-MHC had no activity on its own, but at 1 μg/mL completely inhibited growth of *S. aureus* in the presence of subinhibitory berberine. 5′-MHC was also found completely to inhibit MDR-dependent efflux of ethidium bromide and berberine from the cells, and the level of berberine accumulation by cells was sharply increased in the presence of the inhibitor. *S. aureus* is likely to encounter natural cationic antimicrobials like berberines when the microbe is persisting in the environment. *Berberis* species are not known to be infected by bacterial pathogens; apparently owing to the presence of effective antimicrobials like berberine and 5′-MHC. It seems that plants producing antimicrobials might have developed a variety of MDR inhibitors against different MDR pumps of plant pathogens. Such substances, as well as synthetic inhibitors, will become useful additions to the arsenal of antimicrobial agents.

4 CIRCUMVENTING EFFLUX RESISTANCE

Efflux-based resistance can be eliminated either by creating molecules that are not substrates of transporters or by employing pump inhibitors. Both approaches are currently under development.

4.1 Specific Pumps

A new class of semisynthetic tetracyclines, glycylcyclines, has been developed by investigators at American Cyanamid (105). These compounds exhibit potent activity against a broad spectrum of gram-positive and gram-negative bacteria, including those that carry ribosomal protection (Tet[M]) and efflux determinants (Tet [A–D], Tet[K]). Susceptibility to these compounds for tetracycline-resistant strains (MIC >32 μg/mL) ranged from 0.5 to 2.0 μg/mL. It was demonstrated that glycylcyclines overcome efflux-mediated resistance, because they are not recognized by the transporter protein (106). It is noteworthy that glycylcyclines are still effluxed by MDRs; for example, by the MexAB-OprM and MexCD-OprJ from *P. aeruginosa*.

Inhibitors of Tet pumps were identified among certain semisynthetic tetracycline analogues (107–109). It was shown that these compounds inhibited tetracycline efflux in everted vesicles prepared from *E. coli* cells containing the TetB protein. The most potent analogue, 13-CPTC, interfered with tetracycline transport by competitively binding to TetB but itself was transported less efficiently than tetracycline. When combined with doxycycline, 13-CTPC exhibited synergy against *E. coli* strains expressing either TetA or TetB proteins by lowering the MIC by a factor of 2 or greater.

Significant progress has been accomplished in developing new macrolides with enhanced activity. The ketolides are novel semisynthetic 14–membered ring macrolides derived from erythromycin A but characterized by a 3-keto function instead of cladinose moiety. Notable agents are HMR-3647 (Hoechst-Marion-Roussel), ABT-773 (Abbott), and CP-544372 (Pfizer), which are at various stages of development. Although HMR-3647 and ABT-773 are still somewhat affected by the macrolide efflux (Mef) pumps, although to a considerably lesser degree than the older macrolides, CP-544372 (a 4″crolide) appears to not be affected at all (110). Although the ketolides have enhanced activity against macrolide-resistant gram-positive bacteria, their activity against gram-negative bacteria is not superior to azithromycin (111,112). These results infer that these compounds are still subject to efflux by multidrug resistance pumps in gram-negative bacteria.

4.2 MDR Transporters

Research on new fluoroquinolones has been focused on enhancing activity against gram-positive pathogens versus that for ciprofloxacin. These efforts resulted in the discovery of trovafloxacin (Pfizer) (113), clinafloxacin (Parke-Davis), gatifloxacin (Brystol-Myers Squibb), and moxifloxacin (Bayer) (114). It appears that NorA from *S. aureus* and efflux pumps from *S. pneumoniae* confer a lower degree of resistance to these new fluoroquinolones as compared to the older ones (115–119). Direct transport experiments are still needed to clarify whether these pumps do not recognize the new derivatives or whether an increased rate of diffusion of these more lipophilic fluoroquinolones is responsible for overcoming efflux. At the same time, the newer fluoroquinolones are less active against gram-negative bacteria as compared to ciprofloxacin, apparently owing to their more effective efflux by RND pumps (119).

4.3 Efflux Pump Inhibitors

The first inhibitor active against multiple RND transporters in gram-negative bacteria was reported by scientists from Microcide Pharmaceuticals (Mountain View, CA). Empiric screening of small molecule libraries for compounds with efflux pump inhibitory activity resulted in the identification of an inhibitor active against MexAB-OprM, MexCD-OprJ, and MexEF-OprN efflux pumps, which contribute to fluoroquinolone resistance in *P. aeruginosa*. This broad-spectrum inhibitor, MC-207,110 (EPI, efflux pump inhibitor) (see Fig. 2) is also active against RND pumps in many representatives of Enterobacteriaceae, *N. gonorrhoeae*, and *H. influenzae* (120–122). EPI decreased intrinsic resistance to levofloxacin approximately

8-fold in the wild-type strain of *P. aeruginosa*, whereas in the strains that overexpress efflux pumps, the susceptibility was increased 64-fold. As expected, EPI potentiated levofloxacin irrespective of the presence of target-based mutations. Some early SAR of the EPI has recently been published (123).

EPI also dramatically decreased the frequency of selection of resistant bacteria. When the wild-type strain was used for selection experiments at the standard 4 × MIC, the frequency was less than 10^{-11} (vs 10^{-7} without EPI). The EPI diminished the appearance of both efflux-mediated and target-based mutations. Apparently, a single target-based mutation (i.e., in the gyrase) conferred insufficient resistance in the absence of the MexAB-OprM–mediated intrinsic resistance to support the emergence mutants under these selection conditions.

It was found that clinical isolates of *P. aeruginosa* with a wide range of resistant phenotypes showed increased susceptibility to levofloxacin in the presence of EPI (124). In a further survey, it was found that EPI potentiated the antimicrobial action of macrolides in *H. influenzae* and chloramphenicol in *E. coli* and *S. typhimurium*. Therefore, a single compound when combined with different antibiotics may have multiple clinical applications. Importantly, EPI potentiated levofloxacin in several animal models of infection (125), providing in vivo proof-of-principle for the applicability of efflux pump inhibitors.

Inhibitors of the NorA pump from *S. aureus* that potentiate ciprofloxacin against *S. aureus* have been reported by scientists of Influx Inc. (Chicago, IL) (126). Prevention of the emergence of mutants resistant to fluoroquinolones was first demonstrated by inhibition of *S. aureus* NorA (127). As mentioned in the previous section, the plant flavonolignan 5′-MHC is a potent inhibitor of NorA, and there is little doubt that additional natural MDR inhibitors will be identified.

In a number of cases, an MDR pump has been identified in a pathogen, but resistant isolates overexpressing the pump have not been described in a clinical setting. For example, *Mycobacterium smegmatis* harbors an LfrA MDR belonging to the MF family that confers resistance to fluoroquinolones (128,129), and a homologous EfpA MDR is found in *M. tuberculosis* (130). Known quinolone resistant mutants of *M. tuberculosis* carry mutations in DNA gyrase, and not in regulatory elements controlling EfpA. This would suggest that EfpA does not have a role in clinically significant antibiotic resistance. However, the examples of resistance prevention discussed above suggest that even when MDRs are not obviously involved in drug resistance, their role might actually be crucial in development of resistant mutants. Thus one might expect that *M. tuberculosis* lacking a functional EfpA will not produce gyrase mutants with high levels of resistance to fluoroquinolones.

Owing to their broad specificity, MDRs provide a ready-made resistance mechanism for the newest synthetic antibiotics like quinolones. In a sense, various pathogens have already developed a resistance mechanism to current and future antimicrobials. Emulating nature's strategy and potentiating antibiotics with MDR inhibitors can be an effective approach against drug-resistant microorganisms.

REFERENCES

1. Marger MD, Saier MH Jr. A major superfamily of transmembrane facilitators that catalyse uniport, symport and antiport [see comments]. Trends Biochem Sci 1993; 18:13–20.
2. Rubin RA, Levy SB, Heinrikson RL, Kezdy FJ. Gene duplication in the evolution of the two complementing domains of gram-negative bacterial tetracycline efflux proteins. Gene 1990; 87:7–13.
3. McMurry LM, Levy SB. The NH2-terminal half of the Tn10-specified tetracycline efflux protein TetA contains a dimerization domain. J Biol Chem 1995; 270:22752–22757.
4. Varela MF, Sansom CE, Griffith JK. Mutational analysis and molecular modelling of an amino acid sequence motif conserved in antiporters but not symporters in a transporter superfamily [published erratum appears in Mol Membr Biol 1996 Jan-Mar;13(1):66]. Mol Membr Biol 1995; 12:313–319.
5. Roberts MC. Epidemiology of tetracycline-resistance determinants. Trends Microbiol 1994; 2:353–357.
6. Yamaguchi A, Ono N, Akasaka T, Noumi T, Sawai T. Metal-tetracycline/H+ antiporter of *Escherichia coli* encoded by a transposon, Tn10. The role of the conserved dipeptide, Ser65-Asp66, in tetracycline transport. J Biol Chem 1990; 265:15525–15530.
7. Thanassi DG, Suh GS, Nikaido H. Role of outer membrane barrier in efflux-mediated tetracycline resistance of *Escherichia coli*. J Bacteriol 1995; 177:998–1007.
8. Dorman CJ, Foster TJ. Nonenzymatic chloramphenicol resistance determinants specified by plasmids R26 and R55-1 in *Escherichia coli* K-12 do not confer high-level resistance to fluorinated analogs. Antimicrob Agents Chemother 1982; 22:912–914.
9. Bissonnette L, Champetier S, Buisson JP, Roy PH. Characterization of the nonenzymatic chloramphenicol resistance (*cmlA*) gene of the In4 integron of Tn1696: similarity of the product to transmembrane transport proteins. J. Bacteriol 1991; 173:4493–4502.
10. Dorman CJ, Foster TJ, Shaw WV. Nucleotide sequence of the R26 chloramphenicol resistance determinant and identification of its gene product. Gene 1986; 41:349–353.
11. Toro CS, Lobos SR, Calderon I, Rodriguez M, Mora GC. Clinical isolate of a porinless *Salmonella typhi* resistant to high levels of chloramphenicol. Antimicrob Agents Chemother 1990; 34:1715–1719.

12. Ploy MC, Courvalin P, Lambert T. Characterization of In40 of *Enterobacter aerogenes* BM2688, a class 1 integron with two new gene cassettes, *cmlA2* and *qacF*. Antimicrob Agents Chemother 1998; 42:2557–2563.
13. Burns JL, Mendelman PM, Levy J, Stull TL, Smith AL. A permeability barrier as a mechanism of chloramphenicol resistance in *Haemophilus influenzae*. Antimicrob Agents Chemother 1985; 27:46–54.
14. Kim E, Aoki T. Sequence analysis of the florfenicol resistance gene encoded in the transferable R-plasmid of a fish pathogen, *Pasteurella piscicida*. Microbiol Immunol 1996; 40:665–669.
15. Bolton LF, Kelley LC, Lee MD, Fedorka-Cray PJ, Maurer JJ. Detection of multidrug-resistant *Salmonella enterica* serotype *typhimurium* DT104 based on a gene which confers cross-resistance to florfenicol and chloramphenicol. J Clin Microbiol 1999; 37:1348–1351.
16. Poppe C, Smart N, Khakhria R, Johnson W, Spika J, Prescott J. *Salmonella typhimurium* DT104: a virulent and drug-resistant pathogen. Can Vet J 1998; 39:559–565.
17. Glynn MK, Bopp C, Dewitt W, Dabney P, Mokhtar M, Angulo FJ. Emergence of multidrug-resistant *Salmonella enterica* serotype *typhimurium* DT104 infections in the United States. N Engl J Med 1998; 338:1333–1338.
18. Sutcliffe J, Tait-Kamradt A, Wondrack L. *Streptococcus pneumoniae* and *Streptococcus pyogenes* resistant to macrolides but sensitive to clindamycin: a common resistance pattern mediated by an efflux system. Antimicrob Agents Chemother 1996; 40:1817–1824.
19. Clancy J, Dib-Hajj F, Petitpas JW, Yuan W. Cloning and characterization of a novel macrolide efflux gene, *mreA*, from *Streptococcus agalactiae*. Antimicrob Agents Chemother 1997; 41:2719–2723.
20. Tait-Kamradt A, Clancy J, Cronan M, et al. *mefE* is necessary for the erythromycin-resistant M phenotype in *Streptococcus pneumoniae*. Antimicrob Agents Chemother 1997; 41:2251–2255.
21. Giovanetti E, Montanari MP, Mingoia M, Varaldo PE. Phenotypes and genotypes of erythromycin-resistant *Streptococcus pyogenes* strains in Italy and heterogeneity of inducibly resistant strains. Antimicrob Agents Chemother 1999; 43:1935–1940.
22. Arpin C, Canron MH, Noury P, Quentin C. Emergence of *mefA* and *mefE* genes in beta-haemolytic streptococci and pneumococci in France. J Antimicrob Chemother 1999; 44:133–134.
23. Kataja J, Seppala H, Skurnik M, Sarkkinen H, Huovinen P. Different erythromycin resistance mechanisms in group C and group G streptococci. Antimicrob Agents Chemother 1998; 42:1493–1494.
24. Poutanen SM, de Azavedo J, Willey BM, Low DE, MacDonald KS. Molecular characterization of multidrug resistance in *Streptococcus mitis*. Antimicrob Agents Chemother 1999; 43:1505–1507.
25. Neyfakh AA, Bidnenko VE, Chen LB. Efflux-mediated multidrug resistance in *Bacillus subtilis*: similarities and dissimilarities with the mammalian system. Proc Natl Acad Sci USA 1991; 88:4781–4785.
26. Kaatz GW, Seo SM, Ruble CA. Efflux-mediated fluoroquinolone resistance in

Staphylococcus aureus. Antimicrob Agents Chemother 1993; 37:1086–1094.
27. Neyfakh AA, Borsch CM, Kaatz GW. Fluoroquinolone resistance protein NorA of *Staphylococcus aureus* is a multidrug efflux transporter. Antimicrob Agents Chemother 1993; 37:128–129.
28. Ng EY, Trucksis M, Hooper DC. Quinolone resistance mediated by *norA*: physiologic characterization and relationship to *flqB*, a quinolone resistance locus on the *Staphylococcus aureus* chromosome. Antimicrob Agents Chemother 1994; 38:1345–1355.
29. Edgar R, Bibi E. MdfA, an *Escherichia coli* multidrug resistance protein with an extraordinarily broad spectrum of drug recognition [published erratum appears in J Bacteriol 1997 Sep; 179(17):5654]. J Bacteriol 1997; 179:2274–2280.
30. Bohn C, Bouloc P. The *Escherichia coli cmlA* gene encodes the multidrug efflux pump Cmr/MdfA and is responsible for isopropyl-beta-D-thiogalactopyranoside exclusion and spectinomycin sensitivity. J Bacteriol 1998; 180: 6072–6075.
31. Mine T, Morita Y, Kataoka A, Mizushima T, Tsuchiya T. Evidence for chloramphenicol/H+ antiport in Cmr (MdfA) system of *Escherichia coli* and properties of the antiporter. J Biochem (Tokyo) 1998; 124:187–193.
32. Edgar R, Bibi E. A single membrane-embedded negative charge is critical for recognizing positively charged drugs by the *Escherichia coli* multidrug resistance protein MdfA. EMBO J 1999; 18:822–832.
33. Rouch DA, Cram DS, DiBerardino D, Littlejohn TG, Skurray RA. Efflux-mediated antiseptic resistance gene *qacA* from *Staphylococcus aureus*: common ancestry with tetracycline- and sugar-transport proteins. Mol Microbiol 1990; 4:2051–2062.
34. Lomovskaya O, Lewis K. Emr, an *Escherichia coli* locus for multidrug resistance. Proc Natl Acad Sci USA 1992; 89:8938–8942.
35. Colmer JA, Fralick JA, Hamood AN. Isolation and characterization of a putative multidrug resistance pump from *Vibrio cholerae*. Mol Microbiol 1998; 27:63–72.
36. Lee EH, Shafer WM. The *farAB*-encoded efflux pump mediates resistance of gonococci to long-chained antibacterial fatty acids. Mol Microbiol 1999; 33: 839–845.
37. Lewis K. Multidrug resistance pumps in bacteria: variations on a theme. Trends Biochem Sci 1994; 19:119–123.
38. Ma D, Cook DN, Alberti M, Pon NG, Nikaido H, Hearst JE. Molecular cloning and characterization of *acrA* and *acrE* genes of *Escherichia coli*. J Bacteriol 1993; 175:6299–6313.
39. Ma D, Cook DN, Hearst JE, Nikaido H. Efflux pumps and drug resistance in gram-negative bacteria. Trends Microbiol 1994; 2:489–493.
40. Guan L, Ehrmann M, Yoneyama H, Nakae T. Membrane topology of the xenobiotic-exporting subunit, MexB, of the MexA,B-OprM extrusion pump in *Pseudomonas aeruginosa*. J Biol Chem 1999; 274:10517–10522.
41. Fralick JA. Evidence that TolC is required for functioning of the Mar/AcrAB efflux pump of *Escherichia coli*. J Bacteriol 1996; 178:5803–5805.
42. Nikaido H. Multidrug efflux pumps of gram-negative bacteria. J Bacteriol 1996; 178:5853–5859.

43. Zgurskaya HI, Nikaido H. AcrA is a highly asymmetric protein capable of spanning the periplasm. J Mol Biol 1999; 285:409–420.
44. Zgurskaya HI, Nikaido H. Bypassing the periplasm: reconstitution of the AcrAB multidrug efflux pump of *Escherichia coli*. Proc Natl Acad Sci USA 1999; 96:7190–7195.
45. Moore RA, DeShazer D, Reckseidler S, Weissman A, Woods DE. Efflux-mediated aminoglycoside and macrolide resistance in *Burkholderia pseudomallei*. Antimicrob Agents Chemother 1999; 43:465–470.
46. Aires JR, Kohler T, Nikaido H, Plesiat P. Involvement of an active efflux system in the natural resistance of *Pseudomonas aeruginosa* to aminoglycoside. Antimicrob Agents Chemother 1999; 43:2624–2628.
47. Mine T, Morita Y, Kataoka A, Mizushima T, Tsuchiya T. Expression in *Escherichia coli* of a new multidrug efflux pump, MexXY, from *Pseudomonas aeruginosa*. Antimicrob Agents Chemother 1999; 43:415–417.
48. Westbrock-Wadman S, Sherman DR, Hickey MJ, et al. Characterization of a *Pseudomonas aeruginosa* efflux pump contributing to aminoglycoside impermeability. Antimicrob Agents Chemother 1999; 43:2975–2983.
49. Paulsen IT, Skurray RA, Tam R, et al. The SMR family: a novel family of multidrug efflux proteins involved with the efflux of lipophilic drugs. Mol Microbiol 1996; 19:1167–1175.
50. Greener T, Govezensky D, Zamir A. A novel multicopy suppressor of a *groEL* mutation includes two nested open reading frames transcribed from different promoters. EMBO J 1993; 12:889–896.
51. Paulsen IT, Brown MH, Skurray RA. Proton-dependent multidrug efflux systems. Microbiol Rev 1996; 60:575–608.
52. Brown MH, Paulsen IT, Skurray RA. The multidrug efflux protein NorM is a prototype of a new family of transporters. Mol Microbiol 1999; 31:394–395.
53. Morita Y, Kodama K, Shiota S, et al. NorM, a putative multidrug efflux protein, of *Vibrio parahaemolyticus* and its homolog in *Escherichia coli*. Antimicrob Agents Chemother 1998; 42:1778–1782.
54. Ross JI, Eady EA, Cove JH, Cunliffe WJ, Baumberg S, Wootton JC. Inducible erythromycin resistance in staphylococci is encoded by a member of the ATP-binding transport super-gene family. Mol Microbiol 1990; 4:1207–1214.
55. Matsuoka M, Endou K, Kobayashi H, Inoue M, Nakajima Y. A dyadic plasmid that shows MLS and PMS resistance in *Staphylococcus aureus*. FEMS Microbiol Lett 1997; 148:91–96.
56. Milton ID, Hewitt CL, Harwood CR. Cloning and sequencing of a plasmid-mediated erythromycin resistance determinant from *Staphylococcus xylosus*. FEMS Microbiol Lett 1992; 76:141–147.
57. van Veen HW, Venema K, Bolhuis H, et al. Multidrug resistance mediated by a bacterial homolog of the human multidrug transporter MDR1. Proc Natl Acad Sci USA 1996; 93:10668–10672.
58. van Veen HW, Callaghan R, Soceneantu L, Sardini A, Konings WN, Higgins CF. A bacterial antibiotic-resistance gene that complements the human multidrug-resistance P-glycoprotein gene. Nature 1998; 391:291–295.
59. Higgins CF, Gottesman MM. Is the multidrug transporter a flippase? Trends Biochem Sci 1992; 17:8–21.

60. Liu R, Sharom FJ. Site-directed fluorescence labeling of P-glycoprotein on cysteine residues in the nucleotide binding domains. Biochemistry 1996; 35: 11865–11873.
61. Sharom FJ. The P-glycoprotein efflux pump: how does it transport drugs? J Membr Biol 1997; 160:161–175.
62. Ueda K, Taguchi Y, Morishima M. How does P-glycoprotein recognize its substrates? Semin Cancer Biol 1997; 8:151–159.
63. Ruetz S, Gros P. Phosphatidylcholine translocase: a physiological role for the *mdr*2 gene. Cell 1994; 77:1071–1081.
64. van Helvoort A, Smith AJ, Sprong H, et al. MDR1 P-glycoprotein is a lipid translocase of broad specificity, while MDR3 P-glycoprotein specifically translocates phosphatidylcholine. Cell 1996; 87:507–517.
65. Bolhuis H, van Veen HW, Molenaar D, Poolman B, Driessen AJ, Konings WN. Multidrug resistance in *Lactococcus lactis*: evidence for ATP-dependent drug extrusion from the inner leaflet of the cytoplasmic membrane. EMBO J 1996; 15:4239–4245.
66. Bolhuis H, van Veen HW, Brands JR, et al. Energetics and mechanism of drug transport mediated by the lactococcal multidrug transporter LmrP. J Biol Chem 1996; 271:24123–24128.
67. Gramajo HC, White J, Hutchinson CR, Bibb MJ. Overproduction and localization of components of the polyketide synthase of *Streptomyces glaucescens* involved in the production of the antibiotic tetracenomycin C. J Bacteriol 1991: 173:6475–6483.
68. Guilfoile PG, Hutchinson CR. Sequence and transcriptional analysis of the *Streptomyces glaucescens tcmAR* tetracenomycin C resistance and repressor gene loci. J Bacteriol 1992; 174:3651–3658.
69. Caballero JL, Martinez E, Malpartida F, Hopwood DA. Organisation and functions of the *actVA* region of the actinorhodin biosynthetic gene cluster of *Streptomyces coelicolor*. Mol Gen Genet 1991; 230:401–412.
70. Li XZ, Ma D, Livermore DM, Nikaido H. Role of efflux pump(s) in intrinsic resistance of *Pseudomonas aeruginosa*: active efflux as a contributing factor to beta-lactam resistance. Antimicrob Agents Chemother 1994; 38:1742–1752.
71. Li XZ, Nikaido H, Poole K. Role of *mexA-mexB-oprM* in antibiotic efflux in *Pseudomonas aeruginosa*. Antimicrob Agents Chemother 1995; 39:1948–1953.
72. Srikumar R, Li XZ, Poole K. Inner membrane efflux components are responsible for beta-lactam specificity of multidrug efflux pumps in *Pseudomonas aeruginosa*. J Bacteriol 1997; 179:7875–7881.
73. Ma D, Cook DN, Alberti M, Pon NG, Nikaido H, Hearst JE. Genes *acrA* and *acrB* encode a stress-induced efflux system of *Escherichia coli*. Mol Microbiol 1995; 16:45–55.
74. Nikaido H, Basina M, Nguyen V, Rosenberg EY. Multidrug efflux pump AcrAB of *Salmonella typhimurium* excretes only those beta-lactam antibiotics containing lipophilic side chains. J Bacteriol 1998; 180:4686–4692.
75. Shafer WM, Qu X, Waring AJ, Lehrer RI. Modulation of *Neisseria gonorrhoeae* susceptibility to vertebrate antibacterial peptides due to a member of the resistance/nodulation/division efflux pump family. Proc Natl Acad Sci USA 1998; 95:1829–1833.

76. Hsieh PC, Siegel SA, Rogers B, Davis D, Lewis K. Bacteria lacking a multidrug pump: a sensitive tool for drug discovery. Proc Natl Acad Sci USA 1998; 95:6602–6606.
77. Ahmed M, Borsch CM, Neyfakh AA, Schuldiner S. Mutants of the *Bacillus subtilis* multidrug transporter Bmr with altered sensitivity to the antihypertensive alkaloid reserpine. J Biol Chem 1993; 268:11086–11089.
78. Klyachko KA, Schuldiner S, Neyfakh AA. Mutations affecting substrate specificity of the Bacillus subtilis multidrug transporter Bmr. J Bacteriol 1997; 179:2189–2193.
79. Klyachko KA, Neyfakh AA. Paradoxical enhancement of the activity of a bacterial multidrug transporter caused by substitutions of a conserved residue. J Bacteriol 1998; 180:2817–2821.
80. Schwaiger M, Lebendiker M, Yerushalmi H, et al. NMR investigation of the multidrug transporter EmrE, an integral membrane protein. Eur J Biochem 1998;254:610–619.
81. Paulsen IT, Brown MH, Dunstan SJ, Skurray RA. Molecular characterization of the staphylococcal multidrug resistance export protein QacC. J Bacteriol 1995; 177:2827–2833.
82. Grinius LL, Goldberg EB. Bacterial multidrug resistance is due to a single membrane protein which functions as a drug pump. J Biol Chem 1994; 269: 29998–30004.
83. Rosenberg MF, Callaghan R, Ford RC, Higgins CF. Structure of the multidrug resistance P-glycoprotein to 2.5 nm resolution determined by electron microscopy and image analysis. J Biol Chem 1997; 272:10685–10694.
84. Higgins CF, Callaghan R, Linton KJ, Rosenberg MF, Ford RC. Structure of the multidrug resistance P-glycoprotein. Semin Cancer Biol 1997; 8:135–142.
85. Lomovskaya O, Lewis K, Matin A. EmrR is a negative upstream regulator of the *E coli* multidrug resistance pump EmrA. J Bacteriol 1995; 177:2328–2334.
86. Miller PF, Sulavik MC. Overlaps and parallels in the regulation of intrinsic multiple-antibiotic resistance in *Escherichia coli*. Mol Microbiol 1996; 21:441–448.
87. Brooun A, Tomashek JJ, Lewis K. Purification and ligand binding of EmrR, a regulator of a multidrug transporter. J Bacteriol 1999; 181:5131–5133.
88. Zheleznova EE, Markham PN, Neyfakh AA, Brennan RG. Structural basis of multidrug recognition by BmrR, a transcription activator of a multidrug transporter. Cell 1999; 96:353–362.
89. Zheleznova EE, Markham PN, Neyfakh AA, Brennan RG. Preliminary structural studies on the multi-ligand-binding domain of the transcription activator, BmrR, from Bacillus subtilis. Protein Sci 1997; 6:2465–2468.
90. Ahmed M, Borsch CM, Taylor SS, Vazquez-Laslop N, Neyfakh AA. A protein that activates expression of a multidrug efflux transporter upon binding the transporter substrates. J Biol Chem 1994; 269:28506–28513.
91. Markham PN, Ahmed M, Neyfakh AA. The drug-binding activity of the multidrug-responding transcriptional regulator BmrR resides in its C-terminal domain. J Bacteriol 1996; 178:1473–1475.
92. Grkovic S, Brown MH, Roberts NJ, Paulsen IT, Skurray RA. QacR is a repressor protein that regulates expression of the *Staphylococcus aureus* multidrug efflux pump QacA. J Biol Chem 1998; 273:18665–18673.

93. Furukawa H, Tsay JT, Jackowski S, Takamura Y, Rock CO. Thiolactomycin resistance in *Escherichia coli* is associated with the multidrug resistance efflux pump encoded by *emrAB*. J Bacteriol 1993; 175:3723–3729.
94. Poole K. Bacterial multidrug resistance—emphasis on efflux mechanisms and *Pseudomonas aeruginosa*. J Antimicrob Chemother 1994; 34:453–456.
95. Woolridge DP, Vazquez-Laslop N, Markham PN, Chevalier MS, Gerner EW, Neyfakh AA. Efflux of the natural polyamine spermidine facilitated by the *Bacillus subtilis* multidrug transporter Blt. J Biol Chem 1997; 272:8864–8866.
96. Thanassi DG, Cheng LW, Nikaido H. Active efflux of bile salts by *Escherichia coli*. J Bacteriol 1997; 179:2512–2518.
97. Hagman KE, Pan W, Spratt BG, Balthazar JT, Judd RC, Shafer WM. Resistance of *Neisseria gonorrhoeae* to antimicrobial hydrophobic agents is modulated by the *mtrRCDE* efflux system. Microbiology 1995; 141:611–622.
98. Palumbo JD, Kado CI, Phillips DA. An isoflavonoid-inducible efflux pump in *Agrobacterium tumefaciens* is involved in competitive colonization of roots. J Bacteriol 1998; 180:3107–3113.
99. Kolaczkowski M, Goffeau A. Active efflux by multidrug transporters as one of the strategies to evade chemotherapy and novel practical implications of yeast pleiotropic drug resistance. Pharmacol Ther 1997; 76:219–242.
100. Paulsen IT, Sliwinski MK, Nelissen B, Goffeau A, Saier MH Jr. Unified inventory of established and putative transporters encoded within the complete genome of *Saccharomyces cerevisiae*. FEBS Lett 1998; 430:116–125.
101. Skulachev VP. Membrane Bioenergetics. Berlin: Springer-Verlag, 1988.
102. Colombo ML, Bosisio E. Pharmacological activities of *Chelidonium majus* L. (Papaveraceae). Pharmacol Res 1996; 33:127–134.
103. Stermitz FR, Lorenz P, Tawara JN, Zenewicz L, Lewis K. Synergy in a medicinal plant: antimicrobial action of berberine potentiated by 5′-methoxyhydnocarpin, a multidrug pump inhibitor. Proc Natl Acad Sci USA 2000; 97:1433–1437.
104. Norton SA. Useful plants of dermatology. I. Hydnocarpus and chaulmoogra. J Am Acad Dermatol 1994; 31:683–686.
105. Sum PE, Sum FW, Projan SJ. Recent developments in tetracycline antibiotics. Curr Pharm Des 1998; 4:119–132.
106. Someya Y, Yamaguchi A, Sawai T. A novel glycylcycline, 9-(N,N-dimethylglycylamido)-6-demethyl-6-deoxytetracycline, is neither transported nor recognized by the transposon Tn10-encoded metal-tetracycline/H+ antiporter. Antimicrob Agents Chemother 1995; 39:247–249.
107. Nelson ML, Park BH, Andrews JS, Georgian VA, Thomas RC, Levy SB. Inhibition of the tetracycline efflux antiport protein by 13-thio-substituted 5-hydroxy-6-deoxytetracyclines. J Med Chem 1993; 36:370–377.
108. Nelson ML, Park BH, Levy SB. Molecular requirements for the inhibition of the tetracycline antiport protein and the effect of potent inhibitors on the growth of tetracycline-resistant bacteria. J Med Chem 1994; 37:1355–1361.
109. Nelson ML, Levy SB. Reversal of tetracycline resistance mediated by different bacterial tetracycline resistance determinants by an inhibitor of the Tet(B) antiport protein. Antimicrob Agents Chemother 1999; 43:1719–1724.

110. Brennan L, Duignan J, Petitras J, et al. CP-544372: MIC90 studies and killing kinetics against key respiratory tract pathogens. 38th Interscience Conference for Antimicrobial Agents and Chemotherapy, San Diego. 1998.
111. Biedenbach DJ, Barrett MS, Jones RN. Comparative antimicrobial activity and kill-curve investigations of novel ketolide antimicrobial agents (HMR 3004 and HMR 3647) tested against *Haemophilus influenzae* and *Moraxella catarrhalis* strains. Diagn Microbiol Infect Dis 1998; 31:349–353.
112. Jamjian C, Biedenbach DJ, Jones RN. In vitro evaluation of a novel ketolide antimicrobial agent, RU-64004. Antimicrob Agents Chemother 1997; 41:454–459.
113. Brighty KE, Gootz TD. The chemistry and biological profile of trovafloxacin. J Antimicrob Chemother 1997; 39(Suppl)B:1–14.
114. Chopra I. Research and development of antibacterial agents. Curr Opin Microbiol 1998; 1:495–501.
115. Gootz TD, Zaniewski RP, Haskell SL, Kaczmarek FS, Maurice AE. Activities of trovafloxacin compared with those of other fluoroquinolones against purified topoisomerases and *gyrA* and *grlA* mutants of *Staphylococcus aureus*. Antimicrob Agents Chemother 1999; 43:1845–1855.
116. Fukuda H, Hori S, Hiramatsu K. Antibacterial activity of gatifloxacin (AM-1155, CG5501, BMS-206584), a newly developed fluoroquinolone, against sequentially acquired quinolone-resistant mutants and the *norA* transformant of *Staphylococcus aureus*. Antimicrob Agents Chemother 1998; 42:1917–1922.
117. Gill MJ, Brenwald NP, Wise R. Identification of an efflux pump gene, *pmrA*, associated with fluoroquinolone resistance in *Streptococcus pneumoniae*. Antimicrob Agents Chemother 1999; 43:187–189.
118. Pestova E, Beyer R, Cianciotto NP, Noskin GA, Peterson LR. Contribution of topoisomerase IV and DNA gyrase mutations in *Streptococcus pneumoniae* to resistance to novel fluoroquinolones. Antimicrob Agents Chemother 1999; 43:2000–2004.
119. Piddock LJ, Johnson M, Ricci V, Hill SL. Activities of new fluoroquinolones against fluoroquinolone-resistant pathogens of the lower respiratory tract. Antimicrob Agents Chemother 1998; 42:2956–2960.
120. Blais J, Cho D, Tangen K, Ford C, Lee A, Lomovskaya O. Efflux pump inhibitors enhance the activity of antimicrobial agents against a broad selection of bacteria. 39th Interscience Conference for Antimicrobial Agents and Chemotherapy, San Francisco. 1999.
121. Lomovskaya O, Lee A, Warren M, et al. Targeting efflux pumps in *Pseudomonas aeruginosa*. 39th Interscience Conference for Antimicrobial Agents and Chemotherapy, San Francisco. 1999.
122. Renau T, Leger R, Flamme E, et al. Inhibitors of efflux pumps in *Pseudomonas aeruginosa* potentiate the activity of the fluoroquinolone antibacterial levofloxacin. 39th Interscience Conference for Antimicrobial Agents and Chemotherapy, San Francisco. 1999.
123. Renau TE, Leger R, Flamme EM, et al. Inhibitors of efflux pumps in *Pseudomonas aeruginosa* potentiate the activity of the fluoroquinolone antibacterial levofloxacin. J Med Chem 1999; 42:4928–4931.

124. Cho D, Blais J, Tangen K, et al. Prevalence of efflux mechanisms among clinical isolates of fluoroquinolone resistant *Pseudomonas aeruginosa*. 39th Interscience Conference for Antimicrobial Agents and Chemotherapy, San Francisco. 1999.
125. Griffith D, Lomovskaya O, Lee V, Dudley M. Potentiation of levofloxacin by a broad-spectrum efflux pump inhibitor in mouse models of infection caused by *Pseudomonas aeruginosa*. 39th Interscience Conference for Antimicrobial Agents and Chemotherapy, San Francisco. 1999.
126. Markham PN, Westhaus E, Klyachko K, Johnson ME, Neyfakh AA. Multiple novel inhibitors of the NorA multidrug transporter of *Staphylococcus aureus*. Antimicrob Agents Chemother 1999; 43:2404–2408.
127. Markham PN, Neyfakh AA. Inhibition of the multidrug transporter NorA prevents emergence of norfloxacin resistance in *Staphylococcus aureus*. Antimicrob Agents Chemother 1996; 40:2673–2674.
128. Liu J, Takiff HE, Nikaido H. Active efflux of fluoroquinolones in *Mycobacterium smegmatis* mediated by LfrA, a multidrug efflux pump. J Bacteriol 1996; 178:3791–3795.
129. Takiff HE, Cimino M, Musso MC, et al. Efflux pump of the proton antiporter family confers low-level fluoroquinolone resistance in *Mycobacterium smegmatis*. Proc Natl Acad Sci USA 1996; 93:362–366.
130. Doran JL, Pang Y, Mdluli KE, et al. *Mycobacterium tuberculosis efpA* encodes an efflux protein of the QacA transporter family. Clin Diagn Lab Immunol 1997; 4:23–32.

5

Mechanisms of Aminoglycoside Antibiotic Resistance

Gerard D. Wright
McMaster University, Hamilton, Ontario, Canada

Aminoglycoside antibiotics are positively charged carbohydrate-containing molecules that find clinical use for the treatment of infections caused by both gram-negative and gram-positive bacteria. The first aminoglycosides were discovered over 50 years ago and several continue to find important clinical use including gentamicin, tobramycin, amikacin, netilmicin, and streptomycin. These antibiotics target the bacterial ribosome and interfere with protein translation. Unlike other antibiotics that target translation, most aminoglycosides are bactericidal, a desirable feature in an antibiotic. The bactericidal action of aminoglycosides is correlated with the propensity to cause misreading of the mRNA transcript resulting in the production of aberrant proteins.

Resistance to the aminoglycosides can occur through decreased uptake of the drugs, aminoglycoside efflux, and mutations in the rRNA and ribosomal protein. However, it is the presence and action of aminoglycoside-modifying enzymes that are the most relevant in the preponderance of resistant clinical isolates. Three distinct classes of modifying enzyme are known: the phosphotransferases (APHs), the adenyltransferases (ANTs), and the acetyltransferases (AACs). The APHs and ANTs are ATP-dependent enzymes, whereas the AACs require acetylcoenzyme A (acetylCoA). Members of each of these classes of enzyme are known and prevalent in both gram-positive and gram-negative clinical isolates. Several dozen distinct enzymes have been identified and these are designated by the position on the molecule where modification occurs (given by a number in parentheses), the resistance profile, represented by a roman numeral, and the specific gene, indicated by a lower case letter, for example, AAC(6′)-Ia is an aminoglycoside acetyltransferase modifying position 6′.

Research on aminoglycoside-modifying enzymes has greatly benefitted from crystal structures of representative protein from each class: APH(3′)-IIIa, ANT(4′)-Ia, AAC(3)-Ia, and AAC(6′)-Ii. Together with an increasing body of knowledge on the chemical mechanisms of modifying group transfer, and the molecular strategies for aminoglycoside substrate discrimination, the availability of three-dimensional protein structural data is permitting detailed understanding of the basis for aminoglycoside antibiotic resistance and providing insight into the origins of aminoglycoside-modifying enzymes. For example, APHs have been shown to share structural similarities with protein Ser/Thr/Tyr kinases as well as the capacity to phosphorylate proteins and peptides themselves. AACs fall into a growing family of acetyltransferases, which includes protein acyltransferases such as the histone acyltransferases. Furthermore, ANTs are structurally similar to DNA polymerase β and share the same aspects of reaction chemistry.

Knowledge of enzyme mechanism and structure is now fueling research into specific inhibitors of these enzymes and recent results in this area are promising. Understanding of the aminoglycoside-recognition elements utilized by modifying enzymes can be used in the synthesis of aminoglycosides lacking these functionalities, and this has been the basis for much research in new aminoglycoside discovery over the past 30 years. Furthermore, understanding of enzyme mechanism and structure can be used in the design of efficient and specific APH, ANT, and AAC inhibitors. These could find clinical application in reversing the impact of aminoglycoside resistance enzymes through the potentiation of existing aminoglycosides, and possibly the reintroduction of antibiotics no longer in use as the result of the dissemination and impact of aminoglycoside-modifying enzymes.

1 INTRODUCTION

1.1 Aminoglycoside Antibiotics

The aminoglycoside antibiotics are a diverse class of clinically important antimicrobial compounds that have proven to be instrumental in the treatment of infectious diseases since their discovery in the mid 1940s. The aminoglycosides find use in the treatment of infections caused by both gram-positive and gram-negative bacteria (1,2) and, in addition, some protozoa (3,4). In general, they are bactericidal compounds, an important trait especially for treatment of infections in immunocompromised individuals. The aminoglycosides are natural products, derived from bacterial sources, although some clinically important compounds such as amikacin and isepamicin are semisynthetic derivatives. All aminoglycosides contain an aminocyclitol nucleus (a six-carbon ring substituted with alcohol and amino groups) and as such are more formally termed aminoglycoside-aminocyclitol antibiotics; however, this ponderous term is rarely used and the name *aminoglycosides* is generally accepted.

Aminoglycoside antibiotics can be classified into three groups: the 4,5-disubstituted 2-deoxystreptamine group, the 4,6-disubstituted 2-deoxystreptamine group, and a class of "others" which do not fall into the first two groups (Table 1) (Fig. 1). A variety of aminohexoses and/or pentoses substitute the aminocyclitol ring, which gives rise to structural diversity in these molecules. Additional variance in these antibiotics is derived from further substitution by amino (and nonamino)-hexoses, methylation, deoxygenation, and epimerization of various sites on the molecules (Fig. 1). The result is a structurally rich and varied family of compounds, many of which find clinical use as antimicrobial agents. Numbering of the carbon centers, which is essential for deciphering the nomenclature of modifying enzymes, generally follows the rule that the aminocyclitol ring has no suffix, whereas additional rings are labeled with a prime ('), double prime ("), and so on (see Fig. 1). The ubiquitous presence of amino groups confers an overall positive charge to these compounds at physiological pH.

TABLE 1 Aminoglycoside Antibiotics

4,5-Disubstituted 2-deoxystreptamime	4,6-Disubstituted 2-deoxystreptamime	Others
Kanamycin	Neomycin	Streptomycin
Amikacin	Butirosin	Spectinomycin
Tobramicin	Ribostamycin	Fortimicin
Gentamicins	Lividomycin	Apramycin
Isepamicin		Hygromycin
Arbekacin		
Sisomicin		
Netilimicin		

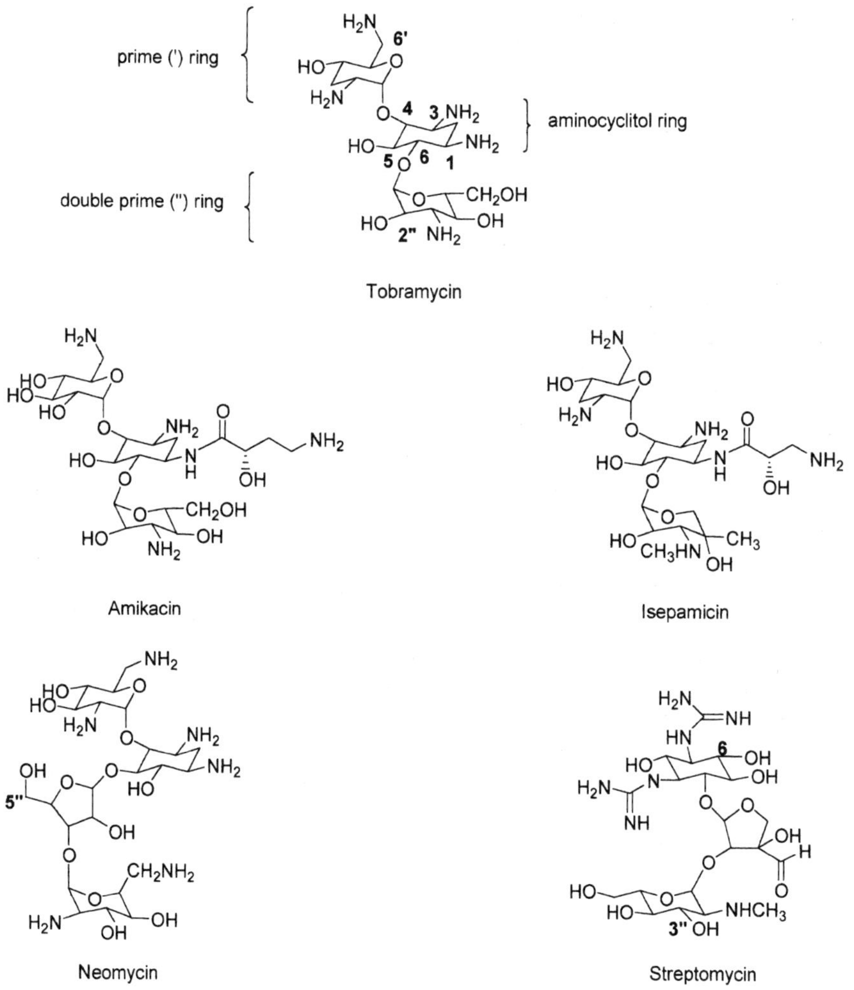

Figure 1 Structures of aminoglycoside antibiotics.

1.2 Interaction with 16S rRNA

The cationic nature of aminoglycosides provides the electronic basis for interaction with the 16S rRNA on the small (30S) ribosomal subunit, which, despite some controversy, remains the generally accepted primary site of action. Specifically, aminoglycosides bind to the region on the ribosome termed the A-site, where the aminoacyl-tRNAs dock and are recognized by the translation machinery and the mRNA is decoded. This affinity for the 16S rRNA is unique among other ribosome-targeted antibiotics such as chloramphenicol, tetracycline, and erythromycin. The ribosome in general and rRNA in particular are the targets of a very significant fraction of known antibiotics. This observation reflects the necessity, ubiquity, and evolutionary conservation of the translational machinery. The rRNA in particular is highly conserved among bacteria, and is therefore a highly attractive antibiotic target that has been exploited a number of times by natural selection.

Although all aminoglycosides target the 16S rRNA, they do not all interact precisely with the same nucleotide bases. Thus, the 2-deoxystreptamine antibiotics have been shown by chemical protection studies to bind adjacent to A1408 and G1494 (*Escherichia coli* 16S rRNA numbering) (5). On the other hand, the aminoglycosides that fall into the others class (see Table 1) have been shown to protect different regions of the 16S rRNA (e.g., streptomycin protects A913, A914, and A915) (5). Dissociation constants of the aminoglycosides kanamycin and tobramycin have been determined to be within 1–10 μM by equilibrium dialysis with intact ribosomes (6) and these values agree reasonably well with K_d values obtained with various aminoglycosides using model A-site RNA oligomers by quantitative chemical footprinting (7,8), competition studies with a fluorescent dye conjugated with paromomycin (9), and surface plasmon resonance (10), where values from 0.1 to .025 μM have been reported.

Recent studies by Puglisi's group have provided molecular insight into the interactions of aminoglycosides paromomycin (11,12), neomycin, ribostamycin, neamine (13), and gentamicin (8) using nuclear magnetic resonance spectroscopy (NMR). Furthermore, solution structures of paromomycin (11) and gentamicin (8) in complex with A-site–derived oligomers have provided the structural basis for the specificity aminoglycoside–16S rRNA interaction and aminoglycoside resistance by chemical modification. These studies have shown that rings I and II of the 4,5-disubstituted (paromomycin) and the 4,5-disubstituted (gentamicin) deoxystreptamine antibiotics occupy the same general binding region, making similar contacts with the A-site rRNA, but that rings III occupy different positions and interact with the rRNA in dissimilar fashions.

These structural studies have also provided the means to evaluate the basis for the specificity of aminoglycosides for bacterial versus eukaryotic rRNA, and it has been determined that the A1408 site (*E. coli* numbering) is critical to this selectivity. In eukaryotes, this position is generally a G, and an A1408 to G mutation in *E. coli* 16S rRNA confers resistance to most aminoglycosides in the mutant bacteria (14,15). The sensitivity of some protozoa to 6′-hydroxyl–containing aminoglycosides such as paromomycin may reflect A C1409-G1491 base pair, which is not found in other eukaryotes but is shared with prokaryotes (15).

Thus, by binding to the decoding region of bacterial ribosomes, aminoglycosides have the potential to interfere with appropriate aminoacyl-tRNA recognition, and this appears to be a key element of the bactericidal action of these antibiotics.

1.3 Aminoglycoside Uptake

Aminoglycoside antibiotics gain entry to the cell through a multiphase process. The initial step in this process is passive accumulation of the positively charged aminoglycosides at the negatively charged cell surface. The antibiotics then gain entry to the bacterial cytosol apparently by diffusion through the plasma membrane. This process is dependent on the electronic potential of the membrane and is thus energy dependent. Support for this model comes from experiments in which inhibitors of membrane potential (such as CCCP and CN^-) prevent aminoglycoside entry (reviewed in Ref. 16). It is generally accepted that this mechanism of aminoglycoside uptake is ubiquitous and does not require a protein component; however, evidence for the participation of a specified protein component in aminoglycoside translocation into the cytosol has been obtained (e.g., oligopeptide binding protein) (17), and this requires additional study.

1.4 Mechanism of Bactericidal Action of Aminoglycosides

Although it is well established that aminoglycosides target the bacterial ribosome, this interaction in and of itself is not sufficient to explain the bactericidal action of these compounds. Other antibiotics that target the translation machinery such as the tetracyclines and chloramphenicol are bacteriostatic rather than bactericidal (18). Once inside the cell, the aminoglycoside antibiotics bind to the decoding region of the ribosomes and cause mistranslation in de novo protein synthesis, resulting in the production of aberrant proteins (19–23). There has also been a long-standing observation that aminoglycosides cause membrane damage as evidenced by the loss of ions from the cell such as K^+ (24–26). It has since been demonstrated that the fate of some of these mistranslated proteins is interaction with the cell membrane, and that this interaction results in altered membrane permeability (27,28). It has been proposed by Davis that aminoglycoside-mediated mistranslation followed by membrane damage caused by perturbation by the altered peptides may account for the breach of membrane integrity which seems to be essential for the bactericidal activity of these antibiotics (29) (spectinomycin, a bacteriostatic aminoglycoside, does not cause mistranslation [23]). Thus, aminoglycosides kill bacteria by pleiotropic means involving ultimate loss of membrane integrity; however, interaction with the ribosome appears to be the primary and critical event.

2 AMINOGLYCOSIDE RESISTANCE

Bacterial resistance to the aminoglycosides can occur through four general mechanisms: 1) altered uptake, 2) antibiotic efflux, 3) target modification, and 4) chemical modification.

2.1 Altered Uptake

Since uptake of aminoglycosides is an energy-requiring phenomenon, mutations that affect the membrane potential can confer aminoglycoside resistance (30,31). Taber and Halfenger (32) isolated multiple aminoglycoside resistant mutants of *Bacillus subtilis* that were deficient in aminoglycoside uptake and one of these was characterized as a menaquinone (a lipophilic quinone required for electron transport) auxotroph. Supplementation of the growth medium with shikimic acid (a menaquinone biosynthesis precursor) restored aminoglycoside sensitivity (33). Similarly, quinone auxotrophs of *Staphylococcus aureus* have an aminoglycoside resistance phenotype that can be abolished by the addition of menaquinone precursors to the medium (34). Furthermore, depletion or mutations in other electron transport components, including cytochrome *aa3* (35) and type γ-subunit of the F1F0 ATPase (36), result in aminoglycoside resistance.

Although electron transport mutations can be readily isolated in the laboratory (and are not the result of exposure to aminoglycosides [32]), they appear to be infrequent sources of resistant organisms in the clinic, possibly because of the potential decreased viability of electron transport mutants in the host.

Another mechanism of aminoglycoside resistance through impaired uptake operates in *Pseudomonas aeruginosa* where overexpression of the a 21.6-kD basic outer membrane protein, OprH, saturates the aminoglycoside-binding sites and thus prevents the first phase of aminoglycoside entry (37). Recently, it has been shown that *E. coli*, which harbor structural or protein expression mutations in the oligopeptide-binding protein, OppA, which is involved in peptide transport across the membrane, show a kanamycin-resistant phenotype (17). These mutants failed to take up [^{14}C]isepamicin and suggests a possible role in aminoglycoside uptake for this protein.

2.2 Aminoglycoside Efflux

Efflux-mediated resistance to aminoglycoside appears to be rare. However, it was recently reported that high-level aminoglycoside resistance in *Burkholderia pseudomallei* is mediated by a multidrug efflux system (AmRAB, OprA) (38). It may be that high-level aminoglycoside resistance observed in other species of *Burkholderia* such as *B. cepacia*, which is a significant pathogen in cystic fibrosis patients, will also be shown to be due to efflux. The recent association of efflux proteins MdfA in *E. coli* (39), MexXY in *P. aeruginosa* (40), and Tap in *Mycobacterium fortuitum* (but not the 83% similar homologue in *Mycobacterium tuberculosis*) (41) with aminoglycoside resistance may point to a broader dispersion of potential efflux mechanisms in bacteria.

2.3 Target Modification

Aminoglycoside resistance through target modification can occur through two mechanisms: 1) point mutation of rRNA or ribosomal proteins or 2) methylation of the 16S rRNA. The latter mechanism is so far found only in actinomycete producers of aminoglycosides where it confers high-level resistance (minimal inhibitory concentration, MIC $> 500\ \mu g/mL$), for example, *Micromonospora purpurea* (gentamicin producer) (42), and *Streptomyces tenabrius* (tobramycin producer) (43).

On the other hand, ribosomal mutation is a clinically important mechanism of resistance in the slow-growing mycobacteria (reviewed in Ref. 44). Resistance to streptomycin can occur through point mutations in the ribosomal protein S12, RpsL (45,46) through an unknown process, although conformational change at the streptomycin-binding site is a likely mechanism. Resistance can also result from mutations in the aminoglycoside target 16S rRNA (*rrs* gene) (45,47). Isolates of *M. tuberculosis* that display resistance to kanamycin and amikacin have mutations in A1400 (48). This base is equivalent to A1408 of the *E. coli* 16S rRNA, which has been shown by structural studies (see Sect. 1.2 above) and mutation analysis (49) to be important to aminoglycoside recognition.

2.4 Modification of Aminoglycosides

Enzymatically catalyzed chemical modification of aminoglycosides remains the most relevant mechanism of resistance in the majority of clinical isolates. Chemical modification can occur through three general mechanisms: *O*-phosphorylation, *O*-adenylation, or *N*-acetylation. All three mechanisms are widespread through both gram-negative and gram-positive bacteria, but the latter appear to have a smaller repertoire of enzymes.

The various aminoglycoside-modifying enzymes are classified by the chemistry of the modifying reaction (phosphoryl, adenyl, or acetyl transfer), their site of aminoglycoside modification (regiospecificity), and by the specific isozyme sequence. Shaw and colleagues have proposed a unifying nomenclature for all aminoglycoside-modifying enzymes where the enzyme is described by type (APH [*O*-phosphotransferase], AAC [*N*-acetyltransferase], or ANT [*O*-adenyltransferase]), the regiospecificity of group transfer in parentheses, for example, (3′), (2″), and so on, followed by a roman numeral indicating a distinct phenotype (these are assigned sequentially as discovered or cloned), and finally a letter indicating the specific gene (50). For example, APH(3′)-Ia is a phosphotransferase which modifies aminoglycosides at position 3′ with a distinct resistance phenotype (in this case, protection against kanamycin, gentamicin B, neomycin, paromomycin, ribostamycin, and lividomycin), and is the first gene cloned with this repertoire (51); on the other hand, APH(3′)-Vc is also an aminoglycoside kinase with the same regiospecificity of phosphoryl transfer (3′-OH), but it has a different resistance phenotype (kanamycin, neomycin, paromomycin, and ribostamycin), and is the third gene cloned with these properties (52). The list of these aminoglycoside-modifying genes continues to grow, but tables of genes, resistance phenotypes, and original references can be found in several extensive reviews (50,53,54). A representative list of clinically relevant enzymes is found in Table 2.

Although genes encoding greater than 70 aminoglycoside-modifying enzymes have already been cloned and a number are being uncovered in whole genome sequencing projects, only a subset of these genes is of significant clinical relevance today given that usage of aminoglycosides is limited to only a few compounds (gentamicin, tobramycin, netilmicin, amikacin, and streptomycin in the United States [1]). For example, ANT(2″)-I, which confers resistance to gentamicin and tobramycin, is common in Enterobacteriaceae worldwide, but depending on aminoglycoside usage patterns, resistance to gentamicin by AAC(3)-II and AAC(3)-VI is also problematic (55,56). Furthermore, combinations of resistance genes such as *aac(6′)*-I and *aac(3)*-II, which result in overall resistance to gentamicin, tobramycin, netilmicin, and amikacin, also are emerging in some countries (56). In gram-positive pathogens such as *S. aureus*, resistance is less complex, and the primary mechanism of gentamicin resistance (>90% of isolates) is a bifunctional enzyme with both aminoglycoside kinase and acetyltransferase activity, AAC(6′)-APH(2″) (57).

Table 2 Representative Aminoglycoside-Modifying Enzymes

Enzyme	Resistance profile	Bacterial source
APH(3′)-Ia	Kan, Neo, Rib, Livid	Enterobacteriaceae
APH(3′)-IIIa	Kan, Amik, Isep, Neo, Rib, But, Livid	Enterococci, staphylococci
APH(3″)-Ib	Strep	Enterobacteriaceae
APH(6)-Id	Strep	Enterobacteriaceae
AAC(6′)-Ib	Kan, Tob, Amik, Neo	Enterobacteriaceae
AAC(6′)-Ii	Kan, Tob, Amik, Neo	*Enterococcus faecium*
AAC(3)-Ia	Kan, Gent, Tob, Fort	Enterobacteriaceae
ANT(2″)-Ia	Kan, Gent, Tob	Enterobacteriaceae
ANT(4′)9Ia	Kan, Tob, Amik, Neo	*Staphylococcus aureus*
ANT(6)-Ia	Strep	*Enterococcus faecalis*
AAC(6′)-(APH2″)		Enterococci, staphylococci
APH activity	Kan, Gent, Amik, Isep, Neo, Rib, But, Livid	
AAC activity	Kan, Amik, Isep, Neo, Rib, But, Livid, Fort (Livid is a poor substrate)	

Abbreviations: Kan, kanamycin; Gent, gentamicin C; Amik, amikacin; Isep, isepamicin; Neo, neomycin; Rib, ribostamycin; But, butirosin; Livid, lividomycin A; Tob, tobramycin; Fort, fortimicin (astromycin); Strep, streptomycin.

2.4.1 O-Phosphotransferases

The aminoglycoside *O*-phosphotransferases, abbreviated APH, are a common resistance mechanism. These enzymes are ATP-dependent kinases of approximately 30 kD, which generate a phosphorylated aminoglycoside and ADP as products. The most prevalent group of aminoglycoside kinases are the APH(3′)s, which confer resistance to kanamycin and neomycin by phosphorylation of the 3′-OH (Fig. 2). Furthermore, some of these enzymes, for example, APH(3′)-Ia, APH(3′)-IIIa, can confer resistance to the 3-deoxyaminoglycoside lividomycin A through phosphoryla-

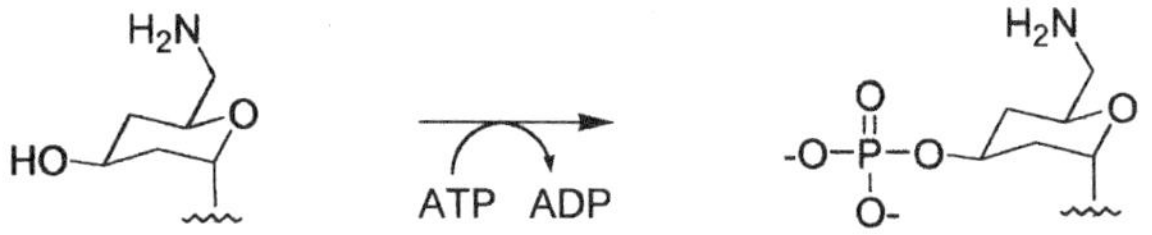

Figure 2 APH aminoglycoside-modifying reaction.

tion of the secondary 5″-alcohol of the pentose ring and in fact this site can be phosphorylated in other 4,5-disubstituted aminoglycosides (58). These enzymes are common in both gram-negative and gram-positive bacteria (59). The prevalence of these resistance elements motivated the search for "resistance-proof" aminoglycosides and prompted the introduction of compounds that lacked the 3′-hydroxyl such as tobramycin. Since these enzymes do not confer resistance to other important 3′-deoxy-aminoglycosides such as gentamicin Cs or isepamicin, the clinical impact of APH(3′)s is now low, although APH(3′)-IIIa does confer resistance to amikacin in gram-positive cocci, and is thus relevant in this contact. Although APH(3′)s no longer are a grave threat to modern aminoglycoside therapy, they have found use as important molecular biological tools where they are frequently used as antibiotic resistance markers; for example, APH(3′)-IIa is the common source of the "neo cassette" found in many cloning plasmids and transposons.

The APH(2″) kinases, on the other hand, are important resistance elements in gram-positive bacteria. The most relevant mechanism is the bifunctional AAC(6′)-APH(2″) that is the primary mechanism of gentamicin C resistance in staphylococci and enterococci. The APH(2″) kinase activity is located to the C-terminus of the enzyme and can efficiently use gentamicin C1, gentamicin C1a, gentamicin C2, isepamicin, netilmicin, sisomicin, and amikacin (among others) as substrates (60–62). The site of 2″-phosphorylation has been confirmed by NMR studies (61), but is not confined to this hydroxyl, and the 3′, 5″, and 3‴ hydroxyls may also be phosphorylated on various aminoglycosides (62). This enzyme activity is quite indiscriminant and therefore a significant challenge for the design of new antibiotics.

Recently, APH(2″) genes have been cloned that are not fused to a 5′-*aac(6′)* gene in *Enterococcus gallinarum* (63) and *E. casseliflavus* (64), indicating that this enzyme activity is increasing in frequency.

Other aminoglycoside kinases have been identified that modify streptomycin (APH[6], APH[3″]), spectinomycin (APH[9]), and hygromycin (APH[4], APH[7″]). With the exception of *strA-StrB* genes found on gram-negative R plasmids such as RSF1010, which respectively encode the streptomycin kinases APH(3″)-Ib and APH(6′)-Id, these kinases are not common mechanism of clinical aminoglycoside resistance.

The three-dimensional structure of one aminoglycoside kinase has been reported—that of APH(3′)-IIIa from gram-positive cocci (65). Since all aminoglycoside kinases share a significant degree of amino acid homology, especially in the active site region, it is likely that the salient issues of enzyme mechanism will be common among these enzymes, although the specific interactions with aminoglycoside substrates, which differ widely

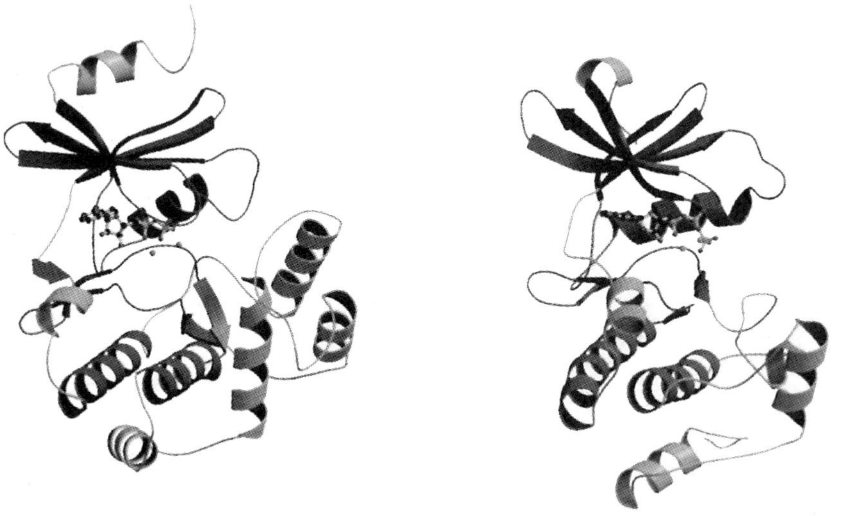

Figure 3 Structures of APH(3′)-IIIa and mouse protein kinase A (cAMP-dependent protein kinase, cAPK).

among enzymes, will be different. The structure of APH(3′)-IIIa bound with ADP is shown in Figure 3. The enzyme has two distinct domains: an N-terminal region consisting largely of β-strands and a C-terminal region that is rich in α-helices. The active site lies at the junction of these domains. The structure revealed two striking features. The first was that the aminoglycoside-binding site was rich in negatively charged amino acid residues. This observation is consistent with the capacity of the enzyme to bind a broad array of positively charged aminoglycosides, which, based on mutagenesis and molecular modeling studies (66), are predicted to bind to the enzyme in a number of distinct conformations.

The second important feature revealed by the three-dimensional structure was the remarkable structural similarity between Ser/Thr/Tyr protein kinases and phosphatidylinositol kinases (see Fig. 3) despite the overall low amino acid homology (<2.5%), suggesting a possible common protein ancestor. This similarity nonetheless prompted an investigation into the potential protein kinase activities of APHs, and indeed APH(3′)-IIIa and the APH activity of the bifunctional AAC(6′)-APH(2″), showed the capacity to act as protein kinases (67). A survey of several known peptide and protein substrates of protein kinases demonstrated that these two antibiotic resistance kinases could phosphorylate some peptides and proteins on Ser residues. The similarity between APHs and protein kinases was further strengthened with the demonstration that several small molecule inhibitors of protein kinases were also inhibitors of APH(3′)-IIIa and APH (2″) (68) (see Sec. 3). Furthermore, site-directed mutagenesis has supported the catalytic importance of active site Asp and Lys residues (Asp190 and Lys44 of APH[3′]-IIIa), which have also been implicated as important to Ser/Thr/Tyr kinase catalysis (65,69).

In summary, aminoglycoside kinases and protein kinases share similarity in protein structure, enzyme mechanism, sensitive to inhibitors, and function. These results then support a common origin for protein and aminoglycoside kinases. Furthermore, other antibiotic resistance kinases such as the erythromycin kinases MPH(2′)-I and MPH(2′)-II (70,71) and viomycin kinase, VPH (72), share sequence similarities within the important active site regions of APHs and protein kinases; thus these enzymes likely form a large superfamily of kinases.

2.4.2 N-Acetyltransferases

The aminoglycoside *N*-acetyltransferases are the largest group of aminoglycoside-modifying enzymes. They are generally 20–25 kD in mass and modify positions 6′, 2′, and 3 of aminoglycosides in an acetylCoA-dependent fashion (Fig. 4). Two AACs with the capacity to modify N-1 have also been reported (73,74). The AAC(3)s, which confer resistance to gentamicin and tobramycin, and the AAC(6′)s, which confer resistance to amikacin and tobramycin, are among the most abundant resistance elements (over 30 isozymes). Not surprisingly then, they are very frequent causes of clinical resistance especially in gram-negative bacteria (56). Furthermore, AAC(6′)-Ie, which forms the N-terminal domain of the AAC(6′)-APH(2″) bifunctional enzyme noted above, is the most frequent source of aminoglycoside resistance in gram-positive organisms.

The AAC(3) and AAC(6′) enzymes are generally encoded on mobile genetic elements such as transposons or plasmids, although some are found in bacterial chromosomes; for example, *aac(6′)-Ii* in *Enterococcus*

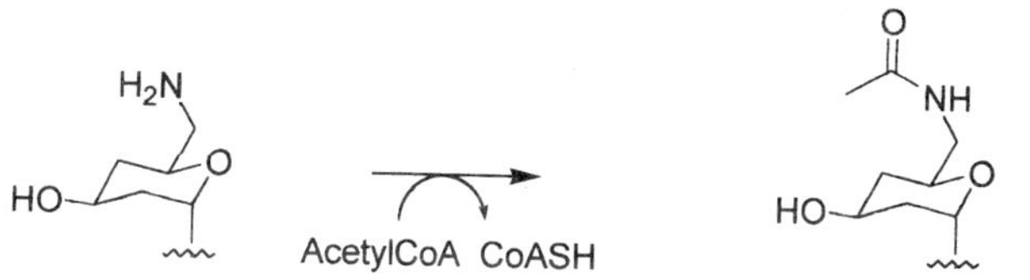

Figure 4 Reaction catalyzed by AACs.

faecium (75). On the other hand, the AAC(2′) enzymes are apparently universally chromosomally encoded: *aac(2′)-Ia* in *Providencia stuartii* (76) and *aac(2′)-Ib-e* in mycobacteria (77,78).

Unlike the case of the emergence of the APH(3′) enzymes, which prompted the replacement of 3′-hydroxyl–containing aminoglycosides such as kanamycin with 3′-deoxy compounds such as tobramycin and gentamicin C, the key importance of NH_2 or OH groups at positions 6′ and 3 for antimicrobial activity has made the presence of AAC(3) and AAC(6′) enzymes highly problematic. Furthermore, as noted above, there are a large number of these enzymes and they are frequently causes of aminoglycoside resistance. The study of AAC enzymes is therefore of key importance, although it has lagged behind research on the mechanism and structure of APHs. Northrop and colleagues reported the kinetic characterization of AAC(3)-I over 20 years ago (79–81) and AAC (6′)-Ib 15 years ago (82,83). These studies demonstrated the broad aminoglycoside substrate specificity of these enzymes and established that they function through a ternary complex mechanism; that is, both acetylCoA and the aminoglycoside need to be present at the enzyme active site for acyl transfer to occur. Although these results do not rule out a mechanism in which the acetyl group is transferred first to the enzyme (acyl-enzyme intermediate) and then to the antibiotic, they more likely support a mechanism in which the acetyl group is transferred directly to the antibiotic from acetylCoA. Consistent with this mechanism, we have not been able to capture an acyl-enzyme intermediate with ^{14}C-acetylCoA and purified AAC(6′)-Ii (K.-A. Draker and G.D. Wright, unpublished data), an enzyme that is chromosomally encoded in all *E. faecium* (75,84).

In addition to this research on the mechanism of AACs, mutagenesis studies have demonstrated that single amino acid substitutions can modulate the aminoglycoside substrate specificity. For example, AAC(6′)-I and AAC(6′)-II share the capacity to modify many aminoglycosides such as kanamycin, but they differ in their propensity to acetylate amikacin and gentamicin C: AAC(6′)-I modifies amikacin but not gentamicin, whereas AAC(6′)-II is incapable of amikacin acetylation but does modify gentamicin C. Shaw and colleagues prepared a series of hybrid AAC(6′) enzymes consisting of various portions of AAC(6′)-Ib and AAC(6′)-IIa and demonstrated that the key elements that conferred amikacin versus gentamicin recognition were in the C-terminus (85). Spontaneous and site-directed mutagenesis studies indicated that modification of amino acid 119 from Ser to Leu could toggle between gentamicin resistance and amikacin sensitivity (85). Similar Ser→Leu mutants resulting from a single C to T transition characterized by amikacin sensitivity and gentamicin resistance have been isolated in *aac(6′)-Ib* from a clinical isolate of *P. aeruginosa*, demonstrating the exquisite balance between antibiotic resistance and sensitivity (86).

The three-dimensional structures of both an AAC(3) and an AAC(6′) have recently been reported. The structure of AAC(3)-Ia, encoded on plasmids in *Serratia marcescens* and other Enterobacteriaceae, was determined to 2.3 Å resolution bound to CoA (87). This enzyme confers resistance to gentamicin C and is widely distributed throughout the world (56). In contrast, AAC(6′)-Ia, the structure of which has been determined to 2.7 Å in complex with acetylCoA (88), is found exclusively in *E. faecium* where it is encoded on the chromosome. Despite low amino acid sequence identity (<11%), there is remarkable conservation in the three-dimensional structure (Fig. 5). Furthermore, reminiscent of the relationship between the structures of APH and protein kinases, there is significant three-dimensional protein structure similarity between the structures of these AACs and other acyltransferases including histone acetyltransferases (89–91) and N-myristoyltransferase (92,93). These proteins had been previously classified as members of the GCN5 superfamily of acyltransferases based on amino acid sequence homologies (94); thus it is gratifying that these three-dimensional structural data have confirmed this homology at the molecular level and point to common evolutionary pathways. This structural similarity has been extended to include function as AAC(6′)-Ii has been shown to have protein acetyltransferase activity in addition to aminoglycoside modification capacity (88). However, this conservation in function and structure has not aided in determining residues important for catalysis, as the invariant amino acids among these proteins and conserved

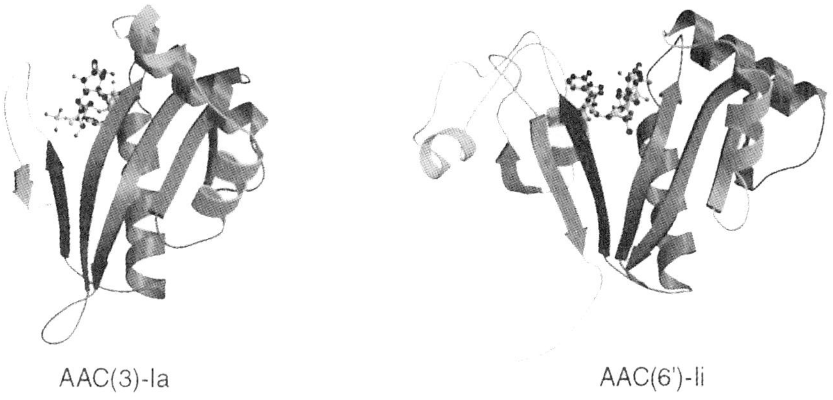

Figure 5 Structures of AAC(3)-Ia and AAC(6′)-Ii.

residues are invariantly hydrophobic in nature and removed from the active site. Thus, unequivocal determination of these enzymes' mechanism awaits additional research.

2.4.3 O-Nucleotidyltransferases

The aminoglycoside *O*-nucleotidyltransferases (ANTs) represent the smallest group of aminoglycoside-modifying enzymes in terms of numbers of reported isozymes (<10), but they have significant impact on clinical aminoglycoside resistance. In particular, ANT(2″)-I (Fig. 6) is a major source of gentamicin and tobramycin resistance in Enterobacteriaceae (56). Unlike the APH family of enzymes, the ANTs are quite diverse at the amino acid level with similarities around 20%, and also differ in predicted molecular mass from ~28 to 38 kD. The most conserved sequence motif, GlySer(Xaa)10-12(Asp or Glu)Zaa(Asp or Glue), where X is any amino acid, is found in the N-terminal region of ANTs. Northrop's group has purified ANT(″)-Ia enzyme from *E. coli* extracts (95), determined the substrate specificity (96), established the kinetic mechanism (97), and the rate limiting step (AMP-aminoglycoside release) (98). Furthermore, the stereochemistry of AMP transfer has been shown to occur with inversion of configuration at the α-phosphorus, implicating a mechanism of direct nucleotidyl transfer to the aminoglycoside hydroxyl, that is, no AMP-enzyme intermediate (99).

These mechanistic results can now be evaluated in light of the crystal line structure of the three-dimensional structure of ANT(4′)-Ia in both the apo and ternary complex forms (Fig. 7) (100,101). This enzyme was originally obtained from *Staphylococcus aureus* where it confers resistance to tobramycin and amikacin (102), and this enzyme shows 27% amino acid homology (10% identity) to the more predominant ANT(2″)-Ia.

ANT(4′)-Ia is a dimer consisting of two identical subunits and reveals two active sites. Each active site is located at the interface of the dimer and each monomer contributes residues that interact with Mg-ATP and the aminoglycoside (kanamycin in the crystal structure) (101). Not

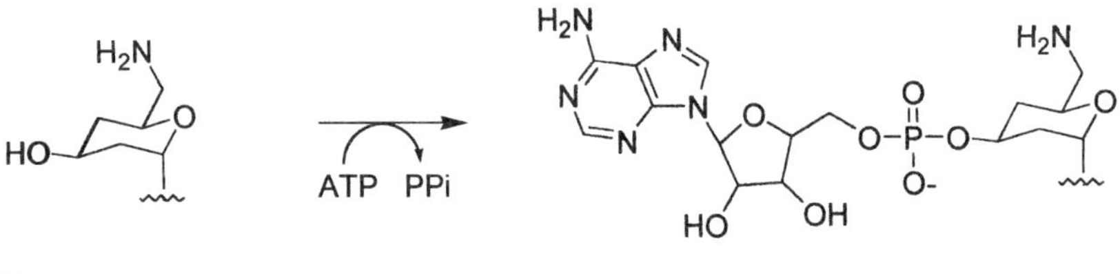

Figure 6 Reaction catalyzed by ANTs.

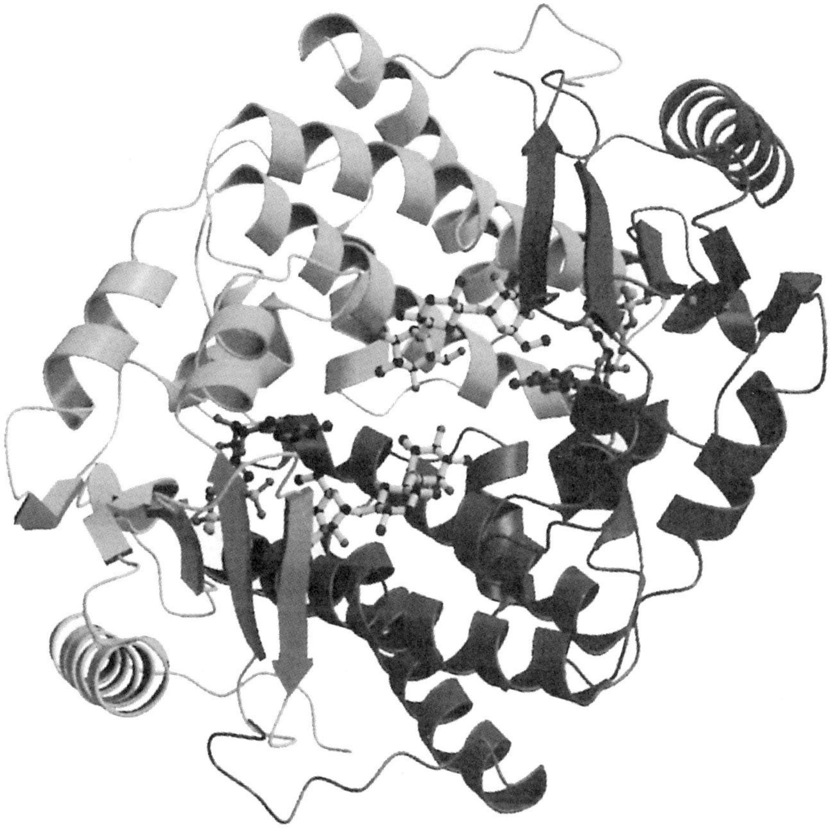

Figure 7 Structure of ANT(4′)-Ia. The dimer bound with two molecules of kanamycin A and AMPCPP is shown. (From Ref. 101.)

surprisingly, the signature motif GlySer(Xaa)10-12(Asp or Glu)Zaa(Asp or Glu) is involved in nucleotide binding where the conserved Ser interacts with the γ-phosphate of ATP and the conserved Asp/Glu residues are Mg^{2+} ligands. The aminoglycoside-binding pocket is lined with negatively charged residues. This general strategy is conserved in all the aminoglycoside resistance enzyme structures determined to date, and is consistent with the requirements for binding a diverse array of positively charged aminoglycoside substrates.

It has been noted that the three-dimensional structure of ANT(4′) is similar to the fold of mammalian DNA polymerase β (103). Recently, it has been proposed that the ANTs form part of a large polymerase β-like superfamily of nucleotidyltransferases and points to the divergence of a minimal nucleotidyltransferase into a variety of important protein groups with diverse function but similar chemical cleavage of NTPs (104).

3 STRATEGIES TO CIRCUMVENT AMINOGLYCOSIDE-MODIFYING ENZYMES

The challenge presented by the dissemination of aminoglycoside resistance elements can be met with two strategies: 1) the discovery of new aminoglycoside antibiotics that are not susceptible to modifying enzymes and 2) the use of inhibitors of modifying enzymes to potentiate the activity of existing aminoglycosides. The first strategy has been the mainstay of the response to resistance over the past several decades. Thus, tobramycin, a 3′-deoxyaminoglycoside, was introduced in the years following the characterization of aminoglycoside modification by APH(3′). These enzymes are incapable of tobramycin modification, and in fact this compound is a good competitive inhibitor of APH(3′) (e.g., see Ref. 105). Similarly, dibekacin (3′,4′-dideoxykanamycin B) was effective against some resistant bacteria as well (106).

The early observation that butirosin, which is derivatized on *N*-1 of the deoxystreptamine ring by an (S)-4-amino-2-hydroxybutyryl (AHB) group, is poorly modified by APH(3′)-I (reviewed in Ref. 107) prompted the synthesis of other AHB aminoglycosides, including amikacin, 1-*N*-AHB kanamycin A (108), which has proven to be an effective and clinically important aminoglycoside antibiotics. Similarly, isepamicin, 1-*N*-(S-3-amino-2-hydroxypropionyl)-gentamicin B, also has found important clinical application (109). Other *N*-1–alkylated aminoglycosides such as netilmicin (1-*N*–ethylsisomicin) (110) have been clinically used (1).

N-alkylated aminoglycosides, including N-6′ derivatives, have been prepared (111,112) and generally evade modification by the abundant AAC(6′)s; however, these derivatives sacrifice antimicrobial activity.

Arbekacin, (1-*N*-(S-3-amino-2-hydroxybutyryl)-3′-4′-kanamycin B) (113), has found clinical use in Japan against aminoglycoside-resistant MRSA. Nonetheless, this compound is a substrate for the bifunctional AAC(6′)-APH(2″) (114). Novel acetylation of the primary amino group of the AHB moiety of arbekacin in cell-free extracts of arbekacin-resistant MRSA has been reported, although not yet associated with a specific resistance enzyme (115). Recently, 2″-amino derivatives of arbekacin have been synthesized and show improved antimicrobial activity against *S. aureus* harboring the bifunctional enzyme (116).

The challenge in these synthetic and semisynthetic approaches to circumvent aminoglycoside resistance by alteration of the sites of enzymatic modification is to preserve antibacterial activity, as it is these sites on the molecules that are frequently important in 16S rRNA interaction. Alkylation of *N*-6′, for example, results in a parallel decrease of affinity for resistance enzymes and antibacterial activity. Mobashery and colleagues have probed the importance of the *N*-2′, *N*-6′, *N*-1, and *N*-3 through the synthesis of deaminated neamine and kanamycin derivatives (117). They have shown that loss of these strategic amines can result in dramatic reduction in enzymatic modification by APH(3′)-Ia and APH(3′)-IIa, whereas in many case still retaining significant antibacterial activity. Although encouraging, these results may not be generally applicable as these same compounds are good substrates for APH(3′)-IIIa (118).

Research in this area continues in several laboratories. For example, Wong's group has prepared several novel aminoglycosides in recent years based on the neamine nucleus (119,120). Some of these compounds show good antibacterial activity and in vitro inhibition of translation (120), although their susceptibility to resistance enzymes is unknown.

The other route to evade aminoglycoside resistance by modifying enzymes is through the use of specific inhibitors of these activities. This approach would rescue the antibacterial properties of pharmacologically well-understood aminoglycosides such as gentamicin C, amikacin, or tobramycin through the coadministration of inhibitors of common resistance enzymes. There is in fact excellent precedent for this approach in the β-lactam field where coadministration of β-lactam antibiotics and β-lactamase inhibitors is now well established in clinical practice. There are several challenges to this approach in the aminoglycoside field however. First is the fact that there are dozens of known aminoglycoside-inactivating enzymes and these use three chemically distinct routes of modification: phosphorylation, adenylation, and acetylation. Since many of the best enzyme inhibitors are based on enzyme mechanism or structure of the predicted transition state, it would be unrealistic to envision an inhibitor that would show activity against all of these mechanisms and enzymes. Nonetheless, all aminoglycoside resistance enzymes share the capacity to bind these structurally diverse molecules, and the available three-dimensional structures of all three classes of modifying enzymes have shown that they all have a highly negatively charged substrate binding site. Therefore, compounds that mimic the structure and charge of aminoglycosides, without the capacity to be modified by these enzymes, could act as "universal" inhibitors.

However, a search for such broad-spectrum compounds may not be necessary. It is known that in fact there are predominant resistance elements in the clinic (55,56), and thus only a few mechanisms need be targeted to achieve significant rescue of aminoglycoside activity. Furthermore, in many cases, resistance is genus- or even species-specific and one could envision cases where targeted molecules would be of great benefit; for example, versus AAC(6′)-APH(2″) in staphylococci and enterococci.

The second challenge to the inhibitor approach is a general requirement for thorough understanding of enzyme mechanism and structure. Modern drug design approaches demand superior knowledge of mechanism, structure, and inhibition to optimize the likelihood of selecting lead compounds with chemotherapeutic potential. Even in high throughput random chemical library screens, downstream optimization of leads by traditional medicinal chemistry or combinatorial methods is greatly facilitated by comprehensive knowledge of enzyme mechanism. As indicated above, this information is now becoming available for all three classes of aminoglycoside-modifying enzyme and examples of new inhibitors have been reported.

The similarity between aminoglycoside kinases and protein kinases has been exploited in a survey of known Ser/Thr/Tyr kinase inhibitors against APH(3′)-IIIa and the kinase activity of AAC(6′)-APH(2″) (68). For example, this screen demonstrated that the flavonoid quercetin was an

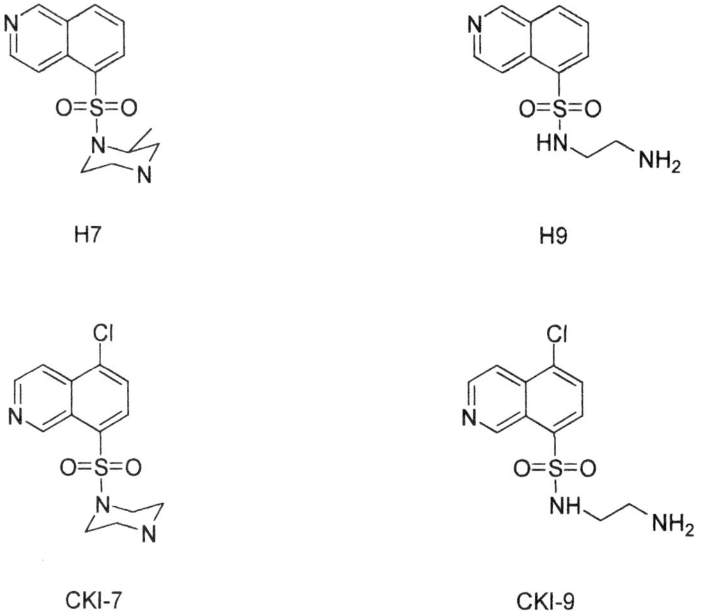

Figure 8 Protein kinase inhibitors that inhibit APHs.

APH(3′)-IIIa inhibitor. The isoquinoline sulfonamide inhibitors such as H-7, H-9, CKI-7, and CKI-9 (Fig. 8) are known to inhibit protein kinases by binding to the ATP-binding site. These compounds behave similarly in APHs, showing competitive inhibition of ATP and noncompetitive inhibition of aminoglycosides, consistent with binding in the ATP site. These inhibitors also exhibit different affinities for APH(3′)-IIIa versus APH(2″) and point to the capacity to engineer APH-specific inhibitors based on the isoquinoline nucleus. Although none of these inhibitors could reverse aminoglycoside resistance in bacterial cultures, these studies provide the proof of principle for screening libraries of protein kinase inhibitors as potentiators of aminoglycoside antibiotics against resistant isolates.

Mobashery's group has designed and synthesized two novel approaches based on the synthesis of aminoglycoside analogues that have the potential to evade the resistance caused by APH(3′)s. The first approach required the preparation of aminoglycosides with a nitro group in position 2′ (121). These compounds were found to be mechanism-based inactivators of APH(3′)-Ia and APH(3′)-IIa, and a mode of action has been proposed that suggests that phosphorylation of the aminoglycoside at position 3′ is followed by elimination of phosphate and the generation of electrophilic nitroalkene in the enzyme active site (Fig. 9). Such compounds readily undergo nucleophilic attack and thus have the potential to alkylate the enzyme through reaction with amino acid side chains; for example, SH of Cys, NH_2 of Lys. These studies provide the only example thus far of compounds with the capacity to irreversibly inhibit APHs.

The second approach involved the synthesis of a kanamycin analogue with a ketone at position 3′, rather than a hydroxyl group (122). In aqueous solution, the ketone is hydrated to form the *gem*-diol. This can act as a substrate for APH(3′), but the phosphate group is unstable in this configuration and the ketone readily regenerated with loss of inorganic phosphate. This compound showed poor biological activity, but the strategy of reversible phosphorylation has been demonstrated and future analogues may prove useful.

ANT(2″)-Ia has been shown to be inhibited by 7-hydroxytropolone (Fig. 10), a natural product produced by *Streptomyces neyagawaensis* (123). This compound was competitive inhibitor of ATP and was identified by its capacity to potentiate tobramycin in ANT(2″)-Ia expressing *E. coli* but not cells expressing AAC(6′), AAC(3), or ANT(3″). These studies indicate that the concept of reversing aminoglycoside resistance through coadministration of resistance enzyme inhibitors is valid.

There are excellent reasons to be optimistic for the discovery of inhibitory compounds that could find clinical use for the reversal of aminoglycoside resistance. The growing understanding of enzyme mechanism

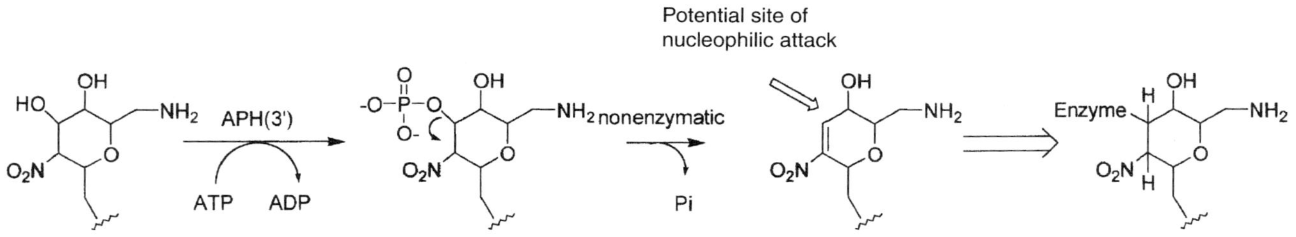

Figure 9 Proposed mechanism of APH inactivation by 2′-NO_2-containing aminoglycosides.

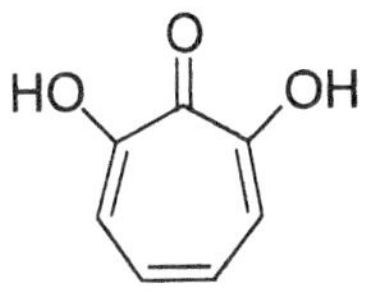

Figure 10 Structure of ANT(2″)-Ia inhibitor 7-hydroxytropolone.

in addition to protein structural data provides the requisite foundation for a concerted effort in this area. The fact that well-established enzyme assays are in place for all three classes of aminoglycoside-modifying enzymes and that these are amenable for high throughput screening methods is also of great benefit. Screens against small molecule libraries may uncover new, nonaminoglycoside leads that may prove to be starting points for inhibitors that can potentiate the activity of aminoglycoside antibiotics in resistant organisms.

4 CONCLUSIONS

Aminoglycosides are clinically important antibiotics that interact with the bacterial ribosome and disrupt proper translation. Aminoglycoside resistance in clinical isolates is largely the result of enzymes that phosphorylate, adenylate, or acetylate the antibiotics. Recent efforts in understanding the mechanisms and structures of these enzymes now opens the possibility for the design of high affinity inhibitors that could reverse resistance and potentiate existing aminoglycoside antibiotics. Several challenges remain to be addressed, however, including issues of the number and diversity of resistance enzymes, transport of inhibitors across cell membranes, and that aminoglycoside usage patterns select for different resistance mechanisms (56). At the same time, genome sequencing efforts have shown that a number of potential or cryptic aminoglycoside resistance genes are located within the genomes of many bacteria, including *M. tuberculosis*, *B. subtilis*, and *P. aeruginosa* (124). The impact of the presence of these elements remains to be assessed, but it speaks to the prevalence and diversity of aminoglycoside resistance within bacterial populations.

ACKNOWLEDGMENTS

I thank Dr. Albert Berghuis for help in preparing protein structure figures. The Medical Research Council of Canada has funded research from my laboratory.

REFERENCES

1. Edson RS, Terrell CL. The aminoglycosides. Mayo Clin Proc 1999; 74:519–528.
2. Gonzalez LS, 3rd, Spencer JP. Aminoglycosides: a practical review. Am Fam Physician 1998; 58:1811–1820.
3. Balana-Fouce R, Reguera RM, Cubria JC, Ordonez D. The pharmacology of leishmaniasis. Gen Pharmacol 1998; 30:435–443.
4. Hoepelman AI. Current therapeutic approaches to cryptosporidiosis in immunocompromised patients. J Antimicrob Chemother 1996; 37:871–880.
5. Moazed D, Noller HF. Interaction of antibiotics with functional sites in 16S ribosomal RNA. Nature 1987; 27:389–394.
6. Teraoka H, Nierhaus KH. Measurement of the binding of antibiotics to ribosomal particles by means of equilibrium dialysis. Methods Enzymol 1979; 59: 862–866.
7. Recht MI, Fourmy D, Blanchard SC, Dahlquist KD, Puglisi JD. RNA sequence determinants for aminoglycoside binding to an A-site rRNA model oligonucleotide. J Mol Biol 1996; 262:421–436.
8. Yoshizawa S, Fourmy D, Puglisi, JD. Structural origins of gentamicin antibiotic action. EMBO J 1998; 17:6437–6448.
9. Wang Y, Hamasaki K, Rando RR. Specificity of aminoglycoside binding to RNA constructs derived from the 16S rRNA decoding region an the HIV-RRE activator region. Biochemistry 1997; 36:768–779.
10. Wong CH, Hendrix M, Priestley ES, Greenberg WA. Specificity of aminoglycoside antibiotics for the A-site of the decoding region of ribosomal RNA. Chem Biol 1998; 5:397–406.
11. Fourmy D, Recht MI, Blanchard SC, Puglisi JD. Structure of the A site of *Escherichia coli* 16S ribosomal RNA complexed with an aminoglycoside antibiotic. Science 1996; 274:1367–1371.
12. Fourmy D, Yoshizawa S, Puglisi JD. Paromomycin binding induces a local conformational change in the A-site of 16 S rRNA. J Mol Biol 1998; 277:333–345.
13. Fourmy D, Recht MI, Puglisi JD. Binding of neomycin-class aminoglycoside antibiotics to the A-site of 16 S rRNA. J Mol Biol 1998; 277:347–362.
14. Recht MI, Douthwaite S, Dahlquist KD, Puglisi JD. Effect of mutations in the A site of 16 S rRNA on aminoglycoside antibiotic-ribosome interaction. J Mol Biol 1999; 286:33–43.
15. Recht MI, Douthwaite S, Puglisi JD. Basis for prokaryotic specificity of action of aminoglycoside antibiotics. EMBO J 1999; 18:3133–3138.
16. Taber HW, Mueller JP, Miller PF, Arrow AS. Bacterial uptake of aminoglycoside antibiotics. Microbiol Rev 1987; 51:439–457.
17. Kashiwagi K, Tsuhako MH, Sakata K, et al. Relationship between spontaneous aminoglycoside resistance in *Escherichia coli* and a decrease in oligopeptide binding protein. J Bacteriol 1998; 180:5484–5488.
18. Gale EF, Cundliffe E, Reynolds PE, Richmond MH, Waring MJ. The Molecular Basis of Antibiotic Action. London: Wiley, 1981:646.
19. Davies J, Gorini L, Davis BD. Misreading of RNA codewords induced by aminoglycoside antibiotics. Mol Pharmacol 1965; 1:93–106.

20. Davies J, Jones DS, Khorana HG. A further study of misreading of codons induced by streptomycin and neomycin using ribopolynucleotides containing two nucleotides in alternating sequence as templates. J Mol Biol 1966; 18: 48–57.
21. Davies J, Davis BD. Misreading of ribonucleic acid code words induced by aminoglycoside antibiotics. The effect of drug concentration. J Biol Chem 1968; 243:3312–3316.
22. Gorini L, Streptomycin and misreading of the genetic code. In: Nomura M, Tissères A, Lengyel P, eds. Ribosomes. Cold Spring Harbor, NY: Cold Spring Harbor Laboratory, 1974:791–803.
23. Lando D, Cousin MA, Privat de Garilhe M. Misreading, a fundamental aspect of the mechanism of action of several aminoglycosides. Biochemistry 1973; 12:4528–4533.
24. Anand N, Davis BD. Damage by streptomycin to the cell membrane of *Escherichia coli*. Nature 1960; 185:22–23.
25. Dubin DT, Davis BD. The effect of streptomycin on potassium flux in *Escherichia coli*. Biochim Biophys Acta 1961; 52:400–402.
26. Hancock R. Early effects of streptomycin on *Bacillus megaterium*. J Bacteriol 1961; 88:633–639.
27. Davis BD, Chen LL, Tai PC. Misread protein creates membrane channels: an essential step in the bactericidal action of aminoglycosides. Proc Natl Acad Sci USA 1986; 83:6164–6168.
28. Busse H-J, Wöstmann C, Bakker EP. The bactericidal action of streptomycin: membrane permeabilization caused by the insertion of mistranslated proteins into the cytoplasmic membrane of *Escherichia coli* and subsequent caging of the antibiotic inside the cells due to degradation of these proteins. J Gen Microbiol 1992; 138:551–561.
29. Davis BD. Mechanism of action of aminoglycosides. Microbiol Rev 1987; 51: 341–350.
30. Bryan LE, Van Den Elzen HM. Effects of membrane-energy mutations and cations on streptomycin and gentamicin accumulation by bacteria: a model for entry of streptomycin and gentamicin in susceptible and resistant bacteria. Antimicrob Agents Chemother 1977; 12:163–177.
31. Muir ME, Wallace BJ. Isolation of mutants of *Escherichia coli* uncoupled in oxidative phosphorylation using hypersensitivity to streptomycin. Biochim Biophys Acta 1979; 547:218–229.
32. Taber H, Halfenger GM. Multiple-aminoglycoside-resistant mutants of *Bacillus subtilis* deficient in accumulation of kanamycin. Antimicrob Agents Chemother 1976; 9:251–259.
33. Taber HW, Sugarman BJU, Halfenger GM. Involvement of menaquinone in the active accumulation of aminoglycosides of *Bacillus subtilis*. J Gen Microbiol 1981; 123:143–149.
34. Miller MH, Edberg SC, Mandel LJ, Behar CF, Steigbigel NH. Gentamicin uptake in wild-type and aminoglycoside-resistant small-colony mutants of *Staphylococcus aureus*. Antimicrob Agents Chemother 1980; 18:722–729.
35. McEnroe AS, Taber HW. Correlation between cytochrome *aa3* concentrations

and streptomycin accumulation in *Bacillus subtilis*. Antimicrob Agents Chemother 1984; 26:507–523.

36. Humbert R, Altendorf K. Defective gamma subunit of ATP synthase (F1F0) from *Escherichia coli* leads to resistance to aminoglycoside antibiotics. J Bacteriol 1989; 171:1435–1444.
37. Young ML, Bains M, Bell A, Hancock RE. Role of *Pseudomonas aeruginosa* outer membrane protein OprH in polymyxin and gentamicin resistance: isolation of an Opr-H deficient mutant by gene replacement techniques. Antimicrob Agents Chemother 1992; 36:2566–2568.
38. Moore RA, DeShazer D, Reckseidler S, Weissman A, Woods DE. Efflux-mediated aminoglycoside and macrolide resistance in *Burkholderia pseudomallei*. Antimicrob Agents Chemother 1999; 43:465–470.
39. Edgar R, Bibi E, MdfA, an *Escherichia coli* multidrug resistance protein with an extraordinarily broad spectrum of drug recognition. J Bacteriol 1997; 179: 2274–2280.
40. Aires JR, Kohler T, Nikaido, H, Plesiat P. Involvement of an active efflux system in the nature resistance of *Pseudomonas aeruginosa* to aminoglycosides. Antimicrob Agents Chemother 1999; 43:2624–2628.
41. Aínsa JA, Blokpoel MC, Otal I, Young DB, De Smet KA, Martin C. Molecular cloning and characterization of Tap, a putative multidrug efflux pump present in *Mycobacterium fortuitum* and *Mycobacterium tuberculosis*. J Bacteriol 1998; 180:5836–5843.
42. Kelemen GH, Cundliffe E, Financsek I. Cloning and characterization of gentamicin-resistance genes from *Micromonospora purpurea* and *Micromonospora rosea*. Gene 1991; 98:53–60.
43. Skeggs PA, Holmes DA, Cundliffe E. Cloning of aminoglycoside-resistance determinants from *Streptomyces tenebrarius* and comparison with related genes from other actinomycetes. J Gen Microbiol 1987; 133:915–923.
44. Musser JM. Antimicrobial agent resistance in mycobacteria: molecular genetic insights. Clin Microbiol Rev 1995; 8:496–514.
45. Finken M, Kirschner P, Meier A, Wrede A, Bottger EC. Molecular basis of streptomycin resistance in *Mycobacterium tuberculosis*: alterations of the ribosomal protein S12 gene and point mutations within a functional 16S ribosomal RNA pseudoknot. Mol Microbiol 1993; 9:1239–1246.
46. Nair J, Rouse DA, Bai GH, Morris SL. The *rpsL* gene and streptomycin resistance in single and multiple drug-resistant strains of *Mycobacterium tuberculosis*. Mol Microbiol 1993; 10:521–527.
47. Douglass J, Steyn LM. A ribosomal gene mutation in streptomycin-resistant *Mycobacterium tuberculosis* isolates. J Infect Dis 1993; 167:1505–1506.
48. Alangaden GJ, Kreiswirth BN, Aouad A, et al. Mechanism of resistance to amikacin and kanamycin in *Mycobacterium tuberculosis*. Antimicrob Agents Chemother 1998; 42:1295–1297.
49. De Stasio EA, Moazed D, Noller HF, Dahlberg AE. Mutations in 16S ribosomal RNA disrupt antibiotic–RNA interactions. EMBO J 1989; 8:1213–1216.
50. Shaw KJ, Rather PN, Hare RS, Miller GH. Molecular genetics of aminoglyco-

side resistance genes and familial relationships of the aminoglycoside-modifying enzymes. Microbiol Rev 1993; 57:138–163.
51. Oka A, Sugisaki H, Takanami M. Nucleotide sequence of the kanamycin resistance transposon Tn*903*. J Mol Biol 1981; 147:217–226.
52. Hoshiko S, Nojiri C, Matsunaga K, Katsumata K, Satoh E, Nagaoka K. Nucleotide sequence of the ribostamycin phosphotransferasegene and of its control region in *Streptomyces ribosidificus*. Gene 1988; 68:285–296.
53. Wright GD, Berghuis AM, Mobashery S. Aminoglycoside antibiotics: structures, function and resistance. In: Rosen BP, Mobashery S, eds. Resolving the Antibiotic Paradox: Progress in Drug Design and Resistance. New York: Plenum Press, 1998:27–69.
54. Wright GD, Thompson PR. Aminoglycoside phosphotransferases: proteins, structures, and mechanism. Front Biosci 1999; 4:D9–D21.
55. Miller GH, Sabatelli FJ, Naples L, Hare RS, Shaw KJ, Groups at ARS. The most frequently occurring aminoglycoside resistance mechanisms-combined results of surveys in eight regions of the world. J Chemother 1995; 7(Suppl.)2: 17–30.
56. Miller GH, Sabatelli FJ, Hare RS, et al. The most frequent aminoglycoside resistance mechanisms—changes with time and geographic area: a reflection of aminoglycoside usage patterns? Clin Infect Dis 1997; 24:S46–S62.
57. Miller GH, Sabatelli FJ, Naples L, Hare RS, Shaw KJ, Groups at ARS. The changing nature of aminoglycoside resistance mechanisms and the role of isepamicin—a new broad-spectrum aminoglycoside. J Chemother 1995; 7 (Suppl.)2:31–44.
58. Thompson PR, Hughes DW, Wright GD. Regiospecificity of aminoglycoside phosphotransferase from *enterococci* and *staphylococci* (APH(3′)-IIIa). Biochemistry 1996; 35:8686–8695.
59. Umezawa H, Okanishi M, Kondo S, et al. Phosphorylative inactivation of aminoglycoside antibiotics by *Escherichia coli* carrying R factor. Science 1967; 157:1559–1561.
60. Ferretti JJ, Gilmore KS, Courvalin P. Nucleotide sequence analysis of the gene specifying the bifunctional 6′-aminoglycoside acetyltransferase 2″-aminoglycoside phosphotransferase enzyme in *Streptococcus faecalis* and identification and cloning of gene regions specifying the two activities. J Bacteriol 1986; 167:631–638.
61. Azucena E, Grapsas I, Mobashery S. Properties of a bifunctional bacterial antibiotic resistance enzyme that catalyzes ATP-dependent 2″-phosphorylation and acetyl-CoA-dependent 6′-acetylation of aminoglycosides. J Am Chem Soc 1997; 119:2317–2318.
62. Daigle DM, Hughes DW, Wright GD. Prodigious substrate specificity of AAC(6′)-APH(2″), an aminoglycoside antibiotic resistance determinant in enterococci and staphylococci. Chem Biol 1999; 6:99–110.
63. Chow JW, Zervos MJ, Lerner SA, et al. A novel gentamicin resistance gene in *Enterococcus*. Antimicrob Agents Chemother 1997; 41:511–514.
64. Tsai SF, Zervos MJ, Clewell DB, Donabedian Sm, Sahm DF, Chow JW. A new

high-level gentamicin resistance gene, aph(2″)-Id, in *Enterococcus* spp. Antimicrob Agents Chemother 1998; 42:1229–1232.
65. Hon WC, McKay GA, Thompson PR, et al. Structure of an enzyme required for aminoglycoside resistance reveals homology to eukariotic protein kinases. Cell 1997; 89:887–895.
66. Thompson PR, Schwartzenhauer J, Hughes DW, Berghuis Am, Wright GD. The COOH terminus of aminoglycoside phosphotransferase (3′)-IIIa is critical for antibiotic recognition and resistance. J Biol Chem 1999; 274:30697–30706.
67. Daigle DM, McKay GA, Thompson PR, Wright GD. Aminoglycoside phosphotransferases required for antibiotic resistance are also serine protein kinases. Chem Biol 1998; 6:11–18.
68. Daigle DM, McKay GA, Wright GD. Inhibition of aminoglycoside antibiotic resistance enzymes by protein kinase inhibitors. J Biol Chem 1997; 272: 24755–24758.
69. Madhusudan, Trafny EA, Xuong N-H, et al. cAMP-dependent protein kinase: crystallographic insights into substrate recognition and phosphotransfer. Prot Sci 1994; 3:176–187.
70. Noguchi N, Emura A, Matsuyama H, O'Hara K, Sasatsu M, Knon M. Nucleotide sequence and characterization of erythromycin resistance determinant that encodes macrolide 2′-phosphotransferase-I in *Escherichia coli*. Antimicrob Agents Chemother 1995; 39:2359–2363.
71. Noguchi N., Katahama J, O'Hara K. Cloning and nucleotide sequence of the *mphB* gene for macrolide 2′-phosphotransferase-II in *Escherichia coli* FEMS Microbiol Lett 1996; 144:197–202.
72. Bibb MJ, Bibb MJ, Ward JM, Cohen SN. Nucleotide sequences encoding and promoting expression of three antibiotic resistance genes indigenous to *Streptomyces*. Mol Gen Genet 1985; 199:26–36.
73. Lovering AM, White LO, Reeves DS. AAC(1): a new aminoglycoside-acetylating enzyme modifying the C1 amino group of apramycin. J Antimicrob Chemother 1987; 20:803–813.
74. Sunada A, Nakajima M, Ikeda Y, Kondo S, Hotta K. Enzymatic 1-*N*-acetylation of paromomycin by an actinomycete strain #8 with multiple aminoglycoside resistance and paromomycin sensitivity. J Antibiot 1999; 52:809–814.
75. Costa Y, Galimand M, Leclercq R, Duval J, Courvalin P. Characterization of the chromosomal *aac(6′)-Ii* gene specific for *Enterococcus faecium*. Antimicrob Agents Chemother 1993; 37:1896–1903.
76. Rather PN, Orosz E, Shaw KJ, Hare R, Miller G. Characterization and transcriptional regulation of the 2′-N-acetyltransferase gene from *Providencia stuartii*. J Bacteriol 1993; 175:6492–6498.
77. Aínsa JA, Martin C, Gicquel B, Gomez-Lus R. Characterization of the chromosomal aminoglycoside 2′-N-acetyltransferase gene from *Mycobacterium fortuitum*. Antimicrob Agents Chemother 1996; 40:2350–2355.
78. Aínsa JA, Pérez E, Pelicic V, Berthet FX, Gicquel B, Martín C. Aminoglycoside 2′-N-acetyltransferase genes are universally present in mycobacteria: characterization of the *aac(2′)-Ic* gene from *Mycobacterium tuberculosis* and the

aac(2')-Id gene from *Mycobacterium smegmatis*. Mol Microbiol 1997; 24: 431–441.
79. Williams JW, Northrop DB. Purification and properties of gentamicin acetyltransferase I. Biochemistry 1976 15:125–131.
80. Williams JW, Northrop DB. Kinetic mechanism of gentamicin acetyltransferase I. J Biol Chem 1978; 253:5902–5907.
81. Williams JW, Northrop DB. Substrate specificity and structure-activity relationships of gentamicin acetyltransferase I. J Biol Chem 1978; 253:5908–5914.
82. Radika K, Northrop D. The kinetic mechanism of kanamycin acetyltransferase derived from the use of alternative antibiotics and coenzymes. J Biol Chem 1984; 259:12543–12546.
83. Radika K, Northrop DB. Substrate specificities and structure-activity relationships for acylation of antibiotics catalyzed by kanamycin acetyltransferase. Biochemistry 1984; 23:5118–5122.
84. Wright GD, Ladak P. Overexpression and characterization of the chromosomal aminoglycoside 6'-*N*-acetyltransferase from *Enterococcus faecium*. Antimicrob Agents Chemother 1997; 41:956–960.
85. Rather PN, Munayyer H., Mann PA, Hare RS, Miller GH, Shaw KJ. Genetic analysis of bacterial acetyltransferases: identification of amino acids determining the specificities of the aminoglycoside 6'-N-acetyltransferase Ib and IIa proteins. J Bacteriol 1992; 175:3196–3203.
86. Lambert T, Ploy M-C, Courvalin P. A spontaneous point mutation in the *aac(6')-Ib'* gene results in altered substrate specificity of aminoglycoside 6'-N-acetyltransferase of a *Pseudomonas fluorescens* strain. FEMS Microbiol Lett 1994; 115:297–304.
87. Wolf E, Vassilev A, Makino Y, Sali A, Nakatani Y, Burley SK. Crystal structure of a GCN5-related N-acetyltransferase: *Serratia marcescens* aminoglycoside 3-*N*-acetyltransferase. Cell 1998; 94:439–449.
88. Wybenga-Groot L, Draker KA, Wright GD, Berghuis AM. Crystal structure of an aminoglycoside 6'-N-acetyltransferase: defining the GCN5-related N-acetyltransferase superfamily fold. Structure 1999; 7:497–507.
89. Dutnall RN, Tafrov ST, Sternglanz R, Ramakrishnan V. Structure of the histone acetyltransferase Hat1: a paradigm for the GCN5-related N-acetyltransferase superfamily. Cell 1998; 94:427–438.
90. Trievel RC, Rojas JR, Sterner DE, et al. Crystal structure and mechanism of histone acetylation of the yeast GCN5 transcriptional coactivator. Proc Natl Acad Sci USA 1999; 96:8931–8936.
91. Clements A, Rojas JR, Trievel RC, Wang L, Berger SL, Marmorstein R. Crystal structure of the histone acetyltransferase domain of the human PCAF transcriptional regulator bound to coenzyme A. EMBO J 1999; 18:3521–3532.
92. Bhatnagar RS, Futterer K, Farazi TA, et al. Structure of N-myristoyltransferase with bound myristoylCoA and peptide substrate analogs. Nat Struct Biol 1998; 5:1091–1097.
93. Weston SA, Camble R, Colls J, et al. Crystal structure of the anti-fungal target N-myristoyl transferase. Nat Struct Biol 1998; 5:213–221.
94. Neuwald AF, Landsman D. GCN5-related histone N-acetyltransferases be-

long to a diverse superfamily that includes the yeast SPT10 protein. Trends Biochem Sci 1997; 22:154–155.
95. Van Pelt JE, Northrop DB. Purification and properties of gentamicin nucleotidyltransferase from *Escherichia coli*: nucleotide specificity, pH optimum, and the separation of two electrophoretic variants. Arch Biochem Biophys 1984; 230:250–263.
96. Gates CA, Northrop DB. Substrate specificities and structure-activity relationships for the nucleotidylation of antibiotics catalyzed by aminoglycoside nucleotidyltransferase 2″-I. Biochemistry 1988; 27:3820–3825.
97. Gates CA, Northrop DB. Alternative substrate and inhibition kinetics of aminoglycoside nucleotidyltransferase 2″-I in support of a Theorell-Chance kinetic mechanism. Biochemistry 1988; 27:3826–3833.
98. Gates CA, Northrop DB. Determination of the rate-limiting segment of aminoglycoside nucleotidyltransferase 2″-I by pH- and viscosity-dependent kinetics. Biochemistry 1988; 27:3834–3842.
99. Van Pelt JE, Iyengar R, Frey PA. Gentamicin nucleodyltransferase. Stereochemical inversion at phosphorus in enzymatic 2′-deoxyadenylyl transfer to tobramycin. J Biol Chem 1986; 261:15995–15999.
100. Sakon J, Liao HH, Kanikula AM, Benning MM, Rayment I, Holden HM. Molecular structure of kanamycin nucleotidyl transferase determined to 3 Å resolution. Biochemistry 1993; 32:11977–11984.
101. Perdersen LC, Benning MM, Holden HM. Structural investigation of the antibiotic and ATP-binding sites in kanamycin nucleotidyltransferase. Biochemistry 1995; 34:13305–13311.
102. Matsumura M, Katakura Y, Imanaka T, Aiba S. Enzymatic and nucleotide sequence studies of a kanamycin-inactivating enzyme encoded by a plasmid from thermophilic bacilli in comparison with that encoded by plasmid pUB110. J Bacteriol 1984; 160:413–420.
103. Holm L, Sander C. DNA polymerase β belongs to an ancient nucleotidyltransferase superfamily. Trends Biol Chem 1995; 20:345–347.
104. Aravind L, Koonin EV. DNA polymerase β-like nucleotidyltransferase superfamily: identification of three new families, classification and evolutionary history. Nucleic Acids Res 1999; 27:1609–1618.
105. McKay GA, Wright GD. Kinetic mechanism of aminoglycoside phosphotransferase type IIIa: evidence for a Theorell-Chance mechanism. J Biol Chem 1995; 270:24686–24692.
106. Umezawa H, Umezawa S, Tsuchiya T, Okazaki Y. 3′,4′-Dideoxykanamycin B active against kanamycin-resistant *Escherichia coli* and *Pseudomonas aeruginosa*. J Antibiot 1971; 24:485–487.
107. Umezawa H, Kondo S. Mechanisms of resistance to aminoglycoside antibiotics. In: Umezawa H, Hooper IR, eds. Aminoglycoside Antibiotics. Berlin: Springer-Verlag, 1982:267–292.
108. Kawaguchi H, Naito T, Nakagowa S, Fuijawa K. BBK8, a new semisynthetic aminoglycoside antibiotic. J Antibiot 1972; 25:695.
109. Nagabhushan TL, Cooper AB, Tsai H, Daniels PJ, Miller GH. The syntheses and biological properties of 1-N-(S-4-amino-2-hydroxybutyryl)-gentamicin

B and 1-N-(S-3-amino-2-hydroxypropionyl)-gentamicin B. J Antibiot 1978; 31: 681–687.
110. Wright JJ. Synthesis of 1-*N*-ethylsisomicin: a broad spectrum semisynthetic aminoglycoside antibiotic. J Chem Soc Chem Commun 1976:206–208.
111. Umezawa H, Nishimura Y, Tsuchiya T, Umezawa S. Syntheses of 6′-N-methylkanamycin and 3′,4′-dideoxy-6′-N-methylkanamycin B active against resistant strains having 6′-N-acetylating enzymes. J Antibiot 1972; 25:743–745.
112. Umezawa H, Iinuma K, Kondo S, Maeda K. Synthesis and antibacterial activity of 6′-N-alkyl derivatives of 1-N-[(S)-4-amino-2-hydroxbutyryl]-kanamycin. J Antibiot (Tokyo) 1975; 28:483–485.
113. Knod S, Iinuma K, Yamamoto H, Maeda K, Umezawa H. Synthese of 1-*N*-(S)-4-amino-2-hydroxybutyryl)-kanamycin B and -3′-4′-dideoxykanamycin B active against kanamycin-resistant bacteria. J Antibiot 1973; 26:412–415.
114. Kondo S, Tamura A, Gomi S, Ikeda Y, Takeuchi T, Mitsuhashi S. Structures of enzymatically modified products of arbekacin by methicillin-resistant *Staphylococcus aureus*. J Antibiot 1993; 46:310–315.
115. Fujimura S, Tokue Y, Takahashi H, et al. A newly recognized acetylated metabolite of arbekacin in arbekacin-resistant strains of methicillin-resistant *Staphylococcus aureus*. J Antimicrob Chemother 1998; 41:495–497.
116. Kondo S, Ikdea Y, Ikeda D, et al. Synthesis of 2″-amino-2″-deoxyarbekacin and its analogs having potent activity against methicillin-resistant *Staphylococcus aureus*. J Antibiot 1994; 47:821–832.
117. Roestamadji J, Grapsas I, Mobashery S. Loss of individual electrostatic interactions between aminoglycoside antibiotics and resistance enzymes as an effective means to overcoming bacterial drug resistance. J Am Chem Soc 1995; 117:11060–11069.
118. McKay GA, Roestamadji J, Mobashery S, Wright GD. Recognition of aminoglycoside antibiotics by enterococcal-staphylococcal aminoglycoside 3′-phosphotransferase type IIIa: role of substrate amino groups. Antimicrob Agents Chemother 1996; 40:2648–2650.
119. Alper PB, Hendrix M, Sears P, Wong C-H. Probing the specificity of aminoglycoside-ribosomal RNA interactions with designed synthetic analogs. J Am Chem Soc 1998; 120:1965–1978.
120. Greenberg WA, Priestley ES, Sears P, et al. Design and synthesis of new aminoglycoside antibiotics containing neamine as an optimal core structure: correlation of antibiotic activity with in vitro inhibition of translation. J Am Chem Soc 1999; 121:6527–6541.
121. Roestamadji J, Grapsas I, Mobashery S. Mechanism-based inactivation of bacterial aminoglycoside 3′-phosphotransferases. J Am Chem Soc 1995; 117:80–84.
122. Haddad J, Vakulenko S, Mobashery S. An antibiotic cloaked by its own resistance enzyme. J Am Chem Soc 1999; 121:11922–11923.
123. Allen NE, Jr, WEA, Jr, Krst HA. 7-Hydroxytropolone: an inhibitor of aminoglycoside-2″-O-adenyltransferase. Antimicrob Agents Chemother 1982; 22: 824–831.
124. Wright GD. Aminoglycoside modifying enzymes. Curr Opin Microbiol 1999; 2:499–503.

6

β-Lactamases and Resistance to β-Lactam Antibiotics

Lakshmi P. Kotra
University of Toronto, Toronto, Ontario, Canada

Jean-Pierre Samama
Centre National de la Recherche Scientifique, Toulouse, France

Shahriar Mobashery
Wayne State University, Detroit, Michigan

The catalytic function of β-lactamases is the primary cause of resistance to β-lactam antibiotics. These enzymes hydrolyze the β-lactam ring of these versatile antibiotics, which is a process that inactivates the drugs. Over 250 β-lactamases are known, which are grouped into four distinct classes (classes A, B, C, and D). The members of each class operate by distinct catalytic mechanisms. A series of recently discovered β-lactamases exhibit a wide breadth for their substrate preferences, which often include penicillins, cephalosporins, and carbapenems, among other substrates. These so-called extended-spectrum β-lactamases (ESBLs) are being identified among enzymes of classes A and D (active-site serine enzymes) and class B β-lactamases (zinc-dependent enzymes). The breadth of phenotypic traits for these enzymes collectively covers all known types of β-lactam antibiotics. The global distribution of the pathogens that harbor these various enzymes is sufficiently different at the present that obsolescence of β-lactam antibiotics has not happened to date. This chapter discusses the various properties of these microbial enzymes, with special emphasis on the structural factors governing their functions.

1 INTRODUCTION

Resistance to antimicrobials is an escalating problem in the clinic, and it is estimated that approximately $30 billion is spent annually in the United States for the treatment of resistant bacterial infections (1). Infections are the leading cause of death on a global scale, and drug resistance is expected to aggravate the already serious situation in the immediate future (2). β-Lactam antibiotics are the most commonly used antibacterial agents in the present chemotherapeutic armamentarium, and β-lactamases, the enzymes that hydrolyze β-lactam antibiotics, are the major cause of resistance to these compounds (3). The fact that we rely so heavily on β-lactams to the present day is remarkable in light of the fact that β-lactamases were discovered before their widespread use clinically (4), and to date over 250 novel β-lactamases have been identified to complicate their therapeutic use (3). The genes for β-lactamases may be chromosomal, plasmidborne, and others may be found on transposable elements. Furthermore, their existence on integrons has also been documented (5,6). Hence, there is ample opportunity for bacteria to share these drug resistance genes, and indeed this has happened extensively (7). It would seem that the diversity in structures of β-lactamases and in mechanisms of genetic dissemination should have put an end to viability of the β-lactam antibacterials in the clinic. The difficulties in treatment of resistant organisms harboring these enzymes are becoming acute, but the demise of these versatile antibacterial agents has not yet happened. In fact, we will remain dependent on β-lactam antibiotics for the foreseeable future. Time will tell if current efforts in the pharmaceutical industry will meet the challenge of bacterial drug resistance by development of novel classes of antibiotics that would ultimately replace β-lactams in therapy (8,9). Meanwhile, it is clear that we need to develop a detailed knowledge of the properties of these enzymes in order to counter their deleterious effects. In this chapter, we will attempt to discuss β-lactamases from the perspective of their mechanisms and structures. We will also explore the means by which random mutation and selection have provided opportunities for these enzymes to extend their substrate specificities such that resistance to virtually any β-lactam antibiotic has been observed.

2 CLASSIFICATION OF β-LACTAMASES

Various classification schemes have been proposed for β-lactamases based on the characteristics of the enzymes and/or their substrate profiles (10,11). Bush proposed a comprehensive functional classification of β-lactamases in 1989, which was expanded in 1995 to include just under 190 β-lactamases (10). This classification system utilized an extensive set of kinetic data on various enzymes and categorized β-lactamases according to the substrate preferences and inhibition characteristics. Four major groups are recognized in this classification. Group 1 consists of cephalosporinases, which are not inhibited by clavulanic acid. Group 2 consists of penicillinases, including broad-spectrum penicillinases that are generally inhibited by the active-site–directed β-lactamase inhibitors. Subgroups of enzymes, namely, 2a, 2b, 2be, 2br, 2c, 2d, 2e and 2f, were defined based on the rates of hydrolysis of carbenicillin, cloxacillin, extended-spectrum β-lactams ceftazidime, cefotaxime, or aztreonam and of inhibition profile by clavulanate, respectively. Enzymes that are inhibited by the metal-chelating agent ethylenediame tetraactic acid (EDTA) are classified as group 3. Group 4 consists of β-lactamases that are not inhibited by clavulanic acid.

However, the classification scheme proposed by Ambler is also commonly used (11,12). The enzymes that hydrolyze β-lactam molecules are grouped into four classes, A, B, C, and D (12–14). Whereas classes A, C, and D have evolved dependence on an active-site serine as their key mechanistic feature, class B enzymes are zinc dependent and hence different. The catalytic process for turnover of the members of the former group involves acylation at the active-site serine by the β-lactam antibiotic followed by deacylation of the acyl-enzyme species. It is noteworthy that these enzymes do not share any sequence homologies, structural similarities, or mechanistic features with serine or zinc-dependent proteases. Class A β-lactamases prefer penicillins as substrates, whereas class C enzymes turn over cephalosporins better (Scheme 1). Class B enzymes can hydrolyze a broad range of substrates, including carbapenems, which resist hydrolysis by most of the other classes of enzymes. Class D β-lactamases, on the other hand, hydrolyze oxacillin-type β-lactams efficiently. Classes A and C of β-lactamase are the most common and the second most common enzymes, respectively (3). The general properties of these enzymes, which operate through distinct mechanisms, will be discussed in the following sections. Our emphasis will be on the structural aspects that impart the given phenotypic consequences. We add that several other reviews on β-lactamases have appeared that complement this chapter in various ways (3, 15–23).

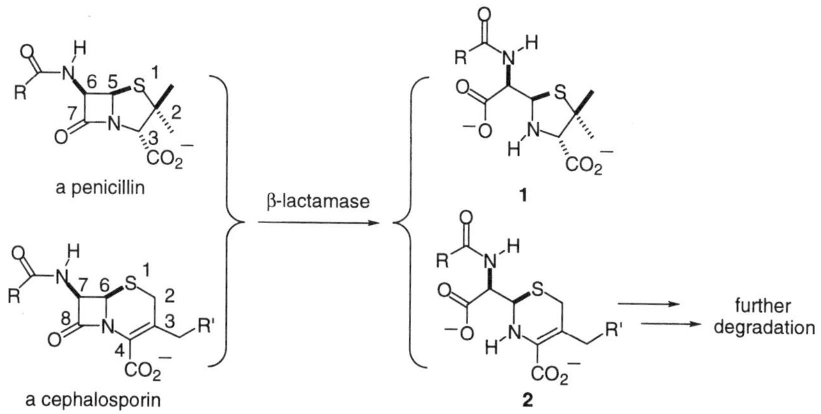

Scheme 1

3 ORIGIN OF β-LACTAMASES

It is now accepted that β-lactamases evolved from penicillin-binding proteins (PBPs) which experience covalent modification by penicillins and other β-lactams. These biosynthetic enzymes assemble and regulate the formation of the bacterial cell wall consisting of cross-linked polymers of peptidoglycan. Certain PBPs carry out the cross-linking reaction, which imparts rigidity to the bacterial cell wall. Penicillins bind to PBPs and acylate the active-site serine. The resultant acyl-enzyme intermediate is sufficiently stable to provide effective inhibition of the biological function of the PBP, and bacteria cannot survive such inhibition.

The kinship of PBPs and β-lactamases is established based on extensive multiple-sequence alignment and structural data (15,17,18). In essence, nature discovered that the same structural motif that binds penicillin (that of PBP) can be used to destroy the drug. Insofar as the resultant acyl-enzyme species between a β-lactam antibiotic and a PBP was relatively stable, evolution of the drug-resistant phenotype had to find a way to make it unstable. For such an evolutionary scheme to be successful, the nascent β-lactamase should be able to experience active-site acylation (inherited from the parental PBP; Scheme 2, species **3**) and deacylation of the acyl-enzyme species. This process, of course, would take place as a consequence of random mutation and selection. It must have taken place in incremental steps to liberate the PBP from inhibition by the β-lactam. It is intuitive that the driving force for liberation from inhibition must have been to make the active PBP available to function again in the cell wall assembly. It is interesting that once acylated by a β-lactam, the modern

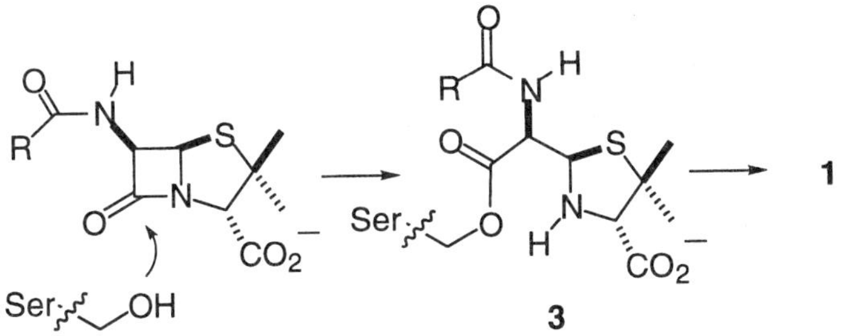

Scheme 2

PBPs undergo slow deacylation with a wide range of deacylation rate constants (24–26) indicative of the diversification among PBPs. Such diversification over evolutionary time scale has introduced substantial sequence divergence among these proteins (15,21).

The evolutionary advent of a PBP that would undergo acylation by β-lactam antibiotics followed by a reasonably rapid deacylation step would have had a clear advantage for the bacterium. Ultimately, the strategy must have been so successful that the PBP that underwent the process of acylation and deacylation fairly effectively started to become more specialized in hydrolysis of the β-lactam antibiotics, so they served the role of bona fide resistance enzymes. Along the way, the bona fide resistance enzyme would detach itself from the surface of the bacterial plasma membrane—the vast majority of PBPs are membrane proteins—so it could serve as a vanguard in fighting the in-coming antibiotics in solution.

Clearly PBPs are ancient proteins, since bacteria came into existence approximately 3.8 billion years ago (27), and the evolution of cell wall must have followed suit at about the same time. But the development of β-lactamases is a relatively recent event, which must have taken place after the evolution of the first biosynthetic pathways for the natural β-lactam antibiotics (3,17,18,28,29). The diversification of function has been impressive in light of various functions for PBPs and the breadth of the substrate profile for β-lactamases (3). The process of evolution for β-lactamases has been accelerated by the extensive use of β-lactams in the clinic over the past 50 years (15,16). As a result, although the degree of sequence homology among these proteins is very low, it would appear from the emerging structural information that the three-dimensional fold of these proteins is preserved (see below) (15). To date, more than 50 structures for PBPs and β-lactamases have been determined using x-ray crystallography (Table 1), giving a wealth of structural information for understanding the function of

Table 1 X-Ray Crystal Structures of β-Lactamases and PBPs

Code	Class	Organism	Mutation	Ref.
1ALQ	A	*S. aureus* PC1	Circularly permuted[a]	125
1BLC	A	*S. aureus* PC1		126
1BLH	A	*S. aureus* PC1		39
1BLP	A	*S. aureus* PC1	D179N	127
1DJA	A	*S. aureus* PC1	A30M, K73H	128
1DJB and 1DJC	A	*S. aureus* PC1	S70A	128
1KGE	A	*S. aureus* PC1	N170M	129
1KGF	A	*S. aureus* PC1	N170Q	129
1OME	A	*S. aureus* PC1	Δ163-178 (ω-loop)	130
1PIO	A	*S. aureus* PC1	Ins(M30), A238S, Del(I239)	131
3BLM	A	*S. aureus* PC1		132
1BSG	A	*S. albus* G		133
1MBL	A	*B. licheniformis* 749/C	E166A	33
2BLM	A	*B. licheniformis* 749/C		134
4BLM	A	*B. licheniformis* 749/C		135
1MFO	A	*M. fortuitum*		136
1AXB	A (TEM-1)	*E. coli* pBR322	V84I, A184V	40
1BTL	A (TEM-1)	*E. coli* pBR322		137
1TEM	A (TEM-1)	*E. coli* pBR322	V84I, A184V	45
1XPB	A (TEM-1)	*E. coli* pBR322	S235A	138
NA[ii]	A (TEM-1)	*E. coli* pBR322	N276D	139
1BT5	A (TEM-1)	*E. coli* pBR322		47
NA	A (TEM-1)	*E. coli* pUC118		140
NA	A (TEM-1)	*E. coli* MV1184	E166N	30
1BUE	A (NMC-A)	*E. cloacae*		72
1BUL	A (NMC-A)	*E. cloacae*		71
NA	A (NMC-A)	*E. cloacae*		46
NA	A (Per-1)	*P. aeruginosa*		141
1BZA	A (Toho-1)	*E. coli* TUH12191	E166A	74
1SHV	A (SHV-1)	*K. pneumoniae* 15571		73
1ZNB	B	*B. fragilis* TAL3636		142
2ZNB	B	*B. fragilis* QMCN3		91
3ZNB	B	*B. fragilis* QMCN3		91
4ZNB	B	*B. fragilis*		92
1BC2	B	*B. cereus* 549/H/9		90
2BC2	B	*B. cereus* 549/H/9		90
3BC2	B	*B. cereus* 549/H/9		90

Table 1 Continued

Code	Class	Organism	Mutation	Ref.
1A7T and 1A8T	B	*B. fragilis*	A171T, D208N	143
1BMC	B	*B. cereus* 569/H		89
1BME and 1BVT	B	*B. cereus* 569/H		90
1BM1 and 2BM1	B	*B. fragilis* TAL2480		144
1SML	B	*S. maltophilia*		145
2BLS and 3BLS	C	*E. coli* K12		112
2BLT	C	*E. cloacae* P99		113
1BLS	C	*E. cloacae* P99		111
S057	C	*C. freundii 1203*		110
Penicillin-binding proteins				
ICEF	DD-transpeptidase	*Streptomyces* sp.		146
1CEG	DD-transpeptidase	*Streptomyces* sp.		146
1PMB	PBP2x	*S. pneumoniae*		147
1SKF	DD-transpeptidase	*Streptomyces* K15		148
3PTE	DD-transpeptidase	*Streptomyces* R61		108

NA, not available.
[a]Circularly permuted with an eight residue linker inserted, which joins original N- and C-terminals and a new N-terminus at residue 254.

these bacterial proteins. Salient features of these structures are discussed here as they relate to the mechanism of hydrolysis of β-lactamases.

4 CLASS A β-LACTAMASES

Enzymes that belong to class A are the most commonly found. These β-lactamases are the best understood in every aspect of their chemistry, biochemistry, and molecular biology. Therefore, we will discuss them in greater depth than enzymes from the other classes.

We recently performed an extensive sequence alignment analysis of over 140 amino acid sequences for members of all classes of β-lactamases and PBPs (15). The sequences are so divergent that there are no significant homologies in general, but a universal identity is shared among *all* these proteins with regard to the motif Ser-X-X-Lys. The serine corresponds to the active-site serine of these proteins—be it a β-lactamase or a PBP—that experiences acylation by the substrate. A lysine at three residues to the carboxyl-terminal side of the serine is clearly important for the functions of both types of enzymes as that too is absolutely conserved. Since the only mechanistic feature that all these proteins share is the active-site serine acylation, the Ser-X-X-Lys sequence is a minimal essential motif for the process of serine acylation.

The mechanism of serine acylation has been an issue of considerable debate in the literature recently, and to date a clear understanding of this process has not emerged. However, once the mechanism of the acylation process is elucidated, it is likely to be broadly shared by all active-site serine β-lactamases and PBPs, as this step is inherited from the primordial PBP that adapted the active-site acylation strategy to begin with.

The contention stems, in part, from lack of a clear knowledge of the titration states of the active-site residues involved in catalysis. Two groups have suggested that Lys-73 is not protonated, and can serve as the general base in activation of Ser-70 during its acylation (30,31). Others believe that a positively charged Lys-73 would serve as an electrostatic anchor in lowering the pK_a of Ser-70 to promote its acylation (32–34), and the suggestion that Glu-166 is the general base for this step has also been put forward (i.e., symmetry in catalysis) (35–37). If Lys-73 is unprotonated, Paetzel and Dalbey have argued that β-lactamase then belongs to the family of enzymes with a Ser/Lys dyad (38). Thus, it was argued that only two residues, that is, the serine and lysine, are essential for the catalytic activity of these proteins, at least for acylation.

A recent comparison of two x-ray structures, one of the *Staphylococcus aureus* PC1 β-lactamase modified by a phosphonate (39), and another of the TEM-1 β-lactamase modified by the same phosphonate (40), indicated a pathway for the formation of the acyl-enzyme intermediate. The phosphonate modified the active-site serine in both enzymes, and the structures mimicked the transition state for the acylation process. However, there are intriguing differences between the two structures. In the *S. aureus* PC1 β-lactamase structure, the side chains of Ser-70 and Lys-73 interact closely, giving the appearance of an interaction of a base (Lys-73) abstracting the proton from serine. On the other hand, the structure for the TEM-1 enzyme shows strong interactions between Ser-130 and the phosphonate oxygen corresponding to the leaving group, indicating that the complex mimics the collapse of the tetrahedral species en route to the formation of the acyl-enzyme intermediate. The process would take place by the transfer of a proton from Ser-130 to the departing amine in the β-lactam substrate. Then a proton would be transferred to Ser-130 from the now-protonated Lys-73 (40). These analyses collectively argue for the existence of Lys-73 in its unprotonated form.

A mechanism for the deacylation of the acyl-enzyme intermediate in class A β-lactamases was proposed by Ishiguro and Imajo that involves the nitrogen atom of the β-lactam moiety (32). According to this mechanism, the hydrolytic water would approach from the *re* face (more commonly known as the β-face) of the ester moiety, and would be activated by the amine of the thiazolidine ring and the C_3 carboxylate moiety of the penicillin substrate. It was argued that the Glu-166 mutant enzymes, which are incapable of undergoing deacylation, would experience a modified hydrogen-bonding pattern by promoting interaction between Ser-130 and the amine of the thiazolidine moiety of the acyl-enzyme intermediate, resulting in a deacylation-deficient species (32). However, a consensus has emerged that argues for residue Glu-166 serving as a general base in promoting a water molecule for the deacylation event (30,33). This assertion is supported by the results from site-directed mutagenesis (41–44) and from studies of β-lactam molecules that acylate the enzymes but resist deacylation (45–49). It is worthy of comment that the water molecule approaches the ester of the acyl-enzyme intermediate from the α-face (Fig. 1).

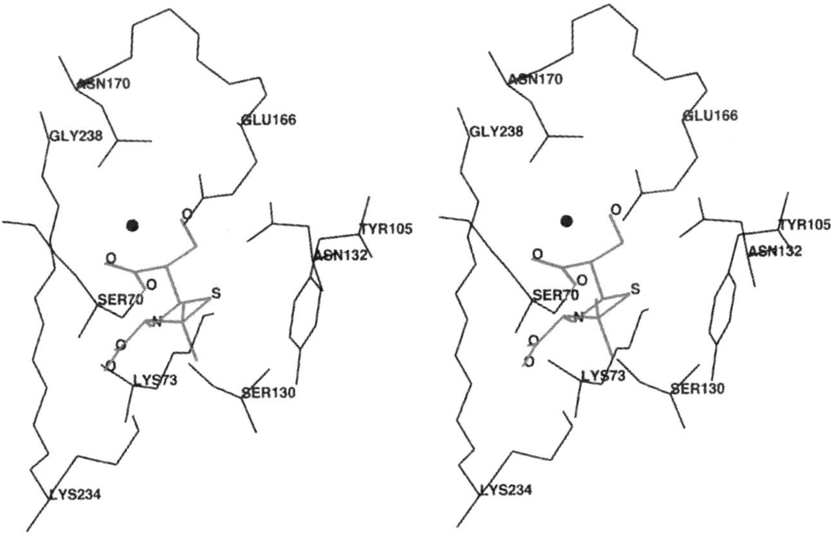

Figure 1 Stereo view of the active site of TEM-1 β-lactamase (class A enzyme) complexed with 6α-hydroxymethyl penicillanic acid (pdb code: 1TEM). The hydrolytic water molecule (shown as a sphere) is approaching the ester moiety from α-face, which is also seen interacting with the hydroxymethyl moiety of the bound inhibitor.

A water molecule in the structure of the *Streptomyces* R61 DD-peptidase/transpeptidase (a PBP) occupies the same spatial location as the hydrolytic water in class A β-lactamases (15). It has been argued that the nascent class A β-lactamase acquired Glu-166 to promote the water molecules that already existed in the structure of the PBP to become deacylation proficient (15). Residue Glu-166 is sequestered on an Ω-loop secondary structural element in the class A enzymes. Despite the fact that there are typically over 20 amino acids in this Ω-loop, only Glu-166 is strictly conserved in all class A enzymes (15). Therefore, the Ω-loop may serve as a structural element, a template for positioning of the residue that promotes the deacylation step (15).

Class A β-lactamases have been argued to be perfect enzymes based on the fact that they operate at the diffusion limit (50). But also, these enzymes catalyze their reaction with the preferred substrates with microscopic rate constants that are large and comparable to the rate of dissection of the product. These analyses for the enzymes from *S. aureus* and *Escherichia coli* indicated that there is no step in catalysis that is solely rate limiting, an observation that was taken to be an indication of fully efficient catalysis by these enzymes (51).

Class A β-lactamases are produced by both gram-positive and gram-negative bacteria. The enzymes of this class are the most heterogeneous with respect to their structures and their kinetic properties. Many of them are plasmidborne, so they have undergone substantial mutational alteration. For example, the prototypic member of this class of β-lactamases from gram negatives, the TEM-1 β-lactamase from *E. coli*, has been known since 1965 (52). To date, 69 variants of this enzyme have been discovered, which show substantial diversity in phenotypic and kinetic properties among themselves.

In general, various members of class A β-lactamases enjoy considerable conservation of the three-dimensional fold regardless of whether they are from gram-negative or gram-positive bacteria. This observation is demonstrated below for the structures of three gram-negative class A β-lactamases from *E. coli* (TEM-1), from *Klebsiella pneumoniae* (SHV-1), from *Enterobacter cloacae* (NMC-A), and for the gram-positive enzyme from *S. aureus* (Fig. 2). Figure 3 shows details of the active site structure for the TEM-1 β-lactamases.

A number of mechanistic properties of class A enzymes are beginning to be elucidated by both biochemical studies and structural information. Two broadly defined phenotypic properties from class A β-lactamases are emerging as a result of recent clinical selection pressures. One is the inhibitor-resistant phenotype, which was first observed in the TEM family, for which the term *inhibitor-resistant TEM* (IRT) was coined. This type of phenotype has now been seen in the SHV family as well (53). The

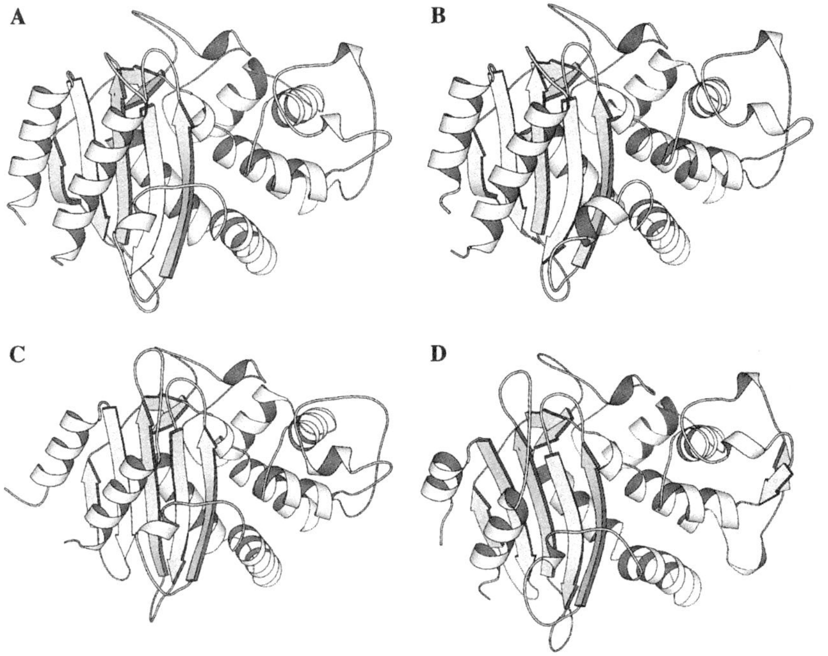

Figure 2 The x-ray crystal structures of the gram-negative class A β-lactamases from *E. coli* (TEM-1) (A), *K. pneumoniae* (SHV-1) (B), and *E. cloacae* (NMC-A) (C) and for the gram-positive enzyme from *Staphylococcus aureus* (D) (pdb codes: 1TEM, 1SHV, 1BUL, and 1BLC, respectively). These figures were prepared using the program MOLSCRIPT (149).

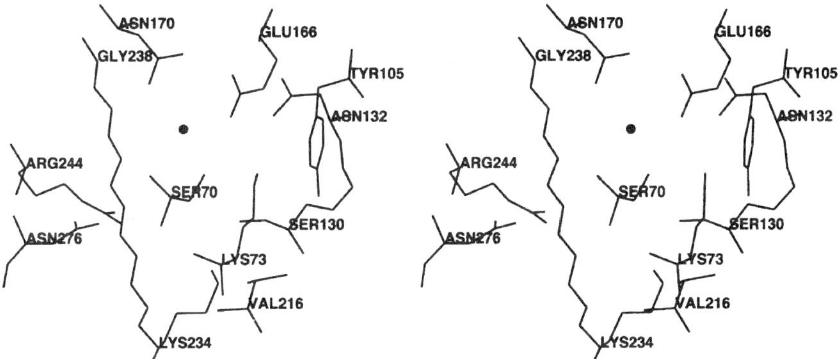

Figure 3 Stereo view of the active site of the TEM-1 β-lactamase, with important residues labeled. The hydrolytic water, coordinated to Glu-166 and Asn-170, is shown as a black sphere.

second phenotype is that of the so-called extended-spectrum β-lactamases (ESBLs). As the name suggests, these variants have broadened their substrate profiles to include such β-lactams as expanded-spectrum cephalosporins and carbapenems. These extended-spectrum β-lactamases are exemplified by the Imi-1, Per-1, Sme-1, Toho-1, and NMC-A β-lactamases, among others (54–57). Both these phenotypes are cause for serious concern in the clinic.

The IRT β-lactamases were first observed in 1992 in France and England (58,59). These mutant variants of the TEM-1 β-lactamase emerged as a consequence of the use of mechanism-based inhibitors for class A β-lactamases, the first of which, clavulanic acid, was introduced to the clinic in 1984 in combination with amoxicillin (17,60). The inhibitors impair the function of the class A β-lactamases, so that the coadministered penicillin would have the opportunity to inhibit the PBP. These mixtures of β-lactamase inhibitors and penicillins have selected the phenotype that resists inhibition of β-lactamase. There were 17 known IRTs as of mid-1999, which were the result of single or multiple mutations of the parental enzyme. However, error-prone polymerase chain reaction (PCR) mutagenesis has identified four mutations of consequence for the IRT β-lactamases (61). These are mutations at positions 69, 130, 244, and 276. Biochemical analyses have shed light on the functions of these amino acids in the mechanism of inhibition of β-lactamases. At position 69, substitution of Met with either Leu, Ile, or Val in the TEM-1 β-lactamase resulted in decreased susceptibility to the β-lactamase inhibitors clavulanate, sulbactam, and tazobactam (62). These mutations were suggested to alter the hydrophobic and/or steric factors near the main-chain nitrogen of Ser-70, leading to the disruption of the interactions at the oxyanion hole within the active site. However, a substitution at position 69 in the Per-1 β-lactamase does not seem to produce a similar effect toward clavulanate (63). Inhibition of the TEM β-lactamases by clavulanic acid commences by acylation of Ser-70, the active-site serine. In the process, Ser-130 is also covalently attached to the acyl-enzyme species (48,64–66). Hence, mutation of Ser-130 to Gly, such as seen in the IRT, prevents this second modification of the active site, resulting in poor inhibition of the enzyme.

The effects of mutations at positions 244 and 276 are related to each other, because their side chains are interacting with one another. However, the mechanistic reasons for the IRT phenotype manifestation for mutations at each site are distinct. The Arg-244 side chain has a number of functions in the TEM-1 β-lactamase. It serves as counterion to the carboxylate of substrates and inhibitors (67). A water molecule critical for the chemistry of inhibition is hydrogen bonded to the side chain of Arg-244 in the TEM-1 β-lactamase (64). Mutation at position 244, such as in Arg-244-Ser, would eliminate this interaction with the water molecule, hence impairing the ability of clavulanic acid to inhibit the enzyme (64). The side chain of Arg-244 is also hydrogen bonded to the side chain of Asn-276. The mutation Asn-276-Asp, seen in IRTs, enhances the strength of interactions between residues at 244 and 276, and as a result the strength of the interaction of Arg-244 with the carboxylate of clavulanic acid is weakened (68). The interaction in the mutant protein also influences the water molecule coordinated to Arg-244, such that the rate constant for inhibition of the enzyme is also affected for the worse. The structural effects with these kinetic consequence are indeed quite subtle, as perceived from the x-ray structure of the Asn-276-Asp variant of the TEM β-lactamase (Fig. 4) (68). This is the only x-ray structure for an IRT enzyme to date.

The structural factors that result in the extended-spectrum phenotypes are quite diverse, and we would not be able to discuss them fully in this chapter, although this subject was covered by a minireview a few

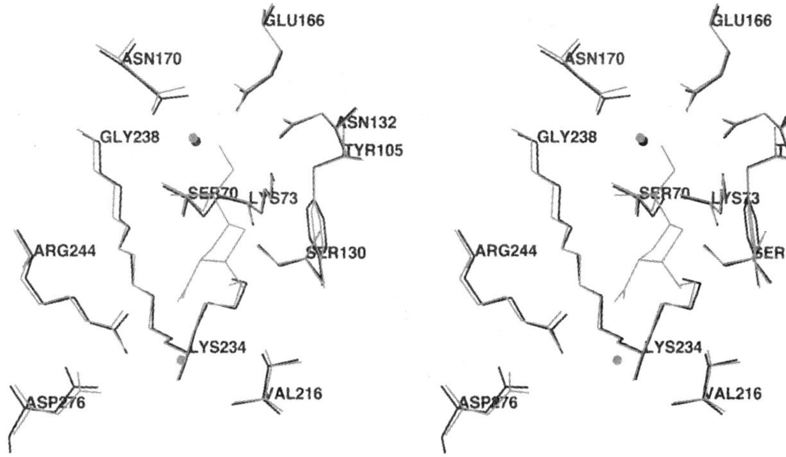

Figure 4 Stereo view of the active sites of wild-type TEM-1 β-lactamase (shown in gray) superimposed with the Asn-176–Asp mutant of TEM-1 β-lactamase (shown in black). Inhibitor of wild-type enzyme, 6α-hydroxymethylpenicillanate, covalently bound to Ser-70 of the wild-type TEM-1 enzyme is shown to illustrate the interactions. The side chain of Asp-276 (at 7 o'clock, in black) is seen shifted toward Arg-244, as is Arg-244 toward Asp-276, resulting in a stronger interaction between these two residues. The coordinated water molecule to Arg-244 (black sphere at 6 o'clock position) is not seen in the crystal structure of the mutant enzyme. The hydrolytic water, however, is seen in both crystal structures (shown here as spheres at 12 o'clock).

years ago (69). We will try to limit our discussion here to cases for which x-ray structure for the enzymes are available. ESBLs with ability to turn over imipenem (**4**), a carbapenem antibiotic, are of special concern. Imipenem is an exceedingly poor substrate for class A β-lactamases. The mechanistic and structural bases for such poor turnover of imipenem by the TEM-1 enzyme have been elucidated recently (47,70). The structural reason behind the poor turnover is the unfavorable interactions of the 6α-hydroxyethyl group of imipenem with the enzyme. These interactions force the acyl-enzyme complex to assume a new conformation, which is no longer predisposed for the deacylation process (47). For a class A enzyme to become adept at turning over imipenem, it has to eliminate the unfavorable interactions of the enzyme with the 6α-hydroxyethyl group of imipenem. This is indeed what has been seen for the NMC-A (nonmetallocarbapenemase of class A) β-lactamase from *E. cloacae* (71). This is also an example of an extremely subtle change in the structure of the enzyme to give a profound phenotypic consequence (71,72). The NMC-A β-lactamase is, in every respect of the catalytic machinery, similar to other prototypic class A enzymes, such as the TEM-1 β-lactamase. However, a close inspection of the structure revealed that the position of one residue, that of Asn-132, has moved away from the active site by a mere 1 Å. This repositioning has enlarged the cavity in which the 6α-hydroxyethyl group of imipenem would fit, and has eliminated the unfavorable steric interactions that were seen in the TEM-1 enzyme (71,72). See Structure 1 and Figure 5.

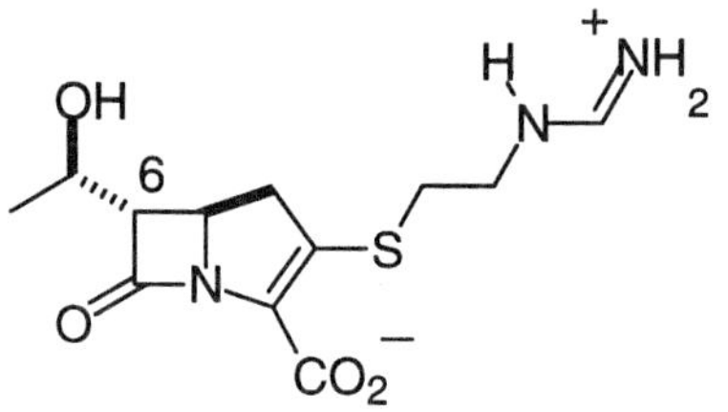

4

The three-dimensional structure of the SHV-1 β-lactamase, which possesses a somewhat broader substrate profile than the TEM-1 enzyme, shows a similar overall fold to that of the TEM-1 enzyme (see Fig. 2). There are, however, few subtle differences that differentiate the active sites of the two enzymes. The Ser-130 to Asn-132 loop and the neighboring Asp-104/Tyr-105 loop in the SHV-1 enzyme have shifted away from the active site by about 0.7–1.2 Å, in comparison to the TEM-1 enzyme, thus widening the active-site (73). A similar shift of Asn-132, seen in the NMC-A β-lactamase, was observed for the SHV-1 enzyme as well (see above).

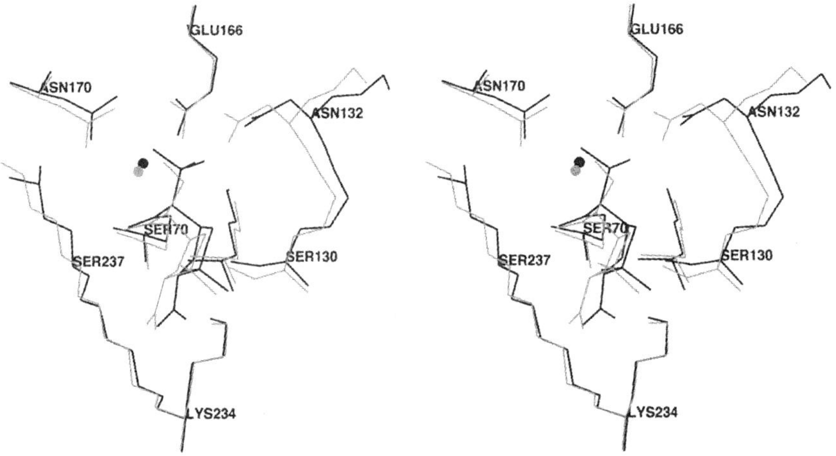

Figure 5 Stereo view of the superimposition of the TEM-1 (shown in gray) and NMC-A (shown in black) β-lactamases complexed with 6α-hydroxymethylpenicillanic acid and 6α-hydroxypropylpenicillanic acid, respectively. The active site of the NMC-A enzyme near the hydrolytic water is expanded by the relocation of the Asn-132 away from the active site, such that this enzyme could accommodate substituents such as the hydroxypropyl moiety in the inhibitor's structure.

The x-ray crystal structure of another ESBL, Toho-1, was recently solved by Ibika et al. (74). The overall structure of this enzyme appears to be similar to that of other class A enzymes. One difference, however, is that the role for Arg-244 in other class A β-lactamases is fulfilled by residue Arg-276 in the Toho-1 β-lactamase. As indicated by the position numbers, these residues are sequestered in different locations in the two enzymes, yet the guanidinium moieties of the side chains in these two enzymes occupy the same location in the active site. This is a difference that may contribute to the ESBL phenotype. Another factor affecting the ESBL phenotype of this enzyme may be the lack of hydrogen-bonding pattern in the Ω-loop that is observed in other crystal structures. Thus, the Ω-loop of Toho-1 differs from that of other ESBLs.

The three-dimensional structure of the Per-1 β-lactamase shows interesting differences when compared to other class A β-lactamases, such as the TEM-1 enzyme (Fig. 6) (J.P. Samama, unpublished results). There are significant differences in the Ω-loop, β3-strand and α2-helix regions compared to the corresponding regions in the TEM-1 β-lactamase. The collective effect of these differences is an enlargement of the active site pocket in

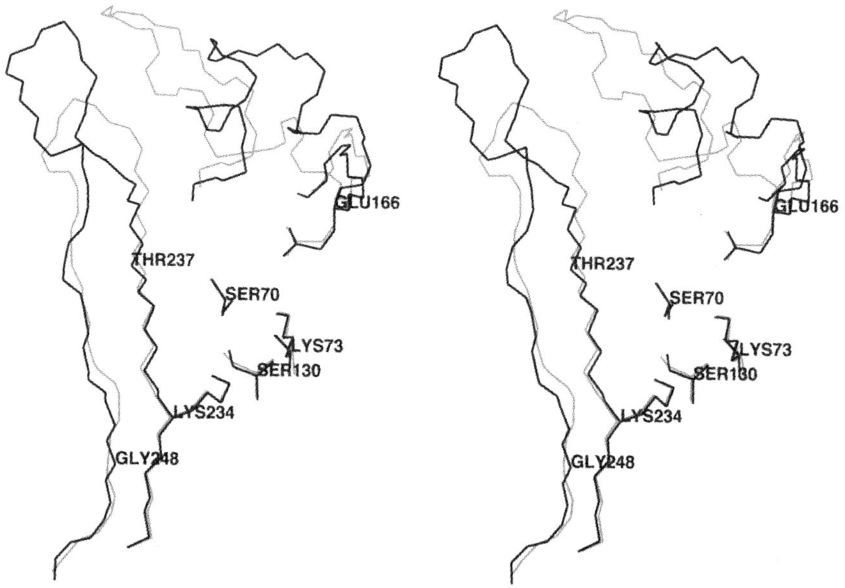

Figure 6 Superimposition of the TEM-1 (shown in gray) and per-1 (shown in black) β-lactamases depicting the differences in the Ω-loop and the β3-sheet regions (top and left regions of the figure, respectively). The region near the Ω-loop is substantially larger in the PER-1 β-lactamase, compared to the TEM-1 enzyme, a factor that is linked to the extended-spectrum activity of the former.

the Per-1 enzyme. Therefore, it would appear that a theme for evolution of the extended-spectrum β-lactamases is enlargement of the active site in order to alleviate the unfavorable steric interactions with certain substrates that have been relatively immune to the deleterious action of β-lactamases.

5 CLASS B β-LACTAMASES

Class B β-lactamases was considered a mere curiosity only a few years ago. First isolated in 1966, its only representative in the 1970s through 1980s was an enzyme from *Bacillus cereus* (75,76). This enzyme required zinc for catalysis (77) and the recent discoveries of zinc-dependent β-lactamase in more than 20 bacterial pathogens that include *Bacteroides fragilis* (78,79), *Pseudomonas aeruginosa* (80), *Xanthomonas maltophilia* (81), *Serratia marcescens* (82), *Klebsiella pneumoniae* (83), *Aeromonas hydrophila* (84), and *Chryseobacterium meningosepticum* (85) have increased interest in these enzymes. The newly discovered class B enzymes, often expressed with other classes of β-lactamases, show an unprecedented breadth of substrate preference, which includes carbapenem antibiotics. This is a cause for serious concern, because carbapenem antibiotics are generally resistant to the action of the other classes of β-lactamases and have enjoyed longevity in the clinic since their introduction in the mid 1980s. The susceptibility of carbapenems to class B enzymes may presage the potential obsolescence of these versatile antibiotics in the future. A significant discovery regarding this class of β-lactamases has been that they may be plasmidborne (86,87). This discovery was first reported for strains from Japan, and it heralds the possibility of facile dissemination of the genes for these enzymes to various organisms (57).

There appears to be substantial structural variation in this family of β-lactamases. Metallo-β-lactamases do not possess a significant homology among themselves, and neither do they show an evolutionary relationship to other classes of β-lactamases (15). However, the active site residues that coordinate to the two zinc metals are conserved among these enzymes. The issue of participation of one zinc ion or two in catalysis is presently the subject of discussion, and a single answer to this question may not be applicable to all enzymes. For example, the active site of the metallo-β-lactamase from *A. hydrophila* AE036 contains the binding sites for two zinc ions, but this enzyme is active with only one bound zinc ion (88). The second zinc-binding site has a lower affinity for the metal, and its occupancy was shown to inhibit the enzyme in a noncompetitive fashion.

Metallo-β-lactamases from *B. fragilis*, *B. cereus*, and *Stenotrophomonas maltophilia* have been crystallized, and their structures have been useful in understanding this class of enzymes and their diversities (Fig. 7) (see Table 1). The structure of the metallo-β-lactamase from *B. cereus* solved by Carfi et al. showed a single zinc ion bound to the active site (89), but subsequent structure analysis of the same enzyme confirmed the binding of a second zinc ion in a low-affinity binding site (90). One of the zinc ions coordinates with three histidine residues, whereas the other one is bound to an aspartic acid, a histidine and a cysteine side chain (Fig. 8). A water molecule coordinated to one of the zinc ions has been suggested to participate in hydrolysis of the β-lactam antibiotic. The Cys-181–Ser mutant of the metallo-β-lactamase from *B. fragilis* has impaired catalytic ability, because the second zinc ion cannot bind to the active site. In further studies, Concha et al. showed that this enzyme can function with a cadmium ion in place of the zinc ion in the active site, and the structure of the cadmium-bound β-lactamase is similar to that of the zinc-bound enzyme (91). However, replacement of cadmium ion with a mercury ion alters the active-site geometry of the enzyme, and thus destroys the catalytic activity of this enzyme. It was concluded that Cys-181 maintains coordination of the metal

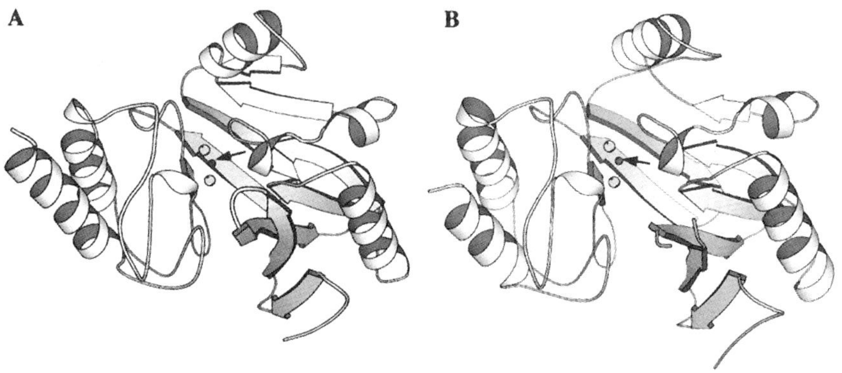

Figure 7 Structures of the zinc-dependent β-lactamases from *B. fragilis* (A) and *B. cereus* 549/H/9 (B) (pdb codes: 1A7T and 1BC2, respectively). Two zinc ions in the active sites of the enzymes are shown as gray spheres and a water molecule (identified by an arrow) is shown as a black sphere coordinated to the two zinc ions (the figures were prepared using the program MOLSCRIPT).

ion, which would be required for activation of the nucleophilic water molecule in this enzyme (92). However, in the case of the *B. cereus* metallo-β-lactamase, the enzyme can function normally with one zinc ion bound to the active site (92).

The nature of the metal dependences of class B β-lactamases appears to vary depending on their sources and structures. Perhaps more struc-

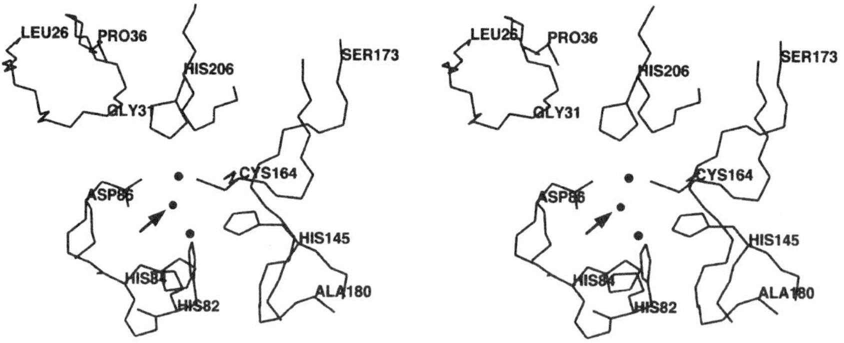

Figure 8 Stereo view of the active site of a class B β-lactamase (from *B. fragilis*; pdb code: 1A7T). Side chains of the residues that coordinate to the zinc ions are shown. A water molecule (identified by an arrow)—in between the two zinc ions—is shown as a sphere that closely interacts with the zinc ions.

tural and kinetic characterizations are needed to understand completely the role of metal ions in catalysis by these enzymes. There are no clinically useful inhibitors for this class of β-lactamases at present, although work in several pharmaceutical companies is making progress in that direction. There is serious need for such inhibitors, since most of these enzymes are capable of hydrolyzing all known β-lactam antibiotics (93). In light of the fact that the various members of this family of enzymes may operate by somewhat different mechanisms, general inhibition of class B β-lactamases by one type of inhibitor—such as achieved by clavulanic acid for class A enzymes—may prove to be difficult. Certainly, a first step toward that goal is an understanding of the mechanisms of these enzymes, which has been a subject for scrutiny recently. It would appear that catalysis does not involve covalent interactions between the enzyme and the substrate. Benkovic and coworkers have shown that the reaction of nitrocefin (a nonclinical compound) with the *B. fragilis* enzyme involves a tightly bound enzyme product complex whose dissociation is rate limiting (94,95).

6 CLASS C β-LACTAMASES

Class C β-lactamases were believed to be exclusively of chromosomal origin until recently when plasmid-borne variants were identified (96–103). This in part explains why they are only found in gram-negative bacteria, and why there are not many mutant variants of the various members of this family of enzymes. Indeed, class C β-lactamases are largely homogeneous as far as their kinetic properties are concerned (18). These enzymes are somewhat larger—approximately 39 kD—than their class A counterparts.

Class C β-lactamases have evolved an entirely distinct mechanism for their deacylation of the acyl-enzyme intermediate. They lack any residue that could correspond to Glu-166 of the class A β-lactamases. However, they possess a conserved tyrosine at position 150, which appears to be a player in the deacylation process (15,16,104). We had argued that if the approach of the hydrolytic water from the α-face were not possible for lack of a general base on that side, then it is likely that the hydrolytic water would approach the ester of the acyl-enzyme intermediate from the opposite β-direction. Such a route for the hydrolytic water would bring it into the coordination sphere of the amine of the acyl-enzyme intermediate—formerly the β-lactam nitrogen—and in contact with the side chain hydroxyl of Tyr-150 (Fig. 9) (105). As such, we have proposed that the collective environment for the hydrolytic water would activate it for the deacylation step. These results are consistent with the attribution of some role in deacylation to Tyr-150, as determined by mutagenesis experiments

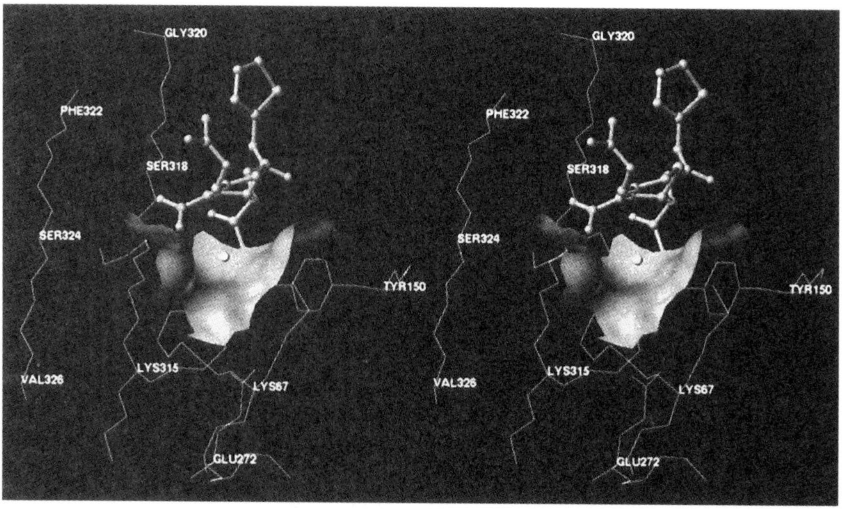

Figure 9 Stereo view of the energy-minimized model for the acyl-enzyme complex of the *E. cloacae* P99 β-lactamase (a class C enzyme; pdb code: 1BLS) with cephalothin. The hydrolytic water is seen coordinated between Tyr-150 and amine of the dihydrothiazine ring, positioned to approach the ester moiety from the β-face. A Connolly water-accessible surface (in gray) is shown for the binding site of the water molecule.

(106,107), and also by studies with β-lactam surrogates with and without the requisite ring amine (105). It is important to note that some PBPs also possess a tyrosine residue spatially equivalent to Tyr-150 in the class C β-lactamases. Therefore, Tyr-150 may have "descended" from the parental PBP, and may not be the product of selection for the deacylation step in evolution of class C β-lactamases. Then, the question becomes why such PBPs do not readily hydrolyze β-lactam antibiotics. The answer to this question is twofold. First, deacylation of the acyl-enzyme species of PBPs modified by β-lactam antibiotics does take place, although the process is slow and the rates vary depending on the substrate and the enzyme. Second, the reason for slow deacylation may be poor activation of the hydrolytic water molecule owing to the presence of the electrophilic Arg-285, as seen in the crystal structure of the *Steptomyes* R61 D-Ala-D-Ala transpeptidase (a PBP) in contrast to the occurrence of Glu-272 in the *E. cloacae* P99 β-lactamase (Fig. 10) (108). It is likely that the collective electrostatic property of the site where the hydrolytic water fits in class C β-lactamases is made favorable by the basic nature of the substrate ring amine, the side chain of Tyr-150 (if it is deprotonated) and the side chain of Glu-272. A

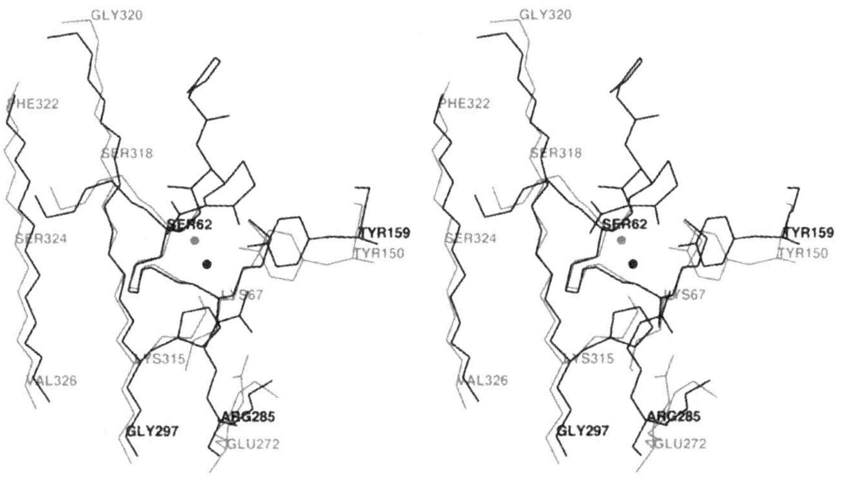

Figure 10 Stereo view of the active site of the *E. cloacae* P99 β-lactamase (in gray, pdb code: 1BLS) superimposed onto the active site of the D-Ala-D-Ala transpeptidase—a PBP from *Streptomyces* R61 (in black; pdb code: 1CEG)—modified covalently by cephalothin (also in black). The modeled hydrolytic water (gray sphere) is shown in the active site of the *E. cloacae* P99 β-lactamase as per Figure 9. The crystallographic water molecule in the structure of D-Ala-D-Ala transpeptidase (shown as a black sphere) is seen coordinated to Arg-285 and Tyr-159; this interaction does not activate the water molecule for hydrolysis of the ester moiety. Arg-285 is replaced by Glu-272 in the β-lactamase structure. There is no opportunity for direct contact between the side chain of Glu-272 and the hydrolytic water.

protonated Tyr-150 would function solely as a ligand in anchoring the hydrolytic water on the β-face of the acyl-enzyme intermediate. The structures of class C β-lactamases show that the surface of the enzymes on the β-face of the acyl-enzyme intermediate has been restructured to allow such an approach for the hydrolytic water (see Fig. 9) in contrast to the PBP situation. Tyr-150 and the ring amine, and possibly Glu-272, contribute to the promotion of the water molecule for the hydrolytic step. We are not suggesting that Glu-272 is a general base here, but that it would merely influence the electrostatic properties of the environment where water binds. Since ring amine—a structural component of the substrate—plays a role in the mechanism of these enzymes (105), class C β-lactamases belong to a group of a handful of enzymes that operate by "substrate-assisted catalysis."

We have shown recently that class C β-lactamases operate at the diffusion limit for turnover of their preferred cephalosporin substrates

(109). Diffusion-controlled catalysis was previously shown for turnover of the preferred penicillin substrates by class A β-lactamases as well (50). Therefore, it would appear that evolution of classes A and C of β-lactamases took entirely different courses, obviously due to different selection pressures; in one case a penicillin, in the other a cephalosporin. The mechanisms for the deacylation step in the two cases are distinct as well. Class A β-lactamases acquired a general base, whereas class C enzymes exploited a substrate-assisted strategy. In the case of class A β-lactamases, the approach of the promoted water is from the α-face of the acyl-enzyme species, whereas that for the class C enzymes is from the β-direction. Furthermore, an extensive sequence alignment of β-lactamases and PBPs indicated that classes A and C of β-lactamases evolved from two different groups of PBPs (15). All these observations collectively and conclusively indicate that the two classes of β-lactamases had different evolutionary experiences, but each reached its full potential by becoming "perfect" (i.e., diffusion controlled) in its catalytic competence.

Three class C β-lactamases have been crystallized to date (Table 1) (Fig. 11). These are the enzymes from *Citrobacter freundii* (110), *E. cloacae* P99 (111), and *E. coli* (112). These enzymes are very similar in their structures, and the similarity is even more pronounced within the active sites. How-

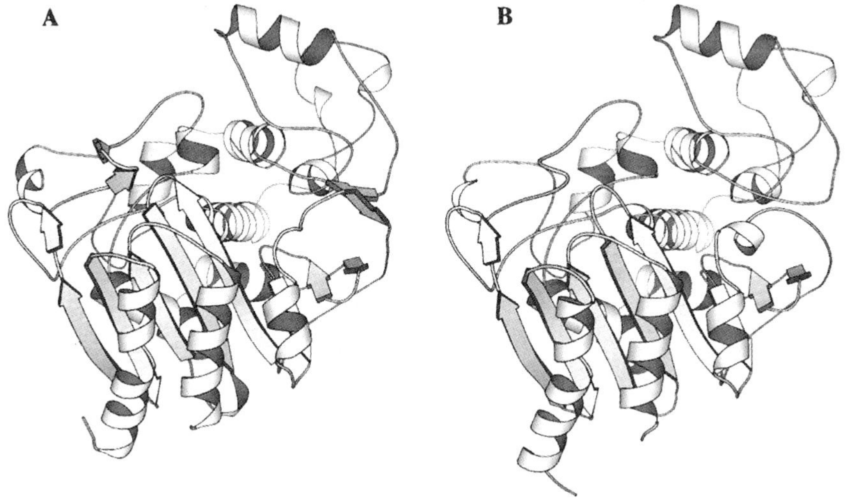

Figure 11 X-ray crystal structures of the class C β-lactamases from *E. coli* K12 (A) and *E. cloacae* P99 (B) (pdb code: 2BLS and 2BLT, respectively) (structures are drawn using the program MOLSCRIPT).

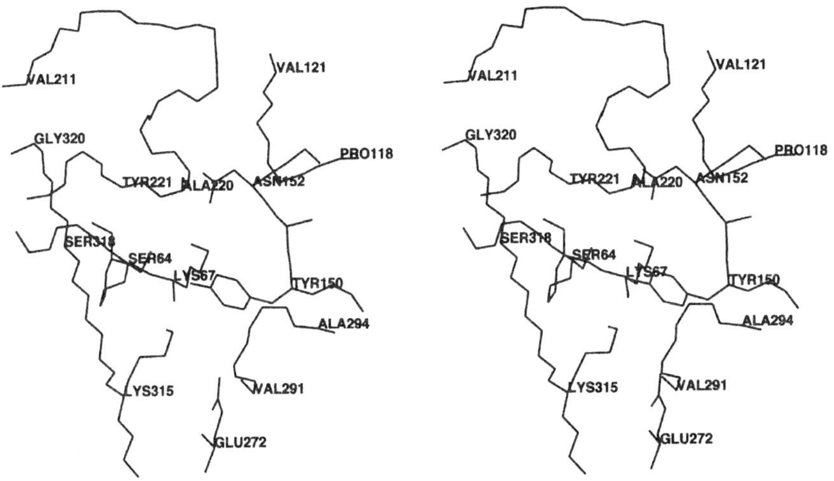

Figure 12 Stereo view of the active site of a class C β-lactamase from *E. cloacae* P99 (pdb code: 2BLT). Important residues, Ser-64, Lys-67, Tyr-150, Lys-315 and Ser-318 are labeled and their side chains are shown.

ever, the active sites of the class C β-lactamases are in general wider than those of class A enzymes, which explains why they readily bind the relatively bulky cephalosporin substrates. The active site of *E. cloacae* P99 is shown in Figure 12.

Oefner and coworkers solved the three-dimensional structure of the chromosomally encoded class C β-lactamase from *C. freundii* covalently bound to the monobactam antibiotic aztreonam and proposed a mechanism for catalysis (110). According to this mechanism, Tyr-150 was suggested to function as a general base during the acylation and deacylation steps (110). The three-dimensional structure of the AmpC β-lactamase, a class C enzyme from *E. coli*, is similar to those from *E. cloacae* and *C. freundii*. The structure of the complex of the AmpC β-lactamase bound to the inhibitor *m*-aminophenylboronic acid resembles the transition-state species for deacylation (112). Similarly, the three-dimensional structure of the complex between a phosphonate inhibitor and the class C enzyme from *E. cloacae* P99 (Fig. 13) was suggested to mimic a transition-state species for the deacylation process.

One of the important features of class C β-lactamases is that they can accommodate cephalosporins in their active site and are not activated by class A β-lactamase inhibitors such as clavulanic acid. Labkovsky et al. suggested that the machinery present in class A β-lactamases to process

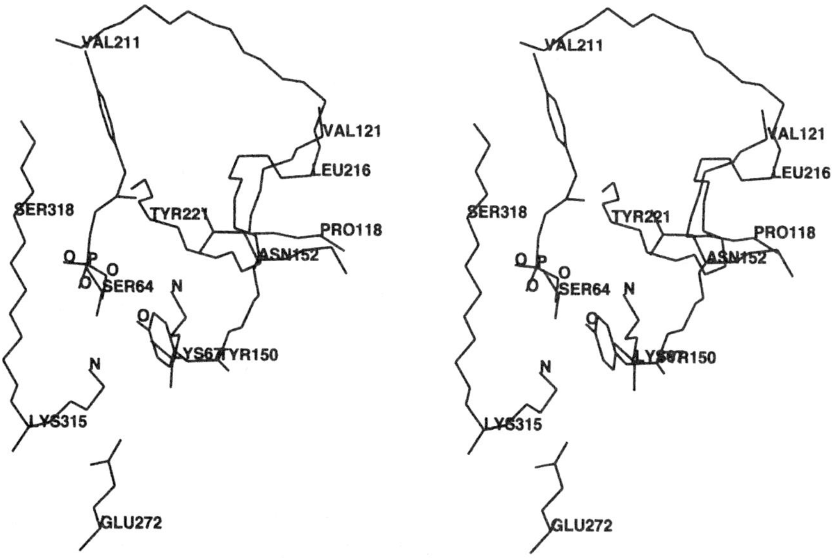

Figure 13 Stereo view of the active site of class C β-lactamase from *E. cloacae* P99 with a phosphonate inhibitor covalently bound to the active site Ser-64, as determined by x-ray crystallography (pdb code: 1BLS). Although one of the oxygens on the phosphonate moiety is in the oxyanion hole formed by the backbone NH groups of Ser-64 and Ser-318, the other oxygen is exposed to the solvent, at 3.4 Å distance from the Tyr-150 hydroxyl group, the closest residue in its vicinity. The hydroxyl group of Tyr-150 is, however, 2.9 and 3.3 Å away from the Oγ of Ser-64 and the Nε of Lys-67, respectively.

the clinical inhibitor clavulanic acid in the course of the inactivation chemistry, namely, a residue such as the Arg-244 of the class A enzymes and the water molecule coordinated to it, does not have a structural counterpart in class C enzymes (113). Perhaps this is the reason for the poor affinity of class C β-lactamases toward the inhibitors of class A enzymes such as clavulanic acid (114).

The catalytic processes of β-lactamases may entail conformational changes. Such has been shown for class A enzymes, the TEM-1 and NMC-A β-lactamases, by x-ray crystallography and molecular dynamics simulations (46,47). Similarly, the class C β-lactamase from *C. freundii* studied by infrared spectroscopy was shown to have multiple carbonyl stretches in the course of turnover of a penicillin. These observations were interpreted to be the result of different conformations for the acyl-enzyme intermediate (115).

7 CLASS D β-LACTAMASES

To date, more than 20 enzymes in class D have been identified (116). These enzymes are becoming important clinically with the discovery of new variants such as Oxa-14, Oxa-15, Oxa-16, and Oxa-18, that show the extended-spectrum property (117–120). The substrate profile for Oxa-17, for example, includes penicillins (oxacillin, amoxicillin, ticarcillin), cephalosporins (cephalothin, ceftazidime, cefotaxime), and the monobactam aztreonam (117). Ceftazidime, cefotaxime, and aztreonam are three of the most important clinical antibiotics.

These enzymes are grouped together, because they seem to prefer oxacillin or cloxacillin as their preferred penicillin substrate (hence, the "Oxa" designation). In addition, class D β-lactamases are the smallest among active-site–serine β-lactamases. For example, the entire sequence of the Oxa-1 class D β-lactamase (including the signal peptide) is 246 amino acids compared to 286 amino acids for the *E. coli* TEM-1 (class A) and 381 amino acids for the *E. cloacae* P99 (class C) β-lactamases. There does not appear to be any striking similarities in their sequences when compared to those of the classes A and C of β-lactamases (15). We have carried out an amino acid sequence alignment for 20 class D enzymes (L.P. Kotra and S. Mobashery, unpublished results) and found a Ser-Thr-Phe-Lys motif that is absolutely conserved and equivalent to the Ser-X-Phe-Lys motif in class A β-lactamases (the motif that includes the active site Ser-70 and Lys-73 residues). A *Tyr*-X-X-X-X-*Tyr*-Gly-X motif was also identified with two strictly conserved tyrosine residues (shown in italics), one of which may potentially correspond to Tyr-150 of class C β-lactamases. There is also a Lys-Thr-Gly signature, which corresponds to the motif containing Lys-234 in class A enzymes, or Lys-315 in class C enzymes. No conserved acidic residue that would correspond to Glu-166 of class A enzymes can be identified based on the sequence alignment, suggesting that the deacylation mechanism for class D enzymes may be different than those for classes A and C. It would thus appear that nature has evolved four distinct mechanisms for hydrolysis of β-lactams by four classes of β-lactamases. No structural information is available for any of the class D enzymes to date. (See "Note Added in Proof" on page 148.)

8 PERSPECTIVES

Introduction of β-lactam antibiotics for use in treatment of infections was one of the most important medical contributions of the twentieth century. Penicillin G was the first β-lactam antibiotic that received widespread clinical application, and its success paved the way for the discovery of other β-lactams and antibiotics of different classes. Despite the presence of the ubiquitous β-lactamases, the clinical importance of β-lactams remains very high. They remain the most commonly used antibiotics in the clinic, and we will have to rely on them for the foreseeable future (93).

The entire genomes of many bacteria have been sequenced, and others are being actively investigated (8,121,122, http://www.tigr.org). Knowledge of these microbial enzymes and the proteins they encode has the potential to revolutionize the pharmaceutical industry in the next few years. The chances are that novel classes of antibiotics will be developed that target enzymes that were previously unknown. But if history is any indication, resistance to any drug will develop in a short time. Indeed, the literature indicates that for the seven major classes of known antibiotics, resistance has developed within 1–4 years from the time of the clinical introduction of the drug (123). This is the vindication of Paul Ehrlich's prophetic statement that "drug resistance follows the drug like a faithful shadow."

We have described in this chapter the different classes of β-lactamases that have arisen in response to the challenge of β-lactam antibiotics. It is important to note that it would appear that the processes of random mutation and selection have resulted in four known classes of β-lactamases, all operating by distinct mechanisms. The many variants of the members of these classes of drug resistance determinants, which essentially cover the full spectrum of phenotypic traits needed to give resistance to all β-lactam antibiotics, document further the power of the evolutionary processes at work in microorganisms. We do not expect to have any obvious replacements for β-lactam antibiotics in the foreseeable future. Only time will tell if the renewed interest in the pharmaceutical industry in development of anti-infectives will meet the clinical challenges before the arrival of what has been presaged as a "postantimicrobial era" (124).

ACKNOWLEDGMENTS

The authors would like to thank Professor James Knox for providing the three-dimensional coordinates of the SHV-1 β-lactamase. This work was supported by the U.S. National Institutes of Health and by the French Ministry of Education and Research.

NOTE ADDED IN PROOF

Since the submission of this manuscript, the OXA-10 β-lactamase, a class D enzyme, has been crystallized (Golemi D, Maveyraud L, Vakulenko S, Tranier S, Ishiwata A, Kotra LP, Samama JP, Mobashery S. The first structural and mechanistic insights for class D β-lactamases: evidence for a novel catalytic process for turnover of β-lactam antibiotics. J Am Chem Soc 2000; 122:6132–5133. Maveyraud L, Golemi D, Kotra LP, Tranier S, Vakulenko S, Mobashery S, Samama JP. Insights into class D β-lactamases are revealed by the crystal structure of the Oxa10 enzyme from *Pseudomonas aeruginosa*. Structure 2000; 8:1289–1298. Paetzel M, Danel F, de Castro L, Mosimann SC, Oage MGP, Strynadka NCJ. Crystal structure of the class D β-lactamase OXA-10. Nat Struct Biol 2000; 7:918–925). This enzyme requires a carbamylated lysine in its active site for its activity. This carbamylated lysine promotes both the acylation and the deacylation steps of the enzyme, hence class D β-lactamases appear to be the only enzymes of this family that utilize symmetry in their catalytic mechanism. Therefore, all four classes of β-lactamases have evolved distinct catalytic mechanisms for their turnover chemistries. This observation validates the proposal by Massova and Mobashery (1998) that different classes of PBPs gave rise to disparate classes of β-lactamases at various evolutionary time points. Furthermore, a publication from the Shoichet group provided structural evidence in favor of the substrate-assisted catalysis by class C β-lactamases (Patera A, Blaszczak LC, Shoichet BK. Crystal structures of substate and inhibitor complexes with AmpC β-lactamase: possible implications for substrate-assisted catalysis. J Am Chem Soc 2000; 122:10504–10512).

REFERENCES

1. Antimicrobial Drug Resistance: Fact Sheet, National Institute of Allergy and Infectious Diseases, National Institutes of Health, Bethesda, MD, March, 1998.
2. The World Health Report 1999. World Health Organization, Geneva, Switzerland, 1999:22.
3. Bush K, Mobashery S. How β-lactamases have driven pharmaceutical drug discovery: From mechanistic knowledge to clinical circumvention. In: Rosen BP, Mobashery S, eds. Resolving the Antibiotic Paradox: Progress in Understanding Drug Resistance and Development of New Antibiotics. New York: Plenum Press, 1998:71–98.
4. Welch H, Lewis CN. Antibiotic Therapy. Washington, DC: Arunde Press, 1951.
5. Mabilat C, Lourencao-Vital J, Goussard S, Courvalin P. A new example of physical linkage between Tn1 and Tn21: The antibiotic multiple-resistance region of plasmid pCFF04 encoding extended-spectrum beta-lactamase TEM-3. Mol Gen Genet 1992; 235:113–121.
6. Stokes HW, Hall RM. The integron In1 in plasmid R46 includes two copies of the oxa2 gene cassette. Plasmid 1992; 28:225–234.

7. Davies J. Inactivation of antibiotics and the dissemination of resistance genes. Science 1994; 264:375–382.
8. Moir DT, Shaw KJ, Hare RS, Vovis GF. Genomics and antimicrobial drug discovery. Antimicrob Agents Chemother 1999; 43:439–446.
9. Knowles DJC, King F. The impact of bacterial genomics on antibacterial drug discovery. In: Rosen BP, Mobashery S, eds. Resolving the Antibiotic Paradox: Progress in Understanding Drug Resistance and Development of New Antibiotics. New York: Plenum Press, 1998:183–195.
10. Bush K, Jacoby GA, Medeiros AA. A functional classification scheme for β-lactamases and its correlation with molecular structure. Antimicrob Agents Chemother 1995; 39:1211–1233.
11. Ambler RP, Coulson AFW, Frère J-M, Ghuysen JM, Joris B, Forsman M, Levesque RC, Tiraby G, Waley SG. A standard numbering scheme for the class A β-lactamases. Biochem J 1991; 276:269–272.
12. Ambler RP, Structure of β-lactamases. Philos Trans Royal Soc London B Biol Sci 1980; 289:321–331.
13. Jaurin B, Grundström T. ampC Cephalosporinase of *Escherichia coli* K-12 has a different evolutionary origin from that of β-lactamases of the penicillinase type. Proc Natl Acad Sci USA 1981; 78:4897–4901.
14. Oullette M, Bissonnette L, Roy PH. Precise insertion of antibiotic resistance determinants into Tn21-like transposons: Nucleotide sequence of the OXA-1 β-lactamase gene. Proc Natl Acad Sci USA 1987; 74:7378–7382.
15. Massova I, Mobashery S. Kinship and diversification of bacterial penicillin-binding proteins and β-lactamases. Antimicrob Agents Chemother 1998; 42: 1–17.
16. Kotra LP, Mobashery S. β-Lactam antibiotics, β-lactamases and bacterial resistance. Bull Inst Pasteur 1998; 96:139– 150.
17. Medeiros AA. Evolution and dissemination of β-lactamases accelerated by generation of β-lactam antibiotics. Clin Infect Dis 1997; 24:S19–S45.
18. Matagne A, Dubus A, Galleni M, Frère J-M. The β-lactamase cycle: A tale of selective pressure and bacterial ingenuity. Nat Prod Rep 1999; 16:1–19.
19. Wright AJ. The penicillins. Mayo Clin Proc 1999; 74:290–307.
20. Maiti SN, Phillips OA, Micetich RG, Livermore DM. β-Lactamase inhibitors: agents to overcome bacterial resistance. Curr Med Chem 1998; 5:441–456.
21. Goffin C, Ghuysen JM. Multimodular penicillin-binding proteins: an enigmatic family of orthologs and paralogs. Microbiol Mol Biol Rev 1998; 62: 1079–1093.
22. Petrosino J, Cantu 3rd C, Palzkill T. β-Lactamases: protein evolution in real time. Trends Microbiol 1998; 6:323–327.
23. Philippon A, Dusart J, Joris B, Frère J-M. The diversity, structure and regulation of β-lactamases. Cell Mol Life Sci 1998; 54:341–346.
24. Nieto M, Perkins HR, Frère J-M, Ghuysen J-M. Fluorescence and circular dichroism studies on the Streptomyces R61 DD-carboxypeptidase-transpeptidase. Biochem J 1973; 135:493–505.
25. Frère J-M, Joris B. Penicillin-sensitive enzymes in peptidoglycan biosynthesis. CRC Crit Rev Microbiol 1985; 11:299–396.

26. Bush K, Smith SA, Ohringer S, Tanaka SK, Bonner DP. Improved sensitivity in assays for binding of novel beta-lactam antibiotics to penicillin-binding proteins of *Escherichia coli*. Antimicrob Agents Chemother 1987; 31:1271–1273.
27. Holland HD. Geo-chemistry evidence for life on earth more than 3850-million years ago. Science 1997; 275:38–39.
28. Martin JF, Gutierrez S. Genes for beta-lactam antibiotic biosynthesis. Antonie Van Leeuwenhoek 1995; 67:181–200.
29. Coque JJ, Liras P, Martin JF. Genes for a beta-lactamase, a penicillin-binding protein and a transmembrane protein are clustered with the cephamycin biosynthetic genes in *Nocardia lactamdurans*. EMBO J 1993; 12:631–639.
30. Strynadka NCJ, Adachi H, Jensen SE, Johns K, Sielecki A, Betzel C, Sutoh K, James MNG. Molecular structure of the acyl-enzyme intermediate in β-lactam hydrolysis at 1.7 Å resolution. Nature 1992; 359:700–705.
31. Swarèn P, Maveyraud L, Guillet V, Masson J-M, Mourey L, Samama JP. Electrostatic analysis of TEM1 beta-lactamase: effect of substrate binding, steep potential gradients and consequences of site-directed mutations. Structure 1995; 3:603–613.
32. Ishiguro M, Imajo S. Modeling study on a hydrolytic mechanism of class A β-lactamases. J Med Chem 1996; 24:2207–2218.
33. Knox JR, Moews PC, Escobar WA, Fink AL. A catalytically-impaired class A β-lactamase: 2 angstroms crystal structure and kinetics of the *Bacillus licheniformis* E166A mutant. Protein Eng 1993; 6:11–18.
34. Madzwick PJ, Waley SG. β-Lactamase I from Bacillus cereus. Structure and site-directed mutagenesis. Biochem J 1987; 248:657–662.
35. Damblon C, Raquet X, Lian LY, Lamotte-Brasseur J, Fonzé E, Charlier P, Roberts GC, Frère J-M. The catalytic mechanism of beta-lactamases: NMR titration of an active-site lysine residue of the TEM-1 enzyme. Proc Natl Acad Sci USA 1996; 93:1747–1752.
36. Lamotte-Brasseur J, Dive G, Dideberg O, Charlier P, Frère J-M, Ghysen JM. Mechanism of acyl transfer by the class A serine beta-lactamase of Streptomyces albus G. Biochem J 1991; 279:213–221.
37. Lamotte-Brasseur J, Jacob-Dubuisson F, Dive, G., Frère J-M, Ghysen JM. Streptomyces albus G serine beta-lactamase. Probing of the catalytic mechanism via molecular modeling of mutant enzymes. Biochem J 1992; 282: 189–195.
38. Paetzel M, Dalbey RE. Catalytic hydroxyl/amine dyads within serine proteases. TIBS 1997; 22:28–31.
39. Chen CCH, Rahil J, Pratt RF, Herzberg O. Structure of a phosphonate-inhibited β-lactamase. An analog of the tetrahedral transition state-intermediate of β-lactam hydrolysis. J Mol Biiol 1993; 234:165–178.
40. Maveyraud L, Pratt RF, Samama JP. Crystal structure of an acylation transition-state analog of the TEM-1 β-lactamase. Mechanistic implications for class A β-lactamases. Biochemistry 1998; 37:2622–2628.
41. Gibson RM, Christensen H, Waley SG. Site-directed mutagenesis of β-lactam I. Single and double mutants of Glu-166 and Lys-73. Biochem J 1990; 272: 613–619.

42. Adachi H, Ohta T, Matsuzawa H. Site-directed mutants, at position 166, of RTEM-1 β-lactamase that form a stable acyl-enzyme intermediate with penicillin. J Biol Chem 1991; 266:3186–3191.
43. Escobar WA, Tan AK, Fink AL. Site-directed mutagenesis of β-lactamase leading to accumulation of a catalytic intermediate. Biochemistry 1991; 30: 10783–10787.
44. Guillauma G, Vanhove M, Lamotte-Brasseur J, Ledent P, Jamin M, Joris B, Frère J-M. Site-directed mutagenesis of glutamate 166 in two beta-lactamases—kinetic and molecular modeling studies. J Biol Chem 1997; 272:5348–5444.
45. Maveyraud L, Massova I, Birck C, Miyashita K, Samama JP, Mobashery S. Crystal structure of 6α-hydroxymethylpenicillanate complexed to the TEM-1 β-lactamase from *Escherichia coli*: evidence on the mechanism of action of a novel inhibitor designed by a computer-aided process. J Am Chem Soc 1996; 118:7435–7436.
46. Mourey L, Kotra LP, Bellettini J, Bulychev A, O'Brien M, Miller MJ, Mobashery S, Samama JP. Inhibition of the broad-spectrum NMC-A β-lactamase from Enterobacter cloacae by monocyclic β-lactams. J Biol Chem 1999; In press.
47. Maveyraud L, Mourey L, Kotra LP, Pédelacq JD, Guillet V, Mobashery S, Samama JP. Structural basis for clinical longevity of carbapenem antibiotics in the face of challenge by the common class A β-lactamases from the antibiotic-resistant bacteria. J Am Chem Soc 1998; 120:9748–9752.
48. Massova I, Mobashery S. Molecular bases for interactions between β-lactam antibiotics and β-lactamases. Acc Chem Res 1997; 30:162–168.
49. Miyashita K, Massova I, Taibi P, Mobashery S. Design, synthesis, and evaluation of a potent mechanism-based inhibitor for the TEM beta-lactamase with implications for the enzyme mechanism. J Am Chem Soc 1995; 117:11055–11059.
50. Hardy LW, Kirsch JF. Diffusion-limited component of reactions catalysed by *Bacillus cereus* β-lactamase I. Biochemistry 1984; 23:1275–1282.
51. Christensen H, Martin MT, Waley SG. β-Lactamases as fully efficient enzymes. Biochem J 1990; 266:853–861.
52. Datta N, Kontomichalou P. Penicillinase synthesis controlled by infectious R factors in *Enterobacteriaceae*. Nature 1965; 208:239–241.
53. Lin S, Thomas M, Shlaes DM, Rudin SD, Knox JR, Anderson V, Bonomo RA. Kinetic analysis of an inhibitor-resistant variant of the OHIO-1 beta-lactamase, an SHV-family class A enzyme. Biochem J 1998; 333:395–400.
54. Nordmann P, Mariotte S, Naas T, Labia R, Nicolas MH. Biochemical properties of a carbapenem-hydrolyzing beta-lactamase from *Enterobacter cloacae* and cloning of the gene into *Escherichia coli*. Antimicrob Agents Chemother 1993; 37:939–946.
55. Naas T, Vandel L, Sougakoff W, Livermore DM, Nordmann P. Cloning and sequence analysis of the gene for a carbapenem-hydrolyzing class A beta-lactamase, Sme-1, from *Serratia marcescens* S6. Antimicrob Agents Chemother 1994; 38:1262–1270.
56. Rasmussen BA, Bush K, Keeney D, Yang Y, Hare R, O'Gara C, Medeiros AA. Characterization of IMI-1 beta-lactamase, a class A carbapenem-hydrolyzing

enzyme from *Enterobacter cloacae*. Antimicrob Agents Chemother 1996; 40: 2080–2086.
57. Rasmussen BA, Bush K. Carbapenem-hydrolyzing β-lactamases. Antimicrob Agents Chemother 1997; 41:223–232.
58. Vedel G, Belaaouaj A, Gilly L, Labia R, Philippon A, Névot P, Paul G. Clinical isolates of *Escherichia coli* producing TRI β-lactamases: novel TEM-enzymes conferring resistance to β-lactamase inhibitors. J Antimicrob Chemother 1992; 30:499–462.
59. Thomson CJ, Amyes SGB. TRC-1: emergence of a clavulanic acid-resistant TEM β-lactamase in a clinical strain. FEMS Microbiol Lett 1992; 91:113–118.
60. Canica MMM, Lu CY, Krishnamoorthy R, Paul GC. Molecular diversity and evolution of bla_{TEM} genes encoding β-lactamases resistant to clavulanic acid in clinical *E. coli*. J Mol Evol 1997; 44:57–65.
61. Vakulenko SB, Geryk B, Kotra LP, Mobashery S, Lerner SA. Selection and characterization of beta-lactam–beta-lactamase inactivator-resistant mutants following PCR mutagenesis of the TEM-1 beta-lactamase gene. Antimicrob Agents Chemother 1998; 42:1542–1548.
62. Chaibi EB, Pedizzi J, Farzaneh S, Barthelemy M, Sirot D, Labia R. Clinical inhibitor-resistant mutants of the β-lactamase TEM-1 at amino-acid position 69—kinetic analysis and molecular modeling. Biochim Biophys Acta 1998; 1382:38–46.
63. Bouthors AT, Jarlier V, Sougakoff W. Amino acid substitutions at positions 69, 165, 244 and 275 of the PER-1 β-lactamase do not impair enzyme inactivation by clavulanate. J Antimicrob Chemother 1998; 42:399–401.
64. Imtiaz U, Bilings E, Knox JR, Manavathu EK, Lerner SA, Mobashery S. Inactivation of class A β-lactamases by clavulanic acid: the role of arginine-244 in a proposed nonconcerted sequence of events. J Am Chem Soc 1993; 115:4435–4442.
65. Brown RP, Aplin RT, Schofield CJ. Inhibition of TEM-2 beta-lactamase from *Escherichia coli* by clavulanic acid: observation of intermediates by electrospray ionization mass spectrometry. Biochemistry 1996; 35:12421–12432.
66. Brown RPA, Aplin RT, Schofield CJ, Frydrych CH. Mass spectrometric studies on the inhibition of TEM-2 beta-lactamase by clavulanic acid derivatives. J Antibiot 1997; 50:184–185.
67. Zafaralla G, Manavathu S, Lerner SA, Mobashery S. Elucidation of the role of arginine-244 in the turnover processes of class A beta-lactamases. Biochemistry 1992; 31:3847–3852.
68. Swarén P, Golemi D, Cabantous S, Bulychev A, Maveyraud L, Mobashery S, Samama JP, X-Ray structure of the Asn276Asp variant of the *Escherichia coli* TEM-1 β-lactamase: direct observation of electrostatic modulation in resistance to inactivation by clavulanic acid. Biochemistry 1999; In press.
69. Knox JR. Extended-spectrum and inhibitor-resistant TEM-type β-lactamases: mutations, specificity, and three-dimensional structure. Antimicrob Agents Chemother 1995; 39:2593–2601.
70. Taibi P, Mobashery S. Mechanism of turnover of imipenem by the TEM β-lactamase revisited. J Am Chem Soc 1995; 117:7600–7605.

71. Mourey L, Miyashita K, Swaren P, Bulychev A, Samama JP, Mobashery S. Inhibition of the NMC-A β-lactamase by a penicillanic acid derivative and the structural bases for the increase in substrate profile of this antibiotic resistance enzyme. J Am Chem Soc 1998; 120:9382–9383.
72. Swaren P, Maveyraud L, Raquet X, Cabantous S, Duez C, Pedelacq JD, Mariotte-Boyer S, Mourey L, Labia R, Nicolas-Chanoine MH, Nordmann P, Frère J-M, Samama JP. X-ray analysis of the NMC-A β-lactamase at 1.64-angstrom resolution, a class A carbapenemase with broad substrate specificity. J Biol Chem 1998; 273:26714–26721.
73. Kuzin AP, Nukaga M, Nukaga Y, Hujer A, Bonomo RA, Knox JR. Structure of SHV-1 β-lactamase. Biochemistry 1999; 38:5720–5727.
74. Ibuka A, Taguchi A, Ishiguro M, Fushinobu S, Ishii Y, Kamitori S, Okuyama K, Yamaguchi K, Konno M, Matsuzawa H. Crystal structure of the E166A mutant of extended-spectrum β-lactamase Toho-1 at 1.8 Å resolution. J Mol Biol 1999; 285:2079–2087.
75. Sabath LD, Abraham EP. Zinc as a cofactor for cephalosporinase from *Bacillus cereus* 569. Biochem J 1966; 98:11C–13C.
76. Kuwabara S, Abraham EP. Some properties of two extracellular β-lactamases from *Bacillus cereus* 569/H. Biochem J 1967; 103:27C–30C.
77. Davies RB, Abraham EP. Metal cofactor requirements of β-lactamase II. Biochem J 1974; 143:129–135.
78. Rasmussen BA, Gluzman Y, Tally FP. Cloning and sequencing of the class B beta-lactamase gene (ccrA) from *Bacteroides fragilis* TAL3636. Antimicrob Agents Chemother 1990; 34:1590–1592.
79. Thompson JS, Malamy MH. Sequencing the gene for an imipenem-cefoxitin–hydrolyzing enzyme (CfiA) from *Bacteroides fragilis* TAL2480 reveals strong similarity between CfiA and *Bacillus cereus* beta-lactamase II. J Bacteriol 1990; 172:2584–2593.
80. Watanabe M, Yiobe S, Inoue M, Mitsuhashi S. Transferable imipenem resistance in *Pseudomonas aeruginosa*. Antimicrob Agents Chemother 1991; 35: 147–151.
81. Saino Y, Kobayashi F, Inoue M, Mitsuhashi S. Purification and properties of inducible penicillin beta-lactamase isolated from *Pseudomonas maltophilia*. Antimicrob Agents Chemother 1982; 22:564–570.
82. Osano E, Arakawa Y, Wacharotayankun R, Ohta M, Horii T, Ito H, Yoshima F, Kato N. Molecular characterization of an enterobacterial metallo beta-lactamase found in a clinical isolate of *Serratia marcescens* that shows imipenem resistance. Antimicrob Agents Chemother 1994; 38:71–78.
83. Yamaguchi H, Nukaga M, Sawait T. EMBL Data Library 1994, D29636.
84. Massida O, Rossolini GM, Satta G. The *Aeromonas hydrophila* cphA gene: molecular heterogeneity among class B metallo-beta-lactamases. J Bacteriol 1991 173:4611–4617.
85. Rossolini GM, Franceschini N, Riccio LM, Mercuri PS, Perilli MG, Galleni M, Frère J-M, Amicosante G. Characterization and sequence of the *Chryseobacterium (Flavobacterium) meningosepticum* carbapenemase: a new molecular

class B beta-lactamase showing a broad substrate profile. Biochem J 1998; 331: 145–152.
86. Watanabe M, Iyobe S, Inoue M, Mitsuhashi S. Transferable imipenem resistance in *Pseudomonas aeruginosa*. Antimicrob Agents Chemother 1991; 35: 147–151.
87. Bandoh K, Watanabe K, Muto Y, Tanaka Y, Kato N, Ueno K. Conjugal transfer of imipenem resistance in *Bacteroides fragilis*. J Antibiot (Tokyo) 1992; 45: 542–547.
88. Valladares MH, Felici A, Weber G, Adolph HW, Zeppezauer M, Rossolinin GM, Amicosante G, Frère J-M, Galleni M. Zinc(II) dependence of the *Aeromonas hydrophila* AE036 metallo-β-lactamase activity and stability. Biochemistry 1997; 36:11534–11541.
89. Carfi A, Pares S, Duee E, Galleni M, Duez C, Frère J-M, Dideberg O. The 3-D structure of a zinc metallo-β-lactamase from *Bacillus cereus* reveals a new type of protein fold. EMBO J 1995; 14:4919–4921.
90. Fabiane SM, Sohi MK, Wan T, Dayne DJ, Bateson JH, Mitchell T, Sutton BJ. Crystal structure of the zinc-dependent β-lactamase from *Bacillus cereus* at 1.9 angstrom resolution: bi-nuclear active site with features of a mono-nuclear enzyme. Biochemistry 1998; 37:12404–12411.
91. Concha NO, Rasmussen BA, Bush K, Herzberg O. Crystal structure of the cadmium- and mercury-substituted metallo-β-lactamase from *Bacteroides fragilis*. Protein Sci 1997; 6:2671–2676.
92. Li Z, Rasmussen BA, Herzberg O. Structural consequences of the active site substitution Cys181→Ser in metallo-β-lactamase from *Bacteroides fragilis*. Protein Sci 1999; 8:249–252.
93. Modugno ED, Felici A. The renewed challenge of β-lactams to overcome bacterial resistance. Curr Opin Anti-Infect Invest Drugs 1999; 1:26–39.
94. Wang ZG, Fast W, Benkovic SJ. Direct observation of an enzyme-bound intermediate in the catalytic cycle of the metallo-beta-lactamase from *Bacteroides fragilis*. J Am Chem Soc 1998; 120:10788–10789.
95. Wang ZG, Benkovic SJ. Purification, characterization, and kinetic studies of a soluble *Bacteroides fragilis* metallo-β-lactamase that provides multiple antibiotic resistance. J Biol Chem 1998; 273:22402–22408.
96. Papanicolaou GA, Medeiros Aa, Jacoby GA. Novel plasmid-mediated beta-lactamase (MIR-1) conferring resistance to oxyimino- and alpha-methoxy beta-lactams in clinical isolates of *Klebsiella pneumoniae*. Antimicrob Agents Chemother 1990; 34:2200–2209.
97. Horii T, Arakawa M, Ohta M, Sugiyama T, Wacharotayankun R, Ito H, Kato N. Characterization of a plasmid-borne and constitutively expressed blaMOX-1 gene encoding AmpC-type beta-lactamase. Gene 1994; 139:93–98.
98. Fosberry AP, Payne DJ, Lawlor EJ, Hodgson JE. Cloning and sequence analysis of blaBIL-1, a plasmid-mediated class C beta-lactamase gene in *Escherichia coli* BS. Antimicrob Agents Chemother 1994; 38:1182–1185.
99. Gonzalez Leiza M, Perez-Diaz JC, Ayala J, Casellas JM, Martinez-Beltran J, Bush K, Baquero F. Gene sequence and biochemical characterization of

FOX-1 from *Klebsiella pneumoniae*, a new AmpC-type plasmid-mediated beta-lactamase with two molecular variants. Antimicrob Agents Chemother 1994; 38:2150–2157.
100. Tzouvelekis LS, Tzelepi E, Mentis AF. Nucleotide sequence of a plasmid-mediated cephalosporinase gene (blaLAT-1) found in *Klebsiella pneumoniae*. Antimicrob Agents Chemother 1994; 38:2207–2209.
101. Baurenfeind A, Stemplinger I, Jungwirth R, Giamarellou H. Characterization of the plasmodic beta-lactamase CMY-2, which is responsible for cephamycin resistance. Antimicrob Agents Chemother 1996; 40:221–224.
102. Gaillot O, Clement C, Simonet M, Philippon A. Novel transferable beta-lactam resistance with cephalosporinase characteristics in *Salmonella enteritidis*. J Antimicrob Chemother 1997; 39:85–87.
103. Bradford PA, Urban C, Mariano N, Projan SJ, Rahal JJ, Bush K. Imipenem resistance in *Klebsiella pneumoniae* is associated with the combination of ACT-1, a plasmid-mediated AmpC beta-lactamase, and the loss of an outer membrane protein. Antimicrob Agents Chemother 1997; 41:563–569.
104. Bush K. The evolution of β-lactamases. In: Antibiotic Resistance: Origin, Evolution, Selection and Spread. Chichester, UK: Wiley, 1997:152–166.
105. Bulychev A, Massova I, Miyashita K, Mobashery S. Nuances of mechanisms and their implications for evolution of the versatile beta-lactamase activity: from biosynthetic enzymes to drug resistance factors. J Am Chem Soc 1997; 119:7619–7625.
106. DuBus A, Normark S, Kania M, Page MG. The role of tyrosine 150 in catalysis of beta-lactam hydrolysis by AmpC beta-lactamase from *Escherichia coli* investigated by site-directed mutagenesis. Biochemistry 1994; 33:8577–8586.
107. Dubus A, Ledent P, Lamotte-Brasseur J, Frère J-M. The roles of residues Tyr150, Glu272, and His314 in class C beta-lactamase. Proteins Struct Funct Genet 1996; 25:473–485.
108. Kelly JA, Knox JR, Zhao H, Frère JM, Ghuysen JM. Crystallographic mapping of beta-lactams bound to a D-alanyl-Dalanine peptidase target enzyme. J Mol Biol 1989; 209:281–295.
109. Bulychev A, Mobashery S. Class C β-lactamases operate at the diffusion limit for turnover of their preferred cephalosporin substrates. Antimicrob Agents Chemother 1999; 43:1743–1746.
110. Oefner C, D'Arcy A, Daly JJ, Gubernator K, Charnas RL, Heinze I, Hubschwerlen C, Winkler FK. Refined crystal structure of β-lactamase from *Citrobacter freundii* indicates a mechanism for β-lactam hydrolysis. Nature 1990; 343:284–288.
111. Lobkovsky E, Billings EM, Moews PC, Rahil J, Pratt RF, Knox JR. Crystallographic structure of a phosphonate derivative of the *Enterobacter cloacae* P99 cephalosporinase: mechanistic interpretation of a β-lactamase transition state analog. Biochemistry 1994; 33:6762–6772.
112. Usher KC, Blaszczak LC, Weston GS, Shoichet BK, Remington SJ. The three dimensional structure of AmpC β-lactamase from *Escherichia coli* bound to a transition state analog: implications for the oxyanion hypothesis and for inhibitor design. Biochemistry 1998; 37:16082–16092.

113. Labkovsky E, Moews PC, Liu H, Zhao H, Frère J-M, Knox JR. Evolution of an enzyme activity: crystallographic structure at 2-Å resolution of cephalosporinase from ampC gene of *Enterobacter cloacae* P99 and comparison with a class A penicillinase. Proc Natl Acad Sci USA 1993; 90:11257–11261.
114. Monnaie D, Frère J-M. Interaction of clavulanate with class C beta-lactamases. FEBS Lett 1993; 334:269–271.
115. Wilkinson A-S, Ward S, Kania M, Page MGP, Wharton CW. Multiple conformations of the acylenzyme formed in the hydrolysis of methicillin by *Citrobacter freundii* β-lactamase: a time-resolved FTIR spectroscopic study. Biochemistry 1999; 38:3851–3856.
116. Naas T, Sougakoff W, Casetta A, Nordmann P. Molecular characterization of OXA-20, a novel class D β-lactamase and its integron from *Pseudomonas aeruginosa*. Antimicrob Agents Chemother 1998; 42:2074–2083.
117. Philippon LN, Naas T, Bouthors ATA, Barakett V, Nordmann P. OXA-18, a class D clavulanic acid–inhibited extended-spectrum beta-lactamase from *Pseudomonas aeruginosa*. Antimicrob Agents Chemother 1997; 41:2188–2195.
118. Danel F, Hall LM, Gur D, Livermore DM. OXA-16, a further extended-spectrum variant of OXA-10 β-lactamase, from two *Pseudomonas aeruginosa* isolates. Antimicrob Agents Chemother 1998; 42:3117–3122.
119. Danel F, Hall LM, Gur D, Livermore DM. OXA-14, another extended-spectrum variant of OXA-10 (PSE-2) β-lactamase from *Pseudomonas aeruginosa*. Antimicrob Agents Chemother 1995; 39:1881–1884.
120. Danel F, Hall LM, Gur D, Livermore, DM. OXA-15, an extended-spectrum variant of OXA-2 beta-lactamase, isolated from a *Pseudomonas aeruginosa* strain. Antimicrob Agents Chemother 1997; 41:785–790.
121. Allsop AL. New antibiotic discovery, novel screens, novel targets and impact of microbial genomics. Curr Opin Microbiol 1998; 1:530–534.
122. Strauss EJ, Falkow S. Microbial pathogenesis: genomics and beyond. Science 1997; 276:707–712.
123. Wong KK, Pmpliano DL. Peptidoglycan biosynthesis: unexploited antibacterial targets within a familiar pathway. In: Rosen BP, Mobashery S, eds. Resolving the Antibiotic Paradox: Progress in Understanding Drug Resistance and Development of New Antibiotics. New York: Plenum Press, 1998:197–217.
124. Cohn MA. Epidemiology of drug resistance: implications for a postantimicrobial era. Science 1992; 257:1050–1054.
125. Pieper U, Hayakawa K, Li Z, Herzberg O. Circularly permeated β-lactamase from *Staphylococcus aureus* PC1. Biochemistry 1997; 36:8767–8774.
126. Chen CCH, Herzberg O. Inhibition of β-lactamase by clavulanate: trapped intermediates in cryocrystallographic studies. J Mol Biol 1992; 224:1103–1113.
127. Herzberg O, Kapadia G, Blanco B, Smith TS, Coulson A. Structural basis for the inactivation of the P54 mutant of β-lactamase from *Staphylococcus aureus* PC1. Biochemistry 1991; 30:9503–9509.
128. Chen CC, Smith TJ, Kapadia G, Wasch S, Zawadzke LE, Coulson A, Herzberg O. Structure and kinetics of the β-lactamase mutants S70A and K73H from *Staphylococcus aureus* PC1. Biochemistry 1996; 35:12251–12258.

129. Zawadzke LE, Chen CC, Banerjee S, Li Z, Wasch S, Kapadia G, Moult J, Herzberg O. Elimination of the hydrolytic water molecule in a class A β-lactamase mutant: crystal structure and kinetics. Biochemistry 1996; 35: 16475–16482.
130. Banerjee S, Pieper U, Kapadia G, Pannell LK, Herzberg O. Role of the omega-loop in the activity, substrate specificity, and structure of class A β-lactamase. Biochemistry 1998; 37:3286–3296.
131. Zawadzke LE, Smith TJ, Herzberg O. An engineered *Staphylococcus aureus* PC1 β-lactamase that hydrolyses third generation cephalosporins. Protein Eng 1995; 8:1275–1285.
132. Herzberg O. Refined crystal structure of β-lactamase from *Staphylococcus aureus* PC1 at 2.0 Å resolution. J Mol Biol 1991; 217:701–719.
133. Dideberg O, Charlier P, Wery JP, Dehottay P, Dusart J, Erpicum T, Frère J-M, Ghuysen JM. The crystal structure of the beta-lactamase of *Streptomyces albus* G at 0.3 nm resolution. Biochem J 1987; 245:911–913.
134. Moews PC, Knox JR, Dideberg O, Charlier P, Frère J-M. β-Lactamase of *Bacillus licheniformis* 749/C at 2 angstroms resolution. Proteins Struct Funct Genet 1990; 7:156–171.
135. Knox JR, Moews PC. β-Lactamase of *Bacillus licheniformis* 749/C. Refinement at 2 angstroms resolution and analysis of hydration. J Mol Biol 1991; 220: 435–455.
136. Sauvage E, Fonze E, Vermeire M, Galleni M, Quinting B, Frère J-M, Charlier P. The crystal structure of the class A β-lactamase *Mycobacterium fortuitum*: structural basis for a broad substrate specificity. Brookhaven Protein Databank ID: IMFO.
137. Jelsch C, Mourey L, Masson JM, Samama JP. Crystal structure of *Escherichia coli* TEM-1 β-lactamase at 1.8 angstroms resolution. Proteins Struct Funct Genet 1993; 16:364–383.
138. Fonze E, Charlier P, Toth Y, Vermeire M, Raquet X, Dubus A. TEM-1 β-lactamase structure solved by molecular replacement and refined structure of the S235A mutant. Acta Crystallogr Sect D 1995; 51:682–694.
139. Samama JP, Mobashery S. Unpublished results.
140. Strynadka NCJ, Jensen SE, Alzari PM, James MNG. A potent new mode of β-lactamase inhibition revealed by the 1.7 Å X-ray crystallographic structure of the TEM-1-BLIP complex. Nature Struct Biol 1996; 3:290–297.
141. Trainer S, Bouthors AT, Maveyraud L, Guillet V, Sougakoff W, Samama JP. The high reduction crystal structure for class A β-lactamase PER-1 reveals the bases for its increase in breadth of activity. J Biol Chem 2000; 275:28075–28082.
142. Concha NO, Rasmussen BA, Bush K, Herzberg O. Crystal structure of the wide-spectrum binuclear zinc β-lactamase from *Bacteroides fragilis*. Structure 1996; 4:823–836.
143. Fitzgerald PMD, Wu JK, Toney JH. Unanticipated inhibition of the metallo-beta-lactamase from *Bacteroides fragilis* by 4-morpholineethanesulfonic acid (Mes): a crystallographic study at 1.85-Å resolution. Biochemistry 1998; 37: 6791–6800.

144. Carfi A, Soto R-P, Duee E, Galleni M, Frère J-M, Dideberg O. X-Ray structure of the Zn(II)-β-lactamase from *Bacteroides fragilis* in an orthorhombic crystal form. Acta Crystallogr D Biol Crystallogr 1998; 54:47–57.
145. Ullah JH, Walsh TR, Taylor IA, Emery DC, Verma CS, Gamblin SJ, Spencer JF. The crystal structure of the L1 metallo-β-lactamase from *Stenotrophomonas maltophilia* at 1.7 angstrom resolution. J Mol Biol 1998; 284:125–136.
146. Kuzin AP, Liu H, Kelly JA, Knox JR. Binding of Cephalothin and Cefotaxime to D-Ala-D-Ala-Peptidase reveals a functional basis of a natural mutation in a low-affinity penicillin-binding protein and in extended-spectrum β-lactamases. Biochemistry 1995; 34:9532–9540.
147. Pares S, Mouz N, Petillot Y, Hakenbeck R, Dideberg O. X-Ray structure of *Streptococcus pneumoniae* PBP2x, a primary penicillin target enzyme. Nature Struct Biol 1996; 3:284–289.
148. Fonze E, Charlier P. Crystal structure of *Streptomyces* K15 DD-transpeptidase. BNL Protein Databank Tracking Number: T16312.
149. Kräulis PJ. MOLSCRIPT: a program to produce both detailed and schematic plots of protein structures. J Appl Crystal 1991; 24:946–950.

7

Target Modification as a Mechanism of Antimicrobial Resistance

David C. Hooper
Harvard Medical School and MassGeneral Hospital, Boston, Massachusetts

Alteration in the target of an antimicrobial drug is a widely used bacterial mechanism of drug resistance and, in addition to reduced drug permeation to its target and drug modification, is one of the three major mechanisms. Resistance by the general mechanism of target modification can be brought about, however, by a remarkable variety of specific means, which have been exploited by different clinically important bacteria. The modification mechanism often results in an altered structure of the original drug target structure that binds the drug poorly or not at all. This alteration in structure can be brought about by naturally occurring spontaneous mutations in the gene(s) encoding the drug target that modify single or limited numbers of amino acids in the target protein, often in the region of a known or putative drug-binding site. Quinolone resistance due to alterations in the target enzymes DNA gyrase and topoisomerase IV involved in DNA synthesis, rifampin resistance due to alterations in the β subunit of the target RNA polymerase involved in RNA synthesis, and low-level penicillin resistance in *Streptococcus pneumoniae* due to alterations in the target transpeptidases (penicillin-binding proteins) involved in cell wall synthesis are examples of this category.

More extensive modifications of a drug target often require other genetic mechanisms. In the case of high-level penicillin resistance in *S. pneumoniae*, more extensive modifications of transpeptidases are possible because of the ability of this organism to exchange DNA segments with related bacterial species, some of which have transpeptidases that bind penicillin poorly, allowing the generation of multiply mosaic transpeptidases with extensively modified regions of these target enzymes in *S. pneumoniae*. In other cases, such as glycopeptide resistance in enterococci and tetracycline and macrolide resistance in many bacteria, the target structures to which these drugs bind (the cell wall for the glycopeptides and the bacterial ribosome for the tetracyclines and macrolides) are exogenously modified by enzymes encoded by DNA acquired on mobile genetic elements, such as plasmids or transposons, which can be transferred between bacteria.

Another novel variation of the altered target mechanism is the overexpression of unmodified drug target–binding sites in such a way that binding of drug to these extra sites limits access of drug to a subset of critical target-binding sites, as is postulated to be the cause of low-level glycopeptide resistance in staphylococci. Finally, in a number of cases, such as resistance to methicillin and other β-lactams in staphylococci, resistance to mupirocin in staphylococci, and resistance to trimethoprim in many species, bacteria have acquired genes, sometimes on mobile genetic elements, that encode an alternative or bypass drug-resistant target enzyme that then provides the functions that would have otherwise been inhibited, allowing growth in the presence of drug. Thus, the creativity of Nature in developing resistance mechanisms under selective pressure has as yet been fully capable of meeting the challenge of new drug development.

1 INTRODUCTION

Modification of targets of antimicrobial agents is one of the three principal mechanisms by which bacteria effect resistance to antimicrobial agents in addition to alteration in drug permeation to its target and drug modification. Within this general mechanism category, however, there is remarkable bacterial diversity in the means by which target modification is accomplished. This chapter discusses six different bacterial strategies (Table 1) for bringing about target modifications that cause antimicrobial resistance, with a focus on specific examples of each strategy that are of general clinical importance.

In many cases, target modification by various means produces alteration of an existing natural target such that it has reduced drug-binding affinity, but in other notable cases, the resistance determinant may block drug access by interaction with the drug target, resistance may occur from target overproduction that sequesters drug thereby limiting its binding to a subset of target molecules at a critical cellular location, or the bacterium may acquire a resistant drug target that can carry out the functions of the natural sensitive target molecule in the presence of drug. Modification of a natural drug target may result from spontaneous chromosomal mutation resulting in single or multiple amino acid substitutions or from homologous recombination with exogenous DNA containing gene segments that encode portions of proteins with reduced drug-binding properties. Analysis of modifications of this type is particularly useful in understanding the structural basis of drug binding to its target. Genes acquired on mobile genetic elements such as plasmids and transposons may also contribute to resistance by target modification by encoding proteins that themselves modify the drug target or block drug access to the target or that act as a drug resistant ("bypass") targets. Thus, multiple variations on the general mechanism of drug target modification have evolved in Nature.

2 SPONTANEOUS MUTATIONAL CHANGES IN TARGET PROTEINS

2.1 Quinolone Resistance

Quinolones are widely used antimicrobial agents with a generally broad spectrum of activity. There are two intracellular targets of the quinolone class of antimicrobials, DNA gyrase and topoisomerase IV, two related enzymes both of which are essential for bacterial DNA replication. Each enzyme is a tetramer composed of two A and two B subunits. In the case of DNA gyrase, the subunits are GyrA and GyrB, and in the case of topoisomerase IV, the subunits are ParC (or GrlA) and ParE (or GrlB). GyrA is homologous to ParC and GyrB is homologous to ParE (1). Quinolones interact with DNA gyrase–DNA complexes and with topoisomerase IV–DNA complexes to trap the enzymes as stabilized reaction intermediates in which broken DNA strands are covalently linked to a tyrosine in the enzyme's active site. These stabilized complexes form a barrier to DNA replication and are necessary but not sufficient for bacterial cell death (2,3). Under some conditions, such as treatment of cells with nalidixic acid together with an inhibitor of protein or RNA synthesis, inhibition of DNA synthesis by the quinolone is unaffected but cell death does not occur as it does in the absence of these inhibitors (4). This dissociation has suggested

TABLE 1 Target Modifications Resulting in Resistance to Antibiotics in Clinical Use

Mechanism	Drug	Targets	Bacteria involved	Location of resistance determinants	Alteration affecting drug interaction with its target
Spontaneous mutational changes in a target protein	Quinolones	DNA gyrase and topoisomerase IV	Multiple gram-negative and gram-positive species	Chromosomal	Reduced target affinity for drug
	Rifamycins	RNA polymerase	Many gram-positive bacteria	Chromosomal	Reduced target affinity for drug
	β-Lactams	Transpeptidases (penicillin-binding proteins)	*Streptococcus pneumoniae*	Chromosomal	Reduced target affinity for drug
Generation of mosaic target proteins by homologous recombination	β-Lactams	Transpeptidases	*Streptococcus pneumoniae*	Chromosomal	Reduced target affinity for drug
Target remodeling by acquired metabolic pathways	Glycopeptides	Cell wall peptide side chains with D-Ala-D-Ala	Enterococci	Plasmids and transposons	Reduced target affinity for drug

Target overproduction as a means to block access to critical target sites	Glycopeptides	Cell wall peptide side chains with D-Ala-D-Ala	*Staphylococcus aureus*	Chromosomal	Sequestration of drug from critical target sites
Target modifications by acquired proteins	Macrolides, lincomycins, streptogramin B	23S RNA of 50S ribosomal subunit	Multiple gram-positive species	Plasmids and transposons	Reduced target affinity for drug
	Tetracyclines	30S ribosomal subunit	Multiple gram-positive and gram-negative species	Plasmids	Reduced target affinity for drug
Acquired bypass targets	β-Lactams	Transpeptidases	Staphylococci	Chromosomal	Alternative target with reduced drug affinity
	Mupirocin	Isoleucyl tRNA synthetase	Staphylococci	Plasmids	Alternative target with reduced drug affinity
	Trimethoprim	Dihydrofolate reductase	Multiple gram-positive and gram-negative bacteria	Plasmids, transposons, and gene cassettes within integrons	Alternative target with reduced drug affinity

that other factors involving new protein or RNA synthesis are necessary for cell death to occur after interaction of the quinolone with its target enzyme–DNA complex. The specific nature of this other factor(s), however, has remained elusive, and for many newer fluoroquinolones inhibition of protein and RNA synthesis has a little or no effect on their bactericidal activity. Release of double-stranded DNA breaks from the quinolone-trapped enzyme-DNA complex, which might occur with different efficiencies for different quinolones, has been postulated to be the ultimate lethal cellular lesion from which cellular DNA repair mechanisms may be able to recover poorly (2). Certain mutants that are bacteriostatically inhibited by otherwise bactericidal antibiotics, such as β-lactams and vancomycin, are also bacteriostatically inhibited by quinolones, suggesting that common cellular pathways mediate the final events that result in cell death (5,6). In the case of *S. pneumoniae*, these pathways have been linked to autolytic activity (6).

Point mutations encoding single amino acid changes in either DNA gyrase or topoisomerase IV can cause quinolone resistance. These resistance mutations have generally been localized to the amino terminal domain of GyrA or ParC and are in proximity to the active site tyrosine (7). This domain has been termed the quinolone resistance determining region (QRDR) of GyrA and ParC (8). The most common site of mutation in GyrA of *Escherichia coli* is at serine-83 (Ser-83) (or a Ser at equivalent positions of GyrA of other species or equivalent positions of ParC), which may be changed to tryptophan (Trp), leucine (Leu), alanine, or other amino acids (9). Ser-83Trp and Ser83Leu mutations of *E. coli* GyrA have been associated with reduced binding of the quinolone norfloxacin and enoxacin to gyrase-DNA complexes (10,11). Many of the common mutations appear to have little effect on the enzyme's catalytic efficiency (12).

Mutations in specific domains of GyrB and ParE have also been shown to cause quinolone resistance (13,14), although these mutations appear to be substantially less common than mutations in GyrA or ParC. GyrB resistance mutations have also been shown to have reduced binding of enoxacin to enzyme-DNA complexes (10). The QRDR of GyrB (or ParE) appears to be distant from the QRDR of GyrA (or ParC) based on the three-dimensional structure inferred from x-ray crystallography of the homologous enzyme yeast topoisomerase II (15). More recent crystalline structures of yeast topoisomerase II, however, have identified other enzyme conformations in which the regions homologous to the QRDRs of GyrA and GyrB are in proximity, suggesting the such a conformation may be important for forming the site of quinolone binding (16). Thus, it appears that mutations in the QRDRs of both GyrA and GyrB act by reducing the affinity of quinolones for the enzyme-DNA complex. Although there are no direct data on quinolone binding to complexes of wild-type and mutant topoisomerase IV, it is presumed that similar mutations in ParC and ParE also reduce quinolone-binding affinity for topoisomerase IV–DNA complexes because of similarity in overall structure and the strong conservation of amino acid sequence in QRDRs of DNA gyrase and topoisomerase IV.

The magnitude of resistance conferred by a single amino acid change in the subunits of DNA gyrase or topoisomerase IV varies both by bacterial species and by quinolone. The variation in the phenotype of a given resistance genotype relates at least in part to the relative sensitivities of DNA gyrase and topoisomerase IV to a given quinolone. Because quinolone interaction with either enzyme target is sufficient to block cell growth and trigger cell death (2), the level of susceptibility of a wild-type bacterium is determined by the most sensitive of the two target enzymes. Interestingly for many quinolones in clinical use, topoisomerase IV is the more sensitive enzyme in gram-positive bacteria, and DNA gyrase is the more sensitive enzyme in gram-negative bacteria (17). Thus, mutations in the subunits of the more sensitive target enzyme generally occur in first-step mutants, providing a genetic definition of the primary target enzyme (14,18,19). The magnitude of the resistance increment from such a first-step mutation may be determined by either the magnitude of the effect of the mutation on enzyme sensitivity or the intrinsic level of sensitivity of the secondary target enzyme, whichever of the two is less. Thus, quinolones that have highly similar activities against both DNA gyrase and topoisomerase IV of a given species may require mutations in a subunit of both enzymes before the mutant bacterium exhibits a substantial resistance phenotype (20,21). Sequential mutations in subunits of both target enzymes have been shown to provide increasing levels of quinolone resistance. In some species in which high-level quinolone resistance is common, such as clinical isolates of methicillin-resistant *Staphylococcus aureus*, mutations in subunits of both enzymes are common (22).

2.2 Rifampin Resistance

Rifampin and related rifamycins, rifabutin and rifapentene, are inhibitors of the essential bacterial enzyme DNA-dependent RNA polymerase and have been used for treatment of mycobacterial and other bacterial infections or colonizations. RNA polymerase appears to be the sole target of rifampin action. Core RNA polymerase is composed of three subunits, β, β', and α. Core polymerase combines with one of several σ subunits to enable specific binding to promoters and initiation of transcription. Rifampin forms a 1:1 complex with RNA polymerase and blocks the initiation of transcription (23,24).

Resistance to rifampin occurs by mutations in the *rpoB* gene that encode amino acid changes in the β subunit of *E. coli* RNA polymerase (25). These mutations are clustered in three highly conserved regions of *rpoB* in the midportion of the gene (cluster I—codons 507–511 and 513–533; cluster II—codons 563–564 and 572; and cluster III—codon 687) (25). These regions appear to be involved in the polymerase antitermination process, because most resistance mutations affected the polymerase read-through of termination signals, although the importance of this occurrence for the fitness of rifampin-resistant mutants in vivo is uncertain (26). It is presumed that these changes reduce the affinity of RNA polymerase for rifampin, although direct binding studies have not been reported.

Resistance to rifampin has been associated with mutations in similar regions of the *rpoB* genes of *Mycobacterium tuberculosis* (27), *M. leprae* (28), *Staph. aureus* (29,30), *S. pneumoniae* (31), and *Neisseria meningitidis* (32). In the case of *M. tuberculosis*, single mutations were associated with high-level resistance to rifampin (27), and single mutations in *Staph. aureus* selected in vitro and in vivo have been associated with both high and low levels of resistance depending on the nature of the amino acid change (29,30). Some clinical isolates of both *Staph. aureus* and *S. pneumoniae* have been found to multiple mutations in the cluster regions (31). Isolates of *Rickettsia typhi* and *R. prowazekii* from patients failing treatment with rifampin have also been found to have homologous mutations, and similar mutations have been found in naturally resistant species of *Rickettsia* (33).

Thus, the ability of single spontaneous mutations to confer high-level rifampin resistance correlates with clinical observations that resistance develops rapidly in clinical settings when rifampin is used alone for therapy of established infections.

2.3 Low-Level Penicillin Resistance in *S. pneumoniae*

β-Lactams target a set of enzymes involved in cell wall biosynthesis, and thus like quinolones have multiple targets within the bacterial cell. These target enzymes are transpeptidases that cross-link the peptidoglycan lattice providing osmotic and structural stability to the cell (34). Because these enzymes bind penicillins, they are commonly referred to as penicillin-binding proteins (PBPs). In *S. pneumoniae*, high molecular weight PBPs 1 (1a, 1b) and 2 (2a, 2b, and 2x) are essential for cell viability.

Single amino acid changes in individual PBPs cause only low-level resistance to penicillins and cephalosporins, and perhaps for this reason have been found only in laboratory mutants. Higher levels of resistance, which have occurred in clinical isolates of *S. pneumoniae*, involve another target modification mechanism that will be discussed in the next section. Amino acid substitutions in PBP 2x have been found in laboratory mutants in domains near the penicillin-binding motifs, and often several amino acid changes are required for substantial reductions in the affinity of PBP 2x for penicillin. Increments in the MIC of penicillin, however, were limited to a change from 0.02 to 0.32 μg/mL even in the presence of as many as four amino acid changes in PBP 2x (35). The need for multiple mutations to reduce drug-binding affinity suggests that there are multiple contact points between penicillins and PBP 2x. In addition, the limited resistance phenotype of PBP 2x mutants even when penicillin affinity of this PBP is reduced suggests that more than one PBP must be changed in order to effect high-level resistance to penicillin by target modification (36). This circumstance is similar in principle to that for fluoroquinolones interacting with dual targets (as discussed above) in which the most sensitive of multiple essential drug targets (be they mutant or wild type) in a given bacterial cell determines the level of drug susceptibility.

3 GENERATION OF MOSAIC TARGET PROTEINS BY HOMOLOGOUS RECOMBINATION

3.1 High-Level Penicillin Resistance in *S. pneumoniae*

As discussed above, individual amino acid changes in the PBP enzyme targets of penicillins and cephalosporins cause only limited levels of resistance to these antimicrobials. In addition, β-lactamases, which degrade penicillin, have not been described as a mechanism of β-lactam resistance in *S. pneumoniae*. This organism, however, has been able to develop substantial resistance to penicillins by a mechanism of transformation and homologous recombination made possible by several distinctive factors. First, *S. pneumoniae* is naturally competent to take up DNA by a process called transformation. If this DNA has sufficient homology to genes on the pneumococcal chromosome, then *S. pneumoniae* can recombine the imported DNA into its chromosomal DNA, creating mosaic genes consisting of segments of both original host and imported DNA. Second, viridans streptococci, many strains of which now contain PBPs with low affinity for penicillins, are genetically related to pneumococci and thus have highly homologous genes encoding PBPs. Third, viridans streptococci are normal inhabitants of the upper respiratory tract, which may also contain pneumococci during periods of colonization or infection, providing a natural opportunity for exchange of genetic information between these organisms.

Clinical isolates of *S. pneumoniae* with reduced susceptibility to penicillin and other β-lactams have been found to have such mosaic genes encoding modified PBP 2x, PBP 2b, and PBP 1a (36). DNA encoding mosaic PBP 2b from penicillin-resistant strains of *S. pneumoniae* is capable of transforming a recipient pneumococcus that has preexisting changes (causing low affinity for penicillin) in PBPs 1a and 2x to higher levels of resistance to penicillin (37). That *S. pneumoniae* without alterations in PBPs 1a and 2x cannot be similarly transformed illustrates the requirements for multiple changes in PBPs necessary to effect high-level resistance. A set of specific amino acid changes at positions 371 and 575–577 found in the mosaic segments of PBP 1a from all resistant clinical isolates from South Africa have been shown genetically to contribute to resistance (38).

In the case of the mosaic gene encoding PBP 2b from a penicillin-resistant pneumococcus, the nonpneumococcal segments of DNA most closely resemble similar segments of DNA from the gene for PBP 2b from penicillin-susceptible strains of *S. mitis*. The mosaic pneumococcal gene, however, also contains other segments of DNA from an unidentified third species (37). Clinical isolates of penicillin-resistant and penicillin-susceptible strains of other viridans streptococci, such as *S. sanguis* and *S. oralis*, as well as *S. pneumoniae* have been found to have mosaic genes for PBPs, highlighting the extent of genetic exchange, but leaving uncertain the directionality of transfer and the original source of the gene segments causing resistance.

The extensive modification of multiple PBPs necessary to cause high-level penicillin resistance in clinical isolates of pneumococci may come at a price. In this regard, the catalytic activity of PBP 2x purified from a resistant clinical isolate has been shown to be substantially lower than that of PBP 2x purified from a susceptible isolate (39). Since penicillin-resistant pneumococci remain capable of colonizing and infecting humans, it is presumed that they have acquired compensatory mechanisms, as yet poorly defined, to assure their fitness in vivo. It has also been suggested that the stringency of these compensatory requirements may be responsible for the limited number of serotypes of penicillin-resistant pneumococci, which have nevertheless been quite successful in spreading throughout the world (34).

4 TARGET REMODELING BY ACQUIRED METABOLIC PATHWAYS

4.1 Glycopeptide Resistance in Enterococci

Vancomycin, teicoplanin, and other glycopeptides bind to components of the bacterial cell wall, which is a peptidoglycan lattice composed of polymers of alternating *N*-acetylglucosamine and *N*-acetylmuramic acid residues that are cross-linked via attached short peptide chains. The peptide chains attached to the polymer backbone are the substrates for the cross-linking reaction, which is catalyzed by transpeptidases (PBPs). Vancomycin binds specifically to the terminal two D-alanine (D-Ala) residues of the peptide side chain. This binding inhibits several reactions in cell wall biosynthesis, including transfer of precursors from a membrane lipid carrier to the peptidoglycan backbone, D,D-carboxypeptidase activity, and transpeptidase activity. To alter the target of vancomycin thus requires remodeling of peptidoglycan structure.

Vancomycin resistance results from substitution of D-lactate (D-Lac) for the terminal D-Ala of the peptide side chain, a change that decreases vancomycin affinity for the cell wall by 1000-fold (40). Cellular transpeptidases, in most cases, remain able to catalyze cross-linking using peptide side chains terminated in D-Ala-D-Lac. The production of a cell wall with peptide side chains terminated in D-Ala-D-Lac is engineered in both *Enterococcus faecium* and *E. faecalis* by acquired clusters of genes located on mobile genetic elements, the best studied of which is transposon Tn1546 (41). Three of the eight genes identified on Tn1546 are necessary for vancomycin resistance (42). *vanA* Encodes a ligase enzyme that catalyzes attachment of D-Lac to D-Ala. *vanH* Encodes a dehydrogenase that catalyzes production of D-Lac from pyruvate. *vanX* Encodes an enzyme that hydrolyzes D-Ala-D-Ala, thereby blocking parallel production of normal peptide side chains with D-Ala-D-Ala termini, which could serve as residual targets for binding of vancomycin (43). Two other genes encode accessory proteins not required for vancomycin resistance. *vanY* Appears to encodes a D,D-carboxypeptidase that hydrolyzes the terminal D-Ala on the cytoplasmic precursor peptide side chain (44). When VanX blocks synthesis of D-Ala-D-Ala–terminated peptides, VanY may have little effect on resistance. In contrast, under conditions in which overall activity of VanX is limited, VanY may additionally remove residual D-Ala-D-Ala peptides as vancomycin targets. The *vanZ* gene appears to contribute to resistance to teicoplanin but not vancomycin when VanH, VanA, and VanX are produced at low levels (45).

In Tn1546, *vanH*, *vanA*, and *vanX* are cotranscribed from a promoter located between two upstream genes, *vanR* and *vanS*, which encode, respectively, response and sensor proteins of a two-component regulatory system (42,46,47). Expression of the *van* gene cluster is thought to be regulated by this sensor-response system, but the signal for induction of expression is not yet known. Induction of expression of high-level glycopeptide resistance by exposure to either vancomycin or teicoplanin is characteristic of the VanA phenotype of the Tn1546 resistance element, but induction can also occur with exposure to structurally unrelated, non-glycopeptide antibiotics (48). Thus, induction is not solely due to the structural features of vancomycin or teicoplanin. A related resistance phenotype, VanB, is distinguished by a lower level of resistance and a lack of inducibility by teicoplanin and is encoded on other mobile elements by a cluster of genes (designated as *vanB* instead of *vanA* for the ligase genes and for the other homologous genes with a subscript B) similar to that found on Tn1546 (49–51). Vancomycin, but not teicoplanin, is an inducer of expression of $vanR_B$-$vanS_B$, thereby providing an explanation for the lack of induction of resistance by teicoplanin in VanB strains (49). VanC, a third phenotype of intrinsic constitutive low-level resistance to glycopeptides, has been found in *E. gallinarum* to be due to peptide side chains terminated in D-Ala-D-serine (52), and naturally highly resistant species such as *Lactobacillus casei*, *Leuconostoc mesenteroides*, and *Pediococcus pentosaceus* have been shown to have D-Ala-D-Lac termini on their peptide side chains (52,53).

The genetic and mechanistic complexity of vancomycin resistance represents a remarkable feat of natural genetic engineering that suggests that bacterial ingenuity given sufficient time and selective advantage is likely to be capable of circumventing the activity of any antimicrobial agent developed for clinical use; that is, resistance is ultimately inevitable and it is only a matter of how rapidly or slowly it emerges, as determined by mechanistic and epidemiological factors. The original source of the vancomycin resistance gene cluster remains uncertain. Homologues of *vanH*, *vanA*, and *vanX* have been found in *Streptomyces toyocaensis*, *Amycolatopsis orientalis* (in the same orientation as in enterococci), and other glycopeptide-producing species of bacteria, suggesting the possibility that the original evolution of such gene clusters might have occurred in these or related species as a means of protection from their own antimicrobial products (54). The G + C content of these genes, however, differs in *Streptomyces toyocaensis* and *A. orientalis* (65%) in comparison to enterococci (44–49%), and in enterococci the G + C content of *vanH*, *vanA*, and *vanX* is 5–10% higher than that of the flanking genes *vanR*, *vanS*, *vanY*, and *vanZ*. Thus, the *vanHAX* gene cluster could have been mobilized en bloc from some donor species, but the exact nature of the donor remains uncertain, as does the origin of a possible intermediate recipient organism containing *vanR*, *vanS*, *vanY*, and *vanZ*. There also may be heterogeneity in such a putative intermediate recipient, since some resistant enterococci lack *vanZ* and the location of *vanY* varies between VanA and VanB resistance elements (49). If the source of *vanHAX* is from one of the glycopeptide-producing species, then considerable evolution of these genes must have occurred before they appear in enterococci, and the organism(s) in which such evolution occurred also remains to be defined.

5 TARGET OVERPRODUCTION AS A MEANS OF BLOCKING ACCESS TO CRITICAL TARGET SITES

5.1 Glycopeptide Resistance in Staphylococci

Although some clinical isolates of other gram-positive bacteria, such as *S. bovis*, have acquired the *vanB* gene of enterococci (55), and *vanA* has been transferred from enterococci to *S. aureus* in the laboratory (56), clinical isolates of staphylococci with reduced susceptibility to glycopeptides have different mechanisms from those in enterococci and do not have the enterococcal resistance genes. The mechanisms of vancomycin resistance in staphylococci may differ between coagulase-negative and coagulase-positive species, and complete details have not been elucidated in any species.

In *Staph. aureus*, some clinical isolates from patients in Japan and the United States who have failed vancomycin therapy have been found to have either (1) heterogeneously expressed low-level resistance in subpopulations of cells that cannot be detected by standard MIC testing methods in clinical laboratories (MIC of vancomycin = 2 μg/mL) (57), or (2) a higher level of resistance (MIC = 8 μg/mL), which is also difficult to detect by disk diffusion but not MIC methods (58,59). Common among all of these strains has been preexisting methicillin resistance and prolonged exposure to vancomycin in vivo. The prototypic Japanese vancomycin-heteroresistant and vancomycin-resistant strains, Mu3 and Mu50, respectively, have been shown to differ from fully vancomycin-susceptible strains in several features, including enhanced incorporation of *N*-acetylglucosamine into the cell wall, an increased pool of the cytoplasmic cell wall precursor monomer UDP-*N*-acetylmuramyl-pentapeptide, enhanced autolysis, and increased production of PBP 2 (60). Mu50, the more resistant strain, exhibited, in addition, a twofold increased thickness of the cell wall, a substantially higher proportion of peptide side chains in which the glutamine residue is nonamidated (side chains composed of L-Ala-*D-Glu*-L-Lys-D-Ala-D-Ala instead of L-Ala-*D-Gln*-L-Lys-D-Ala-D-Ala), and a slight increase in the number of un–cross-linked peptide chains, the latter possibly due to the preference of PBPs for cross-linking the normal rather than the nonamidated peptide side chains (61). The observed increased numbers of un–cross-linked pentapeptide chains, which retain their terminal D-Ala-D-Ala–binding site for vancomycin (cross-linking removes these residues) and an increased thickness of the cell wall has led to proposal of the false target model (62). In this model, vancomycin binding to increased numbers of noncritical D-Ala-D-Ala targets present in the thickened and poorly cross-linked cell wall protects the critical D-Ala-D-Ala target sites at the point of action of the transglycosylase enzymes near to the cell membrane. It has been estimated that Mu50 has a threefold increase in free D-Ala-D-Ala termini, and increased binding of vancomycin to this strain has been reported (62). Thus, vancomycin is sequestered by binding to extra target D-Ala-D-Ala sites far from the site of new cell wall synthesis at the cell membrane.

The genetic determinants of this mechanism of resistance are likely to be multiple and have not yet been defined. Since it is possible to select resistance at the level of Mu50 from Mu3 by growth in the presence of vancomycin in the laboratory, acquired genes do not appear to be necessary for this step(s). Laboratory strains of *Staph. aureus* selected for resistance by repetitive exposure to vancomycin (MIC = 5 μg/mL) have exhibited some of the properties of Mu3 and Mu50, including increased production of PBP 2, an increased muropeptide monomer pool, and thickening of the cell wall (63). These strains, however, lacked nonamidated pentapeptide chains or alterations in cross-linking. *femC* Mutants of *Staph. aureus*, which have reduced glutamine synthetase activity, have been found to have increased numbers of nonamidated peptide side chains (61), so that *femC*-type mutations might be contributory. The *mecA* gene, which is necessary for resistance to methicillin and encodes PBP 2A (see below), appears not to be necessary for vancomycin resistance, since vancomycin resistance persists after excision of *mecA* (62). The association of vancomycin resistance with MRSA strains is likely due to the importance of exposure to vancomycin in selecting resistance, since vancomycin would be used commonly for treatment of patients with MRSA infections and less often for treatment of patients with methicillin-susceptible strains of *Staph. aureus* against which other antibiotics are effective.

There is additional complexity and probably other mechanisms of resistance to glycopeptides in coagulase-negative staphylococci as well as additional properties of teicoplanin resistance that differ from those of vancomycin resistance in *Staph. aureus* that are beyond the scope of this chapter. Thus, multiple and complex mechanisms may occur in different settings, but the model of resistance suggested by the results in Mu3 and Mu50 strains of *Staph. aureus*, represents another potential variation on the altered target mechanism. In this case, in contrast to vancomycin resistance in enterococci in which the target is remodeled, vancomycin resistance in *Staph. aureus* appears to involve target overproduction that titrates vancomycin away for critical target sites.

6 TARGET MODIFICATION BY ACQUIRED PROTEINS

6.1 Macrolide, Streptogramin, and Lincosamide Resistance

Macrolides, streptogramins, and lincosamides inhibit bacterial protein synthesis by binding to the 23S ribosomal RNA (rRNA) component of the bacterial 50S ribosomal subunit (64). One of the most common mechanisms of resistance to these three classes of antimicrobials involves post-transcriptional alteration of a specific base of rRNA that results in ribosomes with reduced drug affinity (65). Specifically, methylation of an adenine at position 2058 of 23S rRNA is accomplished by members of the Erm family of N^6-methyltransferases, the family name being an acronym for *e*rythromycin *r*esistance *m*ethylase. Dimethylated A2058 causes high-level resistance to all generations of macrolides, lincosamides such as clindamycin, and the group B streptogramins represented by pristinamycin (66), a resistance phenotype referred to as MLS_B. This methylated base is located in the peptidyltransferase loop of domain V of the 23S RNA, which is thought to contain at least part of the site of macrolide binding to the ribosome (65).

erm Genes are most often acquired on plasmids and may have arisen from macrolide-producing actinomycetes (67). Enzymatic methylation rather than mutational modification of rRNA at A2058 is important in many bacteria, because they contain multiple copies of rRNA genes, placing a requirement for multiple mutations were resistance to occur by chromosomal mutation alone. Some species, such as *M. intracellulare*, *Helicobacter pylori*, *Mycoplasma* spp., and *Propionobacterium* spp., however, have small numbers of copies of rRNA genes, and resistance in these species has been associated with mutations in rRNA genes that substitute guanine, cytosine, or uracil in place of adenine at position 2058 (68,69) or alter other bases in the peptidyltransferase circle (65).

The expression of Erm methylases is often inducible by low concentrations of erthromycin. Expression of *erm* is negatively regulated by transcriptional and translational attenuator mechanisms related to the secondary structure of the *erm* mRNA leader sequence. In the absence of erythromycin, stem-loop structures of the mRNA mask the first two codons and the ribosome-binding site of *ermC*, thereby reducing the efficiency of translation of ErmC methylase (64). Upstream of *ermC* is an open reading frame encoding a 19–amino acid leader peptide, translation of which stabilizes the stem-loop conformation that masks critical sites for *ermC* translation downstream. In the presence of erythromycin bound to a ribosome, translation of this leader peptide is stalled, destabilizing the blocking stem-loop structures and facilitating a change in the conformation of the attached downstream mRNA that unmasks the sites critical for *ermC* translation. The unmasked *ermC* transcript must then dissociate from the erythromycin-bound ribosome in order for translation to occur on other ribosomes to which erythromycin has not bound. Thus, the low concentrations of erythromycin associated with induction may allow the presence of both erythromycin-bound and erythromycin-unbound ribosomes within the cell. Once sufficient ErmC methylase is translated, then many ribosomes can be modified and high-level resistance to erythromycin ensues.

6.2 Ribosomal Protection from Tetracyclines

Another example of target modification by acquired proteins is one of the two principal mechanisms of resistance to tetracyclines—protection of the ribosomal target. Tetracyclines reversibly inhibit bacterial protein synthesis by disrupting the interaction of aminoacyl-tRNA with the ribosome (70). Tetracyclines bind to a high-affinity site on the 30S ribosomal subunit, although low-affinity sites have been identified on both 30S and 50S subunits. Binding appears to involve the S7 ribosomal protein, but other ribosomal proteins such as S3, S8, S14, and S19 appear to contribute to an optimal drug-binding configuration (71,72). Drug binding is also thought to be in proximity to the 16S rRNA component of the ribosome.

Tetracycline resistance determinants of classes M, O through Q, and S encode related proteins that interact with ribosomes to protect them from the action of tetracycline. These resistance determinants are found on plasmids and transposons and have been identified in gram-positive and gram-negative bacteria (73–75). Best studied of these determinants is *tetM* from streptococci that encodes the TetM protein. Ribosomes isolated from *tetM*-containing cells are resistant to tetracycline in in vitro translation systems if extracted under low-salt but not high-salt conditions, and purified TetM protein confers tetracycline resistance on ribosomes isolated from tetracycline-susceptible cells (76). Binding of tetracycline to the ribosome, however, is not altered by TetM, and resistance occurs even when ribosomes are in substantial excess of TetM in vitro, suggesting that TetM acts catalytically (77). TetM and TetO have structural similarities to elongation factors EF-Tu and EF-G, which bind ribosomes, and also have GTPase activity, but the exact means by which they interact with or modify the ribosome to interfere with tetracycline inhibition of protein synthesis are still unclear. Host factors also appear to be important in that chromosomal *miaA* mutations, which cause defects in Δ^2-isopentenylpyrophosphate transferase, an enzyme that modifies adenosine at position 37 of tRNA, had partial loss of tetracycline resistance in *tetM* cells (77). Modification of base A37 of tRNA is important for accuracy of translation, but it is uncertain if the effects of *miaA* mutations on tetracycline resistance reflect a direct or indirect effect on tetracycline interaction with the ribosome (78). Current hypotheses about the exact mechanism of ribosome protection include an interaction between TetM and the ribosome that inhibits the binding of tetracycline specifically to ribosomes that are actively engaged in protein synthesis or that allows aminoacyl-tRNA to enter the A-site on the ribosome in the presence of tetracycline (79).

Expression of tetracycline resistance by *tetM*, like expression of *erm*-mediated resistance to macrolides, is inducible and regulated by transcriptional and translational attenuation (80). Upstream of *tetM* is an open reading frame (ORF) that encodes two regions of GC-rich RNA inverted repeats flanked by a series of uracil (U) residues, promoting formation of hairpin secondary structures, which cause RNA polymerase to pause. The series of U residues downstream also produces an unstable RNA/DNA hybrid that facilitates destabilization of binding of RNA polymerase. In the absence of tetracycline, translation of the ORF is thought to be retarded because five of the first eight codons require rare aminoacyl-tRNAs. Read-through transcription of *tetM* is more likely to occur if the translating ribosome is in proximity to the transcription complex, thereby destabilizing the hairpin structures that would otherwise retard transcription (81). In this model, in the presence of tetracycline as an inducer, the A- and P-sites on the ribosome are occupied by drug. Transcription and translation are thus delayed and availability of aminoacyl-tRNA is increased, allowing proximity of the ribosome to the transcribing RNA polymerase and subsequent transcription and translation of *tetM*. Other tetracycline resistance determinants, such as *tetO* of *Campylobacter* spp., however, are not inducible by tetracycline and appear to be expressed constitutively (82).

7 BYPASS TARGETS

7.1 Methicillin Resistance in Staphylococci

Methicillin, other semisynthetic penicillins such as oxacillin and nafcillin, and most cephalosporins are not degraded by the common staphylococcal penicillinase enzyme, and thus many of these antibiotics are often used as antistaphylococcal agents. Some strains of staphylococci, including both *Staph. aureus* and coagulase-negative staphylococci, however, are resistant to methicillin by a mechanism that renders them also resistant to all current β-lactam antibiotics as well. Methicillin-resistant strains of staphylococci contain an insertion that occurs at a specific site on the chromosome and includes the *mecA* gene, which encodes a transpeptidase, PBP 2a, which has low affinity for β-lactams and appears to be capable of serving the functions of the essential native transpeptidases of staphylococci. Thus, PBP 2a serves as a single, resistant bypass enzyme for the several normal targets of β-lactams in staphylococci (83).

Expression of methicillin resistance varies among strains of staphylococci. Two regulatory genes, *mecI* and *mecR*, immediately upstream of the *mecA* promoter are transcribed divergently. *mecI* Is homologous to *blaI*, which encodes a repressor of β-lactamase expression, and *mecR* is homologous to *blaR*, which encodes a protein that binds penicillins and leads to transcription of *blaZ*, the structural gene for β-lactamase, and thereby expression of β-lactamase. Thus, MecI and MecR proteins are thought to perform analogous functions to those of the BlaI and BlaR proteins, respectively, in regulating expression of PBP 2a. Mutation and disruption of *mecI* and *mecR* or mutations in the *mecA* promoter are now commonly found in clinical methicillin-resistant isolates of *Staph. aureus* such that PBP 2a production is usually sufficient to provide a resistance phenotype (84). This phenotype, however, can vary owing to other as yet incompletely defined factors that do not correlate simply with the level of PBP 2a production (85,86). BlaI and BlaR may also be involved in regulation of expression of PBP 2a when MecI and MecR are altered (87).

The methicillin resistance phenotype may be either homogeneous or heterogeneous. Heterogeneous resistance results in a varying level of resistance depending on the culture conditions and the β-lactam antibiotic used, and often only 1 in 10^6 cells in a population may express high-level resistance (88,89). This proportion is higher (1 in 10^2) if cells are grown at 30°C or in medium supplemented with NaCl (90). Stably homogeneously resistant strains can be selected from heterogeneous strains by passage on β-lactam antibiotics (see below) (91).

Several chromosomal loci have been identified that are important to allow for expression of methicillin resistance (92). Among these, the *fem* (*f*actors *e*ssential for *m*ethicillin resistance) genes encode proteins involved in the synthesis of the pentaglycine cross-linking chains of the muramyl peptide component of the cell wall. The chains are involved in the transpeptidase reactions catalyzed by PBP2a and other PBPs to generate cross-linking of the peptidoglycan. For example, the *femA* gene encodes an enzyme necessary to add the second and third glycines of the pentaglycine side chain (93). Other mutants such as *fmtA* have in common with *fem* mutants alterations in cell wall structure (94). That *fem* mutants express methicillin resistance poorly or not at all implies that PBP 2a functions poorly (or at least less well than native PBPs) in the setting of such modified cell wall precursor structures. The specific molecular interactions of PBP 2a with its substrates, however, remain to be defined.

The genetic basis for the change from heterogeneous to homogeneous methicillin resistance is beginning to be defined. Stable mutants of *Staph. aureus* with homogeneous resistance can be selected from heterogeneously resistant strains after single exposure to β-lactams (95). These mutants, termed *chr**, have mutation(s) at several non-*fem* and non-*mec* loci. Extragenic revertants of *fem* mutants can also express high-level homogeneous resistance and differ from the *chr** mutants. Thus, multiple types of mutations may allow homogeneous expression of methicillin resistance. In some cases, these mutants have had alterations in genes encoding enzymes with cell wall lytic activity thought to be involved in normal cell wall remodeling (96), including the *lytH* gene (92). The means by which reduced lytic enzymes or other changes contribute to homogeneous methicillin resistance remains to be determined, but it is reasonable to speculate that, as with the *fem* mutants, some may involve cell wall structural changes that improve rather than impair the function of PBP 2a.

The site-specific insertion of *mec* DNA in the chromosome of *Staph. aureus* appears to be an infrequent event, since most clinical MRSA strains appear to have a clonal lineage (97). The original source of *mec* DNA is uncertain but it has been suggested to be coagulase-negative staphylococci (98). As with high-level resistance to penicillin in pneumococci, the infrequency of the genetic events generating new resistant clones has not been a barrier to the persistence and dissemination of methicillin resistance in staphylococci worldwide. Thus, whatever the cost of resistance, highly effective compensatory mechanisms appear to have been developed.

7.2 Mupirocin Resistance in Staphylococci

Mupirocin (pseudomonic acid) is an antibiotic derived from cultures of *Pseudomonas fluorescens* that inhibits isoleucyl tRNA synthetase and thus indirectly inhibits bacterial protein synthesis by depriving the cell of the ability to incorporate a common amino acid into protein. Mupirocin is used in topical preparations to eradicate nasal colonization with *Staph. aureus* and for treatment of certain staphylococcal skin infections or colonizations. High-level resistance to mupirocin in *Staph. aureus* has been found to be encoded on a variety of transferable plasmids (99,100). Two isoleucyl tRNA synthetase activities have been isolated from high-level resistant strains. One was also found in susceptible strains and those with low-level resistance. The other was found only in high-level resistant strains and had substantially reduced sensitivity to mupirocin (101,102). A gene, *ileS2* (originally termed *mupA*), encoding this resistant enzyme, was identified on resistance plasmids and found to encode a protein with 57% identity and

30% similarity in amino acid sequence relative to the native IleS protein (103). The origin of the *ileS2* gene is uncertain. Coagulase-negative staphylococci are generally less susceptible to mupirocin than *Staph. aureus,* and high-level mupirocin resistance has been transferred between coagulase-negative staphylococci and *Staph. aureus* on plasmids (104), but other reservoirs are possible. On some plasmids, *ileS2* is flanked by direct repeats of the insertion sequence IS257 (105), suggesting that it is located on a transposable element. In addition, some but not all strains of *Staph. aureus* with low levels of mupirocin resistance appear to contain the *ile2* gene on the chromosome but not plasmids (106,107).

Thus, an acquired resistant isoleucyl tRNA synthetase that bypasses the sensitive native enzyme allows protein synthesis to proceed in the presence of mupirocin. That the level of resistance to mupirocin is higher when the gene encoding this enzyme is located on plasmids relative to the chromosome suggests that differences in gene expression due either to plasmid copy number or to differences in promoter strength or regulation of expression in the two locations may be responsible for the different levels of resistance.

7.3 Trimethoprim Resistance

Trimethoprim is a synthetic structural analogue of folic acid and a competitive inhibitor of the bacterial enzyme dihydrofolate reductase (DHFR), which in the presence of NADPH converts dihydrofolate to tetrahydrofolate. N^5,N^{10}-methylenetetrahydrofolate is a cofactor for thymidylate synthase, donating a methyl group to convert deoxyuridylate to thymidylate, which is required for DNA synthesis. Although sometimes used alone, trimethoprim has most commonly been combined with a sulfonamide such as sulfamethoxazole, which is an inhibitor of dihydropteroate synthase, the enzyme preceding DHFR in the tetrahydrofolate synthesis pathway that converts *p*-aminobenzoic acid to dihydrofolate. DHFRs are essential enzymes found in all living cells, but the human enzyme is intrinsically resistant to trimethoprim, accounting for the drug's selective antibacterial activity (108).

In clinical settings, acquired bacterial resistance to trimethoprim results most frequently from exogenous acquisition of drug-resistant DHFRs on plasmids or transposons (109). These resistant DHFRs are widely distributed and have been studied extensively. The resistant enzyme exists in the cell in addition to the native sensitive enzyme and is able to provide the necessary bypass enzymatic function to generate tetrahydrofolic acid in the presence of trimethoprim. Although there are numerous types of trimethoprim-resistant DHFRs, these types with some exceptions appear to fall into two principal families. The DHFRs of family 1, comprising types I, V, VI, VII, and Ib, have in common a polypeptide length of 157 amino acids, with the type I enzyme being dimeric. Family 1 enzymes all have increased half-inhibitory concentration (IC_{50}) values for trimethoprim (1–100 μM) (109) relative to those for the chromosomally encoded native enzymes of *E. coli* (0.01 μM) (110) and *Haemophilus influenzae* (0.001 μM) (111). All DHFRs of family 1 mediate resistance to trimethoprim producing MICs of host cells of substantially > 1 mg/mL. Levels of resistance conferred by a particular resistant enzyme may vary owing to different levels of expression of that enzyme from its plasmid- or transposon-encoded gene.

DHFRs of family 2 comprise types IIa, IIb, and IIc and have in common a tetrameric structure of four 78–amino acid subunits (109). All members of this family have exceedingly high levels of resistance to trimethoprim with IC_{50} values of > 100 μM (111,112). The three-dimensional structure of the type IIa–resistant enzyme (113) and the native *E. coli* enzyme (114) have been solved and differ substantially from one another, including in their active sites. Comparisons of these structures with that of the intrinsically resistant avian DHFR (108,114) have led to models in which resistance may now be explained at the molecular level by loss of trimethoprim affinity for the enzyme owing to steric alterations in contact points between trimethoprim and key glutamate and threonine side chains in the DHFR active site as well as by the absence of a key hydrogen bond between the DHFR valine-115 and the 4-amino group of trimethoprim. The structural differences between the *E. coli* and type IIa enzymes are of sufficient magnitude to make implausible any hypothesis that the type IIa–resistant enzyme is evolutionarily derived from the *E. coli* enzyme.

Other DHFRs that are not members of families 1 or 2 (types III, IIIb, IIIc, IV, VIII, IX, X, XII, and S1) are heterogeneous, and tend to confer only low levels of resistance to trimethoprim (109). The origins of resistant DHFRs of any of the types are not certain, but type III is the enzyme most closely related to the native *E. coli* enzyme (~50% identity) (115), and the type S1 is the enzyme most closely related to the *Staph. aureus* native enzyme (80% identity) (116,117).

Resistant DHFRs are generally encoded by genes on mobile genetic elements, including transposons (e.g., *dhfrI* on Tn7, *dhfrVII* on Tn5086, *dhfrIIc* on Tn5090) and plasmids (e.g., *dhfrI* on pLMO150, *dhfrIIa* on R67, *dhfrIIb* on R388) most commonly, and thus may be able to spread readily among bacteria (109). Trimethoprim resistance was in fact the earliest identified example of plasmid-mediated antibiotic resistance by an altered or bypass target mechanism. Transposon Tn7 commonly inserts into a specific site on the chromosome of *E. coli* and other bacteria (118). In addition, many (if not most) *dhfr* genes occur on transferable gene cassettes as part of integrons, which are genetic structures consisting of a series of one or more gene cassettes in tandem and a gene encoding an integrase enzyme that catalyzes the recombination of gene cassettes into the integron structure. Thus, resistance to trimethoprim is often associated with resistance to other antibiotics such as sulfonamides and aminoglycosides, the resistance determinants of which are also often in gene cassettes and often found in integrons.

Less frequently in clinical bacterial isolates, acquired bacterial resistance to trimethoprim can occur by chromosomal mutation (1) leading to thymine auxotrophy, (2) causing an altered native DHFR with reduced affinity for trimethoprim, (3) causing overexpression of the native DHFR, or (4) a combination of mechanisms 2 and 3 (109,119). In the case of thymine auxotrophs, in the presence of exogenous thymine, which is required for growth, the need for the product of DHFR action, tetrahydrofolate, in thymidylate synthesis is bypassed, and thus enzyme inhibition has a minimal effect on cell growth. A combination of both structural changes in the chromosomally encoded enzyme and its overproduction appear to be necessary for high-level trimethoprim resistance in the absence of acquired resistant DHFRs (111). These constraints and the potential fitness disadvantage of thymine auxotrophy may explain why chromosomal trimethoprim resistance is less common than acquired resistant bypass DHFRs in clinical settings.

8 CONCLUSIONS

Although in most cases target modification ultimately results in reduced affinity for drug binding to the target as the ultimate cause of drug resistance (see Table 1), overexpression of natural drug targets does produce resistance on a more limited basis. This limited occurrence of gene overexpression is in contrast to the usual circumstances with resistance mechanisms involving drug modification and active efflux in which overexpression of resistance determinants is a common and mechanistically important occurrence. Current models suggest that vancomycin resistance in *Staph. aureus* occurs by target overexpression (62), and the infrequent occurrence of resistance to trimethoprim mediated by mutant or wild-type chromosomally encoded DHFRs appears to require their overexpression as well (109). In the former case, target overexpression acts to sequester drug from critical target sites. In the latter example, DHFR overexpression may act because of the competitive nature of inhibition by trimethoprim. It is noteworthy that for some drugs, such as quinolones, in which the drug-target complex itself forms the toxic lesion, target enzyme overexpression is predicted to cause increased drug susceptibility rather than resistance.

There are a few other examples of target overexpression mediating resistance, such as resistance to isoniazid, ethambutol, and other drugs in mycobacteria (120). It is argued, however, that this mechanism is used by mycobacteria because in these organisms resistance is almost completely dependent on chromosomal mutation owing to the only limited occurrence of plasmids and transposons in these organisms.

Although target modification mechanisms can in most cases be simplified to reductions in affinity of the target for the drug, the means to that end is impressively diverse among different bacterial pathogens and different drugs used against them. The range of possible resistance mechanisms affects which mechanisms may become dominant in the long run as well as the rapidity with which resistance develops after initial introduction of a drug into clinical settings. For chromosomal mutations, which occur spontaneously in large bacterial populations as the result of low-frequency errors in DNA replication, there is the potential to select for resistance by drug exposure in any bacterium. The likelihood that such spontaneous mutational changes will emerge as the dominant resistance mechanism is in part determined by (1) the magnitude of the increase in resistance possible with single mutations thereby affecting the ability of the mutant bacterium to survive in the presence of drug (2) the consequences of these mutations on the fitness of the mutant bacterium to compete in Nature in both the absence and presence of antibiotic selection pressures, and (3) the presence of more than a single drug target within the cell (121). Thus, the potential level of such mutational resistance is low for penicillin resistance in *S. pneumoniae*, for example, because of the multiple PBPs to which penicillin binds. In this circumstance then other mechanisms may be drawn on by a bacterial population in order to generate successful survivors of drug exposure. The importance of genetic exchange is thus highlighted by the exploitation of transformation and recombination by *S. pneumoniae* to generate alteration in multiple targets and levels of resistance to penicillin not readily possible by chromosomal mutational mechanisms.

The importance of genetic exchange in expanding the range of possibilities for resistance is perhaps best highlighted by the multiplicity of mobile genetic elements that have become important vectors for altered target resistance. Both exogenous target modification mechanisms and bypass targets mediating resistance to macrolides, tetracyclines, mupirocin, and trimethoprim are widely dispersed on plasmids, transposons, and integrons, which have become the dominant source of resistance to these drugs. The rapidity with which resistance may emerge with initial introduction of antibiotics in this circumstance is likely determined by the presence or absence of complete or partial cross-resistance mechanisms existing in natural bacterial populations due to naturally produced antibiotics or in nonhuman reservoirs of bacteria under antibiotic selection pressure that may be introduced into human populations (e.g., antibiotic use in food animals). Risk of such cross resistance may be more likely in antibiotics that are natural products, the presence of which in Nature may have provided pressures for prior evolution of resistance mechanisms that may be incorporated into mobile genetic elements. The mechanism of resistance to vancomycin in enterococci involving multiple genes from apparently diverse sources assembled into a large mobile genetic element is perhaps the most impressive example of exploitation of genetic exchange mechanisms in the service of antibiotic resistance and bodes poorly for any hope of ever developing an antibiotic for which resistance will not emerge ultimately in the presence of persisting opportunities for selection.

REFERENCES

1. Wang JC. DNA topoisomerases. Annu Rev Biochem 1996; 65:635–692.
2. Drlica K, Zhao XL. DNA gyrase, topoisomerase IV, and the 4-quinolones. Microbiol Rev 1997; 61:377–392.
3. Hiasa H, Yousef DO, Marians KJ. DNA strand cleavage is required for replication fork arrest by a frozen topoisomerase-quinolone-DNA ternary complex. J Biol Chem 1996; 271:26424–26429.
4. Dietz WH, Cook TM, Goss WA. Mechanism of action of nalidixic acid on *Escherichia coli*. III. Conditions required for lethality. J Bacteriol 1966; 91: 768–773.
5. Wolfson JS, Hooper DC, Shih DJ, McHugh GL, Swartz MN. Isolation and characterization of an *Escherichia coli* strain exhibiting partial tolerance to quinolones. Antimicrob Agents Chemother 1989; 33:705–709.
6. Novak R, Henriques B, Charpentier E, Normark S, Tuomanen E. Emergence of vancomycin tolerance in *Streptococcus pneumoniae*. Nature 1999; 399:590–593.
7. Cabral JH, Jackson AP, Smith CV, Shikotra N, Maxwell A, Liddington RC. Crystal structure of the breakage-reunion domain of DNA gyrase. Nature 1997; 388:903–906.
8. Yoshida H, Bogaki M, Nakamura M, Nakamura S. Quinolone resistance-determining region in the DNA gyrase *gyrA* gene of *Escherichia coli*. Antimicrob Agents Chemother 1990; 34:1271–1272.
9. Hooper DC. Mechanisms of quinolone resistance. Drug Resist Updates 1999; 2:38–55.
10. Yoshida H, Nakamura M, Bogaki M, Ito H, Kojima T, Hattori H, Nakamura S. Mechanism of action of quinolones against *Escherichia coli* DNA gyrase. Antimicrob Agents Chemother 1993; 37:839–845.
11. Willmott CJ, Maxwell A. A single point mutation in the DNA gyrase A protein greatly reduces binding of fluroquinolones to the gyrase-DNA complex. Antimicrob Agents Chemother 1993; 37:126–127.

12. Aleixandre V, Urios A, Herrera G, Blanco M. New *Escherichia coli gyrA* and *gyrB* mutations which have a graded effect on DNA supercoiling. Mol Gen Genet 1989; 219:306–312.
13. Yoshida H, Bogaki M, Nakamura M, Yamanaka LM, Nakamura S. Quinolone resistance-determining region in the DNA gyrase *gyrB* gene of *Escherichia coli*. Antimicrob Agents Chemother 1991; 35:1647–1650.
14. Breines DM, Ouabdesselam S, Ng EY, Tankovic J, Shah S, Soussy CJ, Hooper DC. Quinolone resistance locus *nfxD* of *Escherichic coli* is a mutant allele of *parE* gene encoding a subunit of topoisomerase IV. Antimicrob Agents Chemother 1997; 41:175–179.
15. Berger JM, Gamblin SJ, Harrison SC, Wang JC. Structure and mechanism of DNA topoisomerase II. Nature 1996; 379:225–232.
16. Fass D, Bogden CE, Berger JM. Quaternary changes in topoisomerase II may direct orthogonal movement of two DNA strands. Nat Struct Biol 1999; 6:322–326.
17. Blanche F, Cameron B, Bernard FX, Maton L, Manse B, Ferrero L, Ratet N, Lecoq C, Goniot A, Bisch D, Crouzet J. Differential behaviors of *Staphylococcus aureus* and *Escherichia coli* type II DNA topoisomerases. Antimicrob Agents Chemother 1996; 40:2714–2720.
18. Trucksis M, Wolfson JS, Hooper DC. A novel locus conferring fluroquinolone resistance in *Staphylococcus aureus*. J Bacteriol 1991; 173:5854–5860.
19. Ng EY, Trucksis M, Hooper DC. Quinolone resistance mutations in topoisomerase IV: relationship of the *flqA* locus and genetic evidence that topoisomerase IV is the primary target and DNA gyrase the secondary target of fluoroquinolones in *Staphylococcus aureus*. Antimicrob Agents Chemother 1996; 40:1881–1888.
20. Pan XS, Fisher LM. *Streptococcus pneumoniae* DNA gyrase and topoisomerase IV: overexpression, purification, and differential inhibition by fluoroquinolones. Antimicrob Agents Chemother 1999; 43:1129–1136.
21. Pan XS, Fisher LM. DNA gyrase and topoisomerase IV are dual targets of clinafloxacin action in *Streptococcus pneumoniae*. Antimicrob Agents Chemother 1998; 42:2810–2816.
22. Schmitz FJ, Jones ME, Hofmann B, Hansen B, Scheuring S, Lückefahr MFA, Verhoef J, Hadding U, Heinz HP, Köhrer K. Characterization of *grlA*, *grlB*, *gyrA*, and *gyrB* mutations in 116 unrelated isolates of *Staphylococcus aureus* and effects of mutations on ciprofloxacin MIC. Antimicrob Agents Chemother 1998; 42:1249–1252.
23. Kumar KP, Reddy PS, Chatterji D. Proximity relationship between the active site of *Escherichia coli* RNA polymerase and rifampicin binding domain: a resonance energy-transfer study. Biochemistry 1992; 31:7519–7526.
24. Kumar KP, Chatterji D. Differential inhibition of abortive transcription initiation at different promoters catalysed by *E. coli* RNA polymerase. Effect of rifampicin on purine or pyrimidine-initiated phosphodiester synthesis. FEBS Lett 1992; 306:46–50.
25. Jin DJ, Gross CA. Mapping and sequencing of mutations in the *Escherichia coli rpoB* gene that lead to rifampicin resistance. J Mol Biol 1988; 202:45–58.

26. Jin DJ, Cashel M, Friedman DI, Nakamura Y, Walter WA, Gross CA. Effects of rifampicin resistant *rpoB* mutations on antitermination and interaction with *nusA* in *Escherichia coli*. J Mol Biol 1988; 204:247–261.
27. Telenti A, Imboden P, Marchesi F, Schmidheini T, Bodmer T. Direct, automated detection of rifampin-resistant *Mycobacterium tuberculosis* by polymerase chain reaction and single-strand conformation polymorphism analysis. Antimicrob Agents Chemother 1993; 37:2054–2058.
28. Honore N, Cole ST. Molecular basis of rifampin resistance in *Mycobacterium leprae*. Antimicrob Agents Chemother 1993; 37:414–418.
29. Wichelhaus TA, Schäfer V, Brade V, Böddinghaus B. Molecular characterization of *rpoB* mutations conferring cross-resistance to rifamycins on methicillin-resistant *Staphylococcus aureus*. Antimicrob Agents Chemother 1999; 43:2813–2816.
30. Aubry-Damon H, Soussy CJ, Courvalin P. Characterization of mutations in the *rpoB* gene that confer rifampin resistance in *Staphylococcus auresu*. Antimicrob Agents Chemother 1998; 42:2590–2594.
31. Padayachee T, Klugman KP. Molecular basis of rifampin resistance in *Streptococcus pneumoniae*. Antimicrob Agents Chemother 1999; 43:2361–2365.
32. Carter PE, Abadi FJ, Yakubu DE, Pennington TH. Molecular characterization of rifampin-resistant *Neisseria meningitidis*. Antimicrob Agents Chemother 1994; 38:1256–1261.
33. Drancourt M, Raoult D. Characterization of mutations in the *rpoB* gene in naturally rifampin-resistant *Rickettsia* species. Antimicrob Agents Chemother 1999; 43:2400–2403.
34. Chambers HF. Penicillin-binding protein-mediated resistance in pneumococci and staphylococci. J Infect Dis 1999; 179:S353–S359.
35. Laible G, Hakenbeck R. Five independent combinations of mutations can result in low-affinity penicillin binding protein 2x of *Streptococcus pneumoniae*. J Bacteriol 1991; 173:6986–6990.
36. Hakenbeck R, Grebe T, Zähner D, Stock JB. β-Lactam resistance in *Streptococcus pneumoniae*: penicillin-binding proteins and non-penicillin-binding proteins. Mol Microbiol 1999; 33:673–678.
37. Dowson CG, Coffey TJ, Kell C, Whiley RA. Evolution of penicillin resistance in *Streptococcus pneumoniae*; the role of *Streptococcus mitis* in the formation of a low affinity PBP2B in *S. pneumoniae*. Mol Microbiol 1993; 9:635–643.
38. Smith AM, Klugman KP. Alterations in PBP 1A essential for high-level penicillin resistance in *Streptococcus pneumoniae*. Antimicrob Agents Chemother 1998; 42:1329–1333.
39. Zhao GS, Yeh WK, Carnahan RH, Flokowitsch J, Meier TI, Alborn WE, Jr, Becker GW, Jaskunas SR. Biochemical characterization of penicillin-resistant and -sensitive penicillin-binding protein 2x transpeptidase activities of *Streptococcus pneumoniae* and mechanistic implications in bacterial resistance to β-lactam antibiotics. J Bacteriol 1997; 179:4901–4908.
40. Bugg TDH, Wright GD, Dutka-Malen S, Arthur M, Courvalin P, Walsh CT. Molecular basis for vancomycin resistance in *Enterococcus faecium* BM4147:

biosynthesis of a depsipeptide peptidoglycan precursor by vancomycin resistance proteins VanH and VanA. Biochemistry 1991; 30:10408–10415.
41. Arthur M, Molinas C, Depardieu F, Courvalin P. Characterization of Tn1546, a Tn3-related transposon conferring glycopeptide resistance by synthesis of depsipeptide peptidoglycan precursors in *Enterococcus faecium* BM4147. J Bacteriol 1993; 175:117–127.
42. Arthur M, Courvalin P. Genetics and mechanisms of glycopeptide resistance in enterococci. Antimicrob Agents Chemother 1993; 37:1563–1571.
43. Reynolds PE, Depardieu F, Dutka-Malen S, Arthur M, Courvalin P. Glycopeptide resistance mediated by enterococcal transposon Tn1546 requires production of VanX for hydrolysis of D-alanyl-D-alanine. Mol Microbiol 1994; 13:1065–1070.
44. Arthur M, Depardieu F, Snaith HA, Reynolds PE, Courvalin P. Contribution of VanY D,D-carboxypeptidase to glycopeptide resistance in *Enterococcus faecalis* by hydrolysis of peptidoglycan precursors. Antimicrob Agents Chemother 1994; 38:1899–1903.
45. Arthur M, Depardieu F, Molinas C, Reynolds P, Courvalin P. The *vanZ* gene of Tn1546 from *Enterococcus faecium* BM4147 confers resistance to teicoplanin. Gene 1995; 154:87–92.
46. Wright GD, Holman TR, Walsh CT. Purification and characterization of VanR and the cytosolic domain of VanS: a two-component regulatory system required for vancomycin resistance in *Enterococcus faecium* BM4147. Biochemistry 1993; 32:5057–5063.
47. Holman TR, Wu Z, Wanner BL, Walsh CT. Identification of the DNA-binding site for the phosphorylated VanR protein required for vancomycin resistance in *Enterococcus faecium*. Biochemistry 1994; 33:4625–4631.
48. Lai MH, Kirsch DR. Induction signals for vancomycin resistance encoded by the *vanA* gene cluster in *Enterococcus faecium*. Antimicrob Agents Chemother 1996; 40:1645–1648.
49. Evers S, Courvalin P. Regulation of VanB-Type vancomycin resistance gene expression by the $VanS_B$-$VanR_B$ two-component regulatory system in *Enterococcus faecalis* V583. J Bacteriol 1996; 178:1302–1309.
50. Quintiliani R, Jr., Courvalin P. Conjugal transfer of the vancomycin resistance determinant *vanB* between enterococci involves the movement of large genetic elements from chromosome to chromosome. FEMS Microbiol Lett 1994; 119:359–364.
51. Carias LL, Rudin SD, Donskey CJ, Rice LB. Genetic linkage and cotransfer of a novel, *vanB*-containing transposon (Tn5382) and a low-affinity penicillin-binding protein 5 gene in a clinical vancomycin-resistant *Enterococcus faecium* isolate. J Bacteriol 1998; 180:4426–4434.
52. Billot-Klein D, Gutmann L, Sable S, Guittet E, Van Heijenoort J. Modification of peptidoglycan precursors is a common feature of the low-level vancomycin-resistant VANB-type *Enterococcus* D366 and of the naturally glycopeptide-resistant species *Lactobacillus casei*, *Pediococcus pentosaceus*, *Leuconostoc mesenteroides*, and *Enterococcus gallinarum*. J Bacteriol 1994; 176:2398–2405.

53. Handwerger S, Pucci MJ, Volk KJ, Liu J, Lee MS. Vancomycin-resistant *Leuconostoc mesenteroides* and *Lactobacillus casei* synthesize cytoplasmic peptidoglycan precursors that terminate in lactate. J Bacteriol 1994; 176:260–264.
54. Marshall CG, Lessard IA, Park IS, Wright GD. Glycopeptide antibiotic resistance genes in glycopeptide-producing organisms. Antimicrob Agents Chemother 1998; 42:2215–2220.
55. Poyart C, Pierre C, Quesne G, Pron B, Berche P, Trieu-Cuot P. Emergence of vancomycin resistance in the genus *Streptococcus*: Characterization of a *vanB* transferable determinant in *Streptococcus bovis*. Antimicrob Agents Chemother 1997; 41:24–29.
56. Noble WC, Virani Z, Cree RGA. Co-transfer of vancomycin and other resistance genes from *Enterococcus faecalis* NCTC 12201 to *Staphylococcus aureus*. FEMS Microbiol Lett 1992; 93:195–198.
57. Hiramatsu K, Aritaka N, Hanaki H, Kawasaki S, Hosoda Y, Hori S, Fukuchi Y, Kobayashi I. Dissemination in Japanese hospitals of strains of *Staphylococcus aureus* heterogeneously resistant to vancomycin. Lancet 1997; 350: 1670–1673.
58. Tenover FC, Lancaster MV, Hill BC, Steward CD, Stocker SA, Hancock GA, O'Hara CM, Clark NC, Hiramatsu K. Characterization of staphylococci with reduced susceptibilities to vancomycin and other glycopeptides. J Clin Microbiol 1998; 36:1020–1027.
59. Smith TL, Pearson ML, Wilcox KR, Cruz C, Lancaster MV, Robinson-Dunn B, Tenover FC, Zervos MJ, Band JD, White E, Jarvis WR. Emergence of vancomycin resistance in *Staphylococcus aureus*. N Engl J Med 1999; 340:493–501.
60. Hanaki H, Kuwahara-Arai K, Boyle-Vavra S, Daum RS, Labischinski H, Hiramatsu K. Activated cell-wall synthesis is associated with vancomycin resistance in methicillin-resistant *Staphylococcus aureus* clinical strains Mu3 and Mu50. J Antimicrob Chemother 1998; 42:199–209.
61. Strandén AM, Roos M, Berger-Bachi B. Glutamine synthetase and heteroresistance in methicillin-resistant *Staphylococcus aureus*. Microb Drug Resist 1996; 2:201–207.
62. Hiramatsu K. Vancomycin resistance in staphylococci. Drug Resist Updates 1998; 1:135–150.
63. Moreira B, Boyle-Vavra S, DeJonge BL, Daum RS. Increased production of penicillin-binding protein 2, increased detection of other penicillin-binding proteins, and decreased coagulase activity associated with glycopeptide resistance in *Staphylococcus aureus*. Antimicrob Agents Chemother 1997; 41: 1788–1793.
64. Weisblum B. Macrolide resistance. Drug Resist Updates 1998; 1:29–41.
65. Weisblum B. Erythromycin resistance by ribosome modification. Antimicrob Agents Chemother 1995; 39:577–585.
66. el Solh N, Allignet J. Staphylococcal resistance to streptogramins and related antibiotics. Drug Resist Updates 1998; 1:169–175.
67. Graham MY, Weisblum B. 23S rribosomal ribonucleic acid of macrolide-producing streptomycetes contains methylated adenine. J Bacteriol 1979; 137: 1464–1467.

68. Meier A, Kirschner P, Springer B, Steingrube VA, Brown BA, Wallace RJ, Jr, Böttger EC. Identification of mutations in the 23S ribosomal RNA gene of clarithromycin resistant *Mycobacterium intracellulare*. Antimicrob Agents Chemother 1994; 38:381–384.
69. Versalovic J, Osato MS, Spakovsky K, Dore MP, Reddy R. Point mutations in the 23S rRNA gene of *Helicobacter pylori* associated with different levels of clarithromycin resistance. J Antimicrob Chemother 1997; 40:283–286.
70. Epe B, Wooley P, Hornig H. Competition between tetracycline and tRNA at both P and A sites of the ribosome of *Escherichia coli*. FEBS Lett 1987; 213:443–447.
71. Goldman RA, Hasan T, Hall CC, Strycharz WA, Cooperman BS. Photoincorporation of tetracycline into *Escherichia coli* ribosomes. Identification of the major proteins photolabeled by native tetracycline and tetracycline photoproducts and implications for the inhibitory action of tetracycline on protein synthesis. Biochemistry 1983; 22:359–368.
72. Buck MA, Cooperman BS. Single protein omission reconstitution studies of tetracycline binding to the 30S subunit of *Escherichia coli* ribosomes. Biochemistry 1990; 29:5374–5379.
73. Martin P, Tieu-Cuot P, Courvalin P. Nucleotide sequence of the *tetM* tetracycline resistance determinant of the streptococcal conjugative shuttle transposon Tn1545. Nucleic Acids Res 1986; 14:7047–7058.
74. LeBlanc DJ, Lee LN, Titmas BM, Smith CJ, Tenover FC. Nucleotide sequence analysis of tetracycline resistance gene *tetO* from *Streptococcus mutans* DL5. J Bacteriol 1988; 170:3618–3626.
75. Taylor DE. Plasmid-mediated tetracycline resistance in *Campylobacter jejuni*: expression in *Escherichia coli* and identification of homology with streptococcal class M determinant. J Bacteriol 1986; 165:1037–1039.
76. Burdett V. Purification and characterization of Tet(M), a protein that renders ribosomes resistant to tetracycline. J Biol Chem 1991; 266:2872–2877.
77. Burdett V. tRNA modification activity is necessary for Tet(M)-mediated tetracycline resistance. J Bacteriol 1993; 175:7209–7215.
78. Schnappinger D, Hillen W. Tetracyclines: antibiotic action, uptake, and resistance mechanisms. Arch Microbiol 1996; 165:359–369.
79. Taylor DE, Chau A. Tetracycline resistance mediated by ribosomal protection. Antimicrob Agents Chemother 1996; 40:1–5.
80. Widdowson CA, Klugman KP. The molecular mechanisms of tetracycline resistance in the pneumococcus. Microb Drug Resist 1998; 4:79–84.
81. Su YA, He P, Clewell DB. Characterization of the *tet*(M) determinant of Tn916: evidence for regulation by transcriptional attenuation. Antimicrob Agents Chemother 1992; 36:769–778.
82. Taylor DE, Hiratsuka K, Ray H, Manavathu EK. Characterization and expression of a cloned tetracycline resistance determinant from *Campylobacter jejuni* plasmid pUA466. J Bacteriol 1987; 169:2984–2989.
83. Chambers HF. Methicillin resistance in staphylococci: molecular and biochemical basis and clinical implications. Clin Microbiol Rev 1997; 10:781–791.
84. Hiramatsu K, Konodo N, Ito T. Genetic basis for molecular epidemiology of MRSA. J Infect Chemother 1996; 2:117–129.

85. Kuwahara-Arai K, Kondo N, Hori S, Tateda-Suzuki E, Hiramatsu K. Suppression of methicillin resistance in a *mecA*-containing premethicillin-resistant *Staphylococcus aureus* strain is caused by the *mecI*-mediated repression of PBP 2′ production. Antimicrob Agents Chemother 1996; 40:2680–2685.
86. Niemeyer DM, Pucci MJ, Thanassi JA, Sharma VK, Archer GL. Role of *mecA* transcriptional regulation in the phenotypic expression of methicillin resistance in *Staphylococcus aureus*. J Bacteriol 1996; 178:5464–5471.
87. Hackbarth CJ, Chambers HF. *blaI* and *blaR1* regulate beta-lactamase and PBP 2a production in methicillin-resistant *Staphylococcus aureus*. Antimicrob Agents Chemother 1993; 37:1144–1149.
88. Hartman BJ, Tomasz A. Expression of methicillin resistance in heterogeneous strains of *Staphylococcus aureus*. Antimicrob Agents Chemother 1986; 29:85–92.
89. Matthews PR, Stewart PR. Resistance heterogeneity in methicillin-resistant *Staphylococcus aureus*. FEMS Microbiol Lett 1984; 22:161–166.
90. Sabath LD, Wallace SJ. Factors influencing methicillin resistance in staphylococci. Ann NY Acad Sci 1974; 236:258–266.
91. Chambers HF, Hackbarth CJ. Effect of NaC1 and nafcillin on penicillin-binding protein 2a and heterogeneous expression of methicillin resistance in *Staphylococcus aureus*. Antimicrob Agents Chemother 1987; 31:1982–1988.
92. Berger-Bachi B, Tschierske M. Role of Fem factors in methicillin resistance. Drug Resist Updates 1998; 1:325–335.
93. Berger-Bachi B, Barberis-Maino L, Strassle A, Kayser FH. FemA, a host-mediated factor essential for methicillin resistance in *Staphylococcus aureus*: molecular cloning and characterization. Mol Gen Genet 1989; 219:263–269.
94. Komatsuzawa H, Ohta K, Labischinski H, Sugai M, Suginaka H. Characterization of *fmtA*, a gene that modulates the expression of methicillin resistance in *Staphylococus aureus*. Antimicrob Agents Chemother 1999; 43:2121–2125.
95. Ryffel C, Strassle A, Kayser FH, Berger-Bachi B. Mechanisms of heteroresistance in methicillin-resistant *Staphylococcus aureus*. Antimicrob Agents Chemother 1994; 38:724–728.
96. Fujimura T, Murakami K. Increase of methicillin resistance in *Staphylococcus aureus* caused by deletion of a gene whose product is homologous to lytic enzymes. J Bacteriol 1997; 179:6294–6301.
97. Kreiswirth B, Kornblum J, Arbeit RD, Eisner W, Maslow JN, McGeer A, Low DE, Novick RP. Evidence for a clonal origin of methicillin resistance in *Staphylococcus aureus*. Science 1993; 259:227–230.
98. Archer GL, Niemeyer DM, Thanassi JA, Pucci MJ. Dissemination among staphylococci of DNA sequences associated with methicillin resistance. Antimicrob Agents Chemother 1994; 38:447–454.
99. Cookson BD. The emergence of mupirocin resistance: a challenge to infection control and antibiotic prescribing practice. J Antimicrob Chemother 1998; 41: 11–18.
100. Morton TM, Johnston JL, Patterson J, Archer GL. Characterization of a conjugative staphylococcal mupirocin resistance plasmid. Antimicrob Agents Chemother 1995; 39:1272–1280.

101. Farmer TH, Gilbart J, Elson SW. Biochemical basis of mupirocin resistance in strains of *Staphylococcus aureus*. J Antimicrob Chemother 1992; 30:587–596.
102. Gilbart J, Perry CR, Slocombe B. High-level mupirocin resistance in *Staphylococcus aureus*: evidence for two distinct isoleucyl-tRNA synthetases. Antimicrob Agents Chemother 1993; 37:32–38.
103. Hodgson JE, Curnock SP, Dyke KGH, Morris R, Sylvester DR, Gross MS. Molecular characterization of the gene encoding high-level mupirocin resistance in *Staphylococcus aureus* J2870. Antimicrob Agents Chemother 1994; 38:1205–1208.
104. Janssen DA, Zarins LT, Schaberg DR, Bradley SF, Terpenning MS, Kauffman CA. Detection and characterization of mupirocin resistance in *Staphylococcus aureus*. Antimicrob Agents Chemother 1993; 37:2003–2006.
105. Needham C, Rahman M, Dyke KG, Noble WC. An investigation of plasmids from *Staphylococcus aureus* that mediate resistance to mupirocin and tetracycline. Microbiology 1994; 140:2577–2583.
106. Ramsey MA, Bradley SF, Kauffman CA, Morton TM. Identification of chromosomal location of *mupA* gene, encoding low-level mupirocin resistance in staphylococcal isolates. Antimicrob Agents Chemother 1996; 40:2820–2823.
107. Bradley SF, Ramsey MA, Morton TM, Kauffman CA. Mupirocin resistance: clinical and molecular epidemiology. Infect Control Hosp Epidemiol 1995; 16:354–358.
108. Matthews DA, Bolin JT, Burridge JM, Filman DJ, Volz KW, Kraut J. Dihydrofolate reductase. The stereochemistry of inhibitor selectivity. J Biol Chem 1985; 260:392–399.
109. Huovinen P, Sundstrom L, Swedberg G, Sköld O. Trimethoprim and sulfonamide resistance. Antimicrob Agents Chemother 1995; 39:279–289.
110. Amyes SGB, Smith JT. The purification and properties of trimethoprim-resistant dihydrofolate reductase mediated by the R-factor, R 388. Eur J Biochem 1976; 61:597–603.
111. Flensburg J, Sköld O. Massive overproduction of dihydrofolate reductase in bacteria as a response to the use of trimethoprim. Eur J Biochem 1987; 162:473–476.
112. Smith SL, Stone D, Novak P, Baccanari DP, Burchall JJ. R plasmid dihydrofolate reductase with subunit structure. J Biol Chem 1979; 254:6222–6225.
113. Matthews DA, Smith SL, Baccanari DP, Burchall JJ, Oatley SJ, Kraut J. Crystal structure of a novel trimethoprim-resistant dihydrofolate reductase specified in *Escherichia coli* by R-plasmid R67. Biochemistry 1986; 25:4194–4204.
114. Matthews DA, Bolin JT, Burridge JM, Filman DJ, Volz KW, Kaufman BT, Beddell CR, Champness JN, Stammers DK, Kraut J. Refined crystal structures of *Escherichia coli* and chicken liver dihydrofolate reductase containing bound trimethoprim. J Biol Chem 1985; 260:381–391.
115. Joyner SS, Fling ME, Stone D, Baccanari DP. Characterization of an R-plasmid dihydrofolate reductase with a monomeric structure. J Biol Chem 1984; 259:5851–5856.
116. Dale GE, Then RL, Stuber D. Characterization of the gene for chromosomal

trimethoprim-sensitive dihydrofolate reductase of *Staphylococcus aureus* ATCC 25923. Antimicrob Agents Chemother 1993; 37:1400–1405.

117. Rouch DA, Messerotti LJ, Loo LS, Jackson CA, Skurray RA. Trimethoprim resistance transposon Tn4003 from *Staphylococcus aureus* encodes genes for a dihydrofolate reductase and thymidylate synthetase flanked by three copies of IS257. Mol Microbiol 1989; 3:161–175.
118. Lichtenstein C, Brenner S. Site-specific properties of Tn7 transposition into the *E. coli* chromosome. Mol Gen Genet 1981; 183:380–387.
119. Then RL. Mechanisms of resistance to trimethoprim, the sulfonamides, and trimethoprim-sulfamethoxazole. Rev Infect Dis 1982; 4:261–269.
120. Chopra I. Over-expression of target genes as a mechanism of antibiotic resistance in bacteria. J Antimicrob Chemother 1998; 41:584–588.
121. Spratt BG. Resistance to antibiotics mediated by target alterations. Science 1994; 264:388–393.

8

Antibiotic Permeability

Harry W. Taber
Wadsworth Center, New York State Department of Health, and
State University of New York at Albany, Albany, New York

The ability of antimicrobial compounds to enter bacterial cells generally is a prerequisite to their antibacterial action. In order to penetrate the outer layers of the cell—the cell envelope—semipermeable membranes and polymeric cell wall structures must be negotiated. Depending on the chemical nature of the antibiotic (hydrophilic or hydrophobic), penetration may occur by use of transmembrane pores, by localized disorganization of the membrane, by diffusion through the lipid bilayer, or by transport processes involving the coopting of nutrient transport systems. Bacterial species differ widely in their envelope structures, and hence in their intrinsic resistance to antibiotics.

The energetic requirements for antibiotic entry appear to vary widely by antibiotic class. Cationic compounds such as the aminoglycosides appear to respond to a threshold level of membrane potential in order to cross the cytoplasmic membrane, whereas others, such as albomycin, depend on transporters of the ATP-binding cassette (ABC) type.

Permeability of the cell envelope can be modified by exposure of bacterial cells to nonantibiotic (e.g., cationic) compounds, a maneuver that will often significantly reduce the MIC for a particular antibiotic-bacterium combination. However, since intrinsic permeability is a balance between influx (often a passive process) and efflux (usually an active process), mechanistic interpretations of "permeabilizing" treatments often are flawed.

Bacterial cells contained in biofilms appear to occur as unique, physiologically heterogeneous populations with a variety of mechanisms for preventing antibiotic action on individuals within the biofilm. Many of these mechanisms are not yet understood, but in large part do not depend on genetic modifications. It seems likely that decreased permeability accounts for at least some of this resistance.

1 INTRODUCTION

As with other portions of this volume, this chapter does not presume to be comprehensive about its subject even if attention is confined to the most recent literature. Rather citations are made to reviews and to publications in the primary literature that illustrate certain advances in the field or problems that have not yielded to adequate solutions. As a consequence, numerous excellent publications have had to be omitted; however, bibliographies contained within the references that are cited will lead the reader to the very considerable depth that the permeability literature enjoys. Except for those concerned with antimicrobial peptides and with efflux of antimicrobial agents, general reviews on bacterial permeability to these agents have not appeared in recent years. Thorough coverage of the earlier literature can be found in the article by Chopra and Ball (1), and for gram-negative bacteria, in the article by Hancock and Bell (2).

Because of the intimate association between uptake of antibiotics into bacterial cells and the drug efflux pathways that have become the focus of so much study during the past several years, discussions of antibiotic permeability are often found embedded within publications the principal intent of which is to explore contributions of efflux mechanisms to antibiotic resistance. The reader should be alert to this, and carefully probe these publications for relationships between drug influx and efflux phenomena. In the text that follows, several illustrations of these relationships will be found.

2 WHAT DO WE MEAN BY PERMEABILITY?

Permeability in its broadest sense means the properties of a cell (in this case, a bacterial cell) that allow an ion or molecule to traverse one or more of its boundary structures (its *envelope* [see below]) and enter the cell. As applied to antimicrobial agents, this might involve penetration only of the outermost layer of the cell envelope, because the particular agent in question would not have to enter the bacterial cell cytoplasm proper in order to exert its inhibitory action. Alternatively, many antibiotics act only after reaching the cytoplasm (e.g., those that affect DNA, RNA or protein synthesis), and thus penetration of the cytoplasmic membrane is required.

The mechanisms for entry of antimicrobial agents vary widely depending upon the chemical nature of the agent and the characteristics of the envelope structure being penetrated. Some of these mechanisms will be discussed below, but in many cases they are not known. A general term to describe entry is *antibiotic uptake*, which simply states that the molecule is moving inward from the environment. A related term is *antibiotic accumulation*, which carries the implication that concentrations of the molecule inside the bacterial cell are higher than outside; that is, inward movement is occurring against a concentration gradient, and energy is being expended in this accumulation process. Finally, use of the term *transport* indicates that the uptake process is both specific and saturable and in-

TABLE 1 Paths for Antibiotic Uptake by Bacteria

1. Uptake by passive diffusion
   a. Saturable, apparently dependent on a facilitator(s) contained within the cell envelope; this facilitator could be either in the outer or the inner membrane, and would be rate-limiting for uptake.
   b. Nonsaturable, not dependent on interaction with a facilitator. Transit of the outer membrane could occur either through a porin-based structure, or—for lipophilic compounds—through the membrane bilayer. Lipophilic compounds could also readily transit the cytoplasmic membrane bilayer, but charged antimicrobials would encounter a barrier unless some mechanism such as self-promoted uptake (see below) were operating.
2. Energy-requiring uptake
   Accumulation of antimicrobial drugs inside the bacterial cell occurs against a concentration gradient. Operationally, inhibitors of energy metabolism will prevent uptake. Uptake could depend either on ATP or on the proton-motive force as a source of energy.
3. Self-promoted uptake
   The antimicrobial agent interacts with a cell envelope structure in such as way as to increase the permeability of that structure to the agent; this seems to occur for certain positively charged antibiotics as they encounter the outer membrane of gram-negative bacteria.
4. Illicit uptake
   The antimicrobial agent utilizes an uptake system normally used by the bacterial cell for uptake of a metabolite or nutrient.

volves some type of carrier component in one or another of the boundary layers. All of these terms are brought to bear in describing antibiotic permeability in the literature.

Table 1 summarizes modes of uptake of antimicrobial compounds by bacterial cells.

3 PERMEABILITY BARRIERS OF THE BACTERIAL CELL

3.1 Structure of Bacterial Cell Envelopes

The envelope of the bacterial cell comprises all structures external to the cytoplasm. This includes the cytoplasmic membrane (CM), which immediately overlies the cytoplasm; the cell wall, composed largely of the polymer peptidoglycan (cross linked to various degrees in different species); and, in gram-negative bacteria, the outer membrane (OM). Gram-positive bacteria lack an OM, but have in general a much thicker cell wall layer than do gram-negatives. In addition, some bacteria, particularly pathogens, possess a capsule lying outside the OM. As a consequence of their two-membrane envelope gram-negative bacteria have a compartment, called the *periplasmic space*, located between the CM and OM, containing many proteins and certain other macromolecules. Gram-positive bacteria have a somewhat comparable compartment within their cell walls, called appropriately the *wall space*. Depending upon the particular gram-positive species, the wall space also contains proteins, some anchored in the CM, others retained within this space by mechanisms not yet elucidated.

Any antimicrobial compound that requires access to the cell in order to act as an inhibitory or bactericidal agent must traverse the bacterial cell envelope, since it completely surrounds the cell. It is not surprising to find, then, that antibiotic-inactivating enzymes are found in the envelope: β-lactamases and aminoglycoside-modifying enzymes are examples.

3.2 Envelope Components Relevant to Antibiotic Permeability

Lipopolysaccharide (LPS) is a complex molecule characteristic of the gram-negative OM, and maintenance of its integrity is essential to proper functioning of the outer membrane as a permeability barrier. It has been known for some time that this structure provides an effective impediment to free movement of antibiotics into the bacterial cell (see, e.g., Refs. 3–5). Detailed analysis of precisely which constituents are essential for LPS integrity has revealed that they are contained within the oligosaccharide *core region* of the molecule, which is strongly negatively charged. The presence of LPS only in the outer leaflet of the OM, with negative charges partially neutralized by divalent cations such as Mg^{2+} and Ca^{2+}, provides a "carpet" of surface negative charge, contributing to the role of the OM as a selective permeability barrier. The core region is being dissected genetically (see e.g., Ref. 6) in order to find the molecular requirements for LPS integrity; so-called "deep-rough" mutations convey hypersensitivity to certain antibiotics (e.g., novobiocin) and to surface-active agents such as sodium dodecylsulfate.

A second class of constituents of the gram-negative OM that produces its permeability characteristics is the porins. These are proteins capable of forming *trans*-OM channels ("pores") that have an aqueous interior and permit influx of hydrophilic small molecules such as nutrients and efflux of waste products (5,7). However, the pores exclude many antibiotics because of their size and in many cases their lipophilicity. The so-called "classic" porins occur in the outer membrane as trimers of a 36 to 38-kD monomer, with a quite open structure to the channel formed. The naming of porins is often referred to the *Escherichia coli* Omp (outer membrane protein) nomenclature, because they were first studied systematically in this species. Thus, OmpC and OmpF together with PhoE are found in *E. coli*, and similar porins are widely distributed among gram-negative bacteria. In addition to size limitation, the trimeric pores formed have some preference for the type of small molecules that will be admitted, with OmpC and OmpF selecting molecules with positive or no charge and PhoE preferring anions. Other porins are synthesized under special circumstances; for example, the LamB porin of *E. coli*, which has specificity for oligosaccharides, is formed under conditions of carbon limitation. Nikaido (9) has presented a brief overview of porin regulation. Although a discussion of the regulation of the classic porins is beyond the scope of the present discussion, it is worth noting that OmpF, which has a larger pore diameter than OmpC, is downregulated in *E. coli* in response to antibiotic exposure mediated by the global regulator MarA (reviewed in Ref. 10; also see Chap. 3).

4 SELF-INDUCED UPTAKE OF AMINOGLYCOSIDE ANTIBIOTICS THROUGH THE OUTER MEMBRANE OF GRAM-NEGATIVE BACTERIA

Polycationic antibiotics can traverse the outer membrane by means other than diffusion through pores. As originally outlined by Hancock (2,11,12), aminoglycoside antibiotics such as tobramycin and gentamicin enter gram-negative bacteria by a self-promoted pathway involving disruption of LPS-Mg^{2+} cross bridges, which as indicated above, pose the major barrier to antibiotic entry. Disruption depends on the cationic structure of aminoglycosides, can be effected by other polycations such as polymyxin and protamine, and is mechanistically dependent on displacement of Mg^{2+} from cross bridges. The resulting rearrangement of LPS and exposure of the outer membrane bilayer provides sufficient localized destabilization of the bilayer to allow entry of aminoglycoside to the periplasm. Bacterial species such as *Burkholderia cepacia*, which have LPS with a low-phosphate and high-arabinosamine content, are resistant to polycationic antibiotics, apparently because the *B. cepacia* LPS does not bind polycations effectively (13). The general phenomenon of self-induced uptake is discussed by Hancock (5) in the context of modification of outer membrane permeability for enhancement of antibiotic efficacy.

5 RESISTANCE ASSOCIATED WITH SPECIES-SPECIFIC VARIATIONS AND MUTATIONAL ALTERATION IN ENVELOPE STRUCTURE

Specificity of porins for substrates is sometimes extended to antibiotics; for example, the β-lactam imipenem is specifically taken up by *Pseudomonas aeruginosa* through the substrate-specific porin OprD (14). OprD formation is highly regulated by nitrogen and carbon sources (15), and its downregulation results in imipenem resistance (16), as do mutations in OprD (17).

Do altered porins occur in drug-resistant clinical isolates? In some species, the answer would seem to be yes: Mallea et al. (18) described two nosocomial clones of *Enterobacter aerogenes* lacking OmpC and OmpF and a third having thermolabile porin. This collection of clinically derived strains has been augmented by Bornet et al. (19); the emergence of imipenem (and multidrug) resistance associated with decreased porin synthesis could be reversed by cessation of imipenem therapy. It appears that at least with *Enterobacter*, the ease with which porin deficiency–associated multidrug resistance arises, in combination with the presence of extended-spectrum β-lactamase, is already creating major difficulties in treatment of this third-leading cause of nosocomial respiratory tract infections. A similar situation may be presenting itself in therapy of *Klebsiella pneumoniae* (20,21). Clinical isolates of this pathogen now are appearing that have lost the OmpK35 porin (a homologue of *E. coli* OmpF), and like *Enterobacter*, this phenotype is seen in combination with the presence of extended-spectrum β-lactamases.

A common pattern may be starting to dominate the emergence of permeability-associated single-drug or multidrug resistance, at least in gram-negative pathogens: restricted formation of porins combined with the induction or overexpression of efflux pump(s) with narrow or broad specificity.

5.1 Illicit Uptake: The Use of Metabolic Uptake Systems for Antibiotic Entry

If an antibiotic bears a sufficiently close structural relationship to a molecule for which bacteria have specific uptake systems, then the antibiotic may be carried into the cell by that system. Two recent examples of this phenomenon are outlined below.

The work of Raynaud et al. (22) on uptake of the antituberculosis drug pyrazinamide (PZA) by mycobacteria suggests that the drug is transported into the cell by a system normally used for transporting nicotinamide. That this transport system has substantial specificity is supported by the finding that pyrazinoic acid (POA), the intracellularly active form of the drug, has low activity when administered extracellularly; that is, the amide form but not the acid form can utilize the nicotinamide transport system.

Measurement of PZA susceptibility of *Mycobacterium tuberculosis* strains is complicated by the low in vitro activity of the drug at neutral pH, a problem that is obviated by carrying out susceptibility measurements at acidic pH (23). The reason for this has been analyzed recently by Zhang et al. (24) as part of a more extensive study on the biochemical basis of PZA susceptibility in *M. tuberculosis* (largely due to defective efflux). The susceptibility measurement effect involves the equilibration of POA—synthesized intracellularly from accumulated PZA—in its protonated form across the cytoplasmic membrane (POA, in common with most weak organic acids, will diffuse across a lipid bilayer). When the extracellular pH is neutral (i.e., similar to the intracellular pH), protonated POA will continue to diffuse out of the cell in an effort to achieve equilibrium with the much larger extracellular volume, and this diffusion will continue as POA is produced from PZA, preventing sufficient accumulation of POA to have an inhibitory effect. Equilibration of protonated POA occurs at a much lower extracellular concentration when the pH outside the cell is acidic, allowing accumulation inside the cell sufficient for inhibition to occur.

The above discussion illustrates some of the general complexities of antimicrobial permeability; in this particular case, although a specific transport system may allow entry of the (pro)drug, conversion to a different chemical form requires an entirely new study of the transmembrane behavior of the active drug.

It seems intuitively likely that bacterial transport systems for nutrients or metabolites will not be easily coopted for entry by naturally occurring antibiotics, since the availability of antibiotics emanating from producer species would reasonably have created a selective pressure for susceptible species to modify (via evolutionary selection) their transport systems to exclude antibiotics. Indeed, the relatively few examples of illicit transport of antimicrobials appear rather to involve synthetic drugs (such as PZA) with antimicrobial activity. However, if the nutrient is sufficiently essential, evolution of the transport system may be constrained, and its exploitation for uptake by a structurally related, naturally occurring antibiotic will remain intact. Such would seem to be the case for the transport of iron-binding molecules (siderophores) across the outer membrane of gram-negative bacteria (reviewed in Ref. 25). Iron—as Fe^{+3}—is insoluble, and is transported into bacteria as soluble iron complexes, bound to bacterially synthesized, low molecular weight siderophores. Outer membrane proteins with requisite specificity bind siderophore-iron complexes, transport them across the outer membrane by energy-dependent processes, and deposit them in the periplasm, where the complexes bind to soluble binding proteins, which in turn deliver them to ABC (ATP-binding cassette) transporters in the cytoplasmic membrane. Gram-positive bacteria have essentially similar systems but without the outer membrane–binding protein component.

Albomycin is a broad-spectrum antibiotic of the Fe^{+3}-binding sideromycin class produced by a species of *Streptomyces*. It has a low MIC for both gram-positive and gram-negative bacterial species, and this high specific activity depends on illicit transport of albomycin by the FhuA-FhuD-FhuB-FhuC system, the physiological activity of which is to transport ferrichrome-Fe^{+3} complexes into the cell. FhuA is the outer membrane–binding protein, FhuD is the periplasm-binding protein, and FhuB and FhuC form a cytoplasmic membrane protein complex that, when energized by ATP, catalyzes the entry of ferrichrome-Fe^{+3} or albomycin-Fe^{+3} complexes into the cytoplasmic compartment. Active transport across the outer membrane via FhuA is thought to depend on input of energy from the cytoplasmic membrane via the TonB-ExbB-ExbD complex (26). When albomycin-Fe^{+3} enters the cytoplasm, a thioribosyl pyrimidine moiety with cell-inhibitory activity must be cleaved by a peptidase from the parent antibiotic molecule. As stressed by Braun (25), this arrangement provides an approach to the design of antibiotics that can be actively transported into bacteria. Structures of Fe^{+3}-hydroxamate and Fe^{+3}-albomycin cocrystals with FhuA provide detailed molecular parameters to the design of such compounds. Interestingly, a rifamycin derivative (CGP 4832), which is not structurally related to either hydroxamates or albomycin (and does not bind iron), also is actively transported across the outer membrane by the FhuA protein energized by the TonB-ExbB-ExbD system, but not involving the other Fhu proteins. This suggests that, after reaching the periplasm, the antibiotic diffuses across the cytoplasmic membrane down its concentration gradient.

Other antibiotic classes have been conjugated to siderophores in order to utilize the siderophore transport systems for delivering active compounds to the bacterial cytoplasm. These are summarized by Braun (25) and include sulfonamides and β-lactams. Results were somewhat mixed; this could be due to difficulties in obtaining intracellular release of the active moiety or delivery to a cellular compartment that is not optimal for antibiotic action. Although the use of Fe^{+3}-siderophore delivery systems for illicit antibiotic uptake is still in an early phase, the general approach is very promising, for at least two reasons: (1) in infections, circulating Fe^{+3} is commonly extremely low, and pathogen iron uptake systems will be derepressed and attempting to function at high efficiency; and (2) antibiotic resistance by loss of an uptake system would be detrimental to the success of the pathogen. Thus, this would appear to be a promising avenue for the development of semisynthetic compounds based on siderophore structures. Clearly knowledge of the intracellular systems that release moieties with antibiotic activity will have to be gained; those systems that are essential to survival of the targeted pathogen would seem to be the best to explore.

6 ENERGY-DEPENDENT UPTAKE OF ANTIBIOTICS AND RESISTANCE ARISING FROM COMPROMISE OF ENERGY METABOLISM

In accord with the thermodynamics of diffusion processes, bacteria will accumulate an antibiotic in their cytoplasmic compartment at higher concentrations than in the extracellular milieu only if an energy-requiring accumulation process is involved. The exception to this occurs if an intracellular binding site for the antibiotic exists, with a sufficiently strong binding constant such that a significant fraction of intracellular antibiotic is not freely diffusible. In this situation, antibiotic would continue to diffuse into the cell down its concentration gradient.

In general, two modes of providing energy to an antibiotic uptake system can be envisioned: (1) the proton electrochemical gradient and (2) ATP (as described above for siderophores and albomycin). Either mode requires that a mechanism exists in the CM for coupling utilization of energy to transmembrane movement of the antibiotic. This might, as discussed in the previous section, be coupled to movement across the OM as well as the CM.

The dependence of aminoglycoside antibiotic uptake on energy production by bacterial cells was established four decades ago (see Refs. 27–29, and references therein). Uptake of aminoglycosides has been studied more intensively by more investigators than any other antibiotic. This focus of attention probably has been due to the energy-dependent nature of the uptake process across the CM, and the insights that it might provide to antibiotic uptake in general. This energy dependence has been associated with the membrane potential (30–33) and may reflect a response by positively charged antibiotics to that potential, which carries a negative charge on the interior face of the CM. Results of Miller and associates (34) are consistent with diffusion across the CM through a voltage-gated channel, which closes following uptake owing to decreased membrane potential associated with effects of aminoglycosides themselves on the CM. This would suggest that in the absence of a membrane potential, no uptake would occur. However, *Escherichia coli* (35), and perhaps other bacteria, can adapt to loss of membrane potential and take up aminoglycosides by ATP-dependent processes.

Loss of aminoglycoside uptake associated with energetic deficiencies in clinically important pathogens such as *Staphylococcus aureus* has been established for some years (see, e.g., Ref. 36). Proctor and his colleagues (37–39) have revisited this problem with small-colony variant (SCV) clinical isolates of *S. aureus*. They have stressed the importance of the ability of these variants to persist in chronic infections in the face of antibiotic treatment. Whereas reduced uptake of cationic antimicrobials accounts for resistance to aminoglycosides, resistance to other antibiotics (e.g., β-lactams) appears to result from changes in growth rate or other more general physiological alterations.

It seems likely that a variety of uptake systems, energized by either a membrane potential or by ATP, will be found to exist in different bacterial species; for example, as discussed in Chap. 15, the antituberculosis drug pyrazinamide recently has been found to enter mycobacteria via an ATP-dependent transport system. Understanding the energetics of antibiotic efflux will require a comparable understanding of the energetics of influx, since approaches to blocking or reducing efflux must not also compromise uptake. In some situations, such as the uptake of rifampin by *E. coli* and *S. aureus* (40), uptake of the antibiotic appears not to be energy dependent at all, as judged by the lack of any effect of inhibitors such as carbonylcyanide *m*-chlorophenyl hydrazone (CCCP) and dinitrophenol on the process. In this particular example, the saturability of rifampin accumulation appeared to reflect its binding to the intracellular target molecule (RNA polymerase) rather than specific interaction with an envelope component.

This may be the situation for many antimicrobial compounds, but to establish this will require study of each individual antibiotic or antibiotic class.

7 MODIFICATION OF UPTAKE BY ADMINISTRATION OF A SECOND AGENT

Combined antibiotic therapy has of course a venerable history in infectious disease therapy. However, increased understanding of bacterial physiology and structure can provide possibilities for therapeutic use of agents that are not themselves antimicrobial, but have an enhancing effect on the action of a known antibiotic. For example, Rajyaguru and Muszynski (41) showed that antimicrobial susceptibility of *B. cepacia* isolates to several standard antibiotics could be significantly enhanced in vitro (fourfold or greater reduction in MIC) by the cationic compounds chlorpromazine and prochlorperazine. The mechanism by which this apparent change in permeability occurred was not clarified, but did not involve changes in outer membrane composition, nor was outer membrane permeability to the fluorescent probe 1-*N*-phenylnaphthylamine enhanced. However, electron microscopy revealed a widening of the periplasmic space, suggesting that these cationic agents may alter interactions between outer membrane and cytoplasmic membrane.

There are difficulties in this type of analysis because of the probable involvement of drug efflux systems in most intrinsic antibiotic resistance (see Chap. 4). In gram-negative bacteria, the structure of these efflux systems commonly is characterized by membrane-spanning cytoplasmic membrane and outer membrane proteins joined by a periplasmic connecting protein. Thus, modifications of the spatial relationship between cytoplasmic and outer membranes could disrupt the structural integrity of drug efflux systems. The opportunistic pathogen *P. aeruginosa* possesses a highly impermeable outer membrane, which contributes to its intrinsic multidrug resistance. This broad resistance is augmented by several multidrug efflux systems (see, e.g., Ref. 42; also Chap. 4). The interplay between the *P. aeruginosa* outer membrane and these efflux systems recently was described by Poole's group (43). In this study, enhancement of outer membrane permeability (using agents whose action was known) could be separated from reduction in efflux by use of mutants in which the MexA-MexB-OprM efflux pump had been genetically inactivated. The two effects (permeability enhancement and efflux reduction) were found to be synergistic in enhancing antibiotic susceptibility. Interestingly, overproduction of the efflux system by means of multicopy plasmids containing the *mexA-mexB-oprM* genes overcame the permeability agents, increasing drug resistance above that of the wild type. Thus, a response of an efflux system–containing bacterial pathogen population to a combined antibiotic/permeabilizing agent exposure might be to select out mutants in which duplication of the efflux genes has occurred.

The area of outer membrane permeabilization has been reviewed by Vaara (44), with a significant focus on cationic agents such as antimicrobial peptides. More recently, Nikaido (45) has discussed the contributions of outer membrane and efflux pumps to drug resistance in gram-negative bacteria in the context of the possibilities for improving drug access. It is clear from the examples cited above, that although the permeabilization approach has promise, it will be necessary to understand a great deal more about the response of bacterial cells to enhanced access of antimicrobials to the gram-negative periplasmic space and to the cytoplasm of all bacteria. A more promising route to counteracting drug efflux may be to find agents that inhibit the efflux pumps directly, as discussed in Chapter 4.

8 EFFECTS OF BIOFILM FORMATION ON ANTIBIOTIC PERMEATION

Much attention has been focused in recent years on bacterial biofilms based on the recognition that they occur commonly both at infection sites and during environmental growth of bacterial populations (46,47). The genetics and physiology of the development of biofilms is being thoroughly explored (48–50). The role of biofilms in modifying antibiotic susceptibility of bacterial cells within these structures has received its share of attention. Initially, it was supposed that the biofilm acted only as a physical barrier, with the abundant extracellular shielding resident bacterial cells from the action of antibiotics by retarding diffusion of the compounds. However, recent data suggests that other explanations are more likely (see Ref. 51, and references therein; 52).

First, the apparent lack of diffusion of some antibiotics into biofilms can be attributed in some cases to reaction inactivation; that is, the inactivation of the antibiotic by the generally higher concentration of cells at the periphery of the biofilm. An example would be failure of β-lactams to penetrate the biofilm when β-lactamase is present in the biofilm population; mutational loss of the β-lactamase results in ready penetration (51).

Second, it has been suggested repeatedly that biofilms contain a physiologically heterogeneous cell population (see, e.g., Refs. 46 and 49), and direct evidence is available to support this contention (see, e.g., Refs. 49 and 53). This heterogeneity includes *persistors*; that is, a fraction of cells that—in the context of antibiotic killing—are resistant even to high drug concentrations (see Ref. 50 for a discussion). Biofilm cells that survive bactericidal drug treatments are not drug-resistant mutants, instead they remain wild type in their genetic makeup. Biofilms appear to promote the formation of higher fractions of these persistors than do planktonic (non-biofilm) cultures. Extension of in vitro studies to in vivo infections involving biofilm formation by bacterial pathogens can be made as follows: antibiotic treatment will eliminate most genetically susceptible members of the population both inside and outside a biofilm, and cells of the immune system will eliminate the planktonic remainder. However, it is known that immune cells are unable to penetrate the extracellular polysaccharide matrix that characterizes biofilms (54); therefore, if a residuum of persistors is present inside the biofilm, the potential for regrowth of the pathogen population following termination of therapy remains significant.

9 OVERCOMING RESISTANCE ASSOCIATED WITH PERMEABILITY BARRIERS

The use of permeability-enhancing agents to improve drug access was discussed above (see Sec. 7). A second approach might be to circumvent the permeability barrier entirely by presenting antibiotic to the cell interior via a membrane fusion mechanism. For example, in recent studies, drug-resistant mucoid *P. aeruginosa* could be eradicated in an animal model by the use of a suspension of liposomes containing encapsulated bactericidal antibiotic (55). Mechanistic studies suggested that delivery of the antibiotic to the bacterial cytoplasm occurred by liposome-bacteria fusion. This is particularly noteworthy because of the high degree of intrinsic resistance exhibited by *P. aeruginosa* consequent to the low-permeability characteristics of its outer membrane.

REFERENCES

1. Chopra I, Ball P. Transport of antibiotics into bacteria. Adv Microb Physiol 1982; 23:183–240.
2. Hancock REW, Bell A. Antibiotic uptake in Gram-negative bacteria. Eur J Clin Microbiol 1988; 7:713–720.
3. Nikaido H. Outer membrane barrier as a mechanism of antimicrobial resistance. Antimicrob Agents Chemother 1989; 33:1831–1836.
4. Van JCN, Gutmann L. Antibiotic resistance due to reduced permeability in gram-negative bacteria. Presse Med 1994; 23:522–531.
5. Hancock REW. The bacterial outer membrane as a drug barrier. Trends Microbiol 1997; 5:37–42.
6. Yethon JA, Heinrichs DE, Monteiro MA, Perry MB, Whitfield C. Involvement of *waaY*, *waaO*, and *waaP* in the modification of *Escherichia coli* lipopolysac-

charide and their role in the formation of a stable outer membrane. J Biol Chem 1998; 273:26310–26316.

7. Nikaido H. Prevention of drug access to bacterial targets: permeability barriers and active efflux. Science 1994; 264:382–388.
8. Nikaido H. Porins and specific diffusion channels in bacterial outer membranes. J Biol Chem 1994; 269:3905–3908.
9. Nikaido H. Microdermatology: cell surface in the interaction of microbes with the external world. J Bacteriol 1999; 181:4–8.
10. Alekshun MN, Levy SB. The *mar* regulon: multiple resistance to antibiotics and other toxic chemicals. Trends Microbiol 1999; 7:410–413.
11. Hancock REW, Raffle VJ, Nicas TI. Involvement of the outer membrane in gentamicin and streptomycin uptake and killing in *Pseudomonas aeruginosa*. Antimicrob Agents Chemother 1981; 19:777–785.
12. Hancock REW. Alterations in outer membrane permeability. Annu Rev Microbiol 1984; 38:237–264.
13. Moore RA, Hancock REW. Involvement of outer membrane of *Pseudomonas cepacia* in aminoglycoside and polymyxin resistance. Antimicrob Agents Chemother 1986; 30:923–926.
14. Trias J, Nikaido H. Protein D2 channel of the *Pseudomonas aeruginosa* outer membrane has a binding site for basic amino acids and peptides. J Biol Chem 1990; 265:15680–15684.
15. Ochs MM, Lu CD, Hancock REW, Abdelal AT. Aminoacid-mediated induction of the basic amino acid-specific outer membrane porin OprD from *Pseudomonas aeruginosa*. J Bacteriol 1999; 181:5426–5432.
16. Masuda N, Sakagawa E, Ohya S. Outer membrane proteins responsible for multiple drug resistance in *Pseudomonas aeruginosa*. Antimicrob Agents Chemother 1995; 39:645–649.
17. Kohler T, Michea-Hamzehpour M, Epp SF, Pechere JC. Carbapenem activities against *Pseudomonas aeruginosa*: respective contributions of OprD and efflux systems. Antimicrob Agents Chemother 1999; 43:424–427.
18. Mallea M, Chevalier J, Bornet C, Eyraud A, Davin-Regli A, Bollet C, Pages J-M. Porin alteration and active efflux: two in vivo drug resistance strategies used by *Enterobacter aerogenes*. Microbiol 1998; 144:3003–3009.
19. Bornet C, Davin-Regli A, Bosi C, Pages J-M, Bollet C. Imipenem resistance of *Enterobacter aerogenes* mediated by outer membrane permeability. J Clin Microbiol 2000; 38:1048–1052.
20. Martinez-Martinez L, Hernandez-Alles S, Alberti S, Tomas JM, Benedi VJ, Jacoby GA. In vivo selection of porin deficient mutants of *Klebsiella pneumoniae* with increased resistance to cefoxitin and third generation cephalosporins. Antimicrob Agents Chemother 1996; 40:342–348.
21. Hernandez-Alles S, Alberti S, Alvarez D, Domenech-Sanchez A, Martinez-Martinez L, Gil J, Thomas JM, Benedi VJ. Porin expression in clinical isolates of *Klebsiella pneumoniae*. Microbiology 1999; 145:673–679.
22. Raynaud C, Laneele M-A, Senaratne RH, Draper P, Laneele G, Daffe M. Mechanisms of pyrazinamide resistance in mycobacteria: importance of lack of uptake in addition to lack of pyrazinamidase activity. Microbiology 1999; 145:1359–1367.

23. Salfinger M, Heifetz L. Determination of pyrazinamide MICs for *Mycobacterium tuberculosis* at different pHs by the radiometric method. Antimicrob Agents Chemother 1988; 32:1002–1004.
24. Zhang Y, Scorpio A, Nikaido H, Sun Z. Role of acid pH and deficient efflux of pyrazinoic acid in unique susceptibility of *Mycobacterium tuberculosis* to pyrazinamide. J Bacteriol 1999; 181:2044–2049.
25. Braun V. Active transport of siderophore-mimicking antibacterials across the outer membrane. Drug Resist Updates 1999; 2:363–369.
26. Braun V. Energy-coupled transport and signal transduction through the gram-negative outer membrane via TonB-ExbB-ExbD-dependent receptor proteins. FEMS Microbiol Rev 1995; 16:295–307.
27. Hancock REW. Aminoglycoside uptake and mode of action—with special reference to streptomycin and gentamicin. I. Antagonists and mutants. J Antimicrob Chemotherapy 1981; 8:249–276.
28. Hancock REW. Aminoglycoside uptake and mode of action—with special reference to streptomycin and gentamicin. II. Effects of aminoglycosides on cells. J Antimicrob Chemotherapy 1981; 8:429–445.
29. Taber HW, Mueller JP, Miller PF, Arrow AS. Bacterial uptake of aminoglycoside antibiotics. Microbiol Rev 1987; 51:439–457.
30. Bryan LE, Van den Elzen HM. Effects of membrane-energy mutations and cations on streptomycin and gentamicin accumulation by bacteria: a model for entry of streptomycin in susceptible and resistant bacteria. Antimicrob Agents Chemother 1979; 12:163–177.
31. Bryan LE, Kwan S. Roles of ribosomal binding, membrane potential, and electron transport in bacterial uptake of streptomycin and gentamicin. Antimicrob Agents Chemother 1983; 23:835–845.
32. Mates SM, Eisenberg ES, Mandel LJ, Patel L, Kaback HR, Miller MH. Membrane potential and gentamicin uptake in *Staphylococcus aureus*. Proc Natl Acad Sci USA 1982; 79:6693–6697.
33. Eisenberg ES, Mandel LJ, Kaback HR, Miller MH. Quantitative association between electrical potential across the cytoplasmic membrane and early gentamicin uptake and killing in *Staphylococcus aureus*. J Bacteriol 1984; 157: 863–867.
34. Leviton IM, Fraimow HS, Carrasco N, Dougherty TJ, Miller MH. Tobramycin uptake in *Escherichia coli* membrane vesicles. Antimicrob Agents Chemother 1995; 39:467–475.
35. Fraimow HS, Greenman JB, Leviton IM, Dougherty TJ, Miller MH. Tobramycin uptake in *Escherichia coli* is driven by either electrical potential or ATP. J Bacteriol 1991; 173:2800–2808.
36. Miller MH, Edberg SC, Mandel LJ, Behar CF, Steigbigel NH. Gentamicin uptake in wild-type and aminoglycoside-resistant small-colony mutants of *Staphylococcus aureus*. Antimicrob Agents Chemother 1980; 18:722–729.
37. Proctor RA, Peters G. Small colony variants in staphylococcal infections: diagnostic and therapeutic implications. Clin Infect Dis 1998; 27:419–423.
38. Proctor RA, Kahl B, von Eiff C, Vaudaux PE, Lew DP, Peters G. Staphylococcal small colony variants have novel mechanisms for antibiotic resistance. Clin Infect Dis 1998; 27(Suppl 1):S68–74.

39. McNamara PJ, Proctor RA. *Staphylococcus aureus* small colony variants, electron transport and persistent infections. Int J Antimicrob Agents 2000; 14: 117–122.
40. Williams KJ, Piddock LJV. Accumulation of rifampicin by *Escherichia coli* and *Staphylococcus aureus*. J Antimicrob Chemother 1998; 42:597–603.
41. Rajyaguru JM, Muszynski MJ. Enhancement of *Burkholderia cepacia* antimicrobial susceptibility by cationic compounds. J Antimicrob Chemother 1997; 40:345–351.
42. Li X-Z, Nikaido H, Poole K. Role of *mexA-mexB-pprM* in antibiotic efflux in *Pseudomonas aeruginosa*. Antimicrob Agents Chemother 1995; 39:1948–53.
43. Li X-Z, Zhang L, Poole K. Interplay between the MexA-MexB-OprM multidrug efflux system and the outer membrane barrier in the antibiotic resistance of *Pseudomonas aeruginosa*. J Antimicrob Chemother 2000; 45:433–436.
44. Vaara M. Agents that increase the permeability of the outer membrane. Microbiol Rev 1992; 56:395–411.
45. Nikaido H. The role of outer membrane and efflux pumps in the resistance of gram-negative bacteria. Can we improve drug access? Drug Resist Updates 1998; 1:93–8.
46. Costerton JW, Lewandowski Z, Caldwell DE, Korber DR, Lappin-Scott HM. Microbial biofilms. Annu Rev Microbiol 1995; 49:711–745.
47. Costerton JW, Stewart PS, Greenberg EP. Bacterial biofilms: a common cause of persistent infections. Science 1999; 284:1318–1322.
48. O'Toole GA, Pratt LA, Watnick PI, Newman DK, Weaver VB, Kolter R. Genetic approaches to study of biofilms. Methods Enzymol 1999; 310:91–109.
49. Pratt LA, Kolter R. Genetic analyses of bacterial biofilm formation. Curr Opin Microbiol 1999; 2:598–603.
50. Lewis K. Programmed death in bacteria. Microbiol Molec Biol Rev 2000; 64: 503–514.
51. Anderl JN, Franklin MJ, Stewart PS. Role of antibiotic penetration limitation in *Klebsiella pneumoniae* biofilm resistance to ampicillin and ciprofloxacin. Antimicrob Agents Chemother 2000; 44:1818–1824.
52. Brooun A, Liu S, Lewis K. A dose-response study of antibiotic resistance in *Pseudomonas aeruginosa* biofilms. Antimicrob Agents Chemother 2000; 44:640–644.
53. Huang C-T, Xu KD, McFeters GA, Stewart PS. Spatial patterns of alkaline phosphatase expression within bacterial colonies and biofilms in response to phosphate starvation. Appl Environ Microbiol 1998; 64:1526–1531.
54. Hoyle BD, Jass J, Costerton JW. The biofilm glycocalyx as a resistance factor. J Antimicrob Chemother 1990; 26:1–5.
55. Sachetelli S, Khalil H, Chen T, Beulac C, Senechal S, Legace J. Demonstration of a fusion mechanism between a fluid bactericidal liposomal formulation and bacterial cells. Biochim Biophys Acta-Biomembranes 2000; 1463:254–266.

9

Phenotypic Tolerance of Bacteria

Rodger Novak
Vienna Biocenter, Vienna, Austria

Elaine I. Tuomanen
St. Jude Children's Research Hospital, Memphis, Tennessee

Since the earliest studies on the mechanism of action of penicillin in the 1940s, it has been evident that penicillin does not kill nongrowing bacteria (1–4). Within minutes after the onset of deprivation of an essential nutrient, all bacteria develop resistance to killing by antibiotics, a phenomenon termed phenotypic tolerance (5). The term *phenotypic tolerance* was chosen to emphasize that the response develops in all bacteria in response to an environmental change and leads to bacterial survival (as distinct from bacterial growth that characterizes resistance). Phenotypic tolerance also protects bacteria against virtually every antibiotic regardless of mechanisms of action. Only a few experimental antibiotics, such as the penems, can break this rule. Survival necessitates that bacteria sense numerous environmental cues through a signal transduction cascade, provide the stringent response, and thereby downmodulate several different death mechanisms (6–8). The best studied of those mechanisms is the triggering of autolysis, but other death processes also occur and are subject to the stringent response. Thus, bacteria actively cooperate using their own enzymatic death machinery to achieve the final killing outcome. Identification of participants in the death cascade will allow development of new antibiotics and diagnostic tools for tracking tolerance in the clinical laboratory.

1 INTRODUCTION

Tolerance, by definition, implies that antibiotic binding to the bacterium becomes disconnected from the mechanism of killing. Using penicillin as an example, a simple model of the killing pathway for growing bacteria is shown in Figure 1. Penicillin binds to its target, the cell wall synthetic enzymes (penicillin-binding proteins, PBPs), resulting in growth arrest. Then an unknown event occurs (schematic of the black box) and the extracellular autolysin is activated and dissolves the cell wall cytoskeleton (killing). Since antibiotics bind normally to nongrowing bacteria, the pro-

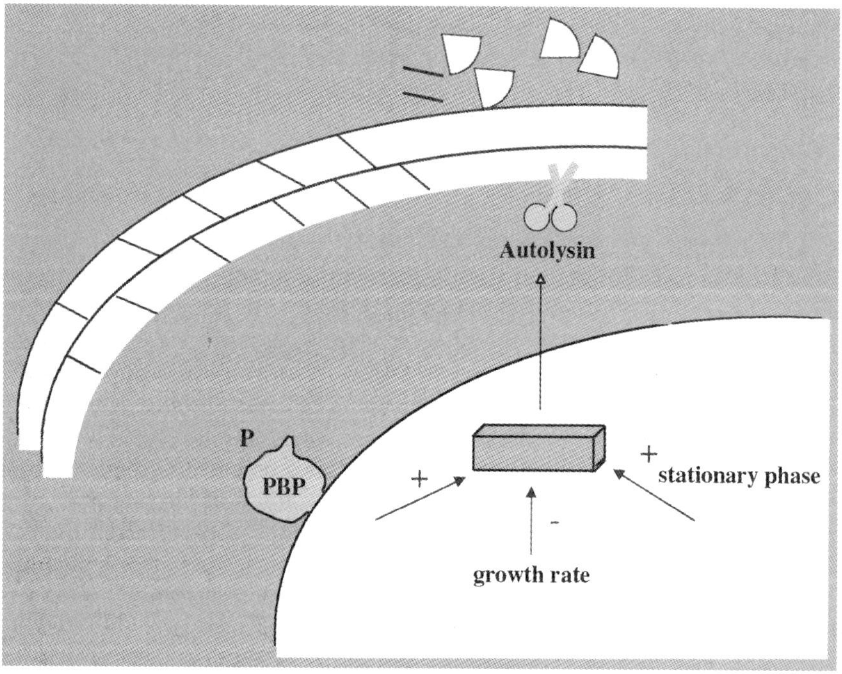

Figure 1 Two-step model of the action of penicillin. Binding of penicillin to PBPs leads to the inhibition of specific steps in cell wall synthesis. This step leads to growth arrest but not cell death. To accomplish death another step is necessary: activation of the autolytic system (scissors). The nature of the signal and the pathway triggering the autolytic enzymes are unknown (black box).

tection from bactericidal consequences that arises within minutes of growth arrest must result from changes in the black box. Changes in any one of three events in nongrowing bacteria could mechanistically explain uncoupling of killing from normal covalent binding of penicillin to the target: 1) A progressive loss of autolysins from the bacterial surface, 2) change in the cell wall composition rendering the peptidoglycan less susceptible to hydrolysis, and 3) activation of a regulatory process decreasing activity of autolysins. Although autolysins are slowly lost from the surface of nongrowing bacteria, the minor role of this event is indicated by the observation that nongrowing bacteria do not lyse upon the addition of exogenous autolysin (9). As for the cell wall composition, studies on the mechanism of phenotypic tolerance in *Escherichia coli* have shown that after 30–60 min of starvation, cell wall structural alterations are made which render the peptidoglycan more resistant to autolysis by antibiotics and detergents (10). Immediately upon growth arrest, the rate of peptidoglycan synthesis decreases to 30% of that of growing *E. coli* but does not stop. The cell wall produced during growth arrest is clearly structurally different and less hydrolyzable by autolysins extracted from normally growing *E. coli* (11,12). The synthesis of this altered cell wall seems to require the activity of penicillin-binding protein 7 (PBP7). However, although important to the generation of phenotypic tolerance, alteration of cell wall composition over the whole cell surface does not occur quickly enough to explain the extraordinarily rapid onset of phenotypic tolerance, a phenomenon starting within minutes after onset of starvation. The rapid onset of phenotypic tolerance is thought to arise by the regulation of hydrolytic enzymes, a process so poorly understood as to be appropriately represented by a black box.

2 THE BIOLOGY OF AUTOLYTIC ENZYMES

To obtain better insight into the physiological mechanism involved in creating antibiotic tolerance, a basic understanding of the nature of cell wall hydrolases is essential. Cell wall hydrolases maintain the peptidoglycan during bacterial growth and split the septum during cell separation. They are constitutively expressed and are always in position to act on the cell wall; that is, they are in the periplasm of gram-negative bacteria and on the cell surface of gram-positive bacteria. Thus, to prevent continuous autolysis they must be strongly negatively regulated. To act as autolysins, these hydrolases must completely deregulate and thus entirely degrade the cell wall (13). The antibacterial effects of β-lactam antibiotics are initiated by the binding of antibiotics to a species-specific group of membrane-located PBPs. This binding inhibits the transglycosylation and transpeptidation steps in cell wall synthesis executed by PBPs and leads to the cessation of bacterial growth (14). Although fundamental to the action of antibiotics, it is unknown how this binding of β-lactams to PBPs leads to deregulation of extracellular autolytic enzymes. It has been proposed that a second process arising from the bacteria itself unleashes the activity of cell wall hydrolases leading to lysis and cell death.

The crucial role of autolysins in mediating cell death was first described for *Streptococcus pneumoniae* based on biochemical and physiological analysis of an antibiotic-tolerant mutant created by chemical mutagenesis. Low specific activity of a cell wall–hydrolyzing enzyme, an *N*-acetylmuramic acid-L-alanine amidase, resulted in lack of killing by penicillin and other cell wall–active antibiotics (15). Further studies demonstrated that antibiotic tolerance arises if either the autolysins are not triggered or the autolysins themselves are not active. For example, in *E. coli*, which has several autolytic enzymes, the degree of tolerance and, in parallel, the degree of lysis is proportional to the number of mutagen-induced defective autolysins (16). In contrast to these laboratory strains, antibiotic tolerance in clinical isolates seems to arise from changes in the control of autolysin enzymatic activity rather than loss of autolysin expression.

3 THE STRINGENT RESPONSE

The most striking example of physiological downregulation of autolysis is the stringent response that occurs during deprivation of an essential nutrient (17). Since nutrient deprivation leads to nongrowing or slowly growing bacteria, the experimental system most frequently used for the study of phenotypic tolerance involves the transfer of bacteria from a nutritionally complete medium to a medium missing an essential amino acid. Within minutes of the onset of starvation, the bacteria stop growing as part of a coordinate reduction in all macromolecular synthetic rates (18). The development of this response is under the control of the *relA* locus, a gene present in all bacteria (19). Mutation of the *relA* gene abolishes the stringent response and phenotypic tolerance. Amino acid deprivation results in inhibition of peptidoglycan, protein, and RNA and DNA synthesis in *relA*$^{+}$ bacteria but not *relA*-deficient mutants. Upon starvation, bacteria rapidly accumulate guanosine 3′,5′-bispyrophosphate (ppGpp), which is synthesized by ppGpp synthetase I encoded by the *relA* gene (20–22). ppGpp in turn coordinately shuts down the synthesis of macromolecules such as DNA, phospholipids (23), and cell wall peptidoglycan (19). This inhibits an early step in the synthesis of the cell wall component UDP-N-acetylmuramyl-pentapeptide (24) as well as a late step in the polymerization of peptidoglycan (25). The precise mechanism involved in the effect of ppGpp on peptidoglycan or phospholipid synthesis is not entirely understood (26). β-Lactam antibiotics bind readily to PBPs of starved bacteria, but during the stringent response, hydrolytic enzymes fail to activate. The mechanism remains completely obscure.

4 PHENOTYPIC TOLERANCE IS NOT RESTRICTED TO NONGROWING BACTERIA

Phenotypic tolerance can be brought about by different modulations of the bacterial environment; for example, by lowering the pH of the medium or by adding proteolytic enzymes or inhibitors of the autolytic enzymes (27). It is known that bacteria growing at acidic pH shift synthesis of certain phospholipids resulting in plasma membranes with an increased resistance to stress (28,29). The change in phospholipid composition correlates with tolerance toward cell wall–active antibiotics (6,8). Membrane alterations are considered to be the basis of tolerance in certain types of bacterial fermentation mutants that overproduce acidic catabolites. Similarly, addition of lipoteichoic acid to the growth medium of pneumococcal cultures causes chain formation, resistance to stationary phase lysis, and penicillin tolerance, suggesting that lipoteichoic acids might be involved in the in vivo control of autolysin activity. This assumption is supported by the observation that lipoteichoic acids appeared to inhibit autolysin activity in several bacterial species, including several streptococci and *Bacillus subtilis* (30–32).

5 GROWTH RATE AND DEATH RATE

If the complete cessation of growth leads to phenotypic tolerance, does growth rate itself correlate with the rate of killing? It has been clearly demonstrated that the time to onset of bacterial lysis is inversely proportional to the rate of bacterial growth prior to antibiotic addition—a faster growth rate correlates with a shorter interval before onset of lysis (33). As early as 1944, Hobby and Dawson suggested that bacteria died more quickly when their growth rate was enhanced by the addition of blood to the media (3). Studies with chemostat-grown cultures formally showed that the rate of killing of *E. coli* grown under conditions of reduced growth rate is a direct function of the generation time of the bacterium (34). Thus, tolerance applies not only to completely dormant, nongrowing bacteria but also to bacterial cells which are multiplying slowly. Since most in vivo fluids are suboptimal bacteriological media, slow growth rate is an important parameter working against bactericidal efficiency in vivo.

The complex relationship between growth and death can be seen in two other contrasting observations. Nongrowing, phenotypically tolerant bacteria do not exhibit a delay in the resumption of growth when the drug concentration falls below the MIC (35). This stands in contrast to the phenomenon of the postantibiotic effect that delays regrowth after antibiotic exposure of growing cells (36).

6 PHENOTYPIC TOLERANCE IN VIVO

The importance of phenotypic tolerance influencing bacterial killing in vivo has been underestimated. The interference of bacteriostatic agents such as sulfadiazine, chloramphenicol, and helvonic acid with the bactericidal activity of antibiotics, including penicillin, streptomycin, and isoniazid, represents one of the earliest clinical examples of phenotypic tolerance (1,2,37,38). The complex, multiple nutritional requirements of important human pathogens such as neisseriae, *S. pneumoniae*, and staphylococci make it likely that these pathogens often encounter suboptimal levels of nutrients in humans. Bacteria observed in natural ecological environments have been shown to grow slowly as a result of nutrient depletion (39). In 1977, Sabath and colleagues described the unsuspected high frequency with which tolerant strains of staphylococci could be isolated from clinical specimens originating in deep-seated infections (40). Infection sites such as cerebrospinal fluid, joint fluid, cardiac vegetations, abscesses, and bone are likely to be deficient in nutrients essential for many bacterial species. Nutrient limitation engenders important physiological changes including decreased drug permeability across the outer or cytoplasmic membranes (41–43), alterations in the structure of peptidoglycan such that it resists degradative enzymes (44–47), or loss of PBPs (48–50). This transition is accompanied by major changes in the cell metabolism, including changes of the composition of the cell wall (51). The identification of stationary phase–inducible promoters and the characterization of *rpoS*, a gene encoding an alternative sigma factor in *E. coli* that controls many genes induced at the onset of stationary phase (52), is only one example illustrating the dynamic adaptations of cells shifting their growth rate.

Since the 1980s, reliable data have become available clearly demonstrating that bacteria grow at rates substantially slower in many animal models than in vitro cultivation (53). In particular, the rabbit meningitis model provided interesting insights into the in vivo growth behavior of bacteria. A direct comparison of a minimal, chemically defined medium and the composition of normal cerebrospinal fluid (CSF) indicated that although CSF contained sufficient glucose, it was more than 100-fold deficient in leucine, isoleucine, cysteine, glycine, valine, serine, magnesium, and manganese (54). It was shown that after inoculation into CSF, pneumococci demonstrated a long lag period before the onset of bacterial growth and the maximal growth rates were significantly slower than those achieved in vitro (55). Both nongrowing and slow-growing populations contributed to phenotypic tolerance of pneumococci entering the CSF during the onset of bacterial meningitis (56,57). Studies using the rabbit model of meningitis demonstrated the importance of bactericidal antimicrobial therapy to achieve cure (58). Treatment with bacteriostatic agents was insufficient, leading to death or persistence of bacteria in the CSF. This study provided the explanation for the dramatic differences in cure rates of pneumococcal meningitis reported by Lepper and Dowling in 1951 (59). They found a 79% cure rate in patients receiving penicillin G (bactericidal) alone but only a 30% cure rate in patients receiving penicillin plus chlortetracycline (bacteriostatic). Arrest of growth by the bacteriostatic drug invokes the stringent response rendering the bacteria insensitive to the action of a bactericidal antibiotic.

7 β-LACTAM ANTIBIOTICS THAT KILL NONGROWING BACTERIA

The deleterious effects of phenotypic tolerance on bactericidal activity of β-lactam antibiotics led to the search for antibiotics that could break the tolerance rule (55). To identify antibacterial agents capable of killing *E. coli* after starvation, an *E. coli* lysine auxotroph was grown in a chemically defined medium and then transferred to the same medium without the essential amino acid lysine (11). At various intervals thereafter, aliquots of the nongrowing culture were exposed to a wide range of antibiotics. Most antibiotics failed completely, including, for example, β-lactams, aminoglycosides, and quinolones. However, three interesting exceptions were found: Penems, nocardicin, and cycloserine were able to kill and lyse *E. coli* after 10 min of starvation. This activity was finally lost 30 min into starvation (Fig. 2). These experiments provided some important mechanistic insights into phenotypic tolerance: 1) The unusual bactericidal activity of these antibiotics on nongrowing bacteria derives from a similar mechanism as in growing cells in that it requires an active autolysin. Autolysin-deficient pneumococci do not lyse with nocardicin whether growing or nongrowing. 2) Lytic activity in nongrowing bacteria is a direct function of the concentration of the antibiotic; that is, the higher the concentration of the antibiotic, the greater the degree of lysis. At extremely high concentrations (100 times the MIC), even benzylpenicillin is effective in lysing *E. coli* starved for brief periods (5 min). 3) Lytic activity is found to vary inversely with the duration of starvation prior to exposure to the antibiotics. The longer the duration of starvation prior to antibiotic challenge, the less

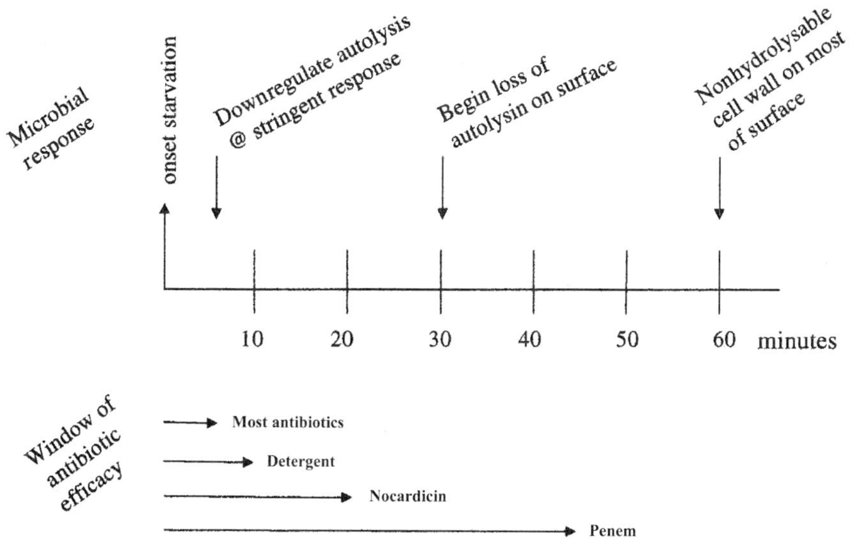

Figure 2 Schematic representation of the relative timing of bacterial adaptations to starvation and their sensitivity to killing by various antibiotics.

effective the antibiotic. Even an effective antibiotic, such as imipenem, is inactive against bacteria prestarved for > 60 min.

7.1 Mechanism

In current models, the antibacterial effects of β-lactam antibiotics are initiated by the binding of antibiotic to PBPs. The PBPs are covalently acylated and thus removed from the metabolically active pool of PBPs. These events explain the arrest of cell growth. By an as yet unknown mechanism, inhibition of cell wall synthesis activates autolysins that lead to cell wall degradation and lysis (60). In the case of phenotypic tolerance of nongrowing bacteria, autolysins are not triggered. Reinitiating growth by supplementing the missing amino acid leads to restoration of the autolytic phenotype, suggesting that starvation of bacteria leads to a downregulation of the activity of the trigger pathway rather than eliminating the autolysin itself. This hypothesis is further substantiated by the fact that autolysin preparations from nongrowing strains retain their hydrolytic activities when transferred to growing cells.

A model of the mechanism of phenotypic tolerance must explain how the development of the stringent response might interfere with the lytic pathway and why or how a few select antibiotics still act temporarily despite the stringent response. Since a *relA*-deficient mutant demonstrates a significant delay in the shutdown of macromolecular synthesis in response to amino acid starvation it is conceivable that RelA is involved directly or indirectly with the generation of phenotypic tolerance. The most obvious explanation for how antibiotics like nocardicin or penems kill nongrowing cells would be that they relax the stringent response, an effect known for chloramphenicol (24). However, nocardicin or penems do not disturb the shutdown of macromolecular synthesis in amino acid–deprived cells. Therefore, lysis of nongrowing bacteria by nocardicin or penems does not directly distort the stringent response, although the possibility exists that these β-lactam antibiotics act on the level of coordinate regulation by an as yet undefined mechanism.

8 NEW POINTS OF VIEW: LESSONS FROM GENOTYPIC TOLERANCE

In contrast to phenotypic tolerance, tolerance even during active growth can result from genetic mutations. The simplest example of tolerance is the loss of function of the autolysin LytA without which pneumococci fail to lyse and die (15). For reasons that are not clear, no clinical isolates have been found bearing loss of function mutation of the autolysin gene. Yet, clinical isolates of pneumococci have been reported that are genetically tolerant (61). Clinical genotypic tolerance appears to arise by genetic alteration at the level of regulation of autolysin activity (62).

In recent study, a pneumococcal mutant was identified that failed to die in the presence of all antibiotics tested, including β-lactams, aminoglycosides, and vancomycin (63). Analysis of the affected gene, *vncS*, revealed it to be a sensor for a two-component regulatory system, VncS-VncR (Fig. 3). VncS-VncR most likely represents an early element in the autolytic triggering pathway, controlling the activity of autolysin via levels of phosphorylation of the response regulator VncR (63). A phosphorylated form of VncR is hypothesized to repress the activation of autolysin, whereas, for lysis, a stimulus sensed by the histidine kinase VncS dephosphorylates VncR and enables triggering of autolysis. It is not clear if this control occurs at the level of transcription, translation, or enzymatic activation. However, in the model, VncS-VncR functions as relay station reacting to extracellular signals; for example, cell density in stationary phase lysis or the binding of antibiotics to PBPs. The nature of the signal and its downstream effects that result in autolysin activation remain to be determined.

Another example for the role of signal transduction in antibiotic action is suggested by the mechanism of β-lactamase induction (64). It had

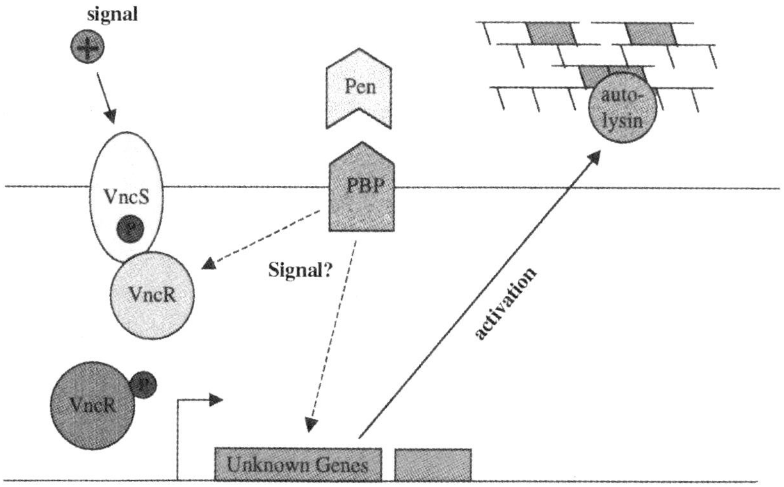

Figure 3 Genes that are known to result in genetic tolerance suggest signal transduction is operative in antibiotic action. Binding of penicillin to PBPs might create a signal directly activating genes involved in regulating the activity of autolysins (uncapping of inhibited autolysin). Mutants lacking autolysin are tolerant (15). Alternatively, a signal might activate the two-component regulatory system, VncS/VncR (63), which seems to represent the relay station for the autolytic trigger pathway in *S. pneumoniae*. The histidine kinase/phosphatase VncS controls the level of phosphorylation of the response regulator VncR. VncR-P represses the activation of autolysin. A stimulus sensed by VncS activates its phosphatase activity, dephosphorylates VncR-P. VncR enables autolysis; e.g., during stationary phase or due to triggering by antibiotics. Genetic loss of VncS traps the system into continued repression of autolysis by VncR-P, resulting in tolerance.

been proposed that during normal bacterial physiology cell wall byproducts are sensed by elements of the chromosomal *amp* regulon, thereby allowing bacteria to change gene transcription as a function of the "health" of the extracellular peptidoglycan. This system is co-opted during β-lactamase induction when the extracellular β lactam stimulates transcription of the chromosomal β-lactamase.

9 PERSPECTIVES

All bacteria share a program linking growth rate and death rate. Antibiotics tap into this endogenous control circuits in order to kill, but this occurs only when bacteria are rapidly growing. The elements in this control circuit for death are unknown, with the exception of the recent discovery that signal transduction is required for the death response. An understanding of bacterial suicidal participants is critical for developing new antibiotics, particularly those that will not fail when the growing gets tough in vivo.

REFERENCES

1. Chain E, Duthie ES. Bactericidal and bacteriolytic action of penicillin on the staphylococcus. Lancet 1945; 1:652–657.
2. Hobby, GL, Dawson MH. The effect of sulfonamides on the action of penicillin. J Bacteriol 1946; 51:447–456.
3. Hobby GL, Dawson MH. Effect of rate of growth of bacteria on action of penicillin. Proc Soc Exp Biol Med 1944; 56:181–184.
4. Hobby GL, Meyer K, Chaffee E. Observations on the mechanism of action of penicillin. Proc Soc Exp Biol 1942; 50:281–285.
5. Tuomanen E, Tomasz A. Mechanism of phenotypic tolerance of nongrowing pneumococci to beta-lactam antibiotics. Scand J Infect Dis 1991; Suppl. 74 (Suppl):102–112.
6. Goodell EW, Lopez R, Tomasz A. Suppression of lytic effect of beta lactams on *Escherichia coli* and other bacteria. Proc Natl Acad Sci USA 1976; 73:3293–3297.
7. Lopez R, C. R-L, Tapia A, Waks SB, Tomasz A. Suppression of the lytic and bactericidal effects of cell-wall inhibitory antibiotics. Antimicrob Agents Chemother 1976; 10:697–706.
8. Horne D, Tomasz A. pH-dependent penicillin tolerance of group B streptococci. Antimicrob Agents Chemother 1976; 20:128–135.
9. Horne D, Tomasz A. Lethal effect of a heterologous murein hydrolase on penicillin-treated *Streptococcus sanguis*. Antimicrob Agents Chemother 1980; 17:235–246.
10. Cozens R, Tuomanen E, Tosch W, Zak O, Suter J, Tomasz A. Evaluation of the bactericidal activity of beta lactam antibiotics on slowly growing bacteria cultured in the chemostat. Antimicrob Agents Chemother 1986; 29:797–802.
11. Goodell E, Tomasz A. Alternation of *E coli* murein during amino acid starvation. J Bacteriol 1980; 144:1009–1016.
12. Tuomanen E, Markiewicz Z, Tomasz A. Autolysis-resistant peptidoglycan of anomalous composition in amino-acid-starved *Escherichia coli*. J Bacteriol 1988; 170:1373–1376.
13. Tomasz A. Murein hydrolases-enzymes in search of a physiological function. In: Hakenbeck R, Holtje J, H. L, eds. The Target of Penicillin. Berlin: Walter de Gruyter, 1983:155–172.
14. Tomasz A, Holtje JV. Murein hydrolases and the lytic and killing action of penicillin. In: Schlesinger D, ed. Microbiology. Washington DC: American Society for Microbiology, 1977:209–215.
15. Tomasz A, Albino A, Zanati E. Multiple antibiotic resistance in a bacterium with suppressed autolytic system. Nature 1970; 227:138–140.

16. Kitano K, Williamson R, Tomasz A. Murein hydrolase defect in the beta lactam tolerant mutants of *Escherichia coli*. FEMS Microbiol Lett 1980; 7:133–136.
17. Cashel M, Gentry DR, Hernandez VJ, Vinella D. The stringent response. In: Neidhardt FC, Curtis III R, Ingraham JL, et al., eds. *Escherichia Coli* and *Salmonella*: Cellular and Molecular Biology. Vol. 1. Washington, DC: ASM Press, 1996;1458–1496.
18. Gallant JA. Stringent control in *Escherichia coli*. Annu Rev Genet 1979; 13: 393–415.
19. Ishiguro EE, Ramey WD. Stringent control of peptidoglycan biosynthesis in *Escherichia coli* K-12. J Bacteriol 1976; 127:1119–1126.
20. Metzger S, Dror IB, Aizenman E, et al. The nucleotide sequence and characterization of the *relA* gene of *Escherichia coli*. J Biol Chem 1988; 263:15699–15704.
21. Schreiber G, Metzger S, Aizenbaum E, Roza S, Cashel M, Glaser G. Overexpression of the *relA* gene in *Escherichia coli*. J Biol Chem 1991; 266:3760–3767.
22. Svitil AL, Cashel M, Zyskind JW. Guanosine tetraphosphate inhibits protein synthesis in vivo. A possible protective mechanism for starvation stress in *Escherichia coli*. J Biol Chem 1993; 268:2307–2311.
23. Sokawa Y, Nakao E, Kaziro Y. On the nature of the control by RC gene in *E. coli*: amino acid–dependent control of lipid synthesis. Biochem Biophys Res Commun 1968; 33:108–112.l
24. Ishiguro EE, Ramey WD, Involvement of the *relA* gene product and feedback inhibition in the regulation of UDP-N-acetylmuramyl-peptide synthesis in *Escherichia coli*. J Bacteriol 1978; 135:766–774.
25. Ramey WD, Ishiguro EE. Site of inhibition of peptidoglycan biosynthesis during the stringent response in *Escherichia coli*. J Bacteriol 1978; 135:71–77.
26. Kusser W, Ishiguro EE. Involvement of the *relA* gene in the autolysis of *Escherichia coli* induced by inhibitors of peptidoglycan synthesis. J Bacteriol 1985; 164:861–865.
27. Handwerger S, Tomasz A. Antibiotic tolerance among clinical isolates of bacteria. Revs Infect Dis 1985; 7:368–386.
28. Carson D, Pieringer RA, Danoe-Moore L. Effect of growth rate on lipid and lipoteichoic acid composition in *Streptococcus faecium*. Biochim Biophys Acta 1979; 575:225–233.
29. Op den Kamp JAF, van Iterson W, van Deenen LLM. Studies on the phospholipids and morphology of protoplasts of *Bacillus megaterium*. Biochim Biophys Acta 1967; 135:862–884.
30. Cleveland RF, Holtje JV, Wicken AJ, Tomasz A, Daneo ML, Shockman GD. Inhibition of bacterial wall lysins by lipoteichoic acids and related compounds. Biochem Biophys Res Commun 1975; 67:1128–1135.
31. Cleveland RF, Wicken AJ, Daneo-Moore L, Shockman GD. Inhibition of wall autolysis in *Streptococcus faecalis* by lipoteichoic acids and lipids. J Bacteriol 1976; 126:192–197.
32. Tomasz A, Waks S. Mechanism of action of penicillin: triggering of the pneumococcal autolytic enzyme of inhibitors of cell wall synthesis. Proc Natl Acad Sci USA 1975; 72:4162–4166.
33. Tuomanen E, Cozens R, Tosch W, Zak O, Tomasz A. The rate of killing of

Escherichia coli by beta lactam antibiotics is strictly proportional to the rate of bacterial growth. J Gen Microbiol 1986; 132:1297–1304.
34. Tuomanen E. Newly made enzymes determine ongoing cell wall synthesis and the antibacterial effects of cell wall synthesis inhibitors. J Bacteriol 1986; 167:535–543.
35. McDonald PJ, Craig WA, Kunin CM. Persistent effect of antibiotics on *Staphylococcus aureus* after exposure for limited periods of time. J Infect Dis 1977; 135:217–223.
36. Hobby GL, Lenert TF. The *in vitro* action of antituberculous agents against multiplying and non-multiplying microbial cells. Am Rev Tuberc 1957; 76: 1031–1048.
37. McQuillen K. Lysis resulting from metabolic disturbance. J Gen Microbiol 1958; 18:498–512.
38. Harder W, Dijkhuizen L. Physiological responses to nutrient limitation. Annu Rev Microbiol 1983; 37:1–23.
39. Sabath LD, Lavadiere M, Wheeler N, Blazevic D, Wilkinson BJ. A new type of penicillin resistance in *Staphylococcus aureus*. Lancet 1977; 1:888–896.
40. Brown MRW. Nutrient depletion and antibiotic susceptibility. Chemotherapy 1977; 3:198–200.
41. Cozens RM, Brown MRM. Effect of nutrient depletion on the sensitivity of *Pseudomonas cepacia* to antimicrobial agents. J Pharm Sci 1983; 72:1363–1365.
42. Gilbert P, Brown MRM. Influence of growth rate and nutrient limitation on the gross cellular composition of *Pseudomonas aeruginosa* and its resistance to 3- and 4-chlorophenol J Bacteriol 1978; 133:1066–1072.
43. Horne D, Tomasz A. Factors affecting sensitivity of group B streptococci to an exogenous murein hydrolase. Can J Microbiol 1985; 31:417–422.
44. Mengin-Lecreulx D, Van Heijenoort J. Effect of growth conditions on peptidoglycan content and cytoplasmic steps of its synthesis in *Escherichia coli*. J Bacteriol 1985; 163:208–212.
45. Pisabarro AG, De Pedro MA, Vazquez D. Structural modifications in the peptidoglycan of *Escherichia coli* associated with changes in the state of growth of the culture. J Bacteriol 1985; 161:238–242.
46. Rosenthal RS, Gfell MA, Folkening WJ. Influence of protein synthesis inhibitors on regulation of extent of O-acetylation of gonococcal peptidoglycan. Infect Immun 1985; 49:7–13.
47. Buchanan CE, Sowell MO. Synthesis of penicillin binding protein 6 by stationary phase *Escherichia coli*. J Bacteriol 1982; 151:491–494.
48. Eudy WW, Burrous SE. Generation times of *Proteus mirabilis* and *Escherichia coli* in experimental infections. Chemotherapy 1973; 19:161–170.
49. Neyman SL, Buchanan CE. Restoration of vegetative penicillin-binding proteins during germination and outgrowth of *Bacillus subtilis* spores: relationship of individual proteins to specific cell cycle events. J Bacteriol 1985; 161: 164–168.
50. Kolter R, Siegele DA, Tormo A. The stationary phase of the bacterial life cycle. Annu Rev Microbiol 1993; 47:855–874.
51. Lange R, Hengge-Aronis R. Growth phase-regulated expression of bolA and

morphology of stationary-phase *Escherichia coli* cells is controlled by the noel sigma factor sigma S (*rpoS*). J Bacteriol 1991; 173:4474–4481.
52. Zak O, Sande MA. Correlation of *in vitro* antimicrobial activity of antibiotics with results of treatment in experimental animal models and human infection. In: Sabath LD, ed. Action of Penicillin in Patients. Bern, Switzerland: Hans Huber, 1981:55–67.
53. Letner Ce. Geigy scientific tables. Basel, Switzerland, 1981:168–169.
54. Tuomanen E. Phenotypic tolerance: the search for β-lactam antibiotics that kill nongrowing bacteria. Res Infect Dis 1986; 8(Suppl 3):279–291.
55. Dacey RG, Sande MA. Effect of probenecid on cerebrospinal fluid concentrations of penicillin and cephalosporin derivatives. Antimicrob Agents Chemother 1974; 6:437–441.
56. Tuomanen E, Tomasz A, Hengstler B, Zak O. The relative role of bacterial cell wall and capsule in the induction of inflammation in pneumococcal meningitis. J Infect Dis 1985; 151:535–540.
57. Scheld MW, Sande MA. Bactericidal versus bacteriostatic antibiotic therapy of experimental pneumococcal meningitis in rabbits. J Clin Invest 1983; 71: 411–419.
58. Lepper MH, Dowling HF. Treatment of pneumococcic meningitis with penicillin compared with penicillin plus aureomycin. Studies including observations on an apparent antagonism between penicillin and aureomycin. Arch Intern Med 1951; 88:489–494.
59. Tomasz A. From penicillin-binding proteins to the lysis and death of bacteria: a 1979 view. Re Infect Dis 1979; 1:434–467.
60. Tuomanen E, Durack DT, Tomasz A. Antibiotic tolerance among clinical isolates of bacteria. Antimicrob Agents Chemother 1986; 30:521–527.
61. Tuomanen E, Pollack H, Parkinson A, et al. Microbiological and clinical significance of a new property of defective lysis in clinical strains of pneumococci. J Infect Dis 1988; 158:36–43.
62. Novak R, Henriques B, Charpentier E, Normark S, Tuomanen E. Emergence of vancomycin tolerance in *Streptococcus pneumoniae*. Nature 1999; 399: 590–593.
63. Tuomanen E, Lindquist S, Sande S, et al. Coordinate regulation of β-lactamase induction and peptidoglycan composition by the *amp* operon. Science 1991; 251:201–204.

10

Resistance as a Worldwide Problem

Rosamund Williams
World Health Organization, Geneva, Switzerland

Despite the major technological advances of the last century, infectious diseases still account for 25% of all deaths and up to 45% of deaths in developing countries (1). As an increasing number of noncommunicable diseases are being linked to infectious agents; for example, atherosclerosis to *Chlamydia* infection and mucosa-associated lymphoid tissue (MALT) lymphoma which has been associated with *Helicobacter pylori* infection. The actual proportion of deaths due to infectious causes may well be even higher. In global terms, six diseases (acute respiratory infections, acquired immunodeficiency syndrome (AIDS), diarrheal diseases, tuberculosis, malaria, and measles) account for about 90% of infectious disease deaths. In addition, other tropical diseases, sexually transmitted infections, and hospital-acquired infections contribute significantly to the global disability burden. Although antimicrobial agents are available to combat the majority of these diseases, this window of opportunity is closing as a result of the emergence of antimicrobial resistance. Antimicrobial resistance increases the mortality and morbidity due to infectious disease and has an important impact on the costs of health care. These effects result from the accumulation of resistance determinants by microbial pathogens, thereby reducing the treatment options and enhancing the spread of resistant strains from person to person. Also significant is the marked decline in the development of novel antimicrobial agents to combat resistant pathogens. In the 1970s, it was widely believed that infectious diseases had been all but conquered and that development of new antibacterial agents was increasingly costly and losing commercial attractiveness. The threats posed by resistance should lead to a reexamination of these views. The aim of this chapter is to provide an overview of resistance in the major global infectious diseases; to highlight the impact of resistance and the challenges in quantifying this; and to summarize the factors contributing to the emergence and spread of resistance and how these may be addressed to contain the resistance threat.

1 A GLOBAL VIEW OF THE MAGNITUDE AND TRENDS IN RESISTANCE

Published data on the magnitude and, where available, trends in resistance are summarized in this section. The term *resistance* is almost invariably defined by in vitro tests, and few studies link microbiological resistance to clinical outcome. Further shortcomings of the laboratory data are discussed in Section 5.

1.1 Acute Respiratory Infections

Streptococcus pneumoniae and *Haemophilus influenzae* constitute the most important bacterial causes of community-acquired infection of the respiratory tract. Since its introduction, penicillin has been the drug of choice for the treatment of *S. pneumoniae* infections, but during the last 20 years there have been increasing numbers of isolates showing decreased susceptibility or resistance to penicillin. The evolution and epidemiology of penicillin and erythromycin resistance are described elsewhere in this book (see Chap. 12). The interpretation of the resistance data has been confused, because pneumococcal pneumonia caused by isolates with intermediate resistance responds to penicillin therapy similarly to that caused by fully susceptible strains. This has tended to lead clinicians to mistrust laboratory results and resistance surveillance data. However, the increasing number of reports of resistance from around the world suggest that for a bacterial species that was for many years completely susceptible to penicillin, a real change is underway. It has also been shown clearly that meningitis caused by pneumococcal strains with intermediate resistance does not respond to penicillin therapy and may also be less responsive than penicillin-susceptible strains to chloramphenicol (2).

Resistance in *S. pneumoniae* to other classes of antimicrobials is also on the rise. Erythromycin-resistant pneumococci in South Africa increased from 1 of 5115 clinical isolates tested in 1987 to 28 of 4735 tested in 1996 (3). In England and Wales, erythromycin resistance rose from 2.8 to 8.6% between 1990 and 1995 (4). Cotrimoxazole is one of the first-line drugs for acute respiratory infection in developing countries, but resistance to sulfamethoxazole appears to be consistently higher than penicillin resistance. Surveillance in Cameroun, Egypt, Kenya, and Nigeria have reported rates exceeding 25% (5), as do studies in Mexico, Uruguay and Brazil (6). In Pakistan, various investigations between 1986 and 1992 described rates of sulfamethoxazole resistance ranging between 65 and 93% (7,8). High-level cotrimoxazole resistance has been associated with multiple resistance in isolates in South Africa, and it has increased from 3.8% in 1985 to 44% in 1991 (2,9).

H. influenzae is less commonly isolated than expected from respiratory tract infections in many parts of the world, but this may be due to inadequate laboratory methods for culture and isolation. A survey of the published literature worldwide revealed large differences in rates of resistance to the drugs previously considered to be first choice for treatment (i.e., ampicillin, chloramphenicol, and cotrimoxazole). Analysis of trends over time in three countries revealed a modest increase in resistance. In many developed countries, the impact of antimicrobial resistance in *H. influenzae* has been significantly reduced owing to the dramatic fall in infections subsequent to the widespread use of Hib vaccine. In the United Kingdom, for instance, the risk of invasive infection fell from 1:600 to 1:30,000 in children less than 5 years old within 2 years of Hib vaccine introduction (10). The dramatic decline in infection has yet to be seen in developing countries, since the cost of the Hib vaccine keeps utilization very low (11).

1.2 Diarrheal Diseases

1.2.1 Dysentery

The diarrheal disease where antimicrobial resistance is probably having greatest impact is bacilliary dysentery caused by *Shigella dysenteriae*. Plasmid-mediated resistance was first recognized in *Shigella*, and over time species in this genus have become resistant to sulfonamides, tetracyline, ampicillin, cotrimoxazole, and nalidixic acid. Multiresistant *Sh. dysenteriae* have been isolated in Latin America, Central Africa, and southern and Southeast Asia (12). For example, a study from Sack and colleagues (13) in Bangladesh showed the degree to which first-line treatment had already become compromised by 1992. Although ciprofloxacin and some of the other fluoroquinolones are now widely used, there is concern that the existence of resistance to nalidixic acid (an early member of the quinolone family) in shigellae heralds rapid development of resistance to ciprofloxacin. Furthermore, fluoroquinolones are not licensed for use in children; that is, the patient population that suffers the major burden of diarrheal disease. In developed countries, infections with multiresistant strains of *Shigella* are associated with foreign travel (14). In contrast to severe dysenteric disease, mild infections, usually caused by *Sh. sonnei* may be multiresistant, but antimicrobial therapy is not indicated.

Microbiologically closely related to *Shigella* and often resistant to antimicrobials various groups of *Escherichia coli* (enterohemorrhagic *E. coli*, e.g., serotype O157, enteroinvasive and enterotoxigenic *E. coli*) are recognized causes of gastrointestinal infection with distinct virulence mechanisms. However, the majority of disease is due to toxin production rather than bacterial multiplication, and antimicrobial therapy is not recommended.

1.2.2 Cholera

The symptoms of cholera are entirely due to the production in the gut of an enterotoxin resulting in severe watery diarrhea. Oral or intravenous rehydration is the cornerstone of the treatment, and antimicrobial therapy does not appear to alter the course of the disease. It does, however, reduce the period of excretion of *Vibrio cholerae* from infected cases, and therefore is important in the management of epidemics. Despite widespread use of tetracycline, *V. cholerae* appeared to remain susceptible until 1977 when multiresistant strains were reported in Tanzania (15). Within an epidemic lasting 5 months. susceptibility to tetracycline decreased from 100 to 24%. Since 1980, several outbreaks of resistant and multiresistant *V. cholerae* have been reported in developing countries (16).

1.2.3 Foodborne Infections

Diarrheal disease associated with foodborne infection is most commonly associated with *Salmonella* spp. (other than *Sal. typhi*) and *Campylobacter*. Although these bacteria have been found to carry an increasing number of resistance genes over time (believed by many to be due to the widespread use of antimicrobials as feed additives and mass prophylaxis of food animals), the risks of resistance to human health have still to be evaluated. The disease produced is almost always mild and self-limiting, and antimicrobials are contraindicated, since they tend to prolong carriage. However, a small proportion of patients develop invasive infections that require therapy, and there have been reported cases of treatment failure owing to resistance such as an outbreak of quinolone-resistant *Sal. typhimurium* DT104 resulting in treatment failures in hospitalized patients in Denmark (17). Following the introduction of fluoroquinolones for use in poultry, there has been a dramatic rise in the prevalence of fluoroquinolone-resistant *Campylobacter jejuni* isolated from live poultry, poultry meat, and infected humans, whereas previously resistant strains were not reported in individuals who had not had prior exposure to quinolones (18). Because of their broad antibacterial spectrum, fluoroquinoles are often used for empirical treatment pending laboratory results. One of the clinical effects of fluoroquinolone resistance in *Campylobacter* is extended duration of illness (19).

1.2.4 Typhoid

Like other species of *Salmonella*, *Sal. typhi* has been adept at accumulating resistance determinants, and over recent years a series of typhoid outbreaks caused by multiresistant strains has been described (20). Of additional concern is the finding that infections caused by strains with reduced susceptibility to ciprofloxacin (MICs ranging from 0.25 to 1.0 mg/L and that may be reported by the laboratory as susceptible) do not respond adequately to ciprofloxacin therapy. Recent findings suggest that such strains are now endemic in India and Pakistan (21). For such multiresistant infections, the only available treatment is a third-generation cephalosporin (ceftriaxone or cefotaxime).

1.3 Sexually Transmitted Infections

AIDS (including AIDS patients dying of tuberculosis) currently ranks second among the infectious causes of death worldwide (1). It is beyond the scope of this chapter to review the impact of resistance on antiretroviral therapy. The vast majority of AIDS patients do not currently have access to such combination therapy. However, the emergence of resistance is a growing problem that may compromise therapeutic options even for those who have access to these drugs (22).

Concurrent sexually transmitted infections (STIs) increase the likelihood of acquisition of human immunodeficiency virus (HIV) from an infected partner manyfold, thus the identification and treatment of STIs is a paramount importance in populations at risk of HIV. STIs also have a major impact on reproductive health. The proportion of gonococcal infections caused by penicillin-resistant *Neisseria gonorrhoeae* has increased internationally. High rates of resistance have been documented in countries in the western pacific region with penicillinase-producing *N. gonorrhoeae* (PPNG) exceeding 80% in isolates tested from Korea and the Philippines and 60–80% in Singapore, Malaysia, and China (23). Little has been documented about the trends in PPNG and other resistance in sub-Saharan

Africa, but studies have revealed rates as high as 73% in Abidjan, 67% in Kinshasa (24), and 80% in Ibadan (25).

Tetracyline-resistant strains of *N. gonorrhoeae* (TRNG) are also common and more likely to also be PPNG. Hence, in many developing countries, neither tetracycline nor penicillin is useful in treating gonococcal infections. In contrast, *Treponema pallidum*, the causative agent of syphilis, remains completely susceptible to penicillin treatment. The emergence of quinolone-resistant gonorrhea (QRNG) has now been reported in all continents, but clinical failure has been observed primarily in Asia (26). In Australia, the proportion of QRNG rose slowly between 1984 and 1995 (27).

1.4 Tuberculosis

Although resistance to each new antituberculosis agent developed after the drugs were introduced, recent dramatic outbreaks of multi–drug-resistant tuberculosis (MDR-TB) in HIV-infected patients in the United States and in Europe focused international attention on *Mycobacterium tuberculosis* and the potential threat of drug resistance to TB control. MDR-TB is defined as resistance to both rifampicin and isoniazid, and such strains are often resistant to other agents as well. The second global report from the World Health Organization (WHO) reviewed survey data from 72 countries and found drug resistance in all countries. Several "hot spots" were identified where the prevalence of resistance was alarmingly high (28); probably due to the lack of properly managed national tuberculosis programs, lack of availability of drugs, or poor drug quality. Because of the prolonged nature of the infection, resistance readily develops during treatment if a single agent is used, and combination therapy (with three or four drugs) has long been recognized as essential for cure and is the mainstay of the WHO DOTS (directly observed therapy short-course) strategy. This strategy is effective at controlling tuberculosis, but at present only about 16% of the world's population has access to DOTS. This coverage needs to be rapidly and significantly expanded before MDR-TB and coinfection with HIV make TB control impossible (29).

1.5 Nosocomial Infections

Although not ranking among the leading infectious causes of death (1), hospital-acquired infections are major contributors to mortality and morbidity and significantly increase the cost of hospital care. Such infections are usually caused by multiresistant organisms that have been selected over time in the hospital environment where antimicrobial use is intense. Methicillin-resistant (multiresistant) *Staphylococcus aureus* has become the bane of hospitals worldwide, with some centers reporting up to 80% of

Staph. aureus isolates to be MRSA. Staphylococci (*Staph. aureus* and coagulase-negative *Staphylococcus* spp.) and enterococci cause approximately 30% of all nosocomial infections and 47–50% of nosocomial bloodstream infections in the United States (30). In Europe, similar patterns were observed in a point prevalence study in 17 countries where staphylococci accounted for 49.2% of all infections in ICU patients followed by Enterobacteriaceae (34.4%) and *Pseudomonas aeruginosa* (28.7%) (31). Resistance in such hospital pathogens as *Staph. aureus*, coagulase-negative staphylococci, enterococci, and opportunist gram-negative rods is described in detail in other chapters of this book.

Although resistance in hospital pathogens has been a particular focus of concern in developed countries, it is important to recognize that this is a global problem adding to the heavy burden of resistance in community-acquired infections in many parts of the world. What is more, hospital pathogens are increasingly being recognized as causing infections in the community (e.g., MRSA infections [32]) and in aborigine populations; personal communication).

1.6 Malaria

As a protozoal infection, malaria falls outside the scope of this book. However, no review of the public health impact of antimicrobial resistance would be complete without a mention of the growing challenge of resistance to antimalarial drugs. Resistance to chloroquine and to sulfadoxine-pyrimethamine (Fansidar) has been reported throughout the areas of malaria transmission. A recent review of results from 33 study sites in 10 African countries revealed rates of early treatment failure (i.e., within 3 days of onset and likely to be due to infection with resistant parasites) ranging from 3.2 to 82.6% (WHO, unpublished). Multidrug resistance is a growing concern, particularly in areas such as the Mekong Valley.

2 THE GLOBALIZATION OF RESISTANCE

At the local level, the spread of resistance can be "compartmentalized" into spread in the community (e.g., person to person, animal/food to person) and in hospitals. In the past, these two compartments have tended to harbor different pathogens and different resistance problems. For example sepsis caused by multiresistant gram-negative bacteria is found almost exclusively in hospital patients, whereas antimicrobial-resistant sexually transmitted infections and diarrheal disease are community problems. Increasingly, the separation between the two compartments is blurred owing to, for example, early discharge of hospital patients into the community and increasing spread of resistant "community" pathogens in other "closed" situations (e.g., day care centers, prisons).

From the global viewpoint, the enormous increase in international travel and trade has enabled the rapid globalization of resistance. The numbers of infections in travellers returning to Europe and the United States from developing countries is well recognized; the risk that these are antimicrobial-resistant infections, reflecting the resistance situation in the country visited, is increasing as resistance in community pathogens increases worldwide. The secondary local spread of imported resistant strains has been clearly described (e.g., penicillin-resistant pneumococci in Iceland (33) and penicillinase-producing and quinolone-resistant gonococci in Australia [27]). The developing world countries also contribute significantly to the world food trade, but little is known about their use of antimicrobials in food animal rearing, and thus the potential risks to human health from this increasingly global trade.

3 IMPACT OF RESISTANCE

Although it is widely held that resistant infections carry a greater mortality and morbidity than those caused by susceptible strains of the same pathogens, there is a lack of data accurately to document these differences (34). These higher rates are almost certainly due not to an intrinsically greater virulence of resistant strains (although resistance markers and virulence markers have been described on the same plasmids) but to delays in recognizing and treating the resistant infections. There are few examples to date of pathogens that are completely resistant to all known antimicrobials (although vancomycin-intermediate *Staph. aureus* (VISA) and vancomycin-resistant enterococci (VRE) and multiresistant *Sal. typhi* are approaching). However, there are many situations, particularly in the developing world, where pathogens are resistant to all the locally available antimicrobials and are therefore effectively untreatable. A further consequence of the increased morbidity and prolonged illness is the greater opportunity for resistant organisms to spread to new hosts, thereby multiplying the effects of resistance.

The impact of resistance on the cost of treatment should be somewhat easier to document, but there is little literature on the topic. Prolonged illness increases hospital costs and has social costs due to, for example, longer time away from work and the family. For treatment of pneumonia, meningitis, and some STIs, switching from first-line to second- or third-line treatments can increase drug costs by up to 90-fold. This may have significant impact on the patient and on the health system. For example, about 100 patients with TB can be treated for the cost of treating one case of

MDR-TB. Further costs are incurred if the switch to second-line treatment also involves a switch from oral to intramuscular or intravenous therapy.

Lack of information and/or fear of resistance also tend to drive prescribers to choose newer, broad-spectrum agents (which are also more costly) even when the majority of infections may respond to first-line treatment. The need for surveillance and the interpretation and actions resulting from surveillance data to improve prescribing and supply of appropriate drugs (see below) also imply financial commitments.

Lack of data on resistance, antimicrobial consumption, clinical outcome, and linkage between these data mean that, to date, the true impact of resistance is impossible to quantify. However, significant efforts are now being put toward establishing appropriate data-gathering mechanisms to address these issues.

4 ISSUES IN THE DETERMINATION OF THE MAGNITUDE AND IMPACT OF ANTIMICROBIAL RESISTANCE

The resistance data summarized above are drawn from laboratory-based studies of antimicrobial resistance performed on bacterial isolates from patients (or, where stated, healthy carriers). One of the complex issues surrounding interpretation of laboratory-based resistance surveillance is the accuracy with which in vitro tests predict in vivo response to therapy. For some bacteria and sites of infection, the correlation between laboratory result and clinical response is good, whereas for others it appears to be poor. Therefore, questions are often raised concerning the relevance of resistance in the clinical setting.

In addition, there is no international standard method for performing antimicrobial susceptibility tests and different methods have gained popularity in different parts of the world. As a result, bacteria considered susceptible in one laboratory may be reported as resistant elsewhere. Sometimes there is considerable variation in methods within, as well as between, countries. This makes for difficulties in comparing results from different studies. On reviewing the literature, it is clear that the methods and resistance breakpoints are often incompletely described by the investigators, thereby adding to the difficulties interpreting data on magnitude and trends in resistance. Although essential to ensure the validity of the data, the use of quality assurance methods is hard to ascertain from published literature. For instance, of 111 published studies of *S. pneumoniae* resistance in 40 developing countries, only 20 reported the use of quality control strains or participation in quality control programs.

As well as the lack of international standards for laboratory tests, rigorous epidemiological methods are not applied in most resistance surveillance studies. The term *prevalence* in published studies of resistance usually refers to the proportion of resistant organisms among all organisms tested. However, "incidence" and "prevalence" are often used interchangeably. From a public health standpoint, one of the goals of surveillance is to detect the number of patients infected with resistant strains as a proportion of the total number of infections in the population over time. However, the organisation of laboratory-based surveillance often fails to link laboratory data to clinical outcome. Further bias may be introduced by the fact that in vitro tests for resistance are only performed on isolates from the subset of patients from which specimens are collected.

5 FACTORS CONTRIBUTING TO THE EMERGENCE AND SPREAD OF RESISTANCE

The emergence of resistance is the natural response of microbes to the presence of antimicrobials, and it is widely accepted that the greater the use of antimicrobials, the greater will be the emergence of resistance. Thus, although significant gains have been achieved in the fight against infectious diseases as a result of the availability of antimicrobials, their unnecessary use contributes to the resistance problem without concurrent health benefit. Studies (35,36) from a number of countries have shown that as much as 50% of antimicrobial prescribing is inappropriate. In a study in 12 developing countries, antimicrobials constituted 25–63% of all medications prescribed (37) (also see Chap. 17).

Many factors influence inappropriate prescribing, dispensing, and use of antimicrobials in human medicine. In addition, since antimicrobials are used in areas outside human medicine, these uses may contribute to the resistance problem in human infections. The roles that prescribers, dispensers, patients, governments, and health systems play in contributing to inappropriate use of antimicrobials and the possible contribution of uses of antimicrobials outside human medicine are summarized below.

5.1 Prescribers and Dispensers

Prescribers who lack knowledge through lack of education or access to updated information tend to prescribe less rationally (38). Facilities may be inadequate to make a proper clinical diagnosis, and there is often a lack of, or poor utilization of, laboratory services to provide microbiological diagnoses and tests for antimicrobial susceptibility. However, the prescriber may recognize that the patient's symptoms do not call for antimicrobial therapy but nevertheless writes a prescription "just in case" of a secondary infection or on account of perceived or actual pressure from the patient. Writing a prescription is often quicker than explaining why a prescription is not necessary, and individual prescribers, particularly in community settings, rarely see the evidence of their overprescribing in the form of patients presenting on a subsequent occasion with resistant infections. Economic incentives and enticements from pharmaceutical companies further pressure the prescriber to write unnecessary or inappropriate prescriptions.

It is common practice in many countries for antimicrobials to be dispensed without a prescription. Patients buy their antimicrobials at the pharmacy (or in the marketplace) without medical advice, and the guidance provided by the dispenser may be rudimentary (due to lack of education) and/or biased by economic incentives. Lack of ability to pay often results in the purchase of incomplete courses of therapy or cheaper drugs that may be substandard (or counterfeit).

5.2 Patients

In addition to the points raised above, patients themselves contribute to the emergence of resistance through poor compliance to the prescribed course of treatment, especially if the treatment course is long, difficult, and costly (e.g., anti-TB, anti-HIV) or if their symptoms are mild and resolve quickly. Failure to complete a full course of treatment for tuberculosis is strongly linked to the emergence of resistance (28). However, the contribution to resistance of poor compliance in acute, short-lived infections is unclear (39), and further research is needed to define the dose and duration of therapy that optimizes clinical cure while minimizing resistance emergence.

5.3 Hospitals

Although the majority of antimicrobial use occurs in the community, the most intense use is in hospitals. This strong selective pressure together with the multitude of opportunities for resistant strains to spread from patient to patient mean that hospital-acquired infections are mainly caused by multiresistant bacteria or opportunist fungal pathogens that are favored by their intrinsic resistance to antibacterial drugs. Early discharges from hospitals either through changes in practice (e.g., cost reduction in developed countries) or lack of patients' ability to pay (particularly in developing countries) help to disseminate multiresistant strains such as MRSA into the community.

5.4 Governments

Lack of information about prevalent resistance problems or poor supply chain management may result in lack of availability of essential antimicrobials in the public sector in developing countries. Weaknesses in legislation or its enforcement contribute to resistance by allowing the circulation in the market of substandard or counterfeit antimicrobials. Advertising and promotion of drugs undoubtedly increases sales, and poor regulation of these activities encourages unnecessary use of antimicrobials (as noted above). National health systems should ensure that appropriate antimicrobials are accessible to those in need. Reimbursement systems may influence prescribing or dispensing if, for example, prescription costs for some classes of antimicrobials are fully reimbursed, whereas not for others. Furthermore, lack of training and certification of prescribers and dispensers may be due to poor provision or regulation by governments.

5.5 Contribution of Nonhuman Uses of Antimicrobials

Large volumes of antimicrobial agents are used throughout the world for purposes outside human medicine. There is currently much debate about the impact of these uses on the global resistance problem in human infections. The use of antimicrobials, at subtherapeutic concentrations, as growth promoters in food animal rearing and in herd or flockwide prophylaxis is widely believed to have selected for resistance in foodborne pathogens such as *Salmonella* and *Campylobacter* spp. and probably in some opportunist zoonotic bacteria such as enterococci. As in human medicine, appropriate therapeutic use in animals also contributes to the emergence of resistance in animal pathogens, but these risks must be balanced against the value of antimicrobials in treating sick animals and delivering healthy animals to market and the food chain.

The use of antimicrobials in production of other foods, for example, fruit crops, aquaculture, and rice growing contributes to the global selective pressure favoring resistance emergence but usually in pathogens far removed from those causing human infection. However, the extent to which resistance genes circulate between different ecosystems needs further elucidation.

6 ADDRESSING THE RESISTANCE THREAT

The containment of antimicrobial resistance requires interventions directed to slowing emergence of resistance and controlling spread of resistant strains. Although it is widely accepted that reducing the selective pressure by reducing inappropriate use of antimicrobials should result in a reduction in the emergence of resistance, there are significant gaps in our understanding of the relationship between use and resistance. Interventions to control the spread of infections per se should also reduce the spread of resistance, but behavioral change and better use of existing tools are needed to translate available scientific knowledge into effective practice.

6.1 Appropriate Antimicrobial Use

As discussed above, the reasons behind misuse of antimicrobials are many and varied; thus, improving use will require a combination of interventions tailored to local and national realities. A greater understanding of the behavioral, social, and cultural issues; why physicians prescribe antimicrobials and why patients ask for them is an important prelude to designing appropriate educational programs. Sharing information on the effects of interventions both those that work and those that do not, is needed. Treatment guidelines and algorithms can provide useful tools for the prescriber and boost confidence to "withhold" unnecessary prescriptions. For the individual patient, improved diagnostic facilities and more rapid diagnostics could play an important role in reducing unnecessary prescriptions, but the cost-benefit analyses need to be done to convince the deliverers of health care. In some countries it is the need for enforcement of regulations concerning the availability and quality control of antimicrobial products that should be paramount.

Appropriate antimicrobial use also depends on having the antimicrobials available and accessible to those in need. Because of the global spread of resistance, this may result in the need for newer, more expensive drugs in countries whose health services cannot afford them.

6.2 Public Health Infrastructure

A strategy to contain resistance needs to be based on a sound public health infrastructure (40). The resistance problem could be diminished if fewer people became infected in the first place. Thus, the basic needs for clean water, adequate sanitation, and improved living conditions cannot be ignored. Furthermore, active programs for disease prevention (e.g., immunization, health education) should be strengthened. The public health system must be capable of detecting outbreaks of infectious disease and responding rapidly and effectively.

Surveillance of antimicrobial resistance and transformation of the data collected into information for action is critical to monitoring trends in resistance, detecting new resistance problems, and evaluating the effects of interventions (41). Surveillance data should be used to update local and national treatment guidelines, formularies, and essential drug lists, and thus to guide the provision of appropriate antimicrobials in the health system. Although many local and national surveillance programs currently exist, flow of information to guide policy decisions needs to be strengthened. Furthermore, guidelines need to be developed to assist prescribers and policy makers to formulate criteria for "switching" from first- to second- or third-line drugs. Surveillance of antimicrobial usage is still very scarce but is desirable to monitor prescribing patterns and conformation with treatment guidelines.

6.3 Research for Novel Antimicrobials

In part, antimicrobial resistance has become a threat because of the dearth of new molecules effective against resistant pathogens. New techniques being applied in drug discovery (e.g., genomics, combinatorial chemistry, bioinformatics) are anticipated to produce novel lead molecules (42). However, there is a need for active dialogue between the public and private sectors to encourage the development of these agents if controls on prescribing or limitations in access result in unfavorable market conditions. These challenges are all the greater because of the continuing need for appropriate antimicrobials to address the resistance problem globally in the face of widely differing national priorities and resources, diverse health delivery systems and levels of regulation, and availability and use of antimicrobial agents.

DISCLAIMER

The mention of certain manufacturers' products does not imply that they are endorsed or recommended by the WHO in preference to others of a similar nature that are not mentioned.

REFERENCES

1. WHO Report on Infectious Diseases: removing obstacles to healthy development 1999. http://www.who.int/infectious-disease-report.
2. Koornhoff HJ, Wasas A, Klugman KP. Antimicrobial resistance in *Streptococcus pneumoniae*: a South African perspective. Clin Infect Dis 1992; 15:84–94.
3. Widdowson CA, Klugman KP. Emergence of the M phenotype erythromycin-resistant pneumococci in South Africa. Emerg Infect Dis. 1998; 4:277–281.
4. Goldsmith CE, Moore JE, Murphy PG. Pneumococcal resistance in the UK. J Antimicrob Chemother 1997; 40:11–18.
5. Ohene A. Bacterial pathogens and their antimicrobial susceptibility in Kumasi, Ghana. East Afr Med J 1997; 74:450–455.
6. Handsman D. Chloramphenicol-resistant pneumococci in West Africa. Lancet 1978; 1:1102–1103.

7. Mastro TD, Ghafoor A, Nomani NK, et al. Antimicrobial resistance in pneumococci in children with acute lower respiratory tract infection in Pakistan. Lancet 1991; 337:156–159.
8. Mastro TD, Nomani NK, Ishaq Z, et al. Use of nasopharyngeal isolates of *Streptococcus pneumoniae* and *Haemophilus influenzae* from children in Pakistan for surveillance for antimicrobial resistance. Pediatr Infect Dis J 1993; 15:84–94.
9. Akpede O, Abiodun PO, Sykes M, Salami CE. Childhood bacterial meningitis beyond the neonatal period in southern Nigeria: changes in organism/antibiotic susceptibility. East Afr Med J 1994; 71:14–20.
10. Jordens JZ, Slack MPE. *Haemophilus influenzae*: then and now. Eur J Clin Microbiol Infect Dis 1995; 14:935–948.
11. Pecoul B, Chirac P, Trouiller P, Pinel J. Access to essential drugs in poor countries. JAMA 1999; 288:361–367.
12. Shears P. Shigella infections. Ann Trop Med Parasitol 1996; 90:105–114.
13. Sack DB, Rahaman M, Yunus M, Khan EM. Antimicrobial resistance in organisms causing diarrheal disease. Clin Infect Dis 1997; 24(Suppl 1):S102–105.
14. Parsonnet J, Greene KD, Gerber AR, et al. *Shigella dysenteriae* type 1 infections in US travellers to Mexico. Lancet 1989; 2:543–545.
15. Mhalu FS, Muari PW, Ijumba J. Rapid emergence of El Tor *Vibrio cholerae* resistant to antimicrobials during first six months of fourth cholera epidemic in Tanzania. Lancet 1979; 1:345–347.
16. Anand VK, Arora S, Patwari AK, et al. Multidrug resistance in *Vibrio cholerae*. Indian Pediatr 1996; 33:774–777.
17. Molbak K, Baggesen DL, Aarestrup FM, et al. An outbreak of multi-drug resistant, quinolone-resistant *Salmonella enterica* serotype *typhimurium* DT 104. N Engl J Med 1999; 341:1420–1425.
18. Piddock LJ. Fluoroquinolone resistance: overuse of fluoroquinolones in human and veterinary medicine can breed resistance. Br Med J 1998; 317:1029–1030.
19. Smith KE, Besser JM, Hedberg CW, et al. Quinolone-resistant *Campylobacter jejuni* infections in Minnesota, 1992–1998. Investigation Team. N Engl J Med 1999; 340:1525–1532.
20. Rowe B, Ward LR, Threlfall EJ. Multi drug-resistant *Salmonella typhi*: a worldwide epidemic. Clin Infect Dis 1997; 24(Suppl):S106–S109.
21. Threlfall EJ, Ward LR, Skinner JA, et al. Ciprofloxacin-resistant *Salmonella typhi* and treatment failure. Lancet 1999; 353:1590–1591.
22. Yerly S, Kaiser L, Race E, et al. Transmission of antiretroviral-drug-resistant HIV-1 variants. Lancet 1999; 354:729–733.
23. Anon. Surveillance of antibiotic resistance in *Neisseria gonorrhoeae* in the WHO Western Pacific Region. STI/HI/AIDS Surveillance Report, October 1999.
24. Van Dyck E, Crabbe EF, Nzila N, et al. Increasing resistance of *Neisseria gonorrhoeae* in West and Central Africa. Consequence on therapy of gonococcal infection. Sex Transm Dis 1997; 24:32–37.
25. Osoba AO, Johnston NA, Obunbanjo BO, Ochei J. Plasmid profile of *Neisseria gonorrhoeae* in Nigeria and efficacy of spectinomycin in treating gonorrhoea. Genitourin Med 1987; 63:1–5.

26. Anon. Resistance in gonococci isolated in the WHO Western Pacific Region to various antimicrobials used in the treatment of gonorrhoea. Communicable Dis Intell 1995; 19:495–499.
27. Tapsall JW, Limnios EA, Shultz TR. Continuing evolution of the pattern of quinolone resistance in *Neisseria gonorrhoeae* isolated in Sydney, Australia. Sex Transm Dis 1998; 25:415–417.
28. WHO/IUATLD Global Project on anti-tuberculosis drug resistance surveillance. Anti-tuberculosis drug resistance in the world. Report No. 2. World Health Organization WHO/CDS/TB/2000.278.
29. World Health Organization. Status of tuberculosis in the 22 high burden countries. 1999; World Health Organization WHO/CDS/TB/99.271.
30. Cormican MG, Jones RN. Emerging resistance to antimicrobial agents in gram-positive bacteria. Enterococci, staphylococci and non-pneumococcal streptococci. Drugs 1996; 51(Suppl 1):6–12.
31. Vincent JI, Bihar DJ, Suter PM, et al. The prevalence of nosocomial infection in intensive care units in Europe: results of the European Prevalence of Infection in Intensive Care (EPIC) study. JAMA 1995; 274:639–644.
32. Morbidity and Mortality Weekly Report. Four pediatric deaths from community-acquired methicillin-resistant *Staphylococcus aureus*—Minnesota and North Dakota, 1997–1999. MMWR 1999; 48:707–710.
33. Kristinsson KG. Epidemiology of penicillin resistant pneumococci in Iceland. Microb Drug Resist 1995; 1:121–125.
34. Holmberg SD, Solomon SL, Blake PA. Health and economic impacts of antimicrobial resistance. Rev Infect Dis 1987; 9:1065–1078.
35. Gonzales R, Steiner JF, Sande MA. Antibiotic prescribing for adults with cold, upper respiratory tract infections, and bronchitis by ambulatory care physicians. JAMA 1997; 278:901–904.
36. Guillemont D, Carbon C, Vauzelle-Kervoedran F, et al. Inappropriateness and variability of antibiotic prescription among French office-based physicians. J Clin Epidemiol 1998; 51:61–68.
37. Hogerzeil HV, Bino, Ross-Degnan D, et al. Field test for rational drug use in twelve developing countries. Lancet 1993; 342:1408–1410.
38. Kunin CM. Antibiotic armageddon. Clin Infect Dis 1997; 25:240–241.
39. Lambert HP. Don't keep taking the tablets? Lancet 1999; 354:943–945.
40. World Health Organization. The WHO Global Strategy for Containment of Antimicrobial Resistance. WHO/CDS/CSR/DRS/2001.2.
41. Williams RJ, Heymann DL. Containment of antibiotic resistance. Science 1998; 279:1153–1154.
42. Bax RP, Mullen NA. Response of the pharmaceutical industry to antimicrobial resistance. Clin Infect Dis 1999; 5:289–303.

11

Genetic Methods for Detecting Bacterial Resistance Genes

Amalio Telenti
Centre Hospitalier Universitaire Vaudois, University of Lausanne, Lausanne, Switzerland

Fred C. Tenover
Centers for Disease Control and Prevention, Atlanta, Georgia

Identification of resistance genotypes in bacteria can be accomplished through detection of novel genetic material and characterization of mutations in specific genes. When the presence of a specific genetic determinant is a hallmark of antimicrobial resistance, specific probes or amplification of the region by polymerase chain reaction (PCR) or other technique can be used as diagnostic tools. When resistance develops by modification of a gene otherwise present in susceptible strains, a precise analysis for mutations is required. Mutation scanning technologies aim to find unknown mutations in candidate resistance genes. This involves assessment of broad or multiple genetic regions and of each identified polymorphism as relevant to the process investigated. In contrast, mutation screening techniques aim at finding mutations already known to be relevant for the phenotype are under study. The most promising techniques are based on the use of oligonucleotide or DNA arrays and on the application of fluorescent-energy transfer techniques. In the future, two separate families of tools will be required: (1) high-performance automated equipment capable of detecting multiple targets (i.e., a "genetic antibiogram") and (2) simple tools for use in field work or in less technologically demanding situations. However, significant work remains to be accomplished on defining the basic mechanisms of antimicrobial resistance, correlating phenotypes and genotypes, developing labor- and cost-effective equipment and tools for commercial use, and establishing strict quality control and quality assurance guidelines before these tests can become mainstream diagnostic tests.

1 INTRODUCTION

Resistance to antimicrobial agents in bacteria can be medicated by several mechanisms including 1) changes in the permeability of the cell envelope which limit the amount of drug that has access to cellular targets, 2) active efflux of the drug out of the cell, 3) modification of the site of drug action, 4) provision of alternate enzymatic pathways around those blocked by antibacterial therapy, or 5) destruction or inactivation of the antimicrobial agent (1–3). Resistance phenotypes in bacterial agents are usually detected using one of several standardized methods, including agar dilution, broth microdilution, or disk diffusion testing (4,5); however, these methods do not differentiate among resistance mechanisms. Although resistance may occur due to mutations in key genetic loci in the bacterial genome, most resistance to antimicrobial agents is mediated by genes acquired via mobile genetic elements such as plasmids and transposons. Thus, identification of resistance genotypes is accomplished through detection of novel genetic material and characterization of mutations in specific genes.

Genetic tests aimed at the detection of mutations or resistance genes in bacterial isolates have been developed based on the supposition that gene carriage results in a resistance phenotype, although exceptions to this rule have been documented (6–8). Since genetic methods, including DNA probes, PCR, and other amplification techniques, are now used in a variety of clinical laboratories for identification and quantitation of pathogenic microorganisms, and for human genetics, the application of the same genetic methods to detect antimicrobial resistance genes is a natural extension of the molecular diagnostics arena. The new technologies for rapid DNA sequence analysis and the coupling of novel technologies, such as branch-DNA assays (9) and TaqMan reporter systems (10), to the detection of resistance determinants, is significantly enhancing the detection of mutations and genes associated with antibacterial resistance in a variety of microbial pathogens (11,12).

Significant work remains to be accomplished on defining the basic mechanisms of antimicrobial resistance, correlating phenotypes and genotypes, developing labor- and cost-effective equipment and tools for commercial use, and establishing strict quality control and quality assurance guidelines before these tests can become mainstream diagnostic tests. However, the application of genetic techniques to the recognition of determinants of resistance will clearly enhance our understanding of the epidemiology of antimicrobial resistance and improve therapy during the early stages of infection.

2 REASONS FOR USING GENETIC TESTS TO DETECT RESISTANCE GENES

There are three reasons to pursue the identification of antimicrobial resistance genes and mutations associated with resistance (Table 1). First, rapid detection of resistance could be important for effective antimicrobial chemotherapy of bacterial infections, such as sepsis and meningitis. In many cases, anti-infective therapy must be initiated before the identity of the organism and its antimicrobial susceptibility pattern can be established.

Table 1 Advantages and Disadvantages of Genotypic Resistance Testing

Advantages
Rapid detection of resistance can optimize therapy of sepsis and meningitis before culture results are available.
Direct analysis of clinical specimens allows detection of noncultivable or organisms with fastidious growth requirements.
Arbitration of borderline MIC results that can impact therapeutic choices.
Provides gold standard for reevaluating current breakpoints or establishing new breakpoints for novel antimicrobial agents based on mechanisms of resistance.
Disadvantages
Limited sensitivity in the presence of low number of organisms or mixed populations of susceptible and resistance organisms.
Multiple mechanisms of resistance in some organisms limit the types of determinants that can be sought. Lack of a universal technology for testing all resistance genes.
Novel genetic determinants continue to appear. Geographical variation in the prevalence of some resistance genes needs to be explored.
Some genotypic changes detected by mutation screening methods are not associated with a resistance phenotype.
Contamination from amplification products limits specificity of some tests.
Special laboratory facilities required.

Thus, physicians may disregard the results when they become available unless the patient is not responding to the empiric therapy which may be neither clinically nor economically optimal (13).

Second, genetic methods can be applied directly on clinical material, diminishing laboratory biohazard, and can be applied to assess resistance in noncultivable microorganisms, such as *Mycobacterium leprae* (14).

Third, genetic methods are helpful for arbitrating MIC results that are at or near the breakpoint for resistance for bacterial species. For example, isolates of *Staphylococcus aureus* with oxacillin MICs between 2 and 8 μg/mL may contain the *mecA* (methicillin) resistance gene determinant or may produce high levels of β-lactamase that slowly hydrolyze oxacillin (15,16). Although vancomycin would be the drug of choice for the former cases, hyper–β-lactamase producers can be effectively treated with more effective β-lactam/β-lactamase inhibitor combinations (17). A test showing the absence of the *mecA* gene suggests that a physician could use an antimicrobial agent other than vancomycin to treat the infection. Genetic tests also can be used as the gold standard for resistance when evaluating the accuracy of new susceptibility testing methods that use clinical isolates or stock cultures with borderline MICs (18,19).

In summary, rapid and unequivocal identification of resistant organisms directly in clinical specimens may guide therapy early in the course of a patient's disease before culture and susceptibility results are available. For example, PCR assays can detect mutations in the *rpoB* locus associated with rifampin resistance in *Mycobacterium tuberculosis* (20,21). Such mutations indicate that the strain is at least resistant to rifampin and likely to be resistant to multiple drugs. PCR results that indicate the presence of *M. tuberculosis* and mutations in the *rpoB* locus directs the physician to avoid rifampin and to use alternative antimycobacterial agents.

3 STRATEGIES TO DETECT MUTATIONS ASSOCIATED WITH RESISTANCE

The number of molecular technologies capable of identifying specific DNA or RNA sequences, or mutations at particular loci, have increased dramatically over the last several years. Many of these technologies have been adapted for the diagnosis of human diseases or for bacterial and viral species identification. However, because of these advances, there is a tendency to overlook the fact that molecular diagnostics today is in a far from ideal state (22).

When acquisition of a new gene, or the presence of a specific genetic determinant is a hallmark of antimicrobial resistance, specific probes or amplification of the region by PCR or other technique (ligase chain reaction, transcription-mediated amplification, branched-DNA amplification) can be used as diagnostic tools to predict therapeutic outcome. Examples of DNA probes and PCR primer sets that can be used to detect resistance genes have been compiled elsewhere (23). When resistance develops in a microorganism by modification of a gene otherwise present in susceptible strains, a precise analysis for mutations is required. In molecular diagnosis, distinction is made between mutation scanning and screening methodologies (22).

3.1 Mutation Scanning

Scanning technologies aim to find *unknown* mutations in candidate resistance genes (Table 2). This may involve assessment of broad or multiple genetic regions and careful assessment of the identified polymorphisms as relevant or not to the process investigated. Thereafter, epidemiological analysis, assessment of large sample collections, and laboratory selection or construction of mutants carrying specific candidate polymorphisms may be needed to elucidate the role and significance of the observed sequence variations.

Although DNA sequence analysis is the gold standard for detection of point mutations, a number of techniques have been proposed that facilitate less labor-intensive detection of mutations, higher processing rates, or greater potential for automation, such as single-strand conformation polymorphism, denaturing gradient gel electrophoresis, heteroduplex analysis, and chemical or enzymatic cleavage of mismatched DNA.

3.2 Mutation Screening

Screening techniques aim to find *known* mutations, preferably with high throughput. These mutations have already been established as relevant, causal, or useful markers for the phenotype under study. The most relevant technologies are presented in Table 3. The most promising techniques are based on the use of oligonucleotide arrays and on the application of fluorescent-energy transfer techniques (24,25). Indeed, two separate families of tools soon will be required: (1) high-performance automated equipment capable of detecting multiple targets (i.e., a "genetic antibiogram") and (2) simple tools for use in field work or in less technologically demanding situations (e.g., detection of resistance to antituberculosis drugs in developing countries).

Holding the lead for the first category of instruments are those based on microarray technology, where oligomers representing the various wild-type and mutated sequences and oligomers for species confirmation are placed on the same surface and interrogated simultaneously (Fig. 1). Real-

TABLE 2 Mutation Detection (Scanning) Technologies

Complete gene sequences	This is currently the reference method. With the introduction of high-throughput 96-channel capillary sequencers, sequencing is becoming more accessible to diagnostic clinical laboratories.
Single-strand conformation polymorphism (SSCP) (128,129)	SSCP is based on sequence-dependent structural differences in amplification products that are detected by nondenaturing gel electrophoresis. Mutation detection rates of 80–90% can be achieved under optimal conditions.
Denaturing gradient gel electrophoresis (130)	This methods detects sequence differences as changes in partial melting behavior of double-stranded DNA, and hence differences in migration distances of fragments are observed when subjected to increasing denaturing conditions. Mutation detection rates of 90–95% can be achieved under optimal conditions.
Heterduplex analysis (131,132)	Heteroduplex analysis detects mismatches between mixed wild-type and mutated sequences by mobility shifts of bands during electrophoresis. An increasingly popular variant of the technique is denaturing high-performance liquid chromatography. This technology, as applied in the WAVE apparatus (Transgenomic, Santa Clara, CA), detects changes in PCR products at rates of 95–99%. A combination between heteroduplex analysis and denaturing gel electrophoresis appears to improve the respective performance of both techniques (133).
Chemical and enzymatic cleavage of mismatch (134,135)	This technique is based on the heteroduplex analysis but includes cleavage of DNA strands at mismatched positions, which allows the precise location of a mutation. An example of this technology is the Invader assay (Third Wave Technologies, Madison, WI), which is an isothermal, non–PCR-based method. Displacement of the mutated sequence by a complementary oligonucleotide is followed by enzymatic cleavage of the structure and detection of the product.
DNA chips (25)	Very high-density DNA chips were first developed by Affymetrix (Santa Clara, CA). After further system development to reduce unit cost and increase robustness in diagnostic settings, this technology and its variants is thought to be one of the more powerful future technologies for mutation detection.

Source: Adapted from Ref. 22.

TABLE 3 Mutation Detection (Screening) Technology

Allele-specific oligonucleotide hybridization This technique involves hybridization of radioactive- or fluorescence-labeled samples to a set of immobilized oligonucleotides that are specific for the wild-type or mutant sequence. This is a precursor of high-density DNA chip technology.
Oligonucleotide ligation assay (136) This assay is based on the binding of both a labeled reporter oligonucleotide and an immobilized target oligonucleotide to sample DNA which serves as the template. The target and reporter oligonucleotides are chosen to be contiguous and complementary to the mutation-containing region, with the mutation(s) to be detected placed as the last nucleotide in the immobilized oligonucleotide. In the case of mismatch, no ligation will occur, whereas as perfect match will lead to a covalent attachment of the reporter oligonucleotide to the target oligonucleotide. Because of the covalent bond, ligation-based techniques have a lower signal to noise ratio than hybridization-based techniques. The Padlock-probe technique (Amersham Pharmacia Biotech Ltd, Buckinghamshire, UK) is an example of this solid-state array format (137).
Solid-state minisequencing (138) In this technique, a solid support contains oligonucleotides, to which the sample DNA is annealed, that are complementary to the sequence variant to be tested. Four uniquely labeled dNTPs are provided, and the nature of the label incorporated identifies the sequence variant in the sample. Variants of this technique include APEX (arrayed primer extension) and SBE (sequencing by extension) (139).
Fluorescent-energy transfer techniques Amplification reactions can be monitored quantitatively in real time using this method which obviates post-PCR handling, thus reducing the potential for contamination. It is currently used in liquid-based format but has the potential for solid-state multiplexing. 1. TaqMan. This method uses the 5′ activity of the *Thermus aquaticus* DNA polymerase to detect an amplification product through the release of a fluorescence reporter dye which is coupled to a probe containing a quencher dye. The cleavage of a the novel structure releases the reporter from proximity to the quencher (ABI PRISM 7700; Perkin-Elmer Corporation, Foster City, CA) (140). 2. LightCycler. In this technology, a mutation-specific sensor probe carrying a fluorphore hybridizes to the target nucleic acid. An adjacent oligonucleotid carries a second fluorophore with overlapping spectral properties and results in fluorescence energy transfer when hybridization takes place (141). 3. Molecular beacons. Based on a stem-loop–structured oligonucleotide probe, the end of the stem contains a fluorochrome and a quencher. When the probe binds to its target, the stem is opened, moving the quencher away from the label, thus restoring fluorescence (142,143).

Source: Adapted from Ref. 22.

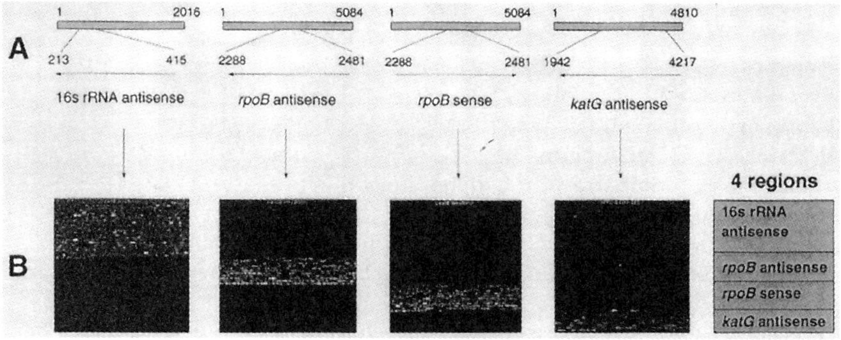

Figure 1 Shown is a *Mycobacterium* probe array for simultaneous species identification and genotypic resistance testing. The *Mycobacterium* probe array is divided into four separate regions, with specific probes for the analysis of either 16S rRNA, *rpoB* (rifampin resistance), and *katG* targets (isoniazid resistance). (A) Diagram of regions analyzed on the array. (B) Hybridization experiments. The fluorescence images were obtained following hybridization of fluorescein-labeled fragmented RNA, generated by in vitro transcription from PCR amplicons. (Adapted from Troesch et al. J Clin Microbiol 1999; 37: 49–55.)

time amplification devices based on fluorochrome accumulation in a closed PCR reaction tube represents the second attractive approach, given the rapidity of results and, most significantly, the absence of postamplification manipulation, which entails significant savings and a critical reduction of the potential for contamination of a laboratory with amplification products (Fig. 2).

4 TARGETS

4.1 Aminoglycoside Resistance Genes

Aminoglycoside resistance genes are common in both gram-positive and gram-negative organisms (26). However, the large number of different types of aminoglycoside resistance genes present in gram-negative organisms (acetyltransferases, adenylyltransferases, and phosphotransferases) and the lack of consensus sequences that would allow detection of multiple types of genes with a single DNA probe or PCR primer set (26) make it difficult to use probes and PCR tests to predict resistance. Rather genetic methods are better suited for classifying new determinants and epidemiological studies (27–30).

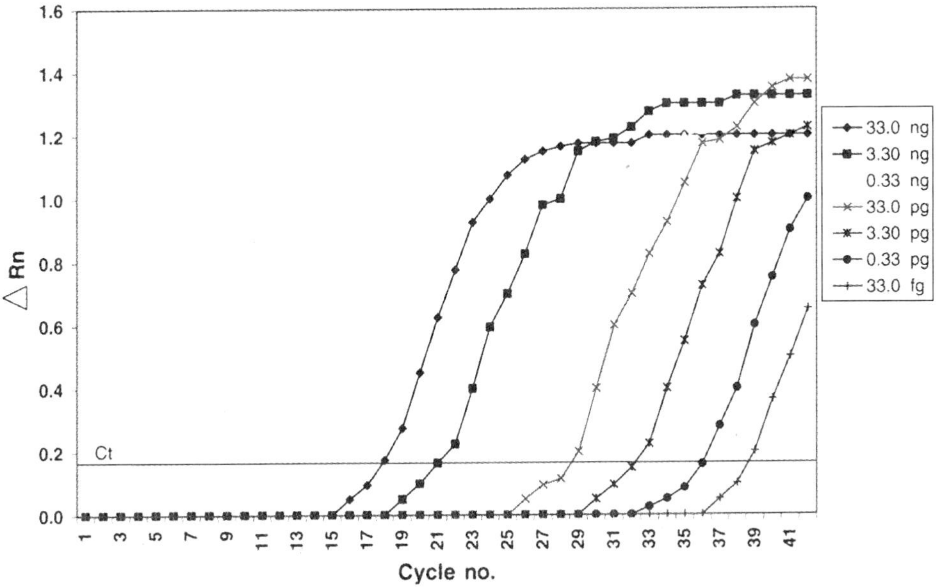

Figure 2 5′-Nuclease assay (TaqMan) for *mecA* resistance gene in *Staphylococcus aureus.* Assay sensitivity using 10× dilution scheme $\Delta R_n = (R_n{}^+) - (R_n{}^-)$, where $(R_n{}^+)$ = (emission intensity of reporter dye)/(emission intensity of passive reference dye) in PCR template and $(R_n{}^-)$ = (emission intensity of reporter dye)/(emission intensity of passive reference dye) in PCR without template or early cycles of real-time reaction. C_t = threshold cycle (i.e., cycle at which a statistically significant increase in ΔR_n is first detected.

Aminoglycoside resistance in gram-positive organisms, however, is more uniform (7). Thus, probes and PCR primers can be used to identify strains that carry genes encoding high-level aminoglycoside resistance. PCR assays and probes have been used, particularly with enterococci, to identify the ANT(6) streptomycin resistance gene and the gene encoding the AAC(6′)/APH(2″) bifunctional enzyme responsible for high-level gentamicin resistance (19,29,31,32).

4.2 β-Lactam Resistance Genes

4.2.1 Methicillin Resistance in Staphylococci

Phenotypic detection of oxacillin resistance in staphylococci, which is primarily mediated by the *mecA* determinant (15,33), continues to be a problem (34), particularly with the coagulase-negative strains (15,35). A *mecA* probe or PCR or branched-DNA assays can differentiate those isolates that are borderline resistant to oxacillin due to the production of large quantities of β-lactamase from those that are resistant owing to the presence of the *mecA* determinant (17,33,34,36,37). The rare strains of *S. aureus* that are resistant to oxacillin by virtue of containing modified penicillin-binding proteins (PBPs) with reduced affinity for oxacillin (the so-called MOD strains) may be misclassified as oxacillin susceptible by the *mecA* gene test, since these strains are truly oxacillin resistant but do not contain the *mecA* gene (38).

4.2.2 β-Lactam Resistance in Pneumococci

Resistance to penicillin and other antimicrobial agents in pneumococci has become a global problem (39–41). Resistance develops when pneumococcal PBPs are remodeled through the acquisition of chromosomal DNA from other pneumococci or other streptococcal species (42,43). Since remodeling of the PBP genes appears to conserve certain motifs, PCR primers have been developed for amplifying regions of the susceptible 2B PBP gene (44). The lack of product in the presence of amplification controls suggest that gene has been remodeled and, therefore, designates resistance. Such an assay does not reliably indicate which strains could be treated with penicillin versus an extended-spectrum cephalosporin, as might be desirable for an assay to be used in a clinical laboratory, but may be used as a screening tool for analyzing large groups of strains for resistance.

4.2.3 β-Lactamase Genes in Gram-Negative Organisms

Several DNA probes and PCR primer sets have been developed to detect the genes encoding the TEM, SHV, OXA, CARB, and ROB β-lactamases present in gram-negative organisms (23). Amplification approaches may be used to detect β-lactamases directly in clinical samples (45). Thus, the utility of direct detection has clearly been demonstrated, although it is rarely, if ever, used in clinical laboratories, primarily because the test is not commercially available.

Nosocomial infections caused by *Klebsiella pneumoniae, K. oxytoca* (46,47), and other Enterobacteriaceae that produce extended-spectrum β-lactamases, plasmid-mediated class C β-lactamases (e.g., AmpC), and other enzymes capable of hydrolyzing cefotaxime, ceftriaxone, ceftazidime, and aztreonam are increasing in the United States and Europe (48–54). Although, DNA sequencing has become the gold standard for analyzing novel β-lactamase genes, better methods are needed for rapid recognition and typing of the resistance determinants. PCR primers, although helpful, are often nonspecific, although PCR primers for genes mediating resistance to imipenem, such as *bla*$_{IMP}$ (55), which appears to be spreading quickly in Japan and the Far East (56), can be helpful.

4.3 Glycopeptide Resistance Genes

Resistance to vancomycin in enterococci has become a major global issue during the last few years. Resistance in enterococci can be mediated by several different genes, including *vanA*, *vanB*, *vanC1*, *vanC2*, *vanC3*, *vanD*, and *vanE* (57–64). DNA probes and PCR assays that detect these genes have been described. PCR has also been used to detect the *vanA* resistance gene in enterococcal DNA purified directly from fecal material recovered on rectal swabs (65). Recently, decreased susceptibility to glycopeptides has been noted in strains of *S. aureus* from Japan and the United States (66,67). However, genetic assays to detect this glycopeptide resistance in staphylococci have yet to be developed, since the mechanisms of resistance remain unknown.

4.4 Macrolide, Lincosamide, and Streptogramin Resistance Genes

The genes mediating resistance to erythromycin have been the targets both of DNA probes and PCR assays. PCR assays that detect erythromycin methylase genes (*erm* genes) that mediate resistance to macrolides, lincosamides (such as clindamycin) and streptogramins (i.e., the MLS resistance phenotype) (68–70); the *msrA* gene that mediates resistance only to macrolides and streptogramins (MS resistance); and the macrolide efflux genes, *mefA* and *mefE* (71–73), have been described as well as a novel efflux mechanism in *S. aureus* (74). Studies by Eady et al. demonstrate that the *msrA* gene is common in isolates of *S. aureus* (74). Studies by Eady et al. demonstrate that the *msrA* gene is common in isolates of *S. aureus* and produces a phenotype of erythromycin resistance but clindamycin susceptibility (69). The *mefE* gene appears to be a common cause of erythromycin resistance in pneumococci in the United States (75). Since most strains of staphylococci that are erythromycin resistant are presumed to be resistant to clindamycin as well, *msrA* probes and PCR assays may be useful in determining the presence of the *msrA* and *ermA* genes if clindamycin therapy was a critical issue.

4.5 Quinolone Resistance Genes

There are two major mechanisms of quinolone resistance: alteration of the target sites, the organisms' gyrase (*gyrA* and *gyrB*) and topoisomerase (*parC* and *parE*) (76–80), and active efflux of the drug out of the cell, which limits access of the drug to the target site. Resistance is usually associated with point mutations in the *gyr* or *par* loci. Since DNA probes are not sufficiently sensitive to detect these changes, investigators have used PCR coupled with direct sequencing of the amplification products to identify changes in the nucleotide sequence of the *gyrA*, *gyrB*, *parC*, and *parE* genes (81–83).

4.6 Tetracycline Resistance Genes

Tetracycline resistance can be mediated by at least 15 different genes (84). DNA probes and PCR have been used for epidemiological studies of the *tet A*, *tet B*, *tet C*, *tet D*, *tet E*, *tet F*, *tet H*, *tet K*, *tet L*, *tet M*, *tet N*, *tet O*, *tet Q*, and *tet S* determinants (84–91). Of these, *tet* M is the most widespread, having been located in a very diverse group of organisms, including staphylococci, streptococci, pneumococci, and gram-negative bacilli. PCR primers have been described for the *tet M*, *tet O*, and *tet Q* genes (92–97). Multiple types of the *tet M* gene also have been recognized through sequence analysis (98). A novel efflux mechanism of tetracycline resistance *tet H* has been described in *Pasteurella multocida* (99) as well as what appears to be a novel ribosomal protection mechanism of resistance encoded by *tetU* in *Enterococcus faecium* (100).

4.7 Other Agents

Genetic testing is possible for genes encoding chloramphenicol acetyltransferases (CAT) that mediate resistance to chloramphenicol (101) in gram-negative organisms (*catI*, *catII*, *catIII*) and in gram-positive anaerobes, including *Clostridium perfringens* and *C. difficile* (*catP*, *catQ*, and *catD*) (102–104). Additional genes that show relatively little DNA sequence homology with those mentioned above are present in staphylococci, streptococci, and aerobic gram-positive bacilli.

Mupirocin is an antistaphylococcal agent that is used to reduce carriage of staphylococci among infected patients and hospital personnel. Recently, a PCR assay was described that can detect high-level mupirocin resistance (105). However, the practical value of the assay has not been assessed in a clinical laboratory setting.

There are two major sulfonamide resistance genes, *sul*I and *sul*II. Both have been cloned and sequenced and probes have been described for each gene (106).

The number of genes capable of mediating trimethoprim resistance in bacteria continues to expand (107,108). DNA probes have proven to be powerful tools for detecting and classifying novel trimethoprim resistance genes (called *dhfr*) (107,109,110). A novel trimethoprim resistance gene, *fol*H, also has been recognized in *Haemophilus influenzae* (111).

5 DETECTING RESISTANCE IN MYCOBACTERIA

Multidrug resistant strains of *M. tuberculosis* (MDR-TB) have been recognized in hospitals in the United States, Southeast Asia, and in parts of Europe where they constitute a major public health problem (112). *M. tuberculosis* grows slowly (1–4 weeks), and consequently, the rapid identification of resistance by genotypic methods represents a significant achievement. At molecular level, development of resistance in *M. tuberculosis* results from chromosomal mutational events. MDR-TB reflects the stepwise accumulation of individual mutations in several independent genes (113) and not a "block" acquisition of multidrug resistance. A considerable amount of work has been devoted in the last few years to understanding mechanisms of resistance and to identifying the genes involved (114,115). Mutations in *katG, inhA, kasA,* and *ahpC* genes are found in 62–90% of isoniazid-resistant strains, rifampin resistance is associated (> 96%) with *rpoB* mutations, pyrazinamide resistance with *pncA* mutations (72–97%), ethambutol resistance with mutations in *embB* (47–65%), streptomycin resistance with *rrs* or *rpsL* mutations (70%), and fluoroquinolone resistance with *gyrA* substitutions (75–94%). Additional genes and mechanisms may play a role, particularly in association with lower levels of resistance.

The target determinant with perhaps the greatest clinical relevance is *rpoB*. Detection of mutations in a limited region of *rpoB*, which can be performed through a single amplification reaction, has excellent sensitivity for the detection of rifampin resistance worldwide, and it also serves as a marker for multidrug resistance in *M. tuberculosis*. Commercial assays to detect this resistance have been described (116).

In contrast, detection of resistance to isoniazid, which together with rifampin is the other primary antituberculosis agent remains complex. Three genetic loci, *katG, inhA,* and *kasA,* have been associated with isoniazid resistance in *M. tuberculosis* (114,115,117–119), and multiple amplification reactions are necessary to include all possible mutations within those regions or the respective promoters. Limited targeting of some of those regions may be useful for detection of isoniazid resistance in specific geographical settings (120,121).

6 GUIDELINES FOR IMPLEMENTING GENETIC TESTING

Today, DNA probes are rarely used to detect resistance genes. Rather laboratories use RNA or DNA amplification assays, which are more accessible and easier to adapt to clinical laboratory use than DNA probes. The critical issue with amplification assays is the reliability of results. The need for quality control measures, including use of amplification controls such as simultaneous amplification or rRNA or rDNA sequences, or optimally, a low–copy number internal standard, to ensure the availability of amplifiable nucleic acid and the absence of inhibitory substances in the reaction, cannot be stressed enough. The temperatures used in PCR assays optimized for use with purified DNA or DNA from bacterial isolates obtained in pure culture may not be stringent enough to avoid false-positive results when used with clinical samples, such as blood or cerebrospinal fluid, where considerably more nonspecific priming can occur (45,122). It may be necessary to increase the temperatures of the assays, particularly the annealing temperatures, to avoid this problem. Using control reactions containing no template DNA to identify nonspecific products owing to contamination of Taq polymerase with DNA is critical (123). Mechanisms for addressing the current limitations of genotypic testing are shown in Table 4.

One should never assume that PCR primers reported in the literature have undergone rigorous testing. Rather primer sets should be thoroughly tested for specificity, self-complementarity, and dimer formation before use. According to the Clinical Laboratory Improvement Act of 1988, validation of DNA problem and PCR tests by the clinical laboratory in which they are to be used is mandatory before they can be used for analysis of clinical specimens. Methods for validation are published by the National Committee for Clinical Laboratory Standards (124).

Now that DNA probes and PCR assays for detecting and differentiating resistance genes are becoming commercially available, more surveys of resistance mechanisms, such as those described by Eady et al. (69), Kapur et al. (21), Ounissi et al. (7), and Shaw et al. (8), should be under-

Table 4 Proposal for Answering Current Limitations of Genotypic Testing

Problem	Technical solution
Contamination from amplification products	Closed-tube or closed-system technologies
Amplification inhibition	Internal amplification standards
Multiple regions to interrogate	Robotics, microassay sequencing, nanoliter-sized DNA sampling
Limited sensitivity	Improvements in DNA concentration techniques, selective DNA capture and enrichment methods, and enhancement of the cumulative benefit from repetitive analysis of clinical material

taken to determine the reservoirs of resistance genes and how resistance genes disseminate in hospitals and community settings. Such studies would also help to determine the frequency with which organisms carry resistance genes that are not expressed. Although still considered experimental, many of the probe and PCR methods for detecting resistance genes described (23) are already having a positive effect on guiding therapy early in the course of infection and making the treatment of infectious diseases less empiric.

7 FUTURE DIRECTIONS

Laboratory diagnosis of human genetic disorders and bacterial, viral, and parasitic infections will increasingly be performed via genotypic analysis. On the technical side, DNA sequence interrogation will require development of multipurpose hardware and software that will require minimal interventions from the technician. Closed-tube, contamination-free systems will be required to reduce analytic systems to what has been termed "nanoliter chemistry" (125,126). With these innovations, ultrafast testing will become a reality (127).

On the knowledge side, there is a need for more complete understanding of resistance mechanisms, data on geographical distribution of resistance determinants, and extensive validation of phenotype-genotype correlations. The field of mycobacterial resistance clearly has been advanced by population-based prevalence studies of various resistance determinants. The field of HIV resistance is showing the lead through the construction of large genotype-phenotype correlational databases (e.g., VIRCO, Belgium), allowing for the proposal of "virtual phenotypes" from sequence data only. A consensus needs to be reached in standardized reporting of resistance determinants to include species, contribution of the various possible mechanisms to a final phenotype, and epidemiological and geographical data. With this in hand, an strategy for target analysis can be constructed.

ACKNOWLEDGMENTS

We thank J. Kamile Rasheed for his contributions. This work was supported by Swiss National Science Foundation grant 31-47251.96.

REFERENCES

1. Davies J. Inactivation of antibiotics and the dissemination of resistance genes. Science 1994; 264:375–382.

2. Nikaido H. Prevention of drug access to bacterial targets: permeability and active efflux. Science 1994; 264:382–388.
3. Spratt BG. Resistance to antibiotics mediated by target alterations. Science 1994; 264:388–393.
4. National Committee for Clinical Laboratory Standards. 1997. Performance Standards for Antimicrobial Disk Susceptibility Tests. 6th ed. Vol 17, No. 1. Approved Standard M2-A6. Wayne, PA: National Committee for Clinical Laboratory Standards, 1997.
5. National Committee for Clinical Laboratory Standards. Methods for Dilution Antimicrobial Susceptibility Tests for Bacteria that Grow Aerobically. 4th ed. Vol 17, No. 2. Approved Standard M7-A4. Wayne, PA: National Committee for Clinical Laboratory Standards, 1997.
6. Flamm RK, Phillips KL, Tenover FC, Plorde JJ. A survey of clinical isolates of *Enterobacteriaceae* using a series of DNA probes for aminoglycoside resistance genes. Mol Cell Probes 1993; 7:139–144.
7. Ounissi H, Derlot E, Carlier C, Courvalin P. Gene homogeneity for aminoglycoside-modifying enzymes in gram-positive cocci. Antimicrob Agents Chemother 1990; 34:2164–2168.
8. Shaw KJ, Hare RS, Sabatelli FJ, Rizzo M, Cramer CA, Miller GH, Verbist L, Van Landuyt H, Glupczynski Y, Catalano M, Woloj M. Correlation between aminoglycoside resistance profiles and DNA hybridization of clinical isolates. Antimicrob Agents Chemother 1991; 35:2253–2261.
9. Kolbert CP, Arruda J, Varga-Delmore P, Zheng X, Lewis M, Kolberg J, Persing DH. Bracnh-DNA assay for detection of the mecA gene in oxacillin resistant and oxacillin-sensitive staphylococci. J Clin Microbiol 1998; 36:2640–2644.
10. Killgore GE, Holloway B, Tenover FC. Comparison of conventional PCR and 5′ nuclease PCR for detecting the *mecA* gene in staphylococci. Abstract C-228, p. 150 In: *In* Program and Abstracts of the 99th General Meeting of the American Society for Microbiology. American Society for Microbiology, Washington, DC, 1999.
11. Musser JM, Kapur V, Williams DL, Kreiswirth BN, van Sooligan D, van Embden JDA. Characterization of the catalase-peroxidase gene (*katG*) and *inhA* locus in isoniazid-resistant and -susceptible strains of *Mycobacterium tuberculosis* by automated DNA sequencing: restricted array of mutations associated with drug resistance. J Infect Dis 1999; 173:196–202.
12. Pease AC. Light-directed oligonucleotide arrays for rapid DNA sequence analysis Proc Natl Acad Sci USA 1994; 91:5022–5026.
13. Bergeron MG, Ouellette M. Preventing antibiotic resistance through rapid genotypic identification of bacteria and of their antibiotic resistance genes in the clinical microbiology laboratory. Antimicrob Agents Chemother 36:2169–2172.
14. Honoré N, Cole S. Molecular basis of rifampin resistance in *Mycobacterium leprae*. Antimicrob Agents Chemother 1993; 37:414–418.
15. Chambers HF. Methicillin-resistant staphylococci. Clin Microbiol Rev 1988; 1: 173–186.
16. Geha DJ, Uhl JR, Gustaferro CA, Persing DH. Multiplex PCR for identifica-

tion of methicillin-resistant staphylococci in the clinical laboratory. J Clin Microbiol 1994; 32:1768–1772.
17. Massanari RM, Pfaller MA, Wakefield DS, Hammons GT, McNutt LA, Woolson FR, Helms CH. Implications of acquired oxacillin resistance in the management and control of *Staphylococcus aureus* infections. J Infect Dis 1988; 158:702–709.
18. Deguchi T, Yasuda M, Asano M, Tada K, Iwata H, Komeda H, Ezaki T, Saito I, Kawada Y. DNA gyrase mutations in quinolone-resistant clinical isolates of *Neisseria gonorrhoeae*. Antimicrob Agents Chemother 1995; 39:561–563.
19. Swenson JM, Ferraro MJ, Sahm DF, Clark NC, Culver DH, Tenover FC, the National Committee for Clinical Laboratory Standards Study Group on Enterococci. Multilaboratory evaluation of screening methods for detection of high-level aminoglycoside resistance in Enterococci. J Clin Microbiol 1995; 33:3008–3018.
20. Telenti A, Imboden P, Marchesi F, Schmidheini T, Bodmer T. Direct, automated detection of rifampicin-resistant *Mycobacterium tuberculosis* by polymerase chain reaction and single-strand confirmation polymorphism analysis Antimicrob Agents Chemother 1993; 37:2054–20.
21. Kapur V, Li LL, Hamrick MR, Plikaytis BB, Shinnick TM, Telenti A, Jacobs WR, Banerjee A, Cole S, Yuen KY, Clarridge JE, Kreiswirth BN, Musser JM. Rapid *Mycobacterium* species assignment and unambiguous identification of mutations associated with antimicrobial resistance in *Mycobacterium tuberculosis* by automated DNA sequencing. Arch Pathol Lab Med 1995; 119:131–138.
22. van Ommen GJB, Bakker E, den Dunnen JT. The human genome project and the future of diagnostics, treatment and prevention. Lancet 1999; 354(Suppl I):5–10.
23. Tenover FC, Rasheed JK. Genetic methods for detecting antibacterial and antiviral resistance genes. In: Murray PR, Baron EJ, Pfaller MA, Tenover FC, Yolken RH, eds. Manual of Clinical Microbiology. 7th ed. Washington, DC: American Society for Microbiology Press, 1999:1578–1592.
24. Kozal MJ, Shah N, Shen N, Yang R, Fucini R, Merigan TC, Richman DD, Morris D, Hubbell E, Chee M, Gingeras TR. Extensive polymorphisms observed in HIV-1 clade B protease gene using high density oligonucleotide arrays. Nat Med 1996; 2:753–759.
25. Gwynne P, Page G. Microarray analysis: the next revolution in molecular biology. Science 1999; 285:911–938.
26. Shaw KJ, Rather PN, Hare RS, Miller GH. Molecular genetics of aminoglycoside resistance genes and familial relationships of the aminoglycoside-modifying enzymes. Microbiol Rev 1993; 57:138–163.
27. Hannecart-Pokorni E, Depuydt F, DeWit L, Van Bossuyt E, Content J, Vanhoof, R. Characterization of the 6′-N-aminoglycoside acetyltransferase gene *aac(6′)-II* associated with a *sulI*-type integron. Antimicrob Agents Chemother 1997; 41:314–318.
28. Ploy MC, Giamarellou H, Bourlioux P, Courvalin P, P., Lambert T. Detection of *aac(6′)-I* genes in amikacin-resistant *Acinetobacter* spp. by PCR. Antimicrob Agents Chemother 1994; 38:2925–2928.

29. van de Klundert JAM, Vliegenthart JS. PCR detection of genes coding for aminoglycoside modifying enzymes. In: Persing DH, Smith TF, Tenover FC, White TJ, eds. Diagnostic Molecular Microbiology. Principles and Applications. Washington, DC: American Society for Microbiology, 1993:547–552.
30. vanHoof R, Content J, Van Bossuyt E, Dewit L, Hannecart-Pokorni E. Identification of the *aadB* gene coding for the aminoglycoside-2″-*O*-nucleotidyltransferase, ANT(2″), by means of the polymerase chain reaction. J Antimicrob Chemother 1992; 29:365–374.
31. Ferretti JJ, Gilmore KS, Courvalin P. Nucleotide sequence analysis of the gene specifying the bifunctional 6′-aminoglycoside acetyl transferase 2″-aminoglycoside phosphotransferase enzyme in *Streptococcus faecalis* and identification and cloning of gene regions specifying the two activities. J Bacteriol 1986; 167:631–638.
32. Kaufhold A, Podbielski A, Horaud T, Ferrieri P. Identical genes confer high-level resistance to gentamicin upon *Enterococcus faecalis*, *Enterococcus faecium*, and *Staphylococcus aureus*. Antimicrob Agents Chemother 1992; 36:1215–1218.
33. Archer GL, Pennell E. Detection of methicillin resistance in staphylococci by using a DNA probe. Antimicrob Agents Chemother 1990; 34:1720–1724.
34. Ubukata K, Nakagami S, Nitta A, Yamane A, Kawakami S, Sugiura M, Konno A. Rapid detection of the *mecA* gene in methicillin-resistant staphylococci by enzymatic detection of polymerase chain reaction products. J Clin Microbiol 1992; 30:1728–1733.
35. Predari SC., Ligozzi M, Fontana R. Genotypic identification of methicillin-resistant coagulase-negative staphylococci by polymerase chain reaction. Antimicrob Agents Chemother 35:2568–2573.
36. Murakami K, Minamide W, Wada K, Nakamura E, Teraoka H, Watanabe S. Identification of methicillin-resistant strains of staphylococci by polymerase chain reaction. J Clin Microbiol 1991; 29:2240–2244.
37. Kolbert CP, Arruda J, Varga-Delmore P, Zheng X, Lewis M, Kolberg J, Persing DH. Bracnh-DNA assay for detection of the *mecA* gene in oxacillin resistant and oxacillin-sensitive staphylococci. J Clin Microbiol 1998; 36:2640–2644.
38. Tomasz A, Drugeon HB, de Lencastre HM, Jabes D, McDougal L, Bille J. New mechanism for methicillin resistance in *Staphylococcus aureus*: clinical isolates that lack the PBP 2a gene and contain normal penicillin-binding capacity. Antimicrob Agents Chemother 1989; 33:1869–1874.
39. Klugman KP. Pneumococcal resistance to antibiotics. Clin Microbiol Rev 1990; 3:171–196.
40. McDougal LK, Rasheed JK, Biddle JW, Tenover FC. Identification of multiple clones of extended-spectrum cephalosporin-resistant *Streptococcus pneumoniae* isolates in the United States. Antimicrob Agents Chemother 1995; 39:2282–2288.
41. Muñoz R, Coffey TJ, Daniels M, Dowson, CG, Laible G, Casal J, Hakenbeck R, Jacobs M, Musser JM, Spratt BG, Tomasz A. Intercontinental spread of a multiresistant clone of serotype 23F *Streptococcus pneumoniae* J Infect Dis 1991; 164:302–306.

42. Coffey TJ, Dowson CG, Daniels M, Zhou J, Martin C, Spratt BG, Musser JM. Horizontal transfer of multiple penicillin-binding protein genes, and capsular biosynthetic genes, in natural populations of *Streptococcus pneumoniae* Mol Microbiol 1991; 5:2255–2260.
43. Dowson CG, Hutchison A, Spratt BG. Extensive remodeling of the transpeptidase domain of penicillin-binding protein 2B of a penicillin-resistant South African isolate of *Streptococcus pneumoniae*. Microbiol 1989; 3:95–102.
44. Ubukata K, Asahi Y, Yamane A, Konno M. Combinational detection of autolysin and penicillin binding protein 2B genes of *Streptococcus pneumoniae* by PCR. J Clin Microbiol 1996: 34:592–596.
45. Tenover FC, Huang MB, Rasheed JK, Persing DH. Development of polymerase chain reaction assays to detect ampicillin resistance genes in cerebrospinal fluid samples containing *Haemophilus influenzae*. J Clin Microbiol 1994; 32:2729–2737.
46. Fournier B, Roy PH. Variability of chromosomally encoded β-lactamases from *Klebsiella oxytoca*. Antimicrob Agents Chemother 1997; 41:1641–1648.
47. Fournier B, Roy PH, Lagrange PH, Philippon A. Chromosomal β-lactamase genes of *Klebsiella oxytoca* are divided into two main groups, *bla*OXY-1 and *bla*OXY2. Antimicrob Agents Chemother 1996; 40:454–459.
48. Mariotte S, Nordmann P, Nicolas MH. Extended-spectrum β-lactamase in *Proteus mirabilis*. J Antimicrob Chemother 1994; 33:925–935.
49. Bauernfeind S. Wagner, Jungwirth R, Schneider I, Meyer D. A novel class C β-lactamase (FOX-2) in *Escherichia coli* conferring resistance to cephamycins. Antimicrob Agents Chemother 41:2041–2046.
50. Bradford PA, Urban C, Jaiswal A, Mariano N, Rasmussen BA, Projan SJ, Rahal JJ, Bush K. SHV-7, a novel cefotaxime-hydrolyzing β-lactamase, identified in *Escherichia coli* isolates from hospitalized nursing home patients. Antimicrob Agents Chemother 1995; 39:899–905.
51. Nuesch-Inderbinen MT, Hachler H, Kayser FH. Detection of genes coding for extended-spectrum SHV beta-lactamases in clinical isolates by a molecular genetic method, and comparison with the E test. Eur J Clin Microbiol Infect Dis 1996; 15:398–402.
52. Rasheed JK, Jay C, Metchock B, Berkowitz F, Weigel L, Crellin J, Steward C, Hill B, Medeiros AA, Tenover FC. Evolution of extended-spectrum β-lactam resistance (SHV-8) in a strain of *Escherichia coli* during multiple episodes of bacteremia. Antimicrob Agents Chemother. 41:647–653.
53. Shlaes DM, Currie-McCumber C, Hull A, Behlau I, Kron M. OHIO-1 β-lactamase is part of the SHV-1 family. Antimicrob Agents Chemother 1990; 34: 1570–1576.
54. Vercauteren E, Descheemaeker P, Ieven M, Sanders CC, Goossens H. Comparison of screening methods for detection of extended-spectrum β-lactamases and their prevalence among blood isolates of *Escherichia coli* and *Klebsiella* spp. In a Belgian teaching hospital. J Clin Microbiol 1997; 35:2191–2197.
55. Senda K, Arakawa Y, Ichiyama S, Nakashima K, Ito H, Ohsuka S, Shimokata

K, Kato N, Ohta M. PCR detection of metallo-β-lactamase gene (*bla*IMP) in gram-negative rods resistant to broad-spectrum β-lactams. J Clin Microbiol 1996; 34:2909–2913.
56. Ito H, Arakawa Y, Ohsuka S, Wacharotayankun R, Kato N, Ohta M. Plasmid-mediated dissemination of the metallo-β-lactamase gene *bla*IMP among clinically isolated strains of *Serratia marcescens*. Antimicrob Agents Chemother 1995; 39:824–829.
57. Clark NC, Cooksey RC, Hill BC, Swenson JM, Tenover FC. Characterization of glycopeptide resistant enterococci from U.S. hospitals. Antimicrob Agents Chemother 1993; 37:2311–2317.
58. Dutka-Malen S, Molinass C, Arthur M, Courvalin P. The VANA glycopeptide resistance protein is related to D-alanyl-D-alanine ligase cell wall biosynthesis enzymes. Mol Gen Genet 1990; 224:364–372.
59. Woodford N, Morrison D, Johnson AP, Briant V, George RC, Cookson B. Application of DNA probes for *rRNA* and *vanA* genes to investigation of a nosocomial cluster of vancomycin-resistant enterococci. J Clin Microbiol 1993; 31:653–658.
60. Dutka-Malen S, Evers S, Courvalin P. Detection of glycopeptide resistance genotypes and identification to the species levels of clinically relevant enterococci by PCR. J Clin Microbiol 1995; 33:24–27.
61. Perichon B, Reynolds P, Courvalin P. VanD-type glycopeptide-resistant *Enterococcus faecium* BM4339. Antimicrob Agents Chemother 1997; 41:2016–2108.
62. Ostrowsky BE, Clark NC, Thauvin-Eliopoulos C, Venkataraman L, Samore MH, Tenover FC, Eliopoulos GM, Moellering RC Jr, Gold HS. A cluster of VanD vancomycin resistant *Enterococus faecium*: molecular characterization and clinical epidemiology. J Infect Dis 1999; 180:1177–1185.
63. Fines M, Perichon B, Reynolds P, Sahm DF, Couvalin P. VanE, a new type of acquired glycopeptide resistance in *Enterocococcus facealis* BM4405. Antimicrob Agents Chemother 43:2161–2164.
64. Sahm DF, Free L, Handwerger S. Inducible and constitutive expression of *van*C-1–encoded resistance to vancomycin in *Enterococcus gallinarum*. Antimicrob Agents Chemother 1995; 39:1480–1484.
65. Satake S, Clark N, Rimland D, Nolte FS, Tenover FC. Detection of vancomycin-resistant enterococci in fecal samples by PCR. J Clin Microbiol 1997; 35:2325–2330.
66. Centers for Disease Control and Prevention. *Staphylococcus aureus* with reduced susceptibility to vancomycin—United States, 1997. MMWR 1997; 46: 765–766.
67. Hiramatsu K, Hanaki H, Ino T, Yabuta K, Oguri T, Tenover FC. Methicillin-resistant *Staphylococcus aureus* clinical strain with reduced vancomycin susceptibility. J Antimicrob Chemother 1997; 40:135–136.
68. Arthur M, Andremont A, Courvalin P. Distribution of erythromycin esterase and rRNA methylase genes in members of the family Enterobacteriaceae highly resistant to erythromycin. Antimicrob Agents Chemother 1987; 31:404–409.

69. Eady EA, Ross JI, Tipper JL, Walters CE, Cove JH, Noble WC. Distribution of genes encoding erythromycin ribosomal methylases and an erythromycin efflux pump in epidemiologically distinct groups of staphylococci. J Antimicrob Chemother 1993; 31:211–217.
70. Thakker-Varia S, Jenssen JW, Mon-McDermott L, Weinstein MP, Dubin DT. Molecular epidemiology of macrolides-lincosamides-streptogramin B resistance in *Staphylococcus aureus* and coagulase-negative staphylococci. Antimicrob Agents Chemother 1987; 31:735–743.
71. Clancy J, Petitpas J, Dib-Hajj F, Yuan W, Cronan M, Kamath AV, Bergeron J, Retsema JA. Molecular cloning and functional analysis of a novel macrolide-resistance determinant, *mef*A, from *Streptococcus pyogenes*. Mol Microbiol 1996; 22:867–879.
72. Shortridge VD, Flamm RK, Ramer N, Beyer J, Tanaka SK. Novel mechanism of macrolide resistance in *Streptococcus pneumoniae*. Diagn Microbiol Infect Dis 1996; 26:73–78.
73. Sutcliffe J, Grebe T, Tait-Kamradt A, Wondrack L. Detection of erythromycin-resistant determinants by PCR. Antimicrob Agents Chemother 1996; 40: 2562–2566.
74. Wondrack L, Massa M, Yang BV, Sutcliffe J. Clinical strain of *Staphylococcus aureus* inactivates and causes efflux of macrolides. Antimicrob Agents Chemother 1996; 40:992–998.
75. Tait-Kamradt A, Clancy J, Cronan M, Dib-Hajj F, Wondrack L, Yuan W, Sutcliffe J. *mef*E is necessary for the erythromycin-resistant M phenotype in *Streptococcus pneumoniae*. Antimicrob Agents Chemother 1997; 41:2251–2255.
76. Everett MJ, Jin YF, Ricci V, Piddock LJV. Contributions of individual mechanisms to fluoroquinolone resistance in 36 *Escherichia coli* strains isolated from humans and animals. Antimicrob Agents Chemother 1996; 40:2380–2386.
77. Hooper DC, Wolfson JS, Ng EY, Schwartz MN. Mechanisms of action of and resistance to ciprofloxacin. Am J Med 1987; 82(Suppl. 4A):12–20.
78. Takiff HW, Salazar L, Guerrero C, Philipp W, Huang WM, Kreiswirth B, Cole ST, Jacobs WR Jr, Telenti A. Cloning and nucleotide sequence of *Mycobacterium tuberculosis gyrA* and *gryB* genes and detection of quinolone resistance mutations. Antimicrob Agents Chemother 1994; 38:773–780.
79. Tankovic J, Perichon B, Duval J, Courvalin P. Contribution of mutations in *gryA* and *parC* genes to fluoroquinolone resistance of mutants of *Streptococcus pneumoniae* obtained in vivo and in vitro. Antimicrob Agents Chemother 1996; 40:2505–2510.
80. Truong QC, Nguyen Van, J-C, Shlaes D, Gutmann L, Moreau NJ. A novel, double mutation in DNA gyrase A of *Escherichia coli* conferring resistance to quinolone antibiotics. Antimicrob Agents Chemother 1997; 41:85–90.
81. Griggs DJ, Gensberg K, Piddock LJV. Mutations in *gyrA* gene of quinolone-resistant Salmonella serotypes isolated from humans and animals. Antimicrob Agents Chemother 1996; 40:1109–1013.
82. Janoir C, Zeller V, Kitzis M-D, Moreau NJ, Gutmann L. High-level fluoroquinolone resistance in *Streptococcus pneumoniae* requires mutations in *parC* and *gyrA*. Antimicrob Agents Chemother 1996; 40:2760–2764.

83. Vila J, Ruiz J, Goni P, Marcos A, De Anta TJ. Mutation in the *gyrA* gene of quinolone-resistant clinical isolates of *Acinetobacter baumannii*. Antimicrob Agents Chemother 1995; 39:1201–1203.
84. Speer BS, Shoemaker NM, Salyers AA. Bacterial resistance to tetracycline: mechanisms, transfer, and clinical significance. Clin Microbiol Rev 1992; 5: 387–399.
85. Abraham LJ, Berryman DI, Rood JI. Hybridization analysis of the class P tetracycline resistance determinant from the *Clostridium perfringens* R-plasmid, pCW3. Plasmid 1988; 19:113–120.
86. Charpentier E, Gerbaud G, Courvalin P. Characterization of a new class of tetracycline-resistance gent *tet*(S) in *Listeria monocytogenes* BM4210. Gene 1993; 131:27–34.
87. Fletcher HM, Macrina FL. Molecular survey of clindamycin and tetracycline resistance determinants in *Bacteroides* species. Antimicrob Agents Chemother 1991; 35:2415–2418.
88. Marshall B, Tachibana C, Levy SB. Frequency of tetracycline resistance determinant classes among lactose-fermenting coliforms. Antimicrob Agents Chemother 1983; 24:835–840.
89. Martin P, Trieu-Cuot P, Courvalin P. Nucleotide sequence of the *tetM* tetracycline resistance determinant of the streptococcal conjugative shuttle transposon Tn*1545*. Nucleic Acids Res 1986; 14:7047–7058.
90. Sougakoff W, Papadopoulou B, Nordmann P, Courvalin P. Nucleotide sequence and distribution of gene *tet*O encoding tetracycline resistance in *Campylobacter coli*. FEMS Microbiol Lett 1987; 44:153–159.
91. Zhao J, Aoki T. Nucleotide sequence analysis of the class G tetracycline resistance determinant from *Vibrio anguillarum*. Microbiol Immunol 1992; 36:1051–1060.
92. Blanchard A, Crabb DM, Dybvig K, Duffy LB, Cassell GH. Rapid detection of *tetM* in *Mycoplasma hominis* and *Ureaplasma urealyticum* by PCR: *tetM* confers resistance to tetracycline but not necessarily to doxycycline. FEMS Microbiol Lett 1992; 95:277–282.
93. De Barbeyrac B, Dupon M, Rodriguez P, Rnaudin H, Bebear C. A Tn1545-like transposon carries the *tet*(M) gene in tetracycline resistant strains of *Bacteroides ureolyticus* as well as *Ureaplasma urealyticum* but not *Neisseria gonorrhoeae*. J Antimicrob Chemother 1996; 37:223–232.
94. Lacroix JM, Walker CB. Detection and prevalence of the tetracycline resistance determinant Tet Q in the microbiota associated with adult periodontitis. Oral Microbiol Immunol 1996; 11:282–288.
95. Lyras D, Rood JI. Genetic organization and distribution of tetracycline resistance determinants in *Clostridium perfringens*. Antimicrob Agents Chemother 1996; 40:2500–2504.
96. Roberts MC, Chung WO, Roe DE. Characterizations of tetracycline and erythromycin resistance determinants in *Treponema denticola*. Antimicrob Agents Chemother 1996; 40:1690–1694.
97. Roberts MC, Pang Y, Riley DE, Hillier SL, Berger RC, Krieger JN. Detection of Tet M and Tet O tetracycline resistance genes by polymerase chain reaction. Mol Cell Probes 1993; 7:387–393.

98. Olsvik B, Tenover FC, Olsen I. Rasheed JK. Three subtypes of the *tet*(M) gene identified in bacterial isolates from periodontal pockets. Oral Microbiol Immunol 1996; 5:299–303.
99. Hansen LM, Blanchard PC, Hirsh DC. Distribution of *tet*(H) among *Pasteurella* isolates from the United States and Canada. Antimicrob Agents Chemother 1996; 40:1558–1560.
100. Ridenhour MB, Fletcher HM, Mortensen JE, Daneo-Moore L. A novel tetracycline-resistant determinant, *tet*(U), is encoded on the plasmid pKQ10 in *Enterococcus faecium*. Plasmid 1996; 35:71–80.
101. Shaw WV. Chloramphenicol acetyltransferases: enzymology and molecular biology. CRC Crit Rev Biochem 1983; 14:1–46.
102. Bolivar F, Rodsriguez RL, Greene PJ, Betlach MC, Heyneker HL, Boyer HW, Crosa JH, Falkow S. Construction and characterization of new cloning vehicles. II. A multipurpose cloning system. Gene 1977; 2:95–113.
103. Rood JI, Jefferson S, Bannam TL, Wilkie JM, Mullany P, Wren BW. Hybridization analysis of three chloramphenicol resistance determinants from *Clostridium perfringens* and *Clostridium difficile*. Antimicrob Agents Chemother 1989; 33:1569–1574.
104. Wren BW, Mullany P, Tabachali S. Molecular cloning and genetic analysis of a chloramphenicol acetyltransferase determinant from *Clostridium difficile*. Antimicrob Agents Chemother 1988; 32:1213–1217.
105. Gilbert J, Perry CE, Slocombe B. High level mupirocin resistance in *Staphylococcus aureus*: evidence for two distinct isoleucyl-tRNA synthetases. Antimicrob Agents Chemother 1993; 37:32–38.
106. Rådström P, Swedberg G, Sköld O. Genetic analyses of sulfonamide resistance and its dissemination in gram-negative bacteria illustrate new aspects of R plasmid evolution. Antimicrob Agents Chemother 1991; 35:1840–1848.
107. Amyes SGB, Towner KJ. Trimethoprim resistance; epidemiology and molecular aspects. J Med Microbiol 1990; 31:1–19.
108. Singh KV, Reves RR, Pickering LK, Murray BE. Identification by DNA sequence analysis of a new plasmid-encoded trimethoprim resistance gene in fecal *Escherichia coli* isolates from children in day-care centers. Antimicrob Agents Chemother 1992; 36:1720–1726.
109. Jansson C, Franklin A, Sköld O. Spread of newly found trimethoprim resistance gene *dhfrIX* among porcine isolates and human pathogens. Antimicrob Agents Chemother 1992; 36:2704–2708.
110. Pulkkinen L, Houvinen P, Vuorio E, Tiovanen P. Characterization of trimethoprim resistance by use of probes specific for transposon Tn7. Antimicrob Agents Chemother 1984; 26:82–86.
111. De Groot R, Sluijter M, De Bruyn A, Campos J, Goessens WHF, Smith AL, Hermans PWM. Genetic characterization of trimethoprim resistance in *Haemophilus influenzae*. Antimicrob Agents Chemother 1996; 40:2131–2136.
112. Pablos-Mendez A, Raviglione MC, Laszlo A, Binkin N, Rieder HL, Bustreo, Cohn DL, Lambregts-van Weezenbeek CS, Kim SJ, Chaulet P, Nunn P. Global surveillance for antituberculosis-drug resistance, 1994–1997. World Health Organization-International Union against Tuberculosis and Lung Disease Working Group on Anti-Tuberculosis Drug Resistance Surveillance [see

comments] [published erratum appears in N Engl J Med 1998 Jul 9;339(2): 139]. N Engl J Med 1998; 338:1641–1649.
113. Heym B, Honoré N, Truffot-Pernot C, Banerjee A, Schurra C, Jacobs WR Jr, van Embden JDA, Grosset JH, Cole ST. Implications of multidrug resistance for the future of short course chemotherapy of tuberculosis: a molecular study. Lancet 1994; 344:293–298.
114. Ramaswamy S, Musser MJ. Molecular genetic basis of antimicrobial agent resistance in *Mycobacterium tuberculosis*; 1998 update. Tubercle Lung Dis 1998 79:3–29.
115. Zhang Y, Telenti A. Genetics of drug resistance in *Mycobacterium tuberculosis*. In: Hatfull G, Jacobs WR Jr, eds. Molecular Genetics of the Mycobacteria. Washington, DC: American Society for Microbiology, 2000.
116. Cooksey RC, Morlock GP, Glickman S, Crawford JT. Evaluation of a line probe assay kit for characterization of *rpo*B mutations in rifampin-resistant *Mycobacterium tuberculosis* isolates from New York City. J Clin Microbiol 1997; 35:1281–1283.
117. Zhang Y, Heym B, Allen B, Young D, Cole S. The catalase-peroxidase gene and isoniazid resistance of *Mycobacterium tuberculosis*. Science 1992; 358:591–593.
118. Banerjee A, Dubnau E, Quemard A, Balasubramanian V, Um SK, Wilson T, Collins D, De Lisle G, Jacobs W Jr. *inhA*, a gene encoding a target for isoniazid and ethionamide in *Mycobacterium tuberculosis*. Science 1994; 263:227–230.
119. Mdluli K, Slayden RA, Zhu YQ. Inhibition of a *Mycobacterium tuberculosis* beta-ketoacyl ACP synthase by isoniazid. Science 1998; 280:1607–1610.
120. Telenti A, Honore N, Bernasconi C, March J, Ortega A, Heym B, Takiff HE, Cole ST. Genotypic assessment of isoniazid and rifampin resistance in *Mycobacterium tuberculosis*: a blind study at reference laboratory level. J Clin Microbiol 1997; 35:719–723.
121. Lee ASG, Lim IHK, Tang LLH, Telenti A, Wong SY. Contribution of *kasA* analysis to detection of isoniazid-resistant *Mycobacterium tuberculosis* in Singapore. Antimicrob Agents Chemother 1999; 43:2087–2089.
122. Tenover FC, Gootz TD, Gordon KP, Plorde JJ. Development of a DNA probe for the structural gene of the 2″-*O*-adenyltransferase aminoglycoside modifying enzyme. J Infect Dis 1984; 159:678–687.
123. Tyler KD, Wang G, Tyler SD, Johnson WM. Factors affecting reliability and reproducibility of amplification based DNA fingerprinting of representative bacterial pathogens. J Clin Microbiol 1997; 35:339–346.
124. National Committee for Clinical Laboratory Standards. Specifications for molecular microbiology methods for infectious diseases. Proposed Guideline, MM3-P Vol 1, No. 1. National Committee for Clinical Laboratory Standards, Villanova, PA, 1994;
125. Burns MA, Johnson BN, Brahmasandra SN, Handique K, Webster JR, Krishnan M, Sammarco TS, Man PM, Jones D, Heldsinger D, Mastrangelo CH, Burke DT. An integrated nanoliter DNA analysis devise. Science 1998; 282:484–487.
126. Kopp MU, Mello AJ, Manz A. Chemical amplification: continuous flow PCR on a chip. Science 1998; 280:1046–1048.

127. Belgrader P, Benett W, Hadley D, Richards, Stratton P, Mariella R Jr. Milanovich F. PCR detection of bacteria in seven minutes. Science 1999; 284:449–450.
128. Orita M, Suzuki Y, Sekiya T, Hayashi K. Rapid and sensitive detection of point mutations and DNA polymorphisms using the polymerase chain reaction. Genomics 1989; 5:874–879.
129. Telenti A, Imboden P, Marchesi F, Schmidheini T, Bodmer T. Direct, automated detection of rifampin-resistant *M. tuberculosis* by PCR and single-strand conformation polymorphism. Antimicrob Agents Chemother 1993; 37:2054–2058.
130. Sheffield VC, Cox DR, Lerman LS, Myers RM. Attachment of a 40bp G+C-rich sequence (GC clamp) to genomic DNA fragments by the polymerase chain reaction results in improved detection of single-base changes. Proc Natl Acad Sci USA 1989; 86:232–236.
131. White MB, Carvalho M, Derse D, O'Brien SJ, Dean M. Detecting single base substitutions as heteroduplex polymorphisms. Genomics 1992; 12:301–306.
132. Williams DL, Waguespack C, Eisenach K, Crawford JT, Portaels F, Salfinger M, Nolan CN, Abe C, Stich-Groh V, Gillis TP. Characterization of rifampin resistance in pathogenic mycobacteria. Antimicrob Agents Chemother 1994; 38:2380–2386.
133. Scarpellini P, Braglia S, Carrera P, Cedri M, Cichero P, Colombo A, Crucianelli R, Gori A, Ferrari M, Lazzarin A. Detection of rifampin resistance in *Mycobacterium tuberculosis* by double gradient-denaturing gradient gel electrophoresis.
134. Cotton RGH. Detection of mutations in DNA and RNA by chemical cleavage. In: Matthew C, ed. Methods in Molecular Biology. Totowa, NJ: Humana Press, 1991:39–49.
135. Youil R, Kemper BW, Cotton RGH. Detection of 81 of 88 known mouse beta-globin promoter mutations with the T4 endonuclease II—the ECM method. Genomics 1996; 32:431–435.
136. Nilsson M, Krejci K, Koch J, Kwiatkowski M, Gustavsson P, Landegren U. Padlock probes rveal single-nucleotide differences, parent of origin, and in situ distribution of centromeric sequences in human chromosome 13 and 21. Nat Genet 1997; 16:252–255.
137. Lizardi PM, Huang X, Zhu Z, Bray-Ward P, Thomas DC, Ward DC. Mutation detection and single molecule counting using isothermal rolling circle amplification. Nat Genet 1998; 19:225–232.
138. Pastinen T, Krug A, Metspalu A, Peltonen L, Syvnen AC. Minisequencing: a specific tool for DNA analysis on oligonucleotide arrays. Genome Res 1997; 7:606–614.
139. Syvnen AC. From gels to chips: "minisequencing" primer extension for analysis of point mutations and single-nucleotide polymorphisms. Hum Mutat 1999; 13:1–10.
140. Holland PM, Abramson RD, Watson R, Gelfand DH. Detection of specific polymerase chain reactiomn product by utilizing the 5′-3′ exonuclease activity of the *Thermus aquaticus* DNA polymerase. Proc Natl Acad Sci USA 1991; 88:7276–7280.

141. Caplin BE, Rasmussen RP, Barnard PS, Wittwer CT. The most direct way to monitor PCR amplification for quantification and mutation detection. Biochimica 1999; 1:5–8.
142. Tyagi SD, Bratu P, Kramer FR. Multicolor molecular beacons for allele discrimination. Nat Biotech 1998; 16:49–53.
143. Piatek A, Tyagi S, Pol AC Telenti A, Miller LM, Kramer FR, Alland D. Molecular beacon sequence analysis for detecting drug resistance in Mycobacterium tuberculosis. Nat Biotechnol 1998; 16:359–363.

12

Evolution and Epidemiology of Antibiotic-Resistant Pneumococci

Christopher Gerard Dowson
University of Warwick, Coventry, England

Krzysztof Trzciński
National Institute of Hygiene, Warsaw, Poland

Streptococcus pneumoniae is still an important human pathogen. The past two decades have witnessed the global spread of resistance to the major groups of antipneumococcal drugs and there are now no countries free of multi–drug-resistant strains. In this naturally transformable organism, horizontal gene transfer, by either intraspecies or interspecies recombination, has played an important role in the evolution of resistance. However, there is also strong evidence for the global spread of multi–drug-resistant clones. Among these the serotype 23F Spanish, 6B Spanish, and French/Spanish 9/14 resistant clones have reached pandemic status. All three emergent clones are penicillin nonsusceptible (PNSP) and are often resistant to tetracyclines, macrolides, chloramphenicol, and cotrimoxazole. The most prevalent of these, the 23F Spanish clone and its five capsular type variants, has been documented in 24 countries around the world. This chapter shows that the prevalence of PNSP among clinical isolates of *S. pneumoniae* is above 40% in 16 of 60 countries surveyed and below 5% in only 3 of them. This chapter also gives an insight into the mechanisms of the antibiotic resistance acquisition in pneumococci, their evolution, and the epidemiology of multi–drug-resistant strains.

1 INTRODUCTION

Streptococcus pneumoniae is the causative agent of pneumonia, otitis media, meningitis, and bacteremia and is a major cause of morbidity and mortality worldwide, particularly among the young, elderly, and immunocompromised (1). Moreover, the past two decades have witnessed the acquisition and global spread of chromosomal and transposon-encoded resistance to the major groups of effective antibiotics (2–5). There is therefore increasing pressure to develop novel therapeutic agents. However, in order fully to understand the current spread of resistance, we need to look jointly at the mechanisms of resistance and the evolutionary processes involved in their acquisition and dissemination. For this we also need a clear picture of the population structure of carried and invasive isolates of this naturally transformable organism.

The past 50 years of selection by a diverse range of antimicrobials has revealed an extensive range of resistance mechanisms, many of which are dealt with in detail elsewhere in this volume. Therefore, in looking ahead to the selection of novel stable targets for chemotherapy or vaccination, we need to take account of the processes involved in the development of resistance during the past decades. The pneumococcus and other naturally transformable organisms such as *Neisseria* spp. that evolve by intraspecies and interspecies recombination are perhaps among the most difficult to deal with. Many of their loci are effectively moving targets (6,7), not only moving freely from one strain to another but being able to evolve by acquiring highly divergent blocks of nucleotides from related species that will generate novel proteins with altered catalytic activities or different antigenic profiles (8–10). The following gives some insight into the role of horizontal gene transfer in the evolution and epidemiology of antibiotic-resistant pneumococci.

2 THE EVOLUTION OF ANTIBIOTIC RESISTANCE

2.1 β-Lactam Resistance

Since its detection in 1967 penicillin resistance in *S. pneumoniae* has become increasingly prevalent worldwide (11). An *S. pneumoniae* isolate is considered to lack susceptibility when the minimal inhibitory concentration (MIC) of penicillin is greater than 0.06 mg/L (12), and is treated as a PNSP. Isolates for which penicillin MICs ranged from 0.12 to 1.0 mg/L fit in the category of intermediate susceptibility (12). High-level resistance to penicillin is seen when the MIC is greater than 1 mg/L (12). With few exceptions, infections caused by strains intermediately susceptible to penicillin can be successfully treated with other antipneumococcal β-lactams, such as amoxicillin or broad-spectrum cephalosporins, cefotaxime, and ceftriaxone (13,14). Highly penicillin-resistant isolates are invariably cross resistant to other β-lactam antibiotics, including the third-generation cephalosporins cefotaxime and ceftriaxone (15). In countries such as Spain, Hungary, and South Africa, lack of susceptibility to penicillin among *S. pneumoniae* is not only found among a high proportion of all pneumococci isolated (16–18), but isolates commonly possess levels of resistance to penicillin of 1–4 mg/L and occasionally up to 8–16 mg/L (19). The spread of highly penicillin-resistant strains is a major concern (20), as pneumococci of this phenotype are frequently nonsusceptible to several other antipneumococcal drugs (21,22).

2.1.1 Role of Penicillin-Binding Proteins in β-Lactam Resistance

Lack of susceptibility to penicillin in clinical isolates of *S. pneumoniae* is due to the presence of high molecular weight penicillin-binding proteins (PBPs) that have a greatly reduced affinity for the β-lactam antibiotics (19,23). Although there are numerous alterations in the genes encoding low-affinity PBPs (24,25) several within the transpeptidase domain of different PBPs have been identified as being important in resistance (26,27). It would appear for PBP2X that resistance is due to amino acid substitutions within a buried cavity near the catalytic site that contains a structural water molecule (28). The examination of β-lactam–resistant laboratory mutants has shown that beside PBPs, mutations in *ciaR*, *ciaH*, and *cpoA* genes could potentially influence resistance, although currently these alternative determinants have not been found to be responsible for increased resistance among clinical isolates (29–31).

The primary target of a β-lactam antibiotic is the essential PBP (32) with the highest affinity for that particular antibiotic, and for many clinically important β-lactams this is PBP2X (33). However, the use of primary target in this context does not presuppose that this is the only killing target but is that which influences MIC owing to the differential affinities of PBPs for different β-lactam antibiotics. For clinical isolates of *S. pneumoniae* challenged by different β-lactams either as the result of treatment of a pneumococcal infection or during asymptomatic carriage, when a different organism is the desired target, there may be selection for the acquisition of different permutations of low-affinity PBPs. High-level resistance to oxacillin requires low-affinity forms of PBPs 2X and 2B (34), cephalosporin-resistance PBPs 2X and 1A (15,35), and penicillin-resistance PBPs 2X, 1A, and 2B (23). The inevitability of highly penicillin-resistant clinical isolates being cross-resistant to other groups of β-lactams now becomes obvious.

2.1.2 Role of Oral Streptococci in the Formation of Mosaic *pbp* Genes

Low-affinity forms of PBP1A, PBP2B, and PBP2X have arisen initially by the horizontal transfer and recombination of homologous chromosomally encoded PBP genes from closely related species of streptococci. *S. mitis* and *S. oralis* have been identified as two of the species responsible for contributing genetic meterial for the formation of a low-affinity PBP2B in many penicillin-resistant isolates of *S. pneumoniae* (26,36). However, analysis of *pbp* genes from a diverse collection of resistant isolates has revealed that several additional, as yet unidentified, species also have been involved in the evolution of these mosaic genes (26). Recent analysis of the population structure of pneumococci and the closely related oral streptococci has revealed that isolates identified as *S. mitis* represent a highly divergent group of organisms. In addition, there is a previously unidentified group of organisms that lie between *S. mitis* and pneumococci (37). These are being investigated as alternative DNA donors involved in the evolution of PBPs and a range of pneumococcal virulence determinants. Experimentally it has been shown that oral streptococci with MICs for penicillin as high as 64 mg/L can transform pneumococci to this level of resistance, although this requires the acquisition of altered forms of PBPs 2A and 1B from *S. oralis* along with 2X, 1A, and 2B (38,39).

Recent work examining the degree of sexual isolation between pneumococci and the related oral streptococci (40) has revealed, as found previously for *Bacillus* (41), a log linear relationship between nucleotide divergence and sexual isolation.

Apart from the acquisition of novel PBPs by recombination, there is now also evidence that mosaic *pbp* genes have further evolved by spontaneous mutation, altering levels of cross resistance to penicillin and cephalosporins, presumably in response to clinical exposure to these different classes of β-lactams (35).

2.2 Development of Multiple Antibiotic Resistance

Resistance of pneumococci to tetracyclines (42–44), chloramphenicol (45), and macrolides (46,47) is due to acquisition of the highly mobile conjugative transposon Tn*1545* or a related transposon which may carry one or more of these and other resistance determinants (48,49). These transposons possess an integration/excision system, encoded by the genes *int/*

xis, and terminally associated (host-derived) coupling sequences (50). Transfer of the transposon from the donor to the recipient chromosome involves excision of the element from the host chromosome, formation of a covalently closed circular intermediate, entry into the recipient cell, and subsequent integration into its chromosome (50). The site of integration may be determined by DNA topology rather than sequence specificity.

The stability of transposon-encoded resistance within pneumococci has not been determined. However, it is clear that members of some multiply resistant pneumococcal clones do differ in their resistance profiles, that many more allelic variants of the tetracycline-resistance gene (*tetM*) are found within pneumococci than previously described (42), and that different *tetM* alleles can be found among members of the same clonal group (C. Dowson, unpublished). In general, the *tetM*-positive isolates are resistant to all clinically available tetracyclines (51); however, isolates with MICs of tetracycline 2–4 mg/L (susceptible or intermediate susceptible [12]) which gave positive hybridization signals with *tetM* probes have been also described (52).

Two mechanisms of resistance to macrolides have been described in pneumococci thus far. Active efflux due to the acquisition of the *mef(A)* gene was identified in isolates expressing a low-level resistance to erythromycin (MICs ranged from 1 to 32 mg/L). Such isolates were once treated as being macrolide resistant but susceptible to lincosamides and streptogramins (M phenotype) (53,54). The second mechanism described is based on ribosomal protection due to acquisition of the *ermB* gene. The *ermB*-positive isolates are resistant to macrolides, lincosamides, and streptogramins B (MLS_B phenotype) (53,54) and exhibit high-level resistance to erythromycin (MICs above 32 mg/L) (52,55). MLS_B and M phenotypes were also described in *ermB*-negative and *mefE*-negative strains, indicating the presence of novel genes or allelic variants of already identified genes (56). Finally, the macrolide-streptogramin–resistant but lincosamide-susceptible *S. pneumoniae* (so called MS phenotype) also has been described (56).

Pneumococcal resistance to trimethoprim and the sulfonamides, which inhibit bacterial purine synthesis, has also been identified (57–60). However, this is clearly chromosomally encoded and involves alterations to housekeeping *sulA* (dihydropteroate synthase) and *dfr* (dihydrofolate reductase) genes within the pneumococcal genome. Although point mutations and codon duplications are frequently associated with resistance, there is also some evidence that interspecies recombination has played a role in the evolution of resistance (57). A similar situation is found in the evolution of pneumococcal resistance to rifampicin, where there is also evidence of resistance arising owing to recombination rather than the more frequently occurring point mutations within the gene encoding the β subunit of RNA polymerase (*rpoB*) (61).

The use of fluoroquinolones in the treatment of pneumococcal infections has resulted in decreased susceptibility (62). This appears to be due to target alterations in DNA gyrase (GyrA) and topoisomerase I (ParC) (63–65) or to the action of an efflux pump encoded by *pmrA* (66–68). Although alterations in GyrA and ParC appear to have evolved by point mutations in *S. pneumoniae*, it is clear that high-level quinolone-resistant viridans streptococci also have evolved (69) and may, if resistance becomes prevalent, act as a source of resistance genes for pneumococci. Recent preliminary investigations do show evidence that interspecies recombination has played a role in the evolution of fluoroquinolone resistance in clinical isolates of *S. pneumoniae* (A. de la Campa, personal communication).

To date, there are no reports of vancomycin resistance in clinical isolates of *S. pneumoniae*. However, it has been shown that loss of function of the VncS histidine kinase of a two-component sensor-regulator system in laboratory strains of *S. pneumoniae* produced tolerance to vancomycin and other classes of antibiotic, indicating that this may be a precursor to the evolution of vancomycin resistance in the community (70,71).

3 EPIDEMIOLOGY OF *S. PNEUMONIAE*

3.1 Population Structure of *S. pneumoniae*

Asymptomatic carriage of pneumococci in the throat or nasopharynx is widespread, with carriage rates being especially high in children (72–74). There is also clear evidence of spread among families (75) and colonization by multiple pneumococcal capsular types has also been reported (72). Some serotypes are particularly associated with disease in children (76) or adults (77) and others with carriage (78) or human immunodeficiency virus (HIV) infection (79). However, it is only just becoming apparent that among isolates associated with invasive disease there are important virulent pneumococcal clones that are responsible for many cases of disease around the world (7), and recently that those clones are also frequently asymptomaticaly carried (6).

There is clear evidence from population genetic analysis that the pneumococcal chromosome is at linkage equilibrium, that is, freely recombining, and that recombination by transformation and possibly transduction may introduce blocks of nucleotides from other *S. pneumoniae* strains or other species ranging in size for 10s of base pairs (80) to 10s of kilobase pairs (81). This can result in alterations to single loci or whole operons.

One, therefore, has to be careful in epidemiological analyses not to rely upon single markers in strain identification, especially if those markers are immunologically reactive and liable to change under the selective pressure of the human immune system.

Capsular serotyping has been the cornerstone of pneumococcal epidemiology for many years. However, this is a fairly blunt instrument when trying to understand the movement and evolution of specific pneumococcal clones, especially now that serotype exchange among clones is well documented (81–83), and the current best estimate suggests that serotype exchange may occur among 4–6% of isolates (6). Therefore, tracking the spread of prevalent susceptible or resistant clones requires the use of techniques such as pulse field gel electrophoresis (PFGE) (84), restriction fragment end labeling (RFEL) with PBP genotyping (85), or the more recently developed multilocus sequence typing (MLST) (7), or multilocus restriction typing (MLRT) (6). Clearly transportability and access to reference strains and composite databases is important for positive strain identification. A database showing clonal variants is especially important for organisms such as pneumococci, in which clones initially sufficiently stable to track do start to break down due to the ongoing process of recombination.

Apart from tracking the clonal spread of organisms, it is also possible to examine the horizontal spread of resistance genes. This has been undertaken successfully for the dissemination of *pbp* genes by RFLP analysis of amplified *pbp* gene fragments (82,86).

3.2 Intercontinental Spread of Resistant Clones

Numerous multi–drug-resistant pneumococcal clones have been identified (7), with three of these shown to be as major pandemic clones (Table 1). The oldest and most prevalent is the serotype 23F Spanish pandemic or 23F Spanish/USA clone. This clone has been reported in 24 countries (Table 1) and on all continents except Australia. Isolates of this clone are usually resistant to a wide range of antipneumococcal drugs, including tetracyclines, cotrimoxazole, chloramphenicol, and often macrolides. MICs for penicillin are generally 1-2 mg/L (87) but may reach 8 mg/L (88,89). This clone has acquired at least five distinct capsular type variants: 3 (90), 6B (88), 9V (87,90), 14 (88,90–93), and 19F (82), with 19F being the most prevalent variant reported (52,87,90–96).

Second in temporal sequence of isolation is the multi–drug-resistant Spanish 6B clone. This clone spread across Western Europe at the end of the 1980s and is now present in North America, Asia, and Australia (see Table 1). Simultaneously, 6B epidemic clones appeared in Finland (94) and the United States (97,98) but were not identified as derivatives of the Spanish clone (97–99).

The most recent of the three dominant multi–drug-resistant pandemic clones is the major penicillin-resistant Spanish serotype 9V clone. This clone was originally intermediately susceptible to penicillin and additionally resistant to cotrimoxazole; however, by the mid-1990s, members of this clone had acquired resistance to macrolides and chloramphenicol (see Table 1). It is now also clear that serotype 14 variants of this clone are widely distributed in France (89), Denmark, Spain, Uruguay (81), Poland (81,100), Portugal (101), the Netherlands (91), Mexico (88), and Colombia (102). Also 23F variants of this clone have been described in Germany (103).

Despite the fact that there is some degree of similarity observed in resistance profiles of particular pandemic clones, different genes or even mechanisms of resistance might be responsible for similar phenotypes. For example, among the 23F pandemic clone isolates collected in the United States in 1996–1997, both *ermB* and *mefE* genes coding for macrolide resistance were observed (92). Isolates of MLS_B and M phenotypes were observed among Taiwanese PNSP of the same clone isolated in 1996–1997 (96). *S. pneumoniae* of clone 23F isolated in Poland showed the M phenotype only (100), as is characteristic of *mefE*-positive isolates. Moreover, Bulgarian (55), Italian (52), and Portuguese (91,101,104) isolates of this pandemic clone were susceptible to macrolides. This might indicate that antibiotic resistance profiles vary in particular clones rather than exhibiting an immutable pandemic pattern. Fluidity in resistance profile would enable strains to respond to local or national variations in prescribing policy.

There are also several currently more geographically restricted national clones of multi–drug-resistant pneumococci, most of them expressing intermediate susceptibility to penicillin (87,100,101,105,106). One of the best described is the 19A Hungarian clone (103,107),which has been responsible for one of the highest frequencies of resistance to penicillin observed worldwide (108). Perhaps surprisingly, the spread of this clone appears to have been restricted to the Czech Republic (107). This was possibly due to the socioeconomic situation in Europe prior to the end of the 1980s when traveling and mass migration was restricted in former Eastern Bloc countries.

Multidrug resistance in pneumococci is not only observed in PNSP. Penicillin-susceptible serotype 3 strains that are resistant to macrolides, lincosamides (MLS_B phenotype), and tetracyclines have been isolated in South Africa (109); and penicillin-susceptible serotype 6 strains resistant to macrolides and lincosamides, tetracyclines, cotrimoxazole, and chloramphenicol have been isolated in Greece (110). Penicillin-susceptible, multiple resistant serotype 5, 6, 11, and 23 strains have also been isolated in Colombia (111), Portugal (101), and Hong Kong (112).

TABLE 1 International Spread of Multi–Drug-Resistant *Streptococcus pneumoniae* Epidemic Clones

	Country	No. of isolates described	Resistance pattern: PEN MIC (mg/L)	CEF	ERY	TET	CHL	SXT	Method of identification	Ref.
6B Spanish clone										
86–91	Spain	9	0.5–2	—*	(R)	R	R	—	MLEE, PBP RFLP	87
86–88	Spain	6	0.5–2	—	—	(R)	R	—	MLEE, PBPs profiling	94
87–89	Spain	12	2	—	(R)	R	R	—	MLEE, PBPs profiling, PFGE (*Sma*I)	98
96–98	Spain	17	—	—	—	—	—	—	MLST	172
89–91	Iceland	57	0.5–2	—	I/R	R	I/R	—	MLEE, PBPs profiling, PFGE (*Sma*I)	98
91–97	France	7	0.5–1	(I)	—	—	—	—	PBP RFLP, ribotyping	89
	United Kingdom	—	—	—	—	—	—	—	PFGE (*Sma*I)	98
92	Germany	1	0.5–1	I	—	R	R	—	MLEE, PBPs profiling	103
92	United States	3	—	—	—	—	—	—	MLST	172
93–94	Thailand	6	—	—	—	—	—	—	MLST	172
94–97	Hong Kong	19	0.25–2	I	R	R	(R)	—	PBP RFLP, PFGE (*Sma*I, *Apa*I)	93
95–96	The Netherlands	3	—	—	—	—	—	—	MLST	172
97	Australia	1	—	—	—	—	—	—	MLST	172
97	Taiwan	1	1	—	R	R	S	—	MLST, PBP RFLP	96

Table 1 Continued

	Country	No. of isolates described	Resistance pattern: PEN MIC (mg/L)	CEF	ERY	TET	CHL	SXT	Method of identification	Ref.
9/14 French/ Spanish clone										
87	Spain	1	0.5–0.75	R	—	S	S	—	MLEE, PBPs profiling	103
88–98	Spain	12	—	—	—	—	—	—	MLST	172
	Spain	1	R	—	S	S	—	R	RFEL, PBP RFLP	91
88	France	1	0.5–0.75	R	—	S	S	—	MLEE, PBPs profiling	103
91–97	France	5	1–8	I/R	—	—	—	—	PBP RFLP, ribotyping	89
92	Germany	1	2	R	—	S	S	—	MLEE, PBPs profiling	103
92–97	Bulgaria	22	0.5–4		(R)	(R)	SS	R	PBP RFLP, PFGE (*Sma*I)	55
93–96	Italy	8	1–2	—	S	S	S	I/R	PFGE (*Sma*I)	52
93–95	Mexico	6	2–4	I/R	S	S	S	R	PFGE (*Sma*I)	88
93–95	Thailand	4	1–2	R	S	S	S	R	BOX PCR, PBP RFLP, RFEL	106
	Thailand	1	R	—	S	S	—	R	PBP RFLP, RFEL	91
93–94	United Kingdom	5	R	R	S	S	S	R	MLEE, PFGE (*Sma*I, *Apa*I)	173
96–97	United Kingdom	3	R	—	—	—	—	—	MLST	7,172
	United Kingdom	1	R	—	S	S	—	R	PBP RFLP, RFEL	91
	United Kingdom	1	R	—	—	—	—	—	MLRT, MLST, REP PCR	6

93–96	Uruguay	4	R	—	—	—	—	—	MLST	7,172
	Uruguay	1	R	—	—	—	—	—	MLRT, MLST, REP PCR	6
94–96	Colombia	7	1–4	(I/R)	S	S	S	R	PFGE (*Sma*I)	105
94–95	Greece	1	R	—	S	S	—	R	PBP RFLP, RFEL	91
94–95	Poland	2	—	—	—	—	—	—	MLST	172
95–96	Poland	7	0.5–2	R	(R)	R	R	R	BOX PCR, PBP RFLP, RFEL	100
95	The Netherlands	17	R	—	(R)	(R)	—	(R)	PCR RFLP, RFEL	120
95–96	The Netherlands	6	R	—	(R)	(R)	—	(R)	PBP RFLP, RFEL	
95–96	Sweden	22	R	—	S	S	—	R	AP PCR, BOX PCR	174
96	Canada	1	R	—	—	—	—	—	MLST	7,172
96–97	Denmark	1	R	—	—	—	—	—	MLST	7,172
96	Portugal	10	R	—	S	S	S	R	PFGE (*Sma*I)	101
96–97	United States	40	1.5–4	—	(R)	S	S	R	PFGE	92
97	Taiwan	1	1.5	—	R	R	S	—	MLST, PBP RFLP	96
23F Spanish pandemic or 23F Spanish/USA clone										
84–88	Spain	12	1–4	—	—	R	R	—	MLEE, PBPs profiling	94
84–94	Spain	31	0.5–2	—	(R)	R	R	—	MLEE, PBP RFLP, REP PCR	87
88–98	Spain	15	—	—	—	—	—	—	MLST	172
89	Spain	6	R	—	S	R	R	R	MLEE, PBP RFLP	3

Table 1 Continued

	Country	No. of isolates described	Resistance pattern: PEN MIC (mg/L)	CEF	ERY	TET	CHL	SXT	Method of identification	Ref.
87	United Kingdom	1	—	—	—	—	—	—	MLST	172
	United Kingdom	1	R	—	R	S	—	R	PBP RFLP, RFEL	91
87	South Africa	2	4	—	—	R	R	—	MLEE, PBPs profiling	94
91	South Africa	6	2	—	R	R	R	R	MLEE, PBP RFLP	175
	South Africa	1	R	—	S	S	—	R	PBP RFLP, RFEL	91
88–92	France	14	1–2	—	—	—	—	—	PBP RFLP, PFGE (*Sma*I), RAPD, ribotyping	176
88–97	France	12	1.5–8	I/R	—	—	—	—	PBP RFLP, ribotyping	89
92–93	France	24	⩾2	—	R	R	R	R	PFGE (*Sma*I, *Apa*I), PBP RFLP	177
89–92	Portugal	20	0.2–1	(I)	S	R	I/R	R	PFGE (*Sma*I)	104
89–92	Portugal	1	R	—	S	R	—	R	PBP RFLP, RFEL	91
96	Portugal	20	R	—	S	R	R	R	PFGE (*Sma*I)	101
89–90	United States (Ohio)	6	R	—	S	R	R	R	MLEE, PBP RFLP	3
89–90	United States	2	—	—	—	—	—	—	MLST	172
96–97	United States	127	1.5–8	—	(R)	R	R	R	PFGE (*Sma*I)	92
	United States	1	R	—	R	R	—	R	PBP RFLP, RFEL	91
90–92	South Korea	25	1–2	I/R	(R)	R	R	R	PFGE (*Sma*I)	95
96–97	South Korea	6	R	—	—	—	—	—	PFGE (*Sma*I)	118
91	Hungary	1	1–1.5	R	—	R	R	—	MLEE, PBPs profiling	103
92–96	Bulgaria	13	2–4	(R)	S	R	R	R	PBP RFLP, PFGE (*Sma*I)	55

92	Germany	1	2	R	—	R	R	—	MLEE, PBPs profiling	103
	Germany	1	R	—	S	R	—	R	RFEL, PBP RFLP	91
92–94	Thailand	10	2	—	R	R	R	R	BOX PCR, PBP RFLP, RFEL	106
93–94	Thailand	4	R	—	(R)	R	—	R	PBP RFLP, RFEL	91
96–97	Thailand	4	R	—	—	—	—	—	PGFE (*Sma*I)	118
93–95	Canada	10	R	R	S	R	R	R	AP PCR, PFGE (*Sma*I)	178
93–96	Italy	7	1–2	—	S	R	R	R	PFGE (*Sma*I)	52
93–95	Mexico	29	2–8	(I/R)	(R)	R	R	R	PFGE (*Sma*I)	88
94 –96	Colombia	8	1–4	(I/R)	S	R	R	I/R	PFGE (*Sma*I)	105
	Croatia	1	R	—	S	R	—	R	PBP RFLP, RFEL	91
94–95	Greece	2	R	—	(R)	(R)	—	R	RFEL, PBP RFLP	91
94–97	Hong Kong	78	1–2	R	R	R	R	—	PBP RFLP, PFGE (*Sma*I, *Apa*I)	93
95	The Netherlands	22	R	—	(R)	(R)	—	R	PBP RFLP, RFEL	120
95–96	The Netherlands	5	R	—	(R)	(R)	—	R	PBP RFLP, RFEL	91
96	Poland	3	2	R	(R)	R	(R)	R	BOX PCR, PBP RFLP, RFEL	100
96–97	Taiwan	21	0.75–2	—	R	R	R	—	MLST, PBP RFLP	96
96–97	Taiwan	14	R	—	—	—	—	—	PFGE (*Sma*I)	118
96–97	Japan	1	R	—	—	—	—	—	PFGE (*Sma*I)	118
96–97	Malaysia	1	R	—	—	—	—	—	PFGE (*Sma*I)	118
96–97	Singapore	9	R	—	—	—	—	—	PFGE (*Sma*I)	118

—, No data.

Antibiotic resistance patterns: R, resistance; I, intermediate susceptibility; S, susceptibility; symbols in brackets indicate occasional resistance or intermediate susceptibility; PEN, penicillin; MICs of penicillin given when available; CEF, third-generation cefalosporines—cefotaxime or ceftriaxone; ERY, erythromycin; TET, tetracycline; CHL, chloramphenicol; SXT, cotrimoxazole; AP PCR, arbitrary primed PCR; BOX PCR, BOX elements PCR fingerprinting; MLEE, multilocus enzyme electrophoresis; MLRT, multilocus restriction typing; MLST, multilocus sequence typing; PFGE, macrorestriction analysis of chromosomal DNA; *Sma*I, *Apa*I, enzymes used in PFGE analysis; PBPs profiling, antibody reaction patterns of PBPs; PBP RFLP, restriction fragment length polymorphism of *pbp* genes; RAPD, randomly amplified polymorphic DNA analysis; REP PCR, repetitive extragenic palindromic PCR genomic profiling; RFEL, restriction fragments end labeling.

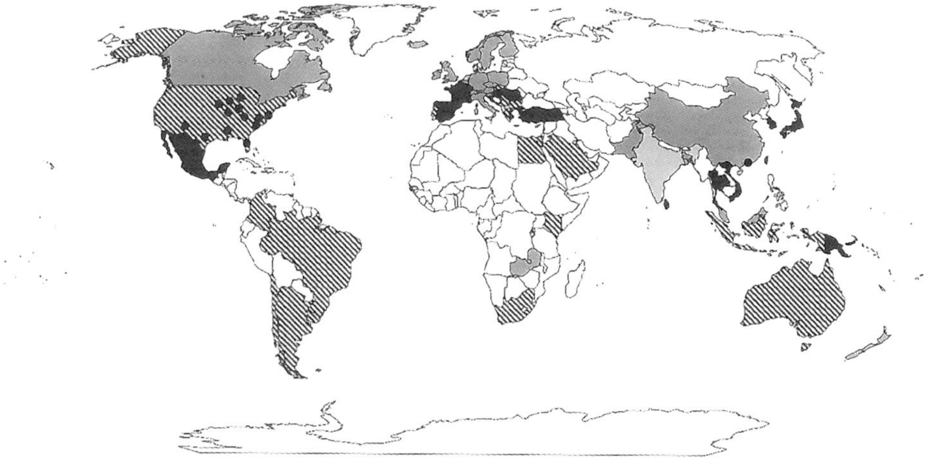

Figure 1 The worldwide prevalence of penicillin-nonsusceptible pneumococci (PNSP, penicillin MIC > 0.1 mg/L). Countries shown in light gray represent those with <5% PNSP (category A), in dark gray those with 5–20% PNSP (category B), in black-white bars with 20–40% PSNP (category C), and black with over 40% PNSP (category D). No data are available for unshaded areas. *Within category A are*: Denmark (116), India (117,118), the Netherlands (119–122); *category B*: Austria (122–124), Bangladesh (125), Belgium (122,126), Canada (114,126–128), China (118,130), Czech Republic (122), Finland (131,132), Germany (119,122,133), Iceland (134), Ireland (121), Italy (122,135), Malaysia (118), New Zealand (136), Norway (137), Pakistan (11), Poland (100,122,138), Slovenia (139), Switzerland (122,140), Sweden (141), United Kingdom (122, 142), Zambia (143); *category C*: Argentina (144), Australia (145), Brazil (122,146, 147), Bulgaria (55,148,149), Chile (150), Colombia (102,105), Egypt (151), Greece (110,119), Indonesia (118), Israel (152), Kenya (153), Portugal (101), Rwanda (154), Saudi Arabia (11), Singapore (118), Slovak Republic (74,155), South Africa (122), Uruguay (156), United States (92,114,122,157,158), Yugoslavia (159); *category D*: Croatia (160), France (89,122), Hungary (108,119,122), Hong Kong (93,122,161), Japan (118), South Korea (118,162), Lebanon (163), Mexico (88, 122), Papua New Guinea (164), Romania (73,165,166), Spain (119,121,122,167, 168), Sri Lanka (118), Thailand (118), Taiwan (118,169), Turkey (170,171), Vietnam (118). Black dots indicate Hong Kong and the United States (Delaware, Florida, Illinois, Iowa, Kentucky, Mississippi, Missouri, Nebraska, New Mexico, North Carolina, Texas, Virginia, Wisconsin) in which reported PNSP prevalence was over 40% (114).

3.3 Prevalence of PNSP Worldwide

A compilation of published data for the prevalence of PNSP in 60 different countries is presented in Figure 1. Surveys do not all cover the same period of time, and for many countries, strains were only collected in one center (e.g., Japan—Nagasaki; South Korea—Seoul; Mexico—Mexico City) or selected areas (mainly highly urbanized) or from populations of particular groups of patients. It is already documented that the prevalence of PNSP depends upon type of infection, patient groups (e.g., HIV positive vs HIV negative), and age group (113). Thus, results of this analysis should be treated with caution, as they do not show a precise picture of PNSP prevalence.

For purposes of this study, prevalence of PNSP was described by four arbitrarily chosen categories based on the percentage of PNSP among clinical or carried *S. pneumoniae* isolates: category A—countries for which the percentage of PNSP was below 5%; category B—prevalence of PNSP ranged from 5 to 20%; category C—PNSP ranged from 20 to 40%; and category D—prevalence of PNSP above 40%. Canada was classified within category B and the United States within category C; however, a study undertaken in 1997 by Doern et al. (114) showed that the percentage of PNSP was higher than that reported by other investigators, and thus both countries may be within higher categories; that is, C for Canada and D for the United States.

For some countries, only sporadic or incomplete data were available. For example, at the beginning of the 1990s, the search for PNSP was undertaken in Russia as part of an eastern European study (73), but only a few isolates have been collected, and since then there has been only one report on the prevalence of PNSP from this country ($<3\%$ PNSP in isolates from sinusitis in Smolensk) (115). As there is a lack of any other current data from this region, it seems unwise to draw any conclusions regarding the percentage of PNSP within Russia.

4 CONCLUSIONS

There has been a substantial increase in antibiotic resistance observed in pneumococci within the last decade. This has been directly connected with the spread of particular pandemic clones of multi–drug-resistant strains and also the development of local epidemic strains. There are no countries that are free of multi–drug-resistant PNSP; however, there are pronounced differences observed in the frequencies of PNSP even between neighboring countries. As to whether this is due to differences in antibiotic usage policies, to vaccination strategies, or to other factors is unclear. Interestingly, the development of multivalent conjugate pneumococcal vaccines may have a profound impact upon the prevalence of antibiotic resistance. Included within the proposed polyvalent vaccines are those childhood-associated serotypes 6B, 9V, 14, and 23F that represent the burden of pandemic multiresistant clones. Eradication of these serotypes from the vaccinated population will hopefully reduce the frequency of their occurrence among other nonvaccinated members of the population. However, this is little comfort to countries unable to afford or implement mass vaccination programs. Moreover, it is unclear in the mid to long term whether restricted valency vaccines will select for new pandemic clones from serotypes beyond the scope of the proposed vaccines or lead to the evolution of novel capsular types. Time will tell whether the commencement of the 21st century ushers in a decline in global pneumococcal infection or just another phase in the evolution of this highly adaptable organism.

REFERENCES

1. Feldman C, Klugman K. Pneumococcal infections. Curr Opin Infect Dis 1997; 10:109–115.
2. Appelbaum PC. Emerging resistance to antimicrobial agents in Gram-positive bacteria—pneumococci. Drugs 1996; 51:1–5.
3. Munoz R, Coffey TJ, Daniels M, Dowson CG, Laible G, Casal J, Hakenbeck R, Spratt BG. Intercontinental spread of a multiresistant clone of serotype-23F *Streptococcus pneumoniae*. J Infect Dis 1991; 164:302–306.
4. McDougal LK, Facklam R, Reeves M, Hunter S, Swenson JM, Hill BC, Tenover FC. Analysis of multiply antimicrobial-resistant isolates of *Streptococcus pneumoniae* from the United States. Antimicrob Agents Chemother 1992; 36:2176–2184.
5. Baquero F. Gram-positive resistance: challenge for the development of new antibiotics. J Antimicrob Chemother 1997; 39:1–6.
6. Muller-Graf CDM, Whatmore AM, King SJ, Trzcinski K, Pickerill AP, Doherty N, Paul J, Griffiths D, Crook D, Dowson CG. Population biology of *Streptococcus pneumoniae* isolated from oropharyngeal carriage and invasive disease. Microbiology. In press.
7. Enright MC, Spratt BG. A multilocus sequence typing scheme for *Streptococcus pneumoniae*: identification of clones associated with serious invasive disease. Microbiology 1998; 144:3049–3060.
8. Smith JM, Dowson CG, Spratt BG. Localized sex in bacteria. Nature 1991; 349: 29–31.
9. Dowson CG, Barcus VA, King SJ, Pickerill AP, Whatmore AM, Yeo M. Horizontal gene transfer and the evolution of resistance and virulence determinants in streptococcus. J Appl Microbiol 1997; 83:S42–S51.
10. Whatmore AM, Barcus VA, Dowson CG. Genetic diversity of the streptococcal competence (com) gene locus. J Bacteriol 1999; 181:3144–3154.

11. Dowson CG, Coffey TJ, Spratt BG. Origin and molecular epidemiology of penicillin-binding–protein-mediated resistance to β-lactam antibiotics. Trends Microbiol 1994; 2:361–366.
12. NCCLS. Methods for dilution antimicrobial susceptibility tests for bacteria that grow aerobically. 4th edition. Approved standards M7-A4. National Committee for Clinical Laboratory Standards, 1997.
13. Gehanno P, Nguyen L, Barry B, Derriennic M, Pichon F, Goehrs JM, Berche P. Eradication by ceftriaxone of *Streptococcus pneumoniae* isolates with increased resistance to penicillin in cases of acute otitis media. Antimicrob Agents Chemother 1999; 43:16–20.
14. Silverstein M, Bachur R, Harper MB. Clinical implications of penicillin and ceftriaxone resistance among children with pneumococcal bacteremia. Pediatr Infect Dis J 1999; 18:35–41.
15. Munoz R, Dowson CG, Daniels M, Coffey TJ, Martin C, Hakenbeck R, Spratt BG. Genetics of resistance to 3rd-generation cephalosporins in clinical isolates of *Streptococcus pneumoniae*. Mol Microbiol 1992; 6:2461–2465.
16. Fenoll A, Bourgon CM, Munoz R, Vicioso D, Casal J. Serotype distribution and antimicrobial resistance of *Streptococcus pneumoniae* isolates causing systemic infections in Spain, 1979–1989. Rev Infect Dis 1991; 13:56–60.
17. Marton A. Pneumococcal antimicrobial resistance—the problem in Hungary. Clin Infect Dis 1992; 15:106–111.
18. Klugman KP, Koornhof HJ. Worldwide increase in pneumococcal antibiotic-resistance. Clin Microbiol Rev 1990; 3:171–196.
19. Markiewicz Z, Tomasz A. Variation in penicillin-binding protein-patterns of penicillin-resistant clinical isolates of pneumococci. J Clin Microbiol 1989; 27:405–410.
20. Ghazal SS, Chowdhury D, Tufenkeji H, AlHowasi M, Ali SM, Aminur M. Ceftriaxone treatment failure of meningitis due to penicillin and ceftriaxone resistant *Streptococcus pneumoniae*. Saudi Med J 1998; 19:510–513.
21. Lister PD. Multiple-resistant pneumococcus—therapeutic problems in the management of serious infections. Eur J Clin Microbiol Infect Dis 1995; 14: S18–S25.
22. McGowan JE, Metchock BG. Penicillin-resistant pneumococci—an emerging threat to successful therapy. J Hosp Infect 1995; 30:472–482.
23. Barcus VA, Ghanekar K, Yeo M, Coffey TJ, Dowson CG. Genetics of high-level penicillin resistance in clinical isolates of *Streptococcus pneumoniae*. FEMS Microbiol Lett 1995; 126:299–303.
24. Dowson CG, Hutchison A, Brannigan JA, George RC, Hansman D, Linares J, Tomasz A, Smith JM, Spratt BG. Horizontal transfer of penicillin-binding protein genes in penicillin-resistant clinical isolates of *Streptococcus pneumoniae*. Proc Natl Acad Sci USA 1989; 86:8842–8846.
25. Martin C, Sibold C, Hakenbeck R. Relatedness of penicillin-binding protein-1a Genes from different clones of penicillin-resistant *Streptococcus pneumoniae* isolated in South-Africa and Spain. EMBO J 1992; 11:3831–3836.
26. Dowson CG, Coffey TJ, Kell C, Whiley RA. Evolution of penicillin resistance in *Streptococcus pneumoniae*—the role of *Streptococcus mitis* in the formation

of a low-affinity PBP2B in *Streptococcus pneumoniae*. Mol Microbiol 1993; 9: 635–643.
27. Smith AM, Klugman KP. Alterations in penicillin-binding protein 2b from penicillin-resistant wild-type strains of *Streptococcus pneumoniae*. Antimicrob Agents Chemother 1995; 39:859–867.
28. Mouz N, Gordon E, DiGuilmi AM, Petit I, Petillot Y, Dupont Y, Hakenbeck R, Vernet T, Dideberg O. Identification of a structural determinant for resistance to β-lactam antibiotics in Gram-positive bacteria. Proc Natl Acad Sci USA 1998; 95:13403–13406.
29. Guenzi E, Gase AM, Sicard MA, Hakenbeck R. A 2-component signal-transducing system is involved in competence and penicillin susceptibility in laboratory mutants of *Streptococcus pneumoniae*. Mol Microbiol 1994; 12: 505–515.
30. Zahner D, Grebe T, Guenzi E, Krauss J, VanDerLinden M, Terhune K, Stock JB, Hakenbeck R. Resistance determinants for β-lactam antibiotics in laboratory mutants of *Streptococcus pneumoniae* that are involved in genetic competence. Micro Drug Resist 1996; 2:187–191.
31. Grebe T, Paik J, Hakenbeck R. A novel resistance mechanism against β-lactams in *Streptococcus pneumoniae* involves CpoA, a putative glycosyltransferase. J Bacteriol 1997; 179:3342–3349.
32. Kell CM, Sharma UK, Dowson CG, Town C, Balganesh TS, Spratt BG. Deletion analysis of the essentiality of penicillin-binding protein-1A, penicillin-binding protein-2B and penicillin-binding protein-2X of *Streptococcus pneumoniae*. FEMS Microbiol Lett 1993; 106:171–175.
33. Grebe T, Hakenbeck R. Penicillin-binding proteins 2B and 2X of *Streptococcus pneumoniae* are primary resistance determinants for different classes of β-lactam antibiotics. Antimicrob Agents Chemother 1996; 40:829–834.
34. Dowson CG, Johnson AP, Cercenado E, George RC. Genetics of oxacillin resistance in clinical isolates of *Streptococcus pneumoniae* that are oxacillin-resistant and penicillin-susceptible. Antimicrob Agents Chemother 1994; 38: 49–53.
35. Coffey TJ, Daniels M, McDougal LK, Dowson CG, Tenover FC, Spratt BG. Genetic analysis of clinical isolates of *Streptococcus pneumoniae* with high-level resistance to expanded-spectrum cephalosporins. Antimicrob Agents Chemother 1995; 39:1306–1313.
36. Sibold C, Henrichsen J, Konig A, Martin C, Chalkley L, Hakenbeck R. Mosaic pbpx genes of major clones of penicillin-resistant *Streptococcus pneumoniae* have evolved from pbpx genes of a penicillin-sensitive *Streptococcus oralis*. Mol Microbiol 1994; 12:1013–1023.
37. Whatmore, AM, et al. Genetic relationships between clinical isolates of *Streptococcus pneumoniae*, *S. oralis* and *S. mitis*: characterisation of 'atypical' pneumococci and organisms allied to *S. mitis* harbouring *S. pneumoniae* virulence factor encoding genes. Infect Immun 2000; 68:1374–1382.
38. Konig A, Reinert RR, Hakenbeck R. *Streptococcus mitis* with unusually high level resistance to β-lactam antibiotics. Microb Drug Resist 1998; 4:45–49.
39. Hakenbeck R, Konig A, Kern I, VanDerLinden M, Keck W, BillotKlein D, Legrand R, Schoot B, Gutmann L. Acquisition of five high-M-r penicillin-

binding protein variants during transfer of high-level β-lactam resistance from *Streptococcus mitis* to *Streptococcus pneumoniae*. J Bacteriol 1998; 180: 1831–1840.

40. Majewski, J, et al. Barriers to genetic exchange between bacterial species: *Streptococcus pneumoniae* transformation. J Bacteriol 2000; 182:1016–1023.
41. Zawadzki P, Roberts MS, Cohan FM. The log-linear relationship between sexual isolation and sequence divergence in Bacillus transformation is robust. Genetics 1995; 140:917–932.
42. Oggioni MR, Dowson CG, Smith JM, Provvedi R, Pozzi G. The tetracycline resistance gene tet(M) exhibits mosaic structure. Plasmid 1996; 35:156–163.
43. Widdowson CA, Klugmann KP, Hanslo D. Identification of the tetracycline resistance gene, tet(O), in *Streptococcus pneumoniae*. Antimicrob Agents Chemother 1996; 40:2891–2893.
44. Widdowson CA, Klugman KP. The molecular mechanisms of tetracycline resistance in the pneumococcus. Microb Drug Resist 1998; 4:79–84.
45. Friedland IR, Klugman KP. Failure of chloramphenicol therapy in penicillin-resistant pneumococcal meningitis. Lancet 1992; 339:405–408.
46. Trieucuot P, Poyartsalmeron C, Carlier C, Courvalin P. Nucleotide sequence of the erythromycin resistance gene of the conjugative transposon Tn1545. Nucleic Acids Res 1990; 18:3660.
47. Widdowson CA, Klugman KP. Emergence of the M phenotype of erythromycin-resistant pneumococci in South Africa. Emerg Infect Dis 1998; 4: 277–281.
48. Linares J, Alonso T, Perez JL, Ayats J, Dominquez MA, Pallares R, Martin R. Decreased susceptibility of penicillin-resistant pneumococci to 24 β-lactam antibiotics. J Antimicrob Chemother 1992; 30:279–288.
49. Clewell DB, Flannagan SE, Jaworski DD. Unconstrained bacterial promiscuity—the Tn916-Tn1545 family of conjugative transposons. Trends Microbiol 1995; 3:229–236.
50. Scott JR, Churchwood GG. Conjugative transposition. Annu Rev Microbiol 1995; 49:367–397.
51. PoyartSalmeron C, TriuCout P, Carlier C, Courvalin P. Nucleotide sequences specific for Tn1545-like conjugative transposon in pneumococci and staphylococci resistant to tetracycline. Antimicrob Agents Chemother 1991; 35: 1657–1660.
52. Marchese A, Ramirez M, Schito GC, Tomasz A. Molecular epidemiology of penicillin-resistant *Streptococcus pneumoniae* isolates recovered in Italy from 1993 to 1996. J Clin Microbiol 1998; 36:2944–2949.
53. Sutcliffe J, TaitKamradt A, Wondrack L. *Streptococcus pneumoniae* and *Streptococcus pyogenes* resistant to macrolides but sensitive to clindamycin: a common resistance pattern mediated by an efflux system. Antimicrob Agents Chemother 1996; 40:1817–1824.
54. Barry AL, Fuchs PC, Brown SD. Antipneumococcal activities of a ketolide (HMR 3647), a streptogramin (quinopritin-dalfopristin), a macrolide (erythromycin), and a lincosamide (clindamycin). Antimicrob Agents Chemother 1998; 42:945–946.
55. Setchanova L, Tomasz A. Molecular characterization of penicillin-resistant

Streptococcus pneumoniae isolates from Bulgaria. J Clin Microbiol 1999; 37: 638–648.
56. Johnston NJ, C. DJ, Kellner JD, Low DE. Prevalence and characterization of the mechanisms of macrolide, lincosamide, and streptogramin resistance in isolates of *Streptococcus pneumoniae*. Antimicrob Agents Chemother 1998; 42:2425–2426.
57. Maskell JP, Sefton AM, Hall LMC. Mechanism of sulfonamide resistance in clinical isolates of *Streptococcus pneumoniae*. Antimicrob Agents Chemother 1997; 41:2121–2126.
58. Pikis A, Keith JM, Rodriquez WJ, Donkersloot JA. Molecular mechanism of trimethoprim (TMP) resistance in *Streptococcus pneumoniae*. Pediat Res 1996; 39:1078.
59. Pikis A, Donkersloot JA, Rodriguez WJ, Keith JM. A conservative amino acid mutation in the chromosome-encoded dihydrofolate reductase confers trimethoprim resistance in *Streptococcus pneumoniae*. J Infect Dis 1998; 178: 700–706.
60. Vinnicombe HG, Derrick JP. Dihydropteroate synthase from *Streptococcus pneumoniae*: characterization of substrate binding order and sulfonamide inhibition. Biochem Biophys Res Commun 1999; 258:752–757.
61. Enright M, Zawadzki P, Pickerill P, Dowson CG. Molecular evolution of rifampicin resistance in *Streptococcus pneumoniae*. Microb Drug Resist 1998; 4:65–70.
62. Doern GV, Pfaller MA, Erwin ME, Brueggemann AB, Jones RN. The prevalence of fluoroquinolone resistance among clinically significant respiratory tract isolates of *Streptococcus pneumoniae* in the United States and Canada—1997 results form the SENTRY Antimicrobial Surveillance Program. Diagn Microbiol Infect Dis 1998; 32:313–316.
63. Choi H, Lee HJ, Lee YH. A mutation in QRDR in the ParC subunit of topoisomerase IV was responsible for fluoroquinolone resistance in clinical isolates of *Streptococcus pneumoniae*. Yonsei Med J 1998; 39:541–545.
64. Taba H, Kusano N. Sparfloxacin resistance in clinical isolates of *Streptococcus pneumoniae*: involvement of multiple mutations in gyrA and parC genes. Antimicrob Agents Chemother 1998; 42:2193–2196.
65. Barry AL, Brown SD, Fuchs PC. Fluoroquinolone resistance among recent clinical isolates of *Streptococcus pneumoniae*. J Antimicrob Chemother 1999; 43:428–429.
66. Brenwald NP, Gill MJ, Wise R. Prevalence of a putative efflux mechanism among fluoroquinolone-resistant clinical isolates of *Streptococcus pneumoniae*. Antimicrob Agents Chemother 1998; 42:2032–2035.
67. Gill MJ, Brenwald NP, Wise R. Identification of an efflux pump gene, pmrA, associated with fluoroquinolone resistance in *Streptococcus pneumoniae*. Antimicrob Agents Chemother 1999; 43:187–189.
68. Markham PN. Inhibition of the emergence of ciprofloxacin resistance in *Streptococcus pneumoniae* by the multidrug efflux inhibitor reserpine. Antimicrob Agents Chemother 1999; 43:988–989.
69. Gonzalez I, Georgiou M, Alcaide F, Balas D, Linares J, DeLaCampa AG.

Fluoroquinolone resistance mutations in the parC, parE, and gyrA genes of clinical isolates of viridans group streptococci. Antimicrob Agents Chemother 1998; 42:2792–2798.
70. Novak R, Henriques B, Charpentier E, Normark S, Tuomanen E. Emergence of vancomycin tolerance in *Streptococcus pneumoniae*. Nature 1999; 399: 590–593.
71. This volume, see Chapter 9.
72. Austrian R. Some aspects of the pneumococcal carrier state. J Antimicrob Chemother 1986; 18:35–45.
73. Appelbaum PC, Gladkova C, Hryniewicz W, Kojouharov B , Kotulova D, Mihalcu F, Schindler J, Setchanova L, Semina N, Trupl J, Tyski S, Urbaskova P, Jacobs MR. Carriage of antibiotic-resistance *Streptococcus pneumoniae* by children in Eastern and Central Europe—a multicenter study with use of standardized methods. Clin Infect Dis 1996; 23:712–717.
74. Sung RYT, Cheng AFB, Chan RCK, Tam JS, Oppenheimer SJ. Epidemiology and etiology of pneumonia in children in Hong-Kong. Clin Infect Dis 1993; 17:894–896.
75. Hendley OJ, Sande MA, Stewart PM, S, Gwaltney JM. Spread of *Streptococcus pneumoniae* in families. I. Carriage rates and distribution of types. J Infect Dis 1975; 132:55–61.
76. Austrian R. Epidemiology of pneumococcal capsular types causing pediatric infections. Pedia Infect Dis J 1989; 8:S21–S22.
77. Scott JAG, Hall AJ, Dagan R, Dixon JMS, Eykyn SJ, Fenoll A, Hortal M, Jette LP, Jorgensen JH, Lamothe F, Latorre C, Macfarlane JT, Shlaes DM, Smart LE, Taunay A. Serogroup-specific epidemiology of *Streptococcus pneumoniae*: associations with age, sex, and geography in 7,000 episodes of invasive disease. Clin Infect Dis 1996; 22:973–981.
78. Hoeprich PD. Bacterial pneumonias. In: Hoeprich PD, Jordan MC, Ronald AR, eds. Infectious Diseases. New York: Lippincott, 1994; 421–433.
79. CreweBrown HH, Karstaedt AS, Saunders GL, Khoosal M, Jones N, Wasas A, Klugman KP. *Streptococcus pneumoniae* blood culture isolates from patients with and without huamn immunodeficiency virus infection: alterations in penicillin susceptibilities and in serogroups or serotypes. Clin Infect Dis 1997; 25:1165–1172.
80. Whatmore AM, Dowson CG. The autolysin encoding gene (lytA) of *Streptococcus pneumoniae* displays restricted allelic variation despite localised recombination events with genes of pneumococcal bacteriophage encoding cell wall lytic enzymes. Infect Immun 1999; 67:4551–4556.
81. Coffey TJ, Daniels M, Enright MC, Spratt BG. Serotype 14 variants of the Spanish penicillin-resistant serotype 9V clone of *Streptococcus pneumoniae* arose by large recombinational replacements of the cpsA-pbpla region. Microbiology 1999; 145:2023–2031.
82. Coffey TJ, Dowson CG, Daniels M, Zhou J, Martin C, Spratt BG, Musser JM. Horizontal transfer of multiple penicillin-binding protein genes, and capsular biosynthetic genes, in natural populations of *Streptococcus pneumoniae*. Mol Microbiol 1991; 5:2255–2260.

83. Coffey TJ, Enright MC, Daniels M, Morona JK, Morona R, Hryniewicz W, Paton JC, Spratt BG. Recombinational exchanges at the capsular polysaccharide biosynthetic locus lead to frequent serotype changes among natural isolates of *Streptococcus pneumoniae*. Mol Microbiol 1998; 27:73–83.
84. Tomasz A, Corso A, Severina EP, EchanizAviles G, Brandileone MCD, Camou T, Castaneda E, Figueroa O, Rossi A, DiFabio JL. Molecular epidemiologic characterization of penicillin-resistant *Streptococcus pneumoniae* invasive pediatric isolates recovered in six Latin-American countries: An overview. Microb Drug Resist 1998; 4:195–207.
85. Hermans PWM, Sluijter M, Hoogenboezem T, Heersma H, VanBelkum A, DeGroot R. Comparative study of 5 different DNA fingerprint techniques for molecular typing of *Streptococcus pneumoniae* strains. J Clin Microbiol 1995; 33:1606–1612.
86. Coffey TJ, Dowson CG, Daniels M, Spratt BG. Horizontal spread of an altered penicillin-binding protein 2B gene between *Streptococcus pneumoniae* and *Streptococcus oralis*. FEMS Microbiol Lett 1993; 110:335–339.
87. Coffey TJ, Berron S, Daniels M, GarciaLeoni ME, Cercenado E, Bouza E, Fenoll A, Spratt BG. Multiply antibiotic-resistant *Streptococcus pneumoniae* recovered from Spanish hospitals (1988–1994): Novel major clones of serotypes 14, 19F and 15F. Microbiology 1996; 142:2747–2757.
88. EchanizAviles G, ValezquezMeza ME, CarnallaBarajas MN, SotoNogueron A, DiFabio JL, SolorzanoSantos F, JimenezTapia Y, Tomasz A. Predominance of the multiresistant 23F international clone of *Streptococcus pneumoniae* among isolates from Mexico. Microb Drug Resist 1998; 4:241–246.
89. Doit C, Loukil C, Fitoussi F, Geslin P, Bingen E. Emergence in France of multiple clones of clinical *Streptococcus pneumoniae* isolates with high-level resistance to amoxicillin. Antimicrob Agents Chemother 1999; 43:1480–1483.
90. Nesin M, Ramirez M, Tomasz A. Capsular transformation of a multidrug-resistant *Streptococcus pneumoniae* in vivo. J Infect Dis 1998; 177:707–713.
91. Hermans PWM, Sluijter M, Dejsirilert S, Lemmens N, Elzenaar K, VanVeen A, Goessens WHF, DeGroot R. Molecular epidemiology of drug-resistant pneumococci: toward an international approach. Microb Drug Resist 1997; 3:243–251.
92. Corso A, Severina EP, Petruk VF, Mauriz YR, Tomasz A. Molecular characerization of penicillin-resistant *Streptococcus pneumoniae* isolates causing respiratory disease in the United States. Microb Drug Resist 1998; 4:325–337.
93. Ip M, Lyon DJ, Yung RWH, Chan C, Cheng AFB. Evidence of clonal dissemination of multidrug-resistant *Streptococcus pneumoniae* in Hong Kong. J Clin Microbiol 1999: 37:2834–2839.
94. Sibold C, Wang JF, Henrichsen J, Hakenbeck R. Genetic relationships of penicillin-susceptible and penicillin-resistant *Streptococcus pneumoniae* strains isolated on different continents. Infect Immun 1992; 60: 4119–4126.
95. Tarasi A, Chong YS, Lee KW, Tomasz A. Spread of the serotype 23F multidrug-resistant *Streptococcus pneumoniae* clone to South Korea. Microb Drug Resist 1997; 3:105–109.
96. Shi ZY, Enright MC, Wilkinson P, Griffiths D, Spratt BG. Identification of three major clones of multiply antibiotic-resistant *Streptococcus pneumoniae* in

Taiwanese hospitals by multilocus sequence typing. J Clin Microb 1998; 36:3514–3519.

97. Versalovic J, Kapur V, Mason EO, Shah U, Koeuth T, Lupski JR, Musser JM. Penicillin-resistant *Streptococcus pneumoniae* strains recovered in Houston—identification and molecular characterization of multiple clones. J Infect Dis 1993; 167:850–856.
98. Soares S, Kristinsson KG, Musser JM, Tomasz A. Evidence for the introduction of a multiresistant clone of serotype-6B *Streptococcus pneumoniae* from Spain to Iceland in the late 1980s. J Infect Dis 1993; 168:158–163.
99. Rudolph KM, Parkinson AJ, Roberts MC. Molecular analysis by pulsed-field gel electrophoresis and antibiogram of *Streptococcus pneumoniae* serotype 6B isolates from selected areas within the United States. J Clin Microbiol 1998; 36:2703–2707.
100. Overweg K, Hermans PWM, Trzcinski K, Sluijter M, DeGroot R, Hryniewicz W. Multidrug-resistant *Streptococcus pneumoniae* in Poland: Identification of emerging clones. J Clin Microbiol 1999; 37:1739–1745.
101. deLencastre H, Kristinsson KG, BritoAvo A, Sanches IS, SaLeao R, Saldanha J, Sigvaldadottir E, Karlsson S, Oliveira D, Mato R, DeSousa MA, Tomasz A. Carriage of respiratory tract pathogens and molecular epidemiology of *Streptococcus pneumoniae* colonization in healthy children attending day care centers in Lisbon, Portugal. Microb Drug Reiss 1999; 5:19–29.
102. Castaneda E, Leal AL, Castillo O, DeLaHoz F, Vela MC, Arango M, Trujillo H, Levy A, Gama ME, Calle M, Valencia ML, Parra W, Agudelo N, Mejia GI, Jaramillo S, Montoya F, Porras H, Sanchez A, Saa D, DiFabio JL, Homma A, Rios AM, Ovalle MV, Serrato J, Navarrete MR, Garcia M, Aristizabal G, Tovar A, Paredes C, Arenas A, Zapata C, Robledo J, Correa N, Suarez C, Garcia V, Gallardo LM, Moreno A, Villamarin N, Bohorquez AL, Lopez P, Guerrero J. Distribution of capsular types and antimicrobial susceptibility of invasive isolates of *Streptococcccus pneumoniae* in Colombian children. Microb Drug Resist 1997; 3:147–152.
103. Reichmann P, Varon E, Gunther E, Reinert RR, Luttiken R, Marton A, Geslin P, Wagner J, Hakenbeck R. Penicillin-resistant *Streptococcus pneumoniae* in Germany—genetic relationship to clones from other European countries. J Med Microbiol 1995; 43:377–385.
104. Pato MVV, DeCarvalho CB, Tomasz A, Ramos MH, Guimaraes C, Teixeira MF, daCosta MN, Sobral MDL, Costa DB, Moreira S, Campose M, Paiva AM, Salgado MJ, Marques JT, Barros RM, Troni MH, Lopes T. Antibiotic susceptibility of *Streptococcus pneumoniae* isolates in Portugal. A multicenter study between 1989 and 1993. Microb Drug Resist 1995; 1:59–69.
105. Castaneda E, Penuela I, Vela MC, Tomasz A. Penicillin-resistant *Streptococcus pneumoniae* in Colombia: presence of international epidemic clones. Microb Drug Resist 1998; 4:233–239.
106. Dejsirilert S, Overweg K, Sluijter M, Saengsuk L, Gratten M, Ezaki T, Hermans PWM. Nasopharyngeal carriage of penicillin-resistant *Streptococcus pneumoniae* among children with acute respiratory tract infections in Thailand: a molecular epidemiological survey. J Clin Microbiol 1999; 37:1832–1838.

107. Figueiredo AMS, Austrian R, Urbanskova P, Teixeira LA, Tomasz A. Novel penicillin-resistant clones of *Streptococcus pneumoniae* in the Czech Republic and in Slovakia. Microb Drug Resist 1995; 1:71–78.
108. Marton A. Epidemiology of resistant pneumococci in Hungary. Microb Drug Resist 1995; 1:127–130.
109. Lawrenson JB, Klugman KP, Eidelman JI, Wasas A, Miller SD, Lipman J. Fatal infection caused by a multiply resistant type-3 pneumococcus. J Clin Microbiol 1988; 26:1590–1591.
110. Syrogiannopoulos GA, Grivea IN, Beratis NG, Spiliopoulou AE, Fasola EL, Bajaksouzian S, Appelbaum PC, Jacobs MR. Resistance patterns of *Streptococcus pneumoniae* from carriers attending day-care centers in southwestern Greece. Clin Infect Dis 1997; 25:188–194.
111. Tamayo M, SaLeao R, Sanches IS, Castaneda E, DeLencastre H. Dissemination of a chloramphenicol-and tetracycline-resistant but penicillin-susceptible invasive clone of serotype-5 *Streptococcus pneumoniae* in Colombia. J Clin Microbiol 1999; 37:2337–2342.
112. Luey KY, Kam KM. Vaccine coverage of *Streptococcus pneumoniae* in Hong Kong with attention to the multiple-antibiotic-resistant strains. Vaccine 1996; 14:1573–1580.
113. Paul J. HIV and pneumococcal infection in Africa—microbiological aspects. Trans R Soc Trop Med Hyg 1997; 91:632–637.
114. Doern GV, Pfaller MA, Kugler K, Freeman J, Jones RN. Prevalence of antimicrobial resistant among respiratory tract isolates of *Streptococcus pneumoniae* in North America: 1997 results from the SENTRY antimicrobial surveillance program. Clin Infect Dis 1998; 27:764–770.
115. Otyagin I, Kamanin E, Voronov E, Chubarova S, Shevkov V, Nekrasova L. Comunity-acquired sinusitis: pathogens and their antimicrobial susceptibility. Clin Microbiol Infect 1999; 5:S3:277.
116. Nielsen SV, Henrichsen J. Incidence of invasive pneumococcal disease and distribution of capsular types of pneumococci in Denmark, 1989–94. Epidemiol Infect 1996; 117:411–416.
117. Thomas K, Lalitha MK, Steinhoff MC, Arora NK, Ratan A, Das B, Awasthi SK, Misra PK, Amita J, Daivanayagam R, Manjula P, Jain D, Agarwal V, Ravindran P, Cherian T, Raghupathy P, Moses P, Jesudason MV, Brahmadathan KN, Jayseelan L, John TJ. Prospective multicentre hospital surveillance of *Streptococcus pneumoniae* disease in India. Lancet 1999; 353:1216–1221.
118. Song JH, Lee NY, Ichiyama S, Yoshida R, Hirakata Y, Fu W, Chongthaleong A, Aswapokee N, Chiu CH, Lalitha MK, Thomas K, Perera J, Yee TT, Jamal F, Warsa UC, Vinh BX, Jacobs MR, Appelbaum PC, Pai CH. Srpead of drug-resistant *Streptococcus pneumoniae* in Asian countries: Asian Network for Surveillance of Resistant Pathogens (ANSORP) study. Clin Infect Dis 1999; 28:1206–1211.
119. Cullmann W. Comparative evaluation of orally active antibiotics against community-acquired pathogens: results of eight European countries. Chemotherapy 1996; 42:11–20.

120. Hermans PWM, Sluijter M, Elzenaar K, vanVeen A, Schonkeren JJM, Nooren FM, vanLeeuwen WJ, deNeeling AJ, vanKlingeren B, Verbrugh HA, deGroot R. Penicillin-resistant *Streptococcus pneumoniae* in the Netherlands: results of a 1-year molecular epidemiologic survey. J Infect Dis 1997; 175:1413–1422.
121. Richard MP, Aguado AG, Mattina R, Marre R. Sensitivity to sparfloxacin and other antibiotics, of *Streptococcus pneumoniae, Haemophilus influenzae* and *Moraxella catarrhalis* strains isolated from adult patients with community-acquired lower respiratory tract infections: a European multicentre study. J Antimicrob Chemother 1998; 41:207–214.
122. Felmingham D, Washington J. Trends in the antimicrobial susceptibility of bacterial respiratory tract pathogens—findings of the Alexander project 1992–1996. J Chemother 1999; 11:5–21.
123. Mittermayer H, Jebelean C, Binder L, Haditsch M, Watschinger R. Antibiotic susceptibility of pneumococci isolated in Austria over a four-year period. Eur J Clin Microbiol Infect Dis 1996; 15:817–820.
124. Georgopoulos A, Buxbaum A, Straschil U, Graninger W. Austrian national survey of prevalence of antimicrobial resistant among clinical isolates of *Streptococcus pneumoniae* 1994–96. Scand J Infect Dis 1998; 30:345–349.
125. Saha SK, Rikitomi N, Ruhulamin M, Masaki H, Hanif M, Islam M, Watanabe K, Ahmed K, Matsumoto K, Sack RB, Nagatake T. Antimicrobial resistance and serotype distribution of *Streptococcus pneumoniae* strains causing childhood infections in Bangladesh, 1993 to 1997. J Clin Microbiol 1999; 37: 798–800.
126. Verhaegen J, Verbist L. In-vitro activity of 21 β-lactam antibiotics against penicillin-susceptible and penicillin-resistant *Streptococcus pneumoniae.* J Antimicrob Chemother 1998; 41:381–385.
127. Lovgren M, Spika JS, Talbot JA. Invasive *Streptococcus pneumoniae* infections: serotype distribution and antimicrobial resistance in Canada, 1992–1995. Can Med Assoc J 1998; 158:327–331.
128. Blondeau JM, Suter M, Borsos S. Determination of the antimicrobial susceptibilities of Canadian isolates of *Haemophilus influenzae, Streptococcus pneumoniae* and *Moraxella catarrhalis.* J Antimicrob Chemother 1999; 43: 25–30.
129. Kellner JD, Ford Jones EL. *Streptococcus pneumoniae* carriage in children attending 59 Canadian child care centers. Arch Pediatr Adoles Med 1999; 153: 495–502.
130. Wang H, Huebner R, Chen MJ, Klugman K. Antibiotic susceptibility patterns of *Streptococcus pneumoniae* in China and comparison of MICs by agar dilution and E-test methods. Antimicrob Agents Chemother 1998; 42:2633–2636.
131. Manninen R, Huovinen P, Nissinen A, Ahonen E, Eerola E, Eskola J, Hirvonen P, Jagerroos H, Katila ML, Kauppinen M, Klossner ML, Kontiainen S, Korpela J, Koskela M, KostialaThompson A, Karkkainen P, Lantto K, Larinkari U, Lehtonen OP, Liimatainen O, Oinonen S, Pietarinen I, Renkonen OV, Sarkkinen H, Schauman K, Sivonen A, Vaara M, Vikberg V. Increasing

antimicrobial resistance in *Streptococcus pneumoniae, Haemophilus influenzae* and *Moraxella catarrhalis* in Finland. J Antimicrob Chemother 1997; 40: 387–392.

132. Manninen R, Leinonen M, Huovinen P, Nissinen A. Reliability of disc diffusion susceptibility testing of *Streptococcus pneumoniae* and adjustment of laboratory-specific breakpoints. J Antimicrob Chemother 1998; 41:19–26.
133. Traub WH, Leonhard B. Antibiotic susceptibility of clinical isolates of *Streptococcus pneumoniae*. Chemotherapy 1996; 42:240–247.
134. Arason VA, Kristinsson KG, Sigurdsson JA, Stefansdottir G, Molstad S, Gudmundsson S. Do antimicrobials increase the carriage rate of penicillin-resistant pneumococci in children? Cross sectional prevalence study. BMJ 1996; 313:387–391.
135. Marchese A, Debbia E, Pesce A, Schito C. Comparative activities of amoxicillin and 10 other oral drugs against penicillin-susceptible and -resistant *Streptococcus pneumoniae* strains recently isolated in Italy. Clin Microbiol Infect 1998; 4:170–172.
136. Brett W, Masters PJ, Lang SD, Ikram RB, Hatch SH, Gordon MS. Antibiotic susceptibility of *Streptococcus pneumoniae* in New Zealand. NZ Med J 1999; 12:74–78.
137. Bergan T, Gaustad P, Hoiby EA, Berdal BP, Furuberg G, Baann J, Tonjum T. Antibiotic resistance of pneumococci in Norway. Int J Antimicrob Agents 1998; 10:77–81.
138. Trzcinski K, Hryniewicz W. Antimicrobial susceptibility of common bacterial pathogens isolated from lower respiratory tract infections in Poland in 1996—the Alexander Project. Med Sci Monitor 1997; 3:714–722.
139. Cizman M, Paragi M, JovanKuhar N, Gubina M, Orazem A, Pokorn M, Kraigher A, Fiser J, Kolman J, Novak D, Drinovec B, Harlander T, Sabotin D, Sobota M, Bozanic V. Antimicrobial resistance of invasive *Streptococcus pneumoniae* in Slovenia, 1993–1995. Scand J Infect Dis 1997; 29:251–254.
140. Wust J, Huf E, Kayser FH. Antimicrobial susceptibilities and serotypes of invasive *Streptococcus pneumoniae* strains in Switzerland. J Clin Microbiol 1995; 33:3159—3163.
141. Nilsson P, Laurell MH. Several different clones present during the penetration phase of resistant *Streptococcus pneumoniae* in the city of Malmo, Sweden. Microb Drug Resist 1999; 5:37–43.
142. Gruneberg RN, Felmingham D, Harding I, Shrimpton SB, Nathwani A. The Nearchus project: antibiotic susceptiblity of respiratory pathogens and clinical outcome in lower respiratory tract infections at 27 centres in the UK. Int J Antimicrob Agents 1998; 10:127–133.
143. Woolfson A, Huebner R, Wasas A, Chola S, GodfreyFaussett P, Klugman K. Nasopharyngeal carrige of community-acquired, antibiotic-resistant *Streptococcus pneumoniae* in a Zambian paediatric population. Bull WHO 1997; 75: 453–462.
144. Rossi A, Corso A, Pace J, Regueira M, Tomasz A. Penicillin-resistant *Streptococcus pneumoniae* in Argentina: frequent occurence of an internationally spread serotype 14 clone. Microb Drug Resist 1988; 4:225–231.

145. Turnidge JD, Bell JM, Collignon PJ. Rapidly emerging antimicrobial resistances in *Streptococcus pneumoniae* in Australia. Med J Aust 1999; 170: 152–155.
146. Levin ASS, Teixeira LM, Sessegolo JF, Barone AA. Resistance of *Streptococcus pneumoniae* to antimicrobials in Sao Paulo, Brazil: clincial features and serotypes. Rev Inst Med Trop Sao Paulo 1996; 38:187–192.
147. Brandileone MCD, Vieira VSD, Casagrande ST, Zanella RC, Guerra M, Bokermann S, DeMoraes JC, Baldacci ER, Chamone CB, Oliveira MAA, DeMatos DGC, Arruda TMC, Coelho MFD, Davila SM, DosSantos AR, DiFabio JL, Vital NC, Leoardio G, Lima MG, Tonelli E, Lopes JMM, Ribeiro JGL, Lainson ZCL, Ramos FLP, Flores RP, Tabosa TJC, Figueiroa A, Souza NF, Silva MJB, Castro JR, Galindo MC, Duque T, Vasconcelos MA, Azevedo AC, Maggi RRS, Araujo C, Rodrigues NG, Oliveira SMF, Pires LJR, Mello ESF, Inoue LK, Nitrini DR, Guanrieri CE, Hidalgo NTR, Fernandes VA, Carnauba EL, Vasconcelos MJ, Rossi A, Lotufo JPB, Hein N, Guida SM, Junciomi MR, Berezin E, Toda EN, Masiero RL, Machado AMO, Afonso MN, Zanatta YO, Baretta MC, Melles CEA. Prevalence of serotypes and antimicrobial resistance of *Streptococcus pneumoniae* strains isolated from Brazilian children with invasive infections. Microb Drug Resist 1997; 3:141–146.
148. Setchanova L. Clinical isolates and nasopharyngeal carriage of antibiotic-resistant *Streptococcus pneumoniae* in Hospital for Infectious Diseases, Sofia, Bulgaria, 1991–1993. Microb Drug Resist 1995; 1:79–84.
149. Girgitzova B, Milanov ST. Etiology of community-acquired pneumonia in hospitalized adults. Clin Microbiol Infect 1999; 5:S3:154.
150. Kertesz DA, DiFabio JL, Brandileone MCD, Castaneda E, EchanizAviles G, Heitmann I, Homma A, Hortal M, Lovgren M, Ruvinksy RO, Talbot JA, Weekes J, Spika JS. Invasive *Streptococccus pneumoniae* infection in Latin American children: results of the Pan American Health Organization surveillance study. Clin Infect Dis 1998; 26:1355–1361.
151. Ostroff SM, Harrison LH, Khallaf N, Assaad MT, Guirguis NI, Harrington S, elAlamy M, Hassan S, Kamal H, Mansour H, Tamam L, Fam S, Hanna N, Shabana F, Sobhy M, Tawfik M, Helmy MF, elSaid MA, elKhaleed M, Lewis A. Resistance patterns of *Streptococcus pneumoniae* and *Haemophilus influenzae* isolates recovered in Egypt from children with pneumonia. Clin Infect Dis 1996; 23:1069–1074.
152. Dagan R, Melamed R, Muallem M, Piglansky L, Yagupsky P. Nasopharyngeal colonization in southern Israel with antibiotic-resistant pneumococci during the first 2 years of life: relation to serotypes likely to be included in pneumococcal conjugate vaccines. J Infect Dis 1996; 174:1352–1355.
153. Kell CM, Jordens JZ, Daniels M, Coffey TJ, Bates J, Paul J, Gilks C, Spratt BG. Molecular epidemiology of penicillin-resistant pneumococci isolated in Nairobi, Kenya. Infect Immun 1993; 61:4382–4391.
154. Bogaerts J, Lepage P, Taelman H, et al. Antimicrobial susceptibility and serotype distribution of *Streptococcus pneumoniae* from Rwanda, 1984–1990. J Infect 1993; 27:157–168.
155. Trupl J, Hupkova H, Balint O, Stankovic I, Pneumococcus Study Group.

Serotype patterns of invasive *Streptococcus pneumoniae* in the Slovak Republic. Clin Microbiol Infect 1999; 5:S3:311.
156. Hortal M, Algorta G, Bianchi I, Borthagaray G, Cestau I, Camou T, Castro M, DeLosSantos M, Diez R, DellAcqua L, Galiana A, Giordano A, Giordano P, LopezGhemi G, Milanese N, Mogdasy C, Palacio R, Pedreira W, Pisano A, Piel L. Capsular type distribution and susceptibility to antibiotics of *Streptococccus pneumoniae* clinical strains isolated from Uruguayan children with systemic infections. Microb Drug Resist 1997; 3: 159–163.
157. Thornsberry c, Ogilvie P, Kahn J, Mauriz Y. Surveillance of antimicrobial resistance in *Streptococcus pneumoniae, Haemophilus influenzae*, and *Moraxella catarrhalis* in the United States in 1996–1997 respiratory season. Diagn Microbiol Infect Dis 1997; 29:249–257.
158. Jones RN, Pfaller MA, Doern GV. Comparative antimicrobial activity of trovafloxacin tested against 3049 *Streptococcus pneumoniae* isolates from the 1997–1998 respiratory infection season. Diagn Microbiol Infect Dis 1998; 32:119–126.
159. Petreska-Sabinovska D, Mroavic M, Lazarevic G. The development of resistance of *Streptococcus pneumoniae* strains islated from children. Clin Microbiol Infect 1999; 5:S3:167.
160. GoicBarisic I, SiskoKraljevic K, Tonkic M, Sardelic S. Susceptibility of respiratory bacterial pathogens to azithromycin. Clin Microbiol Infect 1999; 5:S3:82.
161. Ho PL, Que TL, Tsang DNC, Ng TK, Chow KH, Seto WH. Emergence of fluoroquinolone resistance among multiply resistant strains of *Streptococcus pneumoniae* in Hong Kong. Antimicrob Agents Chemother 1999; 43:1310–1313.
162. Kim SN, Kim SW, Choi IH, Pyo SN, Rhee DK. High incidence of multidrug-resistant *Streptococcus pneumoniae* in South Korea. Microb Drug Resist 1996; 2:401–406.
163. Uwaydah M, Jradeh M, Shihab Z. Antimicrobial resistance of clinical isolates of *Streptococcus pneumoniae* in Lebonon. J Antimicrob Chemother 1996; 38: 283–286.
164. Lehmann D. Epidemiology of acute respiratory-tract infections, especially those due to *Haemophilus influenzae*, in Papua-New-Guinean children. J Infect Dis 1992; 165:S20–S25.
165. Mihalcu F, Pana M, Burcea D, Mitache I, Coman G. Antibiotic resistance study of invasive *Streptococcus pneumoniae*. R Arch Microbiol Immunol 1996; 55:133–143.
166. Dorobat O, Biolan T, Burtea M. Antibiotic-resistance in *Streptococcus pneumoniae* isolated in a clinical hospital in 1997–1998. Clin Microbiol Infect 1999; 5:S3:320.
167. Fenoll A, Jado I, Vicioso D, Perez A, Casal J. Evolution of *Streptococcus pneumoniae* serotypes and antibiotic resistance in Spain: Update (1990 to 1996). J Clin Microbiol 1998; 36:3447–3454.
168. Baquero F, GarciaRodriguez JA, deLomas JG, Aguilar L. Antimicrobial resistance of 1,113 *Streptococcus pneumoniae* isolates from patients with respiratory

tract infections in Spain: results of a 1-year (1996–1997) multicenter surveillance study. Antimicrob Agents Chemother 1999; 43:357–359.
169. Hsueh PR, Teng LJ, Lee LN, Yang PC, Ho SW, Luh KT. Dissemination of high-level penicillin-, extended-spectrum cephalosporin-, and erythromycin-resistant *Streptococcus pneumoniae* clones in Taiwan. J Clin Microbiol 1999; 37:221–224.
170. Sener B, Gunalp A. Trends in antimicrobial resistance of *Streptococcus pneumoniae* in children in a Turkish hospital. J Antimicrob Chemother 1998; 42: 381–384.
171. Sener B, Arikan S, Ergin MA, Gunalp A. Rate of carriage, serotype distribution and penicillin resistance of *Streptococcus pneumoniae* in healthy children. Z Bakteriol-Int J Med Microbiol Virol Parasitol Infect Dis 1998; 288: 421–428.
172. http://mlst.zoo.ox.ac.uk
173. Hall LMC, Whiley RA, Duke B, George RC, Efstratiou A. Genetic relatedness within and between serotypes of *Streptococcus pneumoniae* from the United Kingdom: analysis of multilocus enzyme electrophoresis, pulsed-field gel electrophoresis, and antimicrobial resistance patterns. J Clin Microbiol 1996; 34:853–859.
174. Melander E, Ekdahl K, Hansson HB, Kamme C, Laurell M, Nilsson P, Persson K, Soderstrom M, Molstad S. Introduction and clonal spread of penicillin- and trimethoprim/sulfamethoxazole-resistant *Streptococcus pneumoniae*, serotype 9V, in Southern Sweden. Microb Drug Resist 1998; 4:71–78.
175. Klugman KP, Coffey TJ, Smith A, Wasas A, Meryers M, Spratt BG. Cluster of an erythromycin-resistant variant of the Spanish multiply resistant 23F clone of *Streptococcus pneumoniae* in South-Africa. Eur J Clin Microbiol Infect Dis 1994; 13:171–174.
176. Doit C, Denamur E, Picard B, Geslin P, Elion J, Bingen E. Mechanisms of the spread of penicillin resistance in *Streptococcus pneumoniae* strains causing meningitis in children in France. J Infect Dis 1996; 174:520–528.
177. Ferroni A, Nguyen L, Gehanno P, Boucot I, Berche P. Clonal distribution of penicillin-resistant *Streptococcus pneumoniae* 23F in France. J Clin Microbiol 1996; 34:2707–2712.
178. Louie M, Louie L, Papia G, Talbot J, Lovgren M, Simor AE. Molecular analysis of the genetic variation among penicillin-susceptible and penicillin-resistant *Streptococcus pneumoniae* serotypes in Canada. J Infect Dis 1999; 179:892–900.

13

Resistance Problems Associated with the Enterococcus

George M. Eliopoulos
Harvard Medical School and Beth Israel Deaconess Medical Center, Boston, Massachusetts

In clinical medicine, enterococci have long been recognized as organisms which are relatively resistant to inhibition or killing by a number of antimicrobials commonly used against other gram-positive bacteria. As a result, treatment options are often limited and, to achieve bactericidal activity, synergistic therapy with a cell wall–active antibiotic combined with an aminoglycoside has been required. Resistance to synergistic killing may arise from ribosomal mutation conferring high-level resistance to streptomycin, from the presence of an intrinsic 6′-acetyltransferase in *Enterococcus faecium* which renders several deoxystreptamine aminoglycosides inactive, or from the acquisition by various enterococcal species of other aminoglycoside-modifying enzymes borne on transferable resistance elements. Recently, increases in levels of resistance to β-lactam antibiotics have been noted in *E. faecium*. Such resistance is mediated by low-affinity penicillin–binding proteins, and genetic determinants for these and *(pbp5)* also have been shown to be potentially transferable. Glycopeptide resistance in this genus has emerged over the past decade. Several gene clusters have been identified which ultimately result in the formation of altered peptidoglycan precursors, terminating in either D-alanine-D-lactate or D-alanine-D-serine to which vancomycin binds less well than to the native cell wall component, D-alanine-D-alanine. Other transferable resistance genes have been identified in enterococci, conferring resistance to a number of unrelated classes of antibiotics, including tetracyclines, chloramphenicol, macrolides, lincosamides, and streptogramins A and B. In addition, resistance to fluoroquinolones may result from mutations in *gyrA* or *parC* genes of DNA gyrase or topoisomerase IV, respectively, or from active drug efflux. As a result of this broad repertoire of resistance mechanisms, strains of enterococci now exist which are resistant to virtually all clinically available antimicrobials.

1 INTRODUCTION

Enterococci have long stood out among gram-positive bacteria because of their intrinsic relative resistance to numerous antimicrobials. Moreover, they have the capacity to acquire or to develop traits that render them resistant to even higher concentrations of various antibiotics (1). The emergence of vancomycin resistance among strains of *Enterococcus* in the late 1980s was a devastating development, severely constraining options for treatment of infections caused by such organisms (2). As normal inhabitants of the human gastrointestinal tract, enterococci, including strains resistant to multiple antibiotics, may harmlessly colonize humans and farm animals (3). On the other hand, colonized individuals can be at risk to develop severe and even life-threatening infections due to these organisms when surgery, instrumentation, or immunosuppression broaches or suppresses normal host defenses.

Multiply antibiotic–resistant enterococci are now a serious concern in U.S. hospitals. After coagulase-negative staphylococci and *Staphylococcus aureus*, enterococci were the third most common cause of nosocomial bloodstream infection for the 3-year period ending in April 1998 (4). Vancomycin-resistant strains accounted for more than 20% of nosocomial enterococcal bloodstream infections in the northeastern United States (5). With resistance to multiple agents being common among vancomycin-resistant isolates (6), the significance of such data becomes evident. In several studies, mortality associated with vancomycin-resistant enterococcal bacteremia exceeded that of comparator groups (7,8), although the severity of underlying illnesses of those infected with resistant isolates accounts for much of this excess mortality (9). The additional costs of hospital infection control measures and of more complex antibiotic therapies associated with infection due to multiply resistant enterococci place increased stress on health care budgets.

This chapter will review antibiotic resistance in enterococci, with emphasis on those issues of greatest importance in current clinical practice.

2 RESISTANCE TO SYNERGISTIC KILLING

Enterococci, particularly *E. faecalis*, are a major cause of infective endocarditis, an infection for which bactericidal therapy is warranted. Unfortunately, penicillins, vancomycin, and other cell wall–active antimicrobials as a rule inhibit but do not kill enterococci (10). Early studies documented synergistic killing of enterococci in vitro, as well as enhanced clinical efficacy, when agents acting on the bacterial cell wall were combined with aminoglycosides (11,12). Although enterococci are relatively resistant to aminoglycosides intrinsically, experiments with radiolabeled streptomycin demonstrated enhanced uptake of the aminoglycoside in the presence of penicillin (13). Once inside the bacterial cell, the aminoglycoside exerts bactericidal effects through binding to the ribosome. Thus, the combination of a bacteriostatic agent (penicillin) with an agent without activity at clinically achievable concentrations (streptomycin) resulted in synergistic killing of the enterococcus.

2.1 Ribosomal Resistance to Synergistic Killing

Attainment of synergistic bactericidal activity against enterococci is predicated upon the ability of the aminoglycoside to interact with the bacterial ribosome in a manner leading to lethal effects. Some strains of enterococci are not only resistant to streptomycin at clinically achievable concentrations, but they also exhibit resistance to very high concentrations of the aminoglycoside, with minimal inhibitory concentrations (MICs) > 2000 μg/mL. Mutants derived in vitro demonstrating (high-level) ribosomal resistance to streptomycin were resistant to synergistic killing as described above even though enhanced uptake of a radiolabeled compound still occurred in the presence of penicillin (13). Ribosomal resistance to streptomycin also occurs among clinical isolates of enterococci (14). However, by far the more common mechanism of high-level resistance to aminoglycosides, and resistance to synergistic killing by aminoglycosides in combination with a cell wall–active agent, involves enzymatic modification of the aminoglycoside.

2.2 Aminoglycoside-Modifying Enzymes

High-level resistance to streptomycin (MIC > 2000 μg/mL) in enterococci can result from adenylylation of the aminoglycoside which renders it inactive against the ribosomal target (15). The gene (*aadE*) mediating production of streptomycin 6′-adenylylating enzyme may be plasmidborne and thus potentially transferable between strains (15). Enzymatic modification of streptomycin may also be mediated by *aadA*; the enzyme, ANT(3″) (9) also confers resistance to spectinomycin (16). High-level resistance to streptomycin based on enzymatic modification negates synergistic killing when the aminoglycoside is combined with cell wall–active agents.

High-level resistance to kanamycin (MIC > 2000 μg/mL) was already fairly common among enterococci by the 1970s (17). Resistance was attributable to an enzyme capable of phosphorylating 2-deoxystreptamine aminoglycosides at the 3′-hydroxy position (15). The enzyme is active against a number of aminoglycosides, including kanamycin, neomycin, lividomycin (subject to 5″-*O*-phosphorylation), and amikacin, although the latter is a poor substrate (18). The APH(3′)-IIIa phosphorylating enzyme renders organisms highly resistant to kanamycin and abolishes synergism. Strains possessing this enzyme are also resistant to penicillin-amikacin synergism despite the fact that MICs of amikacin are similar to those of enzyme-negative strains, usually well below 2000 μg/mL (19). Against enzyme-producing strains, amikacin may actually antagonize any bactericidal activity observed with penicillin alone (19,20). Tobramycin, lacking a 3′-OH group, is not susceptible to modification by this enzyme.

Inactivation of kanamycin by another enzyme, 4′,4″-nucleotidyltransferase, has also been described (21). The gene mediating production of the enzyme was plasmid mediated in the strain of *E. faecium* from which it was initially identified. The plasmid was self-transferable and also mediated resistance to MLS_B antibiotics. This aminoglycoside-modifying enzyme rendered strains resistant to synergistic killing when kanamycin, tobramycin, or amikacin was used in combination with a cell wall agent (21).

Characteristic of *E. faecium* (and not found in other enterococci) is the presence of a chromosomally determined enzyme, designated AAC(6′)-Ii, which confers resistance to synergism between cell wall–active agents and kanamycin, tobramycin, netilmicin, and other agents susceptible to modification by acetylation of 6′-amino groups (22). Production of the enzyme occurs at a low level and is difficult to detect by assays of aminoglycoside modification (22). Thus, strains of *E. faecium* are considered to be resistant to synergistic killing by combinations with these aminoglycosides even in the absence of high-level resistance to a specific compound (17). The enzyme has been overexpressed in *Escherichia coli* and purified for detailed analysis (23). Although amikacin possesses a 6′-amino group, and thus is theoretically susceptible to modification, the group is effectively blocked by the N-1 substituent (2-aminohydroxybutyryl) of this compound. Thus, amikacin retains the potential for synergistic interaction with cell wall agents in the absence of other coexisting aminoglycoside-modifying enzymes (17). Because gentamicin C1 is not a substrate for this enzyme, combinations of cell wall–active agents with gentamicin retain synergistic bactericidal activity against *E. faecium* despite the presence of the chromosomal AAC(6′) enzyme (22).

High-level resistance to gentamicin in *E. faecalis* was first reported in 1979 (24). Subsequent studies showed resistance to result from an enzyme with bifunctional activity; that is, a single protein capable of 2″-*O*-phosphorylation and 6′-*N*-acetylation (25). The broad activity of the AAC(6′)-APH(2″) enzyme confers resistance to synergism for combinations including all 2-deoxystreptamine aminoglycosides used in the United States. Kinetic studies performed on purified enzyme indicate impressive activity against a wide range of substrates (26). In addition to the predicted modes of substrate modification, the enzyme also catalyzes 6′-*O*-acetylation (lividomycin A) and phosphorylates hydroxy groups on several aminoglycoside rings. However, in a substantial proportion of isolates, the presence of enzyme is not sufficient to raise gentamicin MICs above 2000 μg/mL (27). Thus, for gentamicin, an MIC in excess of 500 μg/mL is generally considered to represent high-level resistance. Plasmid-mediated high-level gentamicin resistance due to this bifunctional enzyme has now been encountered in other enterococcal species, including *E. faecium* (28), *E. avium*, *E. raffinosus*, and *E. hirae* (29).

The *aac6-aph2* gene of *E. faecalis* HH22 residing on the conjugative plasmid pBEM10 is carried on a transposon (designated Tn*5281*) which is structurally related to transposons mediating bifunctional enzymatic resistance in *Staphylococcus aureus* (Tn*4001*) or *S. epidermidis* (Tn*4031*) (30). A diversity of Tn*4001*-like structures, distinguished by the presence and number of complete or partial IS*256* flanking elements, has been reported on conjugative and nonconjugative plasmids of *E. faecalis* isolated in France (31). Enterococcal plasmids which carry *aac6-aph2* determinants are themselves heterogeneous when strains from diverse geographical areas are examined (32). Transposable elements mediating production of the bifunctional enzyme have also been found in chromosomal locations (33–35). One of these, Tn*5384*, is a composite element mediating resistance to erythromycin as well as to high levels of gentamicin via *aac6-aph2* flanked by two copies of IS*256* (34).

High-level gentamicin resistance has also been reported in the absence of the AAC(6′)-APH(2″) enzyme. Originally described in a strain of *E. casseliflavus*, but subsequently detected in several vancomycin-resistant strains of *E. faecium*, the gene designated *aph(2″)-Id* mediates production of a phosphotransferase resulting in high-level resistance to gentamicin, kanamycin, and tobramycin (among others) but not to neomycin, netilmicin, or amikacin (36). Interestingly, although synergism between ampicillin and amikacin (or neomycin, but not netilmicin) could be demonstrated in the prototype strain, bactericidal synergism between ampicillin and amikacin could *not* be demonstrated against two isolates of *E. faecium* containing this gene (36). The full significance of this observation remains to be explored.

An enzyme initially detected in a strain of *E. gallinarum*, designated APH(2″)-Ic, negates synergistic killing when ampicillin is combined with gentamicin, yet does not cause high-level resistance to the aminoglycoside (37). Synergistic killing of the isolate was observed when ampicillin was combined with netilmicin or amikacin; however, this did not appear to be a consistent phenomenon when other strains were examined (27). Against the original isolate of *E. gallinarum*, the MIC of gentamicin was 256 μg/mL, and for two strains of *E. faecium* and one of *E. faecalis* which contained the gene, the MICs of gentamicin were 256 or 512 μg/mL.

2.3 Clinical Consequences

Because aminoglycosides do not have sufficient activity against enterococci to be useful as single therapeutic agents, the significance of resistance to this class relates to the inability to attain synergistic bactericidal activity (in combination with a cell wall–active agent) which is required for the optimal treatment of enterococcal endocarditis. Within 10 years of the original report of high-level gentamicin resistance in *E. faecalis* from France, in many U.S. medical centers, approximately 25% or more of isolates exhibited this trait (1). In Boston, the first report of a clinical isolate of *E. faecium* with high-level resistance to gentamicin appeared in 1988 (28). However, of strains collected within the next 2 years, more than 60% were highly resistant to gentamicin (38). Generally, high-level resistance to gentamicin precludes synergistic activity of other 2-deoxystreptamine aminoglycosides. However, as the mechanisms of resistance to streptomycin are unrelated, some strains exhibiting high-level resistance to gentamicin will be susceptible to high concentrations of streptomycin, which thus remains an option for therapy (12).

Recognition of novel aminoglycoside resistance mechanisms in enterococci complicates use of time-honored algorithms for predicting the potential for bactericidal synergism based upon knowledge of species and results of screening for resistance to high levels of streptomycin, kanamycin, and gentamicin. Specifically, high-level resistance to gentamicin may not necessarily preclude synergistic activities of other 2-deoxystreptamine aminoglycosides (36), although whether such synergism would be clinically useful has not been proven. On the other hand, absence of high-level resistance to gentamicin does not guarantee synergistic activity with this agent, because some enzymes prevent synergistic activity without raising aminoglycoside MICs above 500 μg/mL (27,37). One instance has been reported of specific lack of synergistic activity of gentamicin (but not tobramycin) against a strain of *E. faecalis* in the absence of detectable modifying enzymes or elevated MIC of gentamicin (39). In this instance, resistance to synergism appeared to be due to a specific defect in the intracellular uptake of gentamicin.

3 RESISTANCE TO CELL WALL–ACTIVE AGENTS

3.1 β-Lactam Antibiotics

3.1.1 Resistance Due to Low-Affinity Penicillin–Binding Proteins

It has been appreciated for years that strains of *E. faecalis* are relatively resistant to penicillins as compared with streptococci. Nevertheless, approximately 90% of isolates are inhibited by ampicillin at concentrations of 1–2 μg/mL or less (40). In turn, *E. faecium* organisms are approximately 10-fold or higher more resistant to penicillins than are *E. faecalis* organisms (1). Binding affinities of radiolabeled penicillin G to cell membranes of enterococci or group D streptococci correlate with the MIC of penicillin against the isolates (41). The major determinant of high levels of resistance to penicillin in *E. faecium* (or *E. hirae*) was shown to be a low-affinity, or slow-binding, penicillin–binding protein (PBP) designated PBP5 (42,43). It is hypothesized that this PBP can carry out critical functions related to peptidoglycan synthesis even in the presence of penicillin at concentrations which saturate (thus inactivate) other PBPs.

From a penicillin-resistant (MIC = 16 μg/mL) strain of *E. hirae*, workers in Belgium identified a low-affinity, 77-kD PBP (termed PBP 3^r), which was found to be immunochemically related to the low-affinity 71-kD PBP5 of *E. hirae* ATCC 9790 (44). At a peptide level, PBP 3^r aligned with sequences of PBP2′, the low-affinity PBP responsible for methicillin-resistance in staphylococci. Substantial similarity was also found based upon comparison of conserved structural motifs (45). In the strain studied, the PBP 3^r gene, as well as *aadE* (streptomycin resistance) and *ermAM* (macrolide resistance), were linked and plasmidborne (46). Later studies demonstrated that the amino acid sequence of plasmidborne PBP3^r of *E. hirae* was 99.8% homologous with that of PBP5 of *E. faecium* (47). This group also found similarity between low-affinity PBP5 of *E. hirae* and staphylococcal PBP2′ (48). Polyclonal antibody raised in rabbits against low-affinity PBP5 of *E. hirae* R40, a resistant derivative of *E. hirae* ATCC

9790, reacted with membrane fractions of all *E. faecium* tested (or their more highly penicillin-resistant derivatives). Reactivity was also seen against those strains of *E. faecalis* from which stable derivatives with higher penicillin resistance levels could be obtained (49). Thus, PBPs conferring resistance to penicillin appear to be common and of relatively conserved structure in this genus.

Synthesis of PBP5 in *E. hirae* appears to be under negative control, as high levels of this protein leading to penicillin resistance in *E. hirae* R40 have been ascribed to an 87–base pair deletion in a sequence (designated *psr*) 1 kb upstream of *pbp5* (50). The *psr* gene product (PBP5 synthesis repressor) may have additional properties related to the physiology of the bacterial cell surface. Strains with truncated repressor regions demonstrate altered spontaneous autolysis in phosphate buffer and increased sensitivity to lysozyme in comparison with strains containing intact *psr* genes (51).

Studies of *E. faecium* isolates with different penicillin MICs have demonstrated the following. Greater amounts of low-affinity PBP5 are produced as levels of resistance increase to a certain point (MICs ≅ 64 μg/mL). When resistance increases to higher levels, smaller amounts of a PBP of even lower affinity are produced (47,52,53). Point mutations in regions of the *pbp5* gene encoding in the C-terminus of the protein are potentially responsible for changes in binding affinity (47,53,54). Transfer of *pbp5* and *psr* (together with genetically linked vancomycin resistance determinants) between trains of *E. faecium* by conjugation has been demonstrated. This process involves transfer of a large segment of chromosomal DNA (55).

As mentioned above, *E. faecalis* organisms are substantially more susceptible to penicillin or ampicillin than are *E. faecium*. Typically, strains of the former species are inhibited by ampicillin at ≤ 2 μg/mL, and almost always at ≤ 4 μg/mL (1). However, strains of *E. faecalis* have been encountered with ampicillin MICs of 32–64 μg/mL. Resistance has been related to increased production of PBP5 and decreased binding of penicillin to PBPs 1 and 6 (56). *E. raffinosus* is another species with characteristically reduced penicillin susceptibility, which is likely due to a low-affinity PBP in this species as well (57).

3.1.2 β-Lactamase Production

Except in very rare instances, resistance to β-lactam antibiotics among enterococci cannot be ascribed to production of β-lactamases. One β-lactamase–producing strain of *E. faecalis* (designated HH22), which was isolated in Houston in the early 1980s, has been extensively characterized (58–61). β-Lactamase production is mediated on a transferable, pheromone response plasmid which also mediates high-level gentamicin resistance. The β-lactamase gene (*blaZ*) was determined to be of staphylococcal origin (60). Constitutive β-lactamase production in this and several other strains of *E. faecalis* (in contrast to inducible production in *S. aureus*) is attributable to the absence or alteration of regulatory genes (*blaI, blaRI*) in most strains examined (62). However, transfer of genes from a constitutive β-lactamase–producing *E. faecalis* with an intact regulatory system into *S. aureus* resulted in inducible β-lactamase production in the latter; strongly suggesting a role for additional, unknown species-related host factors as determinants of enzyme production (63). β-Lactamase production was found to be chromosomally mediated in one cluster of isolates (33). Subsequent work identified genetic determinants for β-lactamase production on a large (approximately 65 kb) chromosomal element, Tn*5385*, which confers resistance to multiple additional antibiotics, including erythromycin (*ermAM*), gentamicin (*aac6'-aph2"*), streptomycin (*aadE*), tetracyclines (*tetM*), and to mercuric chloride (*merRAB*) (64). Strikingly, this composite element consists of staphylococcal as well as enterococcal resistance genes, insertion sequences, and transposons (64).

3.1.3 Resistance Trends and Clinical Significance

Although there is little evidence for change over time in the level of resistance to penicillins among *E. faecalis*, there has been a substantial increase in the level of resistance among strains of *E. faecium*. For strains of this species collected in the 20-year period up to 1988, the MIC_{90} of penicillin was 64 μg/mL. For isolates recovered in the subsequent 2-year period, the MIC_{90} of penicillin was 512 μg/mL (38). Many strains of vancomycin-resistant *E. faecium* encountered today are inhibited by ampicillin only at concentrations of 100 μg/mL or higher; a fact which substantially compromises treatment options for infections due to these isolates (6).

Sporadic isolates of β-lactamase–producing *E. faecalis* have been recovered in the eastern United States, from Lebanon, and from Argentina (65,66). It is important to note that levels of enzyme produced by the strains of *E. faecalis* have generally not raised MICs of penicillin or ampicillin above those of usual isolates (58). Elevated MICs may be detected by high-inoculum testing, but the more readily available technique involves direct detection of β-lactamase activity using the chromogenic substrate nitrocefin. One isolate of a β-lactamase–producing *E. faecium* has been reported thus far (67), but given the high intrinsic resistance to penicillins among current strains of this species, the production of enzyme is not likely to have a major impact. Institutional outbreaks of colonization or colonization with infection have occurred in Boston, Massachusetts, and in Richmond, Virginia (68,69). Although those examples illustrate unequivocally the potential for intrahospital transmission of β-lactamase–producing *E. faecalis*, it is unclear why this does not appear to be an ongoing problem.

3.2 Glycopeptides

Beginning in the mid-1980s, investigators in Europe and in the United States began to encounter strains of enterococci which exhibited resistance to vancomycin or to both vancomycin and teicoplanin (a glycopeptide which is not available for clinical use in the United States). Overlaid on a resistance spectrum that included most currently available antimicrobials, the emergence of vancomycin resistance among enterococci represented a significant negative development.

3.2.1 Original Phenotypic Descriptions of Glycopeptide Resistance Classes

An early practical classification scheme assigned vancomycin-resistant enterococci (VRE) to one of three phenotypic classes. VANA, described strains of *E. faecalis* or *E. faecium* with inducible, relatively high levels of resistance to vancomycin (MIC > 64 μg/mL) and resistance to teicoplanin (>8 μg/mL). Strains of the class VANB were of the same species, generally with lower levels of resistance to vancomycin, but they retained susceptibility to teicoplanin. The VANC class consisted of *E. gallinarum* or *E. casseliflavus* isolates, species considered to be intrinsically resistant to low levels of vancomycin (MIC usually < 64 μg/mL) but susceptible to teicoplanin (70). Subsequently, genetic techniques provided greater power to classify glycopeptide resistance, and revealed shortcomings of the phenotypic classification schemes. First, the range of vancomycin MICs against VANB strains proved to be very broad: from 4 to >1000 μg/mL (71). In addition, mutants resistant to teicoplanin could emerge among VANB strains, which then resemble VANA strains phenotypically (i.e., resistant to both vancomycin and teicoplanin (72). Finally, class C phenotype organisms, *E. gallinarum* or *E. casseliflavus*, can acquire *vanA* or *vanB* resistance genes as well (73–75). Nevertheless, at the present time, genotype is still accurately predicted by phenotypic characteristics for the vast majority of enterococcal isolates (6).

3.2.2 Genotypic Classification and Mechanisms of Resistance

Resistance Due to the vanA Genotype. Studies of a plasmidborne resistance determinant, Tn*1546*, from *E. faecium* BM4147 have produced a detailed understanding of mechanisms responsible for the class A resistance phenotype (76). This transposon contains seven genes involved in glycopeptide resistance (77). The genes *vanS* and *vanR* determine a two-component regulatory system that modulates transcription of downstream elements accounting for inducibility of glycopeptide resistance (78). VanS serves as a histidine kinase sensor protein and VanR as a transcriptional activator (79). Although VanS may mediate phosphorylation of VanR, there is also evidence that in the uninduced state, VanS exerts negative control over VanR (78). Thus, VanS may function not only as a kinase but also as a phosphatase or inhibitor of phosphorylation (80). The specific external signal which leads to induction of glycopeptide resistance is not certain, and various experiments have yielded conflicting results. In one study, several compounds structurally unrelated to vancomycin, including bacitracin, ristocetin, moenomycin, and polymyxin B, demonstrated the capacity to induce the VanS-VanR system in *vanA E. faecium* (81). Others reported that bacitracin, ramoplanin, and penicillin were not inducers for *vanA E. faecalis*, leading them to propose a role for membrane-bound lipid intermediate II in the induction process (82). In a detection system using firefly luciferase under control of the *vanH* promoter, ramoplanin, moenomycin, and bacitracin were inducers, as were various glycopeptides and derivatives irrespective of antimicrobial potency (83). These data supported the possibility that either intermediates or the glycopeptides themselves, or both, could produce an induction signal.

Phosphorylation of VanR increases its affinity of binding to a (*vanH*-) promoter which increases transcription of the *vanHAX* operon (79). The product of *vanA* is a ligase catalyzing synthesis of the depsipeptide D-alanine–D-lactate which, when incorporated into peptidoglycan precursors in place of the normal constituent, D-alanine–D-alanine, yields a vancomycin-binding target which has 1000-fold lower affinity for the glycopeptide than does the normal terminal dipeptide, D-Ala–D-Ala (84). The *vanH*-mediated dehydrogenase contributes to an ample supply of D-lactate for this ligase reaction. The product of *vanX* is a D,D-dipeptidase which degrades competing D-alanine–D-alanine produced by the native ligase, and is essential for expression of vancomycin resistance (85,86).

The level of transcription of the *vanHAX* operon correlates with the level of vancomycin resistance observed (87). Two other genes are not under direct transcriptional control of the *vanH*-promoter. The product of *vanY* is a D,D-carboxypeptidase that can cleave terminal D-alanine from any normal pentapeptide produced; a process which is not essential, but which may contribute modestly to the level of resistance (88,89). The presence of *vanZ* contributes to the full expression of resistance to teicoplanin (87,90).

Induction of vancomycin resistance in enterococci results in morphological changes of the organisms (rod-shaped with increased width) suggesting abnormal autolytic activity (91). Decreased lysis in the presence of penicillin after induction of *vanB* resistance has been described, but this was accompanied only by increased association of D-alanine with lipoteichoic acid, not by changes in the autolytic system of the organism (92). Induced cells of VRE may also display hypersusceptibility to penicillins, and synergistic inhibitory activity of vancomycin–β-lactam combinations has been shown against some strains, suggesting that penicillin-sensitive PBPs are involved in processing the altered peptidoglycan (93).

Examination of VRE collected from U.S. hospitals during the years 1988–1992 and submitted to the Centers for Disease Control and Prevention (CDC) revealed that when *vanA* resistance determinants were present, these were always found on 34- or 60-kb plasmids (16). In a more recent period, examination of vancomycin-resistant *E. faecium* revealed hybridization of a *vanA* probe to the chromosome of 72 strains and to plasmids in only 23; however, this collection was weighted toward clonally related isolates (94). Restriction fragment length polymorphism (RFLP) analysis of amplified Tn*1546*-related elements from Belgian VRE recovered from animals and humans revealed that the (overwhelmingly) predominant types were structurally very closely related (one nucleotide difference in the *vanX* gene) (95). In strains from the northeast United States (96) and from the United Kingdom (97), greater diversity in structure of the Tn*1546*-like element was described.

Resistance Due to the vanB Genotype. The determinant of VANB phenotype resistance are quite analogous to those described for VANA. Studies with *E. faecalis* V583 document the presence of a two-component regulatory system, which is inducible by vancomycin but not teicoplanin (98). The novel ligase protein from this strain, VanB, is 76% homologous at the amino acid level with VanA, and the resulting peptidoglycan also contains D-alanine–D-lactate (99). Analysis of a 630-bp fragment from another strain of *E. faecalis* (SF300) revealed a somewhat different, but closely related, sequence which was designated *vanB2* (100). This fragment shares 96.4% bp homology with the corresponding portion of *vanB*. This resistance trait was transferable in the absence of detectable plasmid transfer. Sequence analyses of long polymerase chain reactions (PCR) amplicons confirm the highly conserved organization of *vanB* (or *vanB1*) and *vanB2* elements (101). Genes homologous with *vanH* and *vanX*, designated $vanH_B$ and $vanX_B$, are present. The gene $vanY_B$ occurs in some, but not all, strains. A gene designated *vanW* is of unknown function. Homologues of *vanZ* are not present in this class (98). The gene cluster resulting in glycopeptide resistance found in *E. faecalis* BM4281 has been designated Tn*1547*, which itself is part of a large, mobile chromosomal element (71,102,103). Plasmid localization of *vanB* resistance determinants has also been documented (104). A *vanB* transposon designated Tn*5382* was found to have inserted close to and downstream of *pbp5* on the chromosome of an isolate of *E. faecium* (55). A surprising and important observation was that vancomycin resistance could be cotransferred to a recipient strain together with penicillin resistance (*pbp5* with *psr*) on a large 130- to 160-kb chromosomal element.

Teicoplanin-resistant mutants selected after exposure to either vancomycin or teicoplanin demonstrate alterations in $VanR_B$-$VanS_B$ sensitivity (82,105). Transconjugants may express altered control of glycopeptide resistance as compared with parental strains. For example, from a constitutively resistant donor, both inducibly and constitutively resistance transconjugants were obtained (106).

vanC Genotypes. Strains of *E. gallinarum* and *E. casseliflavus* demonstrate low-level resistance to vancomycin, but remain susceptible to teicoplanin. MICs of vancomycin are generally ≤ 32 μg/mL, and may actually fall within "susceptible" or "intermediate" ranges (77,107). A ligase gene initially designated *vanC* (now *vanC-1*) was identified in *E. gallinarum* (108). The VanC-1 ligase of *E. gallinarum* results in production not of D-alanine–D-lactate but instead of D-alanine–D-serine, which is also believed to result in weaker binding of vancomycin to the altered pentapeptide precursor (109). D,D-dipeptidase and D,D-carboxypeptidase activities are present as well (109). These dual enzymatic activities are attributable to a single protein, termed $VanXY_c$ (110). Also detected in VANC *E. gallinarum* is *vanT*, a gene mediating production of a serine racemase, which would make available D-serine for use in the ligase reaction (111). VanC-1 ligase coexists in this species with the distantly related native D-alanine–D-alanine ligase (112). Expression of resistance may be inducible or constitutive (113). It is likely that the balance of expression of the two ligase pathways is responsible for the low levels of resistance observed.

The ligase genes *vanC-2* and *vanC-3* have been identified in *E. casseliflavus* and *E. flavescens*, respectively (114). These two genes are nearly identical (>98% sequence identity); they are similar to *vanC-1* of *E. gallinarum* (66% nucleotide identity); and they are found in addition to native ligase genes in their respective species (which were indistinguishable from one another) (114). The VanC-2 ligase, like VanC-1, also favors formation of D-alanine–D-serine (115). Sequence analysis of *vanC-2* amplicons from several isolates revealed up to 7% variation, which is greater than that seen among *vanC-1* isolates (up to 2% variation) (116).

vanD Genotype and Beyond. Another glycopeptide resistance phenotype with constitutive expression of vancomycin resistance (MIC = 64 μg/mL) and with diminished susceptibility to teicoplanin (MIC = 4 μg/mL) was reported in *E. faecium* BM4339 (117). With degenerate primers for ligase genes, a novel gene sequence was identified (*vanD*), the deduced amino acid sequence of which shared 69% identity with VanA and VanB, 43% with VanC, and 34% with the native *E. faecium* ligase. Pentadepsipeptide was detected as the predominant peptidoglycan precursor without evidence of inducibility (117). Organization of the gene cluster was analogous to those other glycopeptide resistance types: $vanR_D$, $vanS_D$, $vanY_D$, $vanH_D$, *vanD*, and $vanX_D$ (although dipeptidase activity was not detected) (118). This strain was also unusual, because a frame shift mutation in the native ligase results in loss of D-alanine–D-alanine ligase activity. Additional *vanD* strains of *E. faecium* have been encountered in Boston (119). These strains were somewhat more resistant to vancomycin (MIC 128–256 μg/mL) and did show evidence of inducible resistance.

Recently, the *vanE* resistance gene has been reported in *E. faecalis* BM4405 (110). This clinical isolate from Chicago exhibited low-level resistance to vancomycin (MIC = 16 μg/mL) and remained susceptible to teicoplanin (MIC = 0.5 μg/mL). Comparison of deduced amino acid sequences of the VanE ligase with those of other glycopeptide resistance ligases revealed closest similarity with VanC (55% identity). Like strains of the VANC class, this *E. faecalis* produced peptidoglycan precursors terminating in D-alanine–D-serine, and exhibited inducible D,D-dipeptidase, D,D-carboxypeptidase, and serine racemase activities (110). As vancomycin-resistent enterococci which defy genotypic classification are still occasionally encountered, it seems likely that as yet undefined resistance determinants exist (120).

The presence of diverse, but related, vancomycin resistance mechanisms raises questions as to the origin of these resistance genes. Homologues of the *vanHAX* gene cluster have been found in glycopeptide-producing stains of *Amycolatopsis* spp. and *Streptomyces toyocaensis* (121, 122). These genes are similar to those of VRE not only in sequence but also in organization. However, differences in G+C content argue against a direct inheritance of VRE genes from the antibiotic producers described.

3.2.3 Clinical Significance

In surveys of U.S. nosocomial bloodstream isolates, 17.7% of the enterococci recovered from 1995 to 1998 were resistant to vancomycin (4). For *E. faecalis*, the most commonly encountered species, the resistance rate was approximately 3%, whereas approximately 50% of *E. faecium* were resistant. Virtually identical results were reported by others (123). The emergence of vancomycin resistance among *E. faecium* is of special concern because of the high prevalence of resistance to multiple agents, including ampicillin, erythromycin, ciprofloxacin, gentamicin, streptomycin, rifampin, and doxycycline (6).

It is possible that some strains of VRE are endowed with specific properties advantageous for establishing endemicity. For example, 3 months after the introduction of a certain clonal type of VRE into a Boston hospital, the new type had become the dominant clone in that institution, as it previously had been in an affiliated institution (124). In the United States, VRE are encountered with substantially greater frequency among hospital inpatient isolates of enterococci, whereas isolates from outpatients are uncommonly (2.5%) vancomycin resistant (125). In northern Europe, *vanA* resistance determinants have been found in meats for sale in markets. In the Netherlands, 70% of retail poultry products were found to contain *vanA* VRE (126). Ingestion of animal VRE by an investigator demonstrated the capacity for animal strains to colonize the human intestine for as long as 3 weeks (127). Even when clonal types of VRE from humans and animals do not overlap, close similarities in *vanA* transposon organization have been observed in some (although not all [97]) studies, indicating the potential for interspecies spread of resistance elements (126,128). By study of RFLP patterns of Tn*1546*-like elements, two transposon types found among isolates from the United States, where there has not been a problem of animal VRE, did not have counterparts isolated from animals (129). Patients who develop gastrointestinal tract colonization through hospital exposure may remain colonized with VRE for many months (130).

The rarity of isolation of VANC type VRE from clinically important specimens and little evidence for transmission have suggested to some that these species should be considered to be distinct from VRE for infection-control purposes (131). However, it is important to note that serious invasive infections by VANC-type isolates to occasionally occur (132).

In addition to the direct threat of infection posed by VRE, their increasing prevalence raises concern that resistance determinants will eventually spread to other gram-positive organisms, including streptococci and staphylococci, that may be inherently more virulent. For example, in the laboratory, vancomycin resistance determinants can be transferred to *Streptococcus pyogenes* and *Listeria monocytogenes* (133), as well as to *S. aureus* (134). Such transfer has not been reported among naturally occurring isolates. On the other hand, a *vanB*-type determinant has been detected in a strain of *S. bovis* (135), and *vanA*-like determinants have been encountered in *Oerskovia turbata* and *Arcanobacterium haemolyticum* (136). A *vanA* gene cluster (*vanRSHAXYZ*) was identified in a clinical isolate of *Bacillus circulans*. This element was distinct, however, from Tn*1546* in that it lacked transposase and resolvase genes (137). Ligase genes homologous with *vanA* and *vanB* (approximately 77 and 69% nucleotide identity, respectively) have been identified in the biopesticide *B. popilliae* (138). The authors had tentatively assigned this gene the designation *vanE*. (Note that this designation was also employed for the gene from *E. faecalis* BM4405 with the VanC-like properties and homologies described above.) Its detection in a specimen from 1945 indicates that it was not introduced from VRE, but might instead reflect a common precursor gene (138).

4 RESISTANCE TO INHIBITORS OF PROTEIN SYNTHESES

4.1 Macrolides, Lincosamides, and Streptogramins

4.1.1 Macrolide Resistance

Erythromycin resistance is very common among current enterococcal isolates, as is cross resistance to the newer macrolides like clarithromycin (139). In a study of 347 strains of *E. faecalis* collected in the United States in the early 1980s, the MIC_{50} and MIC_{75} of erythromycin were 1.56 and 100 μg/mL, respectively (40). More recently, of 403 strains of *E. faecalis* and 90 strains of *E. faecium* collected in Europe during a 1997–1998 surveillance study, only 14.8 and 6.6%, respectively, were susceptible to erythromycin (140). In contrast, erythromycin resistance was found in only 6 of 220 enterococci collected in Washington, DC, in the early 1950s, indicating that resistance is not intrinsic to the genus (141). DNA from the three strains in this collection resistant to both erythromycin and clindamycin hybridized with a probe for the ribosomal methylase gene, *ermAM*, which results in insensitivity to the antibiotics at the 50S ribosomal target site.

MLS_B resistance mediated by *erm* genes in enterococci is inducible in some strains but constitutive in others. Sequence analysis of the *ermAM* region of *E. faecalis* plasmid pAMβ1 revealed deletions in regulatory regions (142). Thus, lack of inducibility in some strains may result from deletions or other alterations leading to truncation of the leader peptide sequence which precedes the *erm* gene and regulates expression of resistance (143). A widely distributed *ermAM*-type MLS_B resistance element, Tn*917* (144,145), has been localized to the bacterial chromosome and to conjugative or nonconjugative enterococcal plasmids (146). An interesting feature of this transposon is that antibiotic exposure promotes transposition. This is postulated to result from antibiotic-promoted transcription extending beyond the antibiotic resistance region to the transposition-related region of the element (144).

Erythromycin resistance determinants can be cotransferred with other antibiotic resistance genes, including those encoding high-level streptomycin and kanamycin resistance (147) or gentamicin resistance (34).

Erythromycin resistance genes (Tn*917*-derived) are also found on multidrug resistance transposons recently identified among enterococci (64).

Other mechanisms of resistance to macrolides have not been completely established in enterococci. Preliminary physiological evidence supports the existence of a macrolide efflux pump, as well as *mef*-like gene sequences, in *E. faecium* (148).

4.1.2 Lincosamide Resistance

E. faecalis organisms are relatively resistant to clindamycin, with most strains being inhibited only at ≥ 16 μg/mL (149). Among isolates collected in Europe during 1997–1998, only 4.4% were susceptible (140). Some strains of *E. faecium* are susceptible to clindamycin. In the study mentioned, 27.5% of *E. faecium* were inhibited at ≤ 0.5 μg/mL; in other collections, approximately 10–15% of *E. faecium* were susceptible (149). Because of the relative resistance of enterococci to lincosamides, it is more difficult to decide whether the MLS_B phenotype in any strain is attributable to *erm* genes, but these are certainly found in enterococci as described above (143).

Resistance due to enzymatic modification of lincosamides has also been described. The gene, *linB*, identified in a strain of *E. faecium* which also contained the *ermAM* gene, encodes a 3-lincosamide *O*-nucleotidyltransferase which adenylates hydroxyl groups on lincomycin and clindamycin (150). This gene was detected by PCR in all 14 (of 508 tested) enterococci which inactivated clindamycin by the Gots' test. All of these were *E. faecium* (150). The resistance determinant could be transferred to *E. faecalis* by conjugation.

4.1.3 Streptogramins

As discussed above, MLS_B resistance attributable to *erm* genes is encountered among enterococci. Ribosomal methylation results in resistance to the streptogramin B antibiotics. Until fairly recently this was primarily of academic interest in view of the limited use of streptogramins in therapy of human infections. The development of quinupristin-dalfopristin, a combination of streptogramin B and streptogramin A components in a 30:70 ratio, changes that situation. This drug is an antimicrobial active against multiply resistant gram-positive organisms. More than 99% of *E. faecium*, including VRE, are susceptible to the agent (151). The two components are not related structurally, but they inhibit ribosomal protein synthesis synergistically (152). Resistance to the streptogramin B component only may affect the level of bactericidal activity of the combination against *E. faecium*, but inhibitory activity of the combination is maintained (153).

Streptogramin B agents are also susceptible to enzymatic inactivation. A gene (*vgb*) initially identified in *S. aureus* (154) and which mediates production of a streptogramin B hydrolase has been encountered in *E.*

faecium (155). In one isolate of *E. faecium*, the *vgb*-like determinant was localized to the bacterial chromosome; in the same strain, the genes *ermAM*, *vanA*, and *satA* (see below) were localized to a large plasmid (156).

Streptogramin A agents are not affected by *erm* genes, but they are susceptible to acetylation mediated by the *satA* gene described in *E. faecium* (157). This gene has been found in animal as well as human isolates of enterococci (158,159). Another streptogramin A acetyltransferase gene, designated *satG*, has been found in quinupristin-dalfopristin–resistant enterococci from sewage, poultry, and hospitals in Germany (160). Efflux mechanisms active against streptogramin A compounds exist in staphylococci but have not yet been described in enterococci. From a small proportion of patients treated with quinupristin-dalfopristin during clinical trials, isolates of *E. faecium* resistant to the agent have been recovered (6). Mechanisms by which resistance emerges under these circumstances have not yet been fully explained. Of note is the fact that *E. faecalis* is typically resistant to quinupristin-dalfopristin (151), and the agent's activity against other enterococcal species is variable.

4.2 Chloramphenicol

Among 152 isolates of *E. faecalis* studied in 1984, the concentrations of chloramphenicol which inhibited growth of 75 and 90% of strains were 6.25 and 12.5 μg/mL, which fall into susceptible and intermediately susceptible ranges, respectively (40). Most strains of vancomycin-resistant *E. faecium* which exhibit resistance to virtually all clinically available antibiotics remain susceptible to chloramphenicol at clinically achievable concentrations (6). As a result, this agent has been used to treat serious VRE infections with modest success (161,162).

Chloramphenicol resistance determinants have been localized to the enterococcal chromosome and to both conjugative and nonconjugative plasmids (146). The drug can be inactivated by enzymatic acetylation (163). Inducibility of enzyme production has been observed (163). At least three distinct classes of *cat* genes mediating production of chloramphenicol acetyltransferases have been recognized on enterococcal plasmids (164).

Evidence has also been presented supporting the existence of active efflux mechanisms for chloramphenicol from *E. faecalis* and *E. faecium* even among strains which were susceptible to the agent by standard criteria (165).

4.3 Tetracyclines

Resistance to tetracyclines is commonly encountered among enterococci. Of recent isolates recovered at our institution, only 33% were susceptible to tetracycline. Resistance to tetracycline and minocycline due to ribosomal protection mechanisms mediated by the genes *tet*(M), *tet*(O) and *tet*(S) has been reported (166). The most common of these is *tet*(M), which is associated with the conjugative transposon Tn*916* (167,168). The transposon can be found on the bacterial chromosome or on plasmids (146,169). Transcription of *tet*(M) is increased by the presence of tetracycline in the growth medium (170). As part of larger transferable resistance units, Tn*916*-like elements may be transferred together with multiple unrelated resistance traits such as vancomycin (*vanB*) and high-level ampicillin resistance in Tn*5382* of *E. faecium* (55).

Enterococci may also demonstrate tetracycline resistance by drug efflux mechanisms (171). Most commonly, resistance due to efflux is mediated by *tet*(L), although *tet*(K) has been reported in the genus (166). Two or three resistance genes (e.g., *tet*[L] + *tet*[M] or *tet*[L] + *tet*[M] + *tet*[S]) can be detected simultaneously in some strains (166). *Tet*(L) may be plasmidborne or chromosomal (172).

5 RESISTANCE TO FLUOROQUINOLONES

The introduction of the fluoroquinolone class of antimicrobials into clinical practice provided an additional option for oral therapy of enterococcal infections. Early studies with ciprofloxacin in vitro indicated that approximately 90% of *E. faecalis* were susceptible at ≤ 2 μg/mL (173,174). However, resistance to this class emerged rapidly. In one French hospital, ciprofloxacin resistance increased from 0% of *E. faecalis* isolated in 1986 to 24% of those isolated in 1992 (175). To a large extent, such observations reflect clonal spread of resistant isolates. Although newer fluoroquinolones with increased potency against gram-positive organisms are under development, strains resistant to the earlier fluoroquinolones generally show reduced susceptibility to the newer agents as well even when MICs remain within the tentatively established susceptible range (176).

Mutations in genes encoding proteins for the A-subunit of DNA gyrase or the ParC-subunit of topoisomerase IV are the most important determinants of fluoroquinolone resistance among bacterial species examined thus far. Mutations in *gyrA* leading to amino acid substitutions at codons 83 (from serine) or 87 (from glutamic acid) have been detected in ciprofloxacin-resistant strains of *E. faecalis* (175,177). With the increasing appreciation of the role of topoisomerase IV in fluoroquinolone resistance, studies examining *parC* of *E. faecalis* demonstrated mutations leading to amino acid substitutions at codons 80 (from serine) or 87 (from glutamic acid), or both, in strains with reduced susceptibility (178). In almost all resistant isolates examined in that study, mutations were detected in both *parC* and *gyrA*. Mutations leading to amino acid substitutions at GyrA codons 83 or 87 *and* at ParC codon 80 were detected in 10 of 11 strains of *E. faecium* resistant to ciprofloxacin and trovafloxacin (179).

The absence of such mutations in some strains with low-level resistance to fluoroquinolones has suggested the possibility that active efflux of these agents may account for resistance in some strains. The active efflux of norfloxacin from *E. faecalis* and *E. faecium* has now been demonstrated (165). MICs of norfloxacin against such strains can fall within the susceptible or intermediately susceptible range. The precise nature of a fluoroquinolone efflux pump in enterococci has not yet been defined, but it seems quite probable that such pumps are common and contribute modestly to resistance in the genus.

6 CONCLUSIONS

Resistance, or at least reduced susceptibility, to multiple antibiotics is so pervasive among enterococci that it is often considered to be an inherent characteristic of the genus. As we have discussed, however, there is clear evidence of increasing resistance over the years. Exactly why resistance is so prevalent in the genus is not known for certain. It is obvious that the existence of mobile, transferable elements bearing multiple unrelated resistance determinants has contributed to the accumulation of resistance traits in these organisms. As the number and different types of resistance determinants accumulate in any clone, it becomes less likely that antibiotic exposures favoring further selection of the clone can be avoided. The unfortunate consequence of such accumulation of resistance traits in enterococci, particularly in *E. faecium*, is that strains now exist which are resistant to most or all of the antibiotics which can be given safely to patients. As a result, increasing hope is being placed on the development of novel antimicrobials with activity against multiply resistant strains.

In 1999, the streptogramin quinupristin-dalfopristin (6) was approved in the United States for use in the treatment of serious or life-threatening infections associated with vancomycin-resistant *E. faecium* bacteremia. Investigational agents with in vitro activity against enterococci and which have been used in humans (although not necessarily to treat enterococcal infections) include the oxazolidinone linezolid (180), evernimicin (SCH 27899) (181), the glycopeptide LY333328 (182,183), and the cyclic lipopeptide daptomycin (184). One or more of these agents might eventually fill the gap left by resistance to penicillins, glycopeptides, and other agents which have been the mainstays of treatment of enterococcal infections. However, even as these novel agents are being investigated, and before widespread exposure of the human flora to the compounds, enterococcal strains with reduced susceptibility to these agents have already been encountered in nature or generated in the laboratory.

NOTE ADDED IN PROOF

Another VanD enterococcus, *E. faecium* BM 4416, has been characterized. MICs of vancomycin and teicoplanin were 128 and 64 μg/mL, respectively. The native ddl ligase was inactive through insertion of IS*19*. In contrast to *E. faecium* BM 4339, in this isolate, $VanX_D$ D,D-dipeptidase activity was measurable (Perichon B, Casadewall B, Reynolds P, Courvalin P. Glycopeptide-resistant *Enterococcus faecium* BM 4416 is a VanD-type strain with an impaired D-alanine:D-alanine ligase. Antimicrob Agents Chemother 2000; 44:1346–1348). The vancomycin resistance determinant tentatively designated *vanE* in reference 138 has been renamed *vanF* (Patel R, Piper K, Cockerill FR III, et al. The biopesticide *Paenibacillus popilliae* has a vancomycin resistance gene cluster homologous to the enterococcal VanA vancomycin resistance gene cluster. Antimicrob Agents Chemother 2000; 44:705–709). A new glycopeptide resistance gene, designated *vanG*, was identified in a strain of *E. faecalis* recovered in Brisbane, Australia (MICs of vancomycin 16 μg/mL, teicoplanin 0.5 μg/mL). The cluster contained genes for a two-component regulatory system, two dipeptidases, and a racemase (McKessar SJ, Berry Am, Bell JM, et al. Genetic characterization of vanG, a novel vancomycin resistance locus of *Enterococcus faecalis*. Antimicrob Agents Chemother 2000; 44:3224–3228).

The putative *E. faecium* efflux determinant described in reference 148 has been named *msrC* based on similarity to the *msrA* efflux determinant of *Staphylococcus epidermidis* (Portillo A, Ruiz-Larrea F, Zaragaza M, et al. Macrolide resistance genes in *Enterococcus* spp. Antimicrob Agents Chemother 2000; 44:967–971). This gene was found in 100% of 233 strains of *E. faecium* Disruption of *msrC* enhanced susceptibility to erythromycin, azithromycin, tylosin, and quinupristin in this species (Singh KV, Malathum K, Murray BE. Disruption of an *Enterococcus faecium* species-specific gene, a homologue of acquired macrolide resistance genes of staphylococci, is associated with an increase in macrolide susceptibility. Antimicrob Agents Chemother 2001; 45:263–266). The acetyltransferase genes *satA* and *satG* conferring dalfopristin resistance have now been renamed *vat*(D) and *vat*(E), respectively. Both are found in some clinical isolates of *E. faecium* (Solatani M, Beighton D, Philpott-Howard, Woodford N. Mechanisms of resistance to quinupristin-dalfopristin among isolates of *Enterococcus faecium* from animals, raw meat and hospital patients in Western Europe. Antimicrob Agents Chemother 2000; 44:433–436). Recent work casts doubt on the role of *psr* as a repressor of *pbp5* transcription (Rice LB, Carias LL, Hutton-Thomas R, et al. Antimicrob Agents Chemother 2001; 45:1480–1486.)

Both quinupritin-dalfopristin and linezolid are now approved by the U.S. Food and Drug Administration. Clinical development of SCH 27899, which had been named evernimicin, has been discontinued. The glycopeptide LY333328 has been named oritavancin.

REFERENCES

1. Eliopoulos GM. Antibiotic resistance in Enterococcus species: an update. Curr Clin Top Infect Dis 1996; 16:21–51.
2. Eliopoulos GM. Vancomycin-resistant enterococci. Mechanism and clinical relevance. Infect Dis Clin North Am 1997; 11:851–865.
3. Devriese LA, Ieven M, Goossens H, Vandamme P, Pot B, Hommez J, Haesebrouck F. Presence of vancomycin-resistant enterococci in farm and pet animals. Antimicrob Agents Chemother 1996; 40:2285–2287.
4. Edmond MB, Wallace SE, McClish DK, Pfaller MA, Jones RN, Wenzel RP. Nosocomial bloodstream infections in United States hospitals: a three-year analysis. Clin Infect Dis 1999; 29:239–244.
5. Jones RN, Marshall SA, Pfaller MA, Wilke WW, Hollis RJ, Erwin ME, Edmond MB, Wenzel RP. Nosocomial enterococcal blood stream infections in the SCOPE Program: antimicrobial resistance, species occurrence, molecular testing results, and laboratory testing accuracy. SCOPE Hospital Study Group. Diagn Microbiol Infect Dis 1997; 29:95–102.
6. Eliopoulos GM, Wennersten CB, Gold HS, Schulin T, Souli M, Farris MG, Cerwinka S, Nadler HL, Dowzicky M, Talbot GH, Moellering RC, Jr. Characterization of vancomycin-resistant *Enterococcus faecalis* isolates from the United States and their susceptibility *in vitro* to dalfopristin-quinupristin. Antimicrob Agents Chemother 1998; 42:1088–1092.
7. Edmond MB, Ober JF, Dawson JD, Weinbaum DL, Wenzel RP, Vancomycin-resistant enterococcal bacteremia: natural history and attributable mortality. Clin Infect Dis 1996; 23:1234–1239.
8. Linden PK, Pasculle AW, Manez R, Kramer DJ, Fung JJ, Pinna AD, Kusne S. Differences in outcomes for patients with bacteremia due to vancomycin-resistant *Enterococcus faecium* or vancomycin-susceptible *E. faecium*. Clin Infect Dis 1996; 22:663–670.
9. Lucas GM, Lechtzin N, Puryear DW, Yau LL, Flexner CW, Moore RD. Vancomycin-resistant and vancomycin-susceptible enterococcal bacteremia: comparison of clinical features and outcomes. Clin Infect Dis 1998; 26:1127–1133.
10. Storch GA, Krogstad DJ. Antibiotic-induced lysis of enterococci. J Clin Invest 1981; 68:639–645.
11. Eliopoulos GM. Increasing problems in the therapy of enterococcal infections [editorial]. Eur J Clin Microbiol Infect Dis 1993; 12:409–412.
12. Eliopoulos GM. Aminoglycoside resistant enterococcal endocarditis. Infect Dis Clin North Am 1993; 7:117–133.
13. Moellering RC, Jr., Weinberg AN. Studies on antibiotic synergism against enterococci. II. Effect of various antibiotics on the uptake of ^{14}C-labeled streptomycin by enterococci. J Clin Invest 1971; 50:2580–2584.
14. Eliopoulos GM, Farber BF, Murray BE, Wennersten C, Moellering RC, Jr.

Ribosomal resistance of clinical enterococcal to streptomycin isolates. Antimicrob Agents Chemother 1984; 25:398–399.
15. Krogstad DJ, Korfhagen TR, Moellering RC, Jr., Wennersten C, Swartz MN. Aminoglycoside-inactivating enzymes in clinical isolates of *Streptococcus faecalis*. An explanation for resistance to antibiotic synergism. J Clin Invest 1978; 62:480–486.
16. Clark NC, Cooksey RC, Hill BC, Swenson JM, Tenover FC. Characterization of glycopeptide-resistant enterococci from U.S. hospitals. Antimicrob Agents Chemother 1993; 37:2311–2317.
17. Moellering RC, Jr., Korzeniowski OM, Sande MA, Wennersten CB. Species-specific resistance to antimicrobial synergism in *Streptococcus faecium* and *Streptococcus faecalis*. J Infect Dis 1979; 140:203–208.
18. McKay GA, Roestamadji J, Mobashery S, Wright GD. Recognition of aminoglycoside antibiotics by enterococcal-staphylococcal aminoglycoside 3′-phosphotransferase type IIIa: role of substrate amino groups. Antimicrob Agents Chemother 1996; 40:2648–2650.
19. Calderwood SB, Wennersten C, Moellering RC Jr. Resistance to antibiotic synergism in *Streptococcus faecalis*: further studies with amikacin and with a new amikacin derivative, 4′-deoxy,6′-N-methylamikacin. Antimicrob Agents Chemother 1981; 19:549–555.
20. Thauvin C, Eliopoulos GM, Wennersten C, Moellering RCJ. Antagonistic effect of penicillin-amikacin combinations against enterococci. Antimicrob Agents Chemother 1985; 28:78–83.
21. Carlier C, Courvalin P. Emergence of 4′,4″-aminoglycoside nucleotidyltransferase in enterococci. Antimicrob Agents Chemother 1990; 34:1565–1569.
22. Costa Y, Galimand M, Leclercq R, Duval J, Courvalin P. Characterization of the chromosomal *aac(6′)-Ii* gene specific for *Enterococcus faecium*. Antimicrob Agents Chemother 1993; 37:1896–1903.
23. Wright GD, Ladak P. Overexpression and characterization of the chromosomal aminoglycoside 6′-N-acetyltransferase from *Enterococcus faecium*. Antimicrob Agents Chemother 1997; 41:956–960.
24. Horodniceanu T, Bougueleret L, El-Solh N, Bieth G, Delbos F. High-level, plasmid-borne resistance to gentamicin in *Streptococcus faecalis* subsp. zymogenes. Antimicrob Agents Chemother 1979; 16:686–689.
25. Ferretti JJ, Gilmore KS, Courvalin P. Nucleotide sequence analysis of the gene specifying the bifunctional 6′-aminoglycoside acetyltransferase 2″-aminoglycoside phosphotransferase enzyme in *Streptococcus faecalis* and identification and cloning of gene regions specifying the two activities. J Bacteriol 1986; 167:631–638.
26. Daigle DM, Hughes DW, Wright GD. Prodigious substrate specificity of AAC(6′)-APH(2″), an aminoglycoside antibiotic resistance determinant in enterococci and staphylococci. Chem Biol 1999; 6:99–110.
27. Chow JW, Donabedian SM, Clewell DB, Sahm DF, Zervos MJ. In vitro susceptibility and molecular analysis of gentamicin-resistant enterococci. Diagn Microbiol Infect Dis 1998; 32:141–146.
28. Eliopoulos GM, Wennersten C, Zighelboim-Daum S, Reiszner E, Goldmann

D, Moellering RC Jr. High-level resistance to gentamicin in clinical isolates of *Streptococcus (Enterococcus) faecium*. Antimicrob Agents Chemother 1988; 32:1528–1532.
29. Straut M, de Cespedes G, Horaud T. Plasmid-borne high-level resistance to gentamicin in *Enterococcus hirae, Enterococcus avium*, and *Enterococcus raffinosus*. Antimicrob Agents Chemother 1996; 40:1263–1265.
30. Hodel-Christian SL, Murray BE. Characterization of the gentamicin resistance transposon Tn5281 from *Enterococcus faecalis* and comparison to staphylococcal transposons Tn4001 and Tn4031. Antimicrob Agents Chemother 1995; 35:1147–1152.
31. Casetta A, Hoi AB, de Cespedes G, Horaud T. Diversity of structures carrying the high-level gentamicin resistance gene (*aac6-aph2*) in *Enterococcus faecalis* strains isolated in France. Antimicrob Agents Chemother 1998; 42: 2889–2892.
32. Patterson JE, Masecar BL, Kauffman CA, Schaberg DR, Hierholzer WJ Jr, Zervos MJ. Gentamicin resistance plasmids of enterococci from diverse geographic areas are heterogeneous. J Infect Dis 1988; 158:212–216.
33. Rice LB, Eliopoulos GM, Wennersten C, Goldmann D, Jacoby GA, Moellering RC, Jr. Chromosomally mediated beta-lactamase production and gentamicin resistance in *Enterococcus faecalis*. Antimicrob Agents Chemother 1991; 35:272–276.
34. Rice LB, Carias LL, Marshall SH. Tn5384, a composite enterococcal mobile element conferring resistance to erythromycin and gentamicin whose ends are directly repeated copies of IS256. Antimicrob Agents Chemother 1995; 39:1147–1153.
35. Thal LA, Chow JW, Clewell DB, Zervos MJ. Tn924, a chromosome-borne transposon encoding high-level gentamicin resistance in *Enterococcus faecalis*. Antimicrob Agents Chemother 1994; 38:1152–1156.
36. Tsai SF, Zervos MJ, Clewell DB, Donabedian SM, Sahm DF, Chow JW. A new high-level gentamicin resistance gene, *aph(2")-Id*, in *Enterococcus* spp. Antimicrob Agents Chemother 1998; 42:1229–1232.
37. Chow JW, Zervos MJ, Lerner SA, Thal LA, Donabedian SM, Jaworski DD, Tsai S, Shaw KJ, Clewell DB. A novel gentamicin resistance gene in Enterococcus. Antimicrob Agents Chemother 1997; 41:511–514.
38. Grayson ML, Eliopoulos GM, Wennersten CB, Ruoff KL, De Girolami PC, Ferraro MJ, Moellering RC, Jr. Increasing resistance to beta-lactam antibiotics among clinical isolates of *Enterococcus faecium*: a 22-year review at one institution. Antimicrob Agents Chemother 1991; 35:2180–2184.
39. Moellering RC, Jr, Murray BE, Schoenbaum SC, Adler J, Wennersten CB. A novel mechanism of resistance to penicillin-gentamicin synergism in *Streptococcus faecalis*. J Infect Dis 1980; 141:81–86.
40. Tofte RW, Solliday JA, Crossley KB. Susceptibilities of enterococci to twelve antibiotics. Antimicrob Agents Chemother 1984; 25:532–533.
41. Williamson R, Calderwood SB, Moellering RC, Jr, Tomasz A. Studies on the mechanism of intrinsic resistance to beta-lactam antibiotics in group D streptococci. J Gen Microbiol 1983; 129:813–822.

42. Fontana R, Cerini R, Longoni P, Grossato A, Canepari P. Identification of a streptococcal penicillin-binding protein that reacts very slowly with penicillin. J Bacteriol 1983; 155:1343–1350.
43. Williamson R, le Bouguenec C, Gutmann L, Horaud T. One or two low affinity penicillin-binding proteins may be responsible for the range of susceptibility of *Enterococcus faecium* to benzylpenicillin. J Gen Microbiol 1985; 131:1933–1940.
44. Piras G, el Kharroubi A, van Beeumen J, Coeme E, Coyette J, Ghuysen JM. Characterization of an *Enterococcus hirae* penicillin-binding protein 3 with low penicillin affinity. J Bacteriol 1990; 172:6856–6862.
45. Piras G, Raze D, el Kharroubi A, Hastir D, Englebert S, Coyette J, Ghuysen JM. Cloning and sequencing of the low-affinity penicillin-binding protein 3r-encoding gene of *Enterococcus hirae* S185: modular design and structural organization of the protein. J Bacteriol 1993; 175:2844–2852.
46. Raze D, Dardenne O, Hallut S, Martinez-Bueno M, Coyette J, Ghuysen JM. The gene encoding the low-affinity penicillin-binding protein 3r in *Enterococcus hirae* S185R is borne on a plasmid carrying other antibiotic resistance determinants. Antimicrob Agents Chemother 1998; 42:534–539.
47. Zorzi W, Zhou XY, Dardenne O, Lamotte J, Raze D, Pierre J, Gutmann L, Coyette J. Structure of the low-affinity penicillin-binding protein 5 PBP5fm in wild-type and highly penicillin-resistant strains of *Enterococcus faecium*. J Bacteriol 1996; 178:4948–4957.
48. el Kharroubi A, Jacques P, Piras G, Van Beeumen J, Coyette J, Ghuysen JM. The *Enterococcus hirae* R40 penicillin-binding protein 5 and the methicillin-resistant *Staphylococcus aureus* penicillin-binding protein 2′ are similar. Biochem J 1991; 280:463–469.
49. Ligozzi M, Aldegheri M, Predari SC, Fontana R. Detection of penicillin-binding proteins immunologically related to penicillin-binding protein 5 of *Enterococcus hirae* ATCC 9790 in *Enterococcus faecium* and *Enterococcus faecalis*. FEMS Microbiol Lett 1991; 67:335–339.
50. Ligozzi M, Pittaluga F, Fontana R. Identification of a genetic element (psr) which negatively controls expression of *Enterococcus hirae* penicillin-binding protein 5. J Bacteriol 1993; 175:2046–2051.
51. Massidda O, Kariyama R, Daneo-Moore L, Shockman GD. Evidence that the PBP 5 synthesis repressor (psr) of *Enterococcus hirae* is also involved in the regulation of cell wall composition and other cell wall-related properties. J Bacteriol 1996; 178:5272–5278.
52. Fontana R, Aldegheri M, Ligozzi M, Lopez H, Sucari A, Satta G. Overproduction of a low-affinity penicillin-binding protein and high-level ampicillin resistance in *Enterococcus faecium*. Antimicrob Agents Chemother 1994; 38:1980–1983.
53. Rybkine T, Mainardi JL, Sougakoff W, Collatz E, Gutmann L. Penicillin-binding protein 5 sequence alterations in clinical isolates of *Enterococcus faecium* with different levels of beta-lactam resistance. J Infect Dis 1998; 178:159–163.
54. Ligozzi M, Pittaluga F, Fontana R. Modification of penicillin-binding protein

5 associated with high-level ampicillin resistance in *Enterococcus faecium*. Antimicrob Agents Chemother 1996; 40:354–357.
55. Carias LL, Rudin SD, Donskey CJ, Rice LB. Genetic linkage and cotransfer of a novel, vanB-containing transposon (Tn5382) and a low-affinity penicillin-binding protein 5 gene in a clinical vancomycin-resistant *Enterococcus faecium* isolate. J Bacteriol 1998; 180:4426–4434.
56. Cercenado E, Vicente MF, Diaz MD, Sanchez-Carrillo C, Sanchez-Rubiales M. Characterization of clinical isolates of beta-lactamase-negative, highly ampicillin-resistant *Enterococcus faecalis*. Antimicrob Agents Chemother 1996; 40:2420–2422.
57. Grayson ML, Eliopoulos GM, Wennersten CB, Ruoff KL, Klimm K, Sapico FL, Bayer AS, Moellering RC, Jr. Comparison of *Enterococcus raffinosus* with *Enterococcus avium* on the basis of penicillin susceptibility, penicillin-binding protein analysis, and high-level aminoglycoside resistance. Antimicrob Agents Chemother 1991; 35:1408–1412.
58. Murray BE, Mederski-Samaroj B. Transferable beta-lactamase. A new mechanism for *in vitro* penicillin resistance in *Streptococcus faecalis*. J Clin Invest 1983; 72:1168–1171.
59. Murray BE, Church DA, Wanger A, Zscheck K, Levison ME, Ingerman MJ, Abrutyn E, Mederski-Samoraj B. Comparison of two beta-lactamase-producing strains of *Streptococcus faecalis*. Antimicrob Agents Chemother 1986; 30:861–864.
60. Murray BE, Mederski-Samoraj B, Foster SK, Brunton JL, Harford P. *In vitro* studies of plasmid-mediated penicillinase from *Streptococcus faecalis* suggest a staphylococcal origin. J Clin Invest 1986; 77:289–293.
61. Murray BE, An FY, Clewell DB. Plasmids and pheromone response of the beta-lactamase producer *Streptococcus (Enterococcus) faecalis* HH22. Antimicrob Agents Chemother 1988; 32:547–551.
62. Tomayko JF, Zschek KK, Singh KV, Murray BE. Comparison of the beta-lactamase gene cluster in clonally distinct strains of *Enterococcus faecalis*. Antimicrob Agents Chemother 1996; 40:1170–1174.
63. Okamoto R, Okubo T, Inoue M. Detection of genes regulating beta-lactamase production in *Enterococcus faecalis* and *Staphylococcus aureus*. Antimicrob Agents Chemother 1996; 40:2550–2554.
64. Rice LB, Carias LL. Transfer of Tn5385, a composite, multiresistance chromosomal element from *Enterococcus faecalis*. J Bacteriol 1998; 180:714–721.
65. Murray BE, Singh KV, Markowitz SM, Lopardo HA, Patterson JE, Zervos MJ, Rubeglio E, Eliopoulos GM, Rice LB, Goldstein FW, et al. Evidence for clonal spread of a single strain of beta-lactamase–producing *Enterococcus (Streptococcus) faecalis* to six hospitals in five states. J Infect Dis 1991; 163:780–785.
66. Murray BE, Lopardo HA, Rubeglio EA, Frosolono M, Singh KV. Intrahospital spread of a single gentamicin-resistant, beta-lactamase–producing strain of *Enterococcus faecalis* in Argentina. Antimicrob Agents Chemother 1992; 36: 230–232.
67. Coudron PE, Markowitz SM, Wong ES. Isolation of a beta-lactamase–

producing, aminoglycoside-resistant strain of *Enterococcus faecium*. Antimicrob Agents Chemother 1992; 36:1125–1126.

68. Rinehart E, Smith NE, Wennersten C, Gorss E, Freeman J, Eliopoulos GM, Moellering RC, Jr, Goldmann DA. Rapid dissemination of beta-lactamase–producing, aminoglycoside-resistant *Enterococcus faecalis* among patients and staff on an infant-toddler surgical ward. N Engl J Med 1990; 323:1814–1818.
69. Markowitz SM, Wells VD, Williams DS, Stuart CG, Coudron PE, Wong ES. Antimicrobial susceptibility and molecular epidemiology of beta-lactamase–producing, aminoglycoside-resistant isolates of *Enterococcus faecalis*. Antimicrob Agents Chemother 1991; 35:1075–1080.
70. Leclercq R. Enterococci acquire new kinds of resistance. Clin Infect Dis 1997; 24 Suppl 1:S80–84.
71. Quintiliani R, Jr, Evers S, Courvalin P. The *vanB* gene confers various levels of self-transferable resistance to vancomycin in enterococci. J Infect Dis 1993; 167:1220–1223.
72. Hayden MK, Trenholme GM, Schultz JE, Sahm DF. *In vivo* development of teicoplanin resistance in a VanB *Enterococcus faecium* isolate. J Infect Dis 1993; 167:1224–1227.
73. Liassine N, Frei R, Jan I, Auckenthaler R. Characterization of glycopeptide-resistant enterococci from a Swiss hospital. J Clin Microbiol 1998; 36:1853–1858.
74. Poulsen RL, Pallesen LV, Frimodt-Moller N, Espersen F. Detection of clinical vancomycin-resistant enterococci in Denmark by multiplex PCR and sandwich hybridization. Apmis 1999; 107:404–412.
75. Dutka-Malen S, Blaimont B, Wauters G, Courvalin P. Emergence of high-level resistance to glycopeptides in *Enterococcus gallinarum* and *Enterococcus casseliflavus*. Antimicrob Agents Chemother 1994; 38:1675–1677.
76. Leclercq R, Derlot E, Duval J, Courvalin P. Plasmid-mediated resistance to vancomycin and teicoplanin in *Enterococcus faecium*. N Engl J Med 1988; 319: 157–161.
77. Leclercq R, Courvalin P. Resistance to glycopeptides in enterococci. Clin Infect Dis 1997; 24:545–554.
78. Arthur M, Depardieu F, Gerbaud G, Galimand M, Leclercq R, Courvalin P. The VanS sensor negatively controls VanR-mediated transcriptional activation of glycopeptide resistance genes of Tn1546 and related elements in the absence of induction. J Bacteriol 1997; 179:97–106.
79. Holman TR, Wu Z, Wanner BL, Walsh CT. Identification of the DNA-binding site for the phosphorylated VanR protein required for vancomycin resistance in *Enterococcus faecium*. Biochemistry 1994; 33:4625–4631.
80. Haldimann A, Fisher SL, Daniels LL, Walsh CT, Wanner BL. Transcriptional regulation of the *Enterococcus faecium* BM4147 vancomycin resistance gene cluster by the VanS-VanR two-component regulatory system in *Escherichia coli* K-12. J Bacteriol 1997; 179:5903–5913.
81. Lai MH, Kirsch DR. Induction signals for vancomycin resistance encoded by

the *vanA* gene cluster in *Enterococcus faecium*. Antimicrob Agents Chemother 1996; 40:1645–1648.
82. Baptista M, Depardieu F, Courvalin P, Arthur M. Specificity of induction of glycopeptide resistance genes in *Enterococcus faecalis*. Antimicrob Agents Chemother 1996; 40:2291–2295.
83. Grissom-Arnold J, Alborn WE, Jr, Nicas TI, Jaskunas SR. Induction of VanA vancomycin resistance genes in *Enterococcus faecalis*: use of a promoter fusion to evaluate glycopeptide and nonglycopeptide induction signals. Microb Drug Resist 1997; 3:53–64.
84. Bugg TD, Wright GD, Dutka-Malen S, Arthur M, Courvalin P, Walsh CT. Molecular basis for vancomycin resistance in *Enterococcus faecium* BM4147: biosynthesis of a depsipeptide peptidoglycan precursor by vancomycin resistance proteins VanH and VanA. Biochemistry 1991; 30:10408–10415.
85. Reynolds PE, Depardieu F, Dutka-Malen S, Arthur M, Courvalin P. Glycopeptide resistance medicated by enterococcal transposon Tn1546 requires production of VanX for hydrolysis of D-alanyl-D-alanine. Mol Microbiol 1994; 13:1065–1070.
86. Wu Z, Wright GD, Walsh CT. Overexpression, purification, and characterization of VanX, a D-D-dipeptidase which is essential for vancomycin resistance in *Enterococcus faecium* BM4147. Biochemistry 1995; 34:2455–2463.
87. Arthur M, Depardieu F, Reynolds P, Courvalin P. Quantitative analysis of the metabolism of soluble cytoplasmic peptidoglycan precursors of glycopeptide-resistant enterococci. Mol Microbiol 1996; 21:33–44.
88. Arthur M, Molinas C, Courvalin P. Sequence of the *vanY* gene required for production of a vancomycin-inducible D,D-carboxypeptidase in *Enterococcus faecium* BM4147. Gene 1992; 120:111–114.
89. Arthur M, Depardieu F, Snaith HA, Reynolds PE, Courvalin P. Contribution of VanY D,D-carboxypeptidase to glycopeptide resistance in *Enterococcus faecalis* by hydrolysis of peptidoglycan precursors. Antimicrob Agents Chemother 1994; 38:1899–1903.
90. Arthur M, Depardieu F, Molinas C, Reynolds P, Courvalin P. The *vanZ* gene of Tn1546 from *Enterococcus faecium* BM4147 confers resistance to teicoplanin. Gene 1995; 154:87–92.
91. Lorian V, Fernandes F. The effect of vancomycin on the structure of vancomycin-susceptible and -resistant *Enterococcus faecium* strains [letter]. Antimicrob Agents Chemother 1997; 41:1410–1411.
92. Gutmann L, Al-Obeid S, Billot-Klein D, Ebnet E, Fischer W. Penicillin tolerance and modification of lipoteichoic acid associated with expression of vancomycin resistance in VanB-type *Enterococcus faecium* D366. Antimicrob Agents Chemother 1996; 40:257–259.
93. al-Obeid S, Billot-Klein D, van Heijenoort J, Collatz E, Gutmann L. Replacement of the essential penicillin-binding protein 5 by high-molecular mass PBPs may explain vancomycin-beta-lactam synergy in low-level vancomycin-resistant *Enterococcus faecium* D366. FEMS Microbiol Lett 1992; 70:79–84.
94. Thal L, Donabedian S, Robinson-Dunn B, Chow JW, Dembry L, Clewell DB, Alshab D, Zervos MJ. Molecular analysis of glycopeptide-resistant *Entero-*

coccus faecium isolates collected from Michigan hospitals over a 6-year period. J Clin Microbiol 1998; 36:3303–3308.
95. Descheemaeker PR, Chapelle S, Devriese LA, Butaye P, Vandamme P, Goossens H. Comparison of glycopeptide-resistant *Enterococcus faecium* isolates and glycopeptide resistance genes of human and animal origins. Antimicrob Agents Chemother 1999; 43:2032–2037.
96. Handwerger S, Skoble J, Discotto LF, Pucci MJ. Heterogeneity of the *vanA* gene cluster in clinical isolates of enterococci from the northeastern United States. Antimicrob Agents Chemother 1995; 39:362–368.
97. Woodford N, Adebiyi AM, Palepou MF, Cookson BD. Diversity of VanA glycopeptide resistance elements in enterococci from humans and nonhuman sources. Antimicrob Agents Chemother 1998; 42:502–508.
98. Evers S, Courvalin P. Regulation of VanB-type vancomycin resistance gene expression by the VanS(B)-VanR (B) two-component regulatory system in *Enterococcus faecalis* V583. J Bacteriol 1996; 178:1302–1309.
99. Evers S, Reynolds PE, Courvalin P. Sequence of the *vanB* and *ddl* genes encoding D-alanine:D-lactate and D-alanine:D-alanine ligases in vancomycin-resistant *Enterococcus faecalis* V583. Gene 1994; 140:97–102.
100. Gold HS, Unal S, Cercenado E, Thauvin-Eliopoulos C, Eliopoulos GM, Wennersten CB, Moellering RC, Jr. A gene conferring resistance to vancomycin but not teicoplanin in isolates of *Enterococcus faecalis* and *Enterococcus faecium* demonstrates homology with *vanB*, *vanA*, and *vanC* genes of enterococci. Antimicrob Agents Chemother 1993; 37:1604–1609.
101. Dahl KH, Simonsen GS, Olsvik O, Sundsfjord A. Heterogeneity in the *vanB* gene cluster of genomically diverse clinical strains of vancomycin-resistant enterococci. Antimicrob Agents Chemother 1999; 43:1105–1110.
102. Quintiliani R, Jr, Courvalin P. Conjugal transfer of the vancomycin resistance determinant *vanB* between enterococci involves the movement of large genetic elements from chromosome to chromosome. FEMS Microbiol Lett 1994; 119:359–363.
103. Quintiliani R, Jr, Courvalin P. Characterization of Tn1547, a composite transposon flanked by the IS16 and IS256-like elements, that confers vancomycin resistance in *Enterococcus faecalis* BM4281. Gene 1996; 172:1–8.
104. Rice LB, Carias LL, Donskey CL, Rudin SD. Transferable, plasmid-mediated *vanB*-type glycopeptide resistance in *Enterococcus faecium*. Antimicrob Agents Chemother 1998; 42:963–964.
105. Baptista M, Rodrigues P, Depardieu F, Courvalin P, Arthur M. Single-cell analysis of glycopeptide resistance gene expression in teicoplanin-resistant mutants of a VanB-type *Enterococcus faecalis*. Mol Microbiol 1999; 32:17–28.
106. Hayden MK, Picken RN, Sahm DF. Heterogeneous expression of glycopeptide resistance in enterococci associated with transfer of *vanB*. Antimicrob Agents Chemother 1997; 41:872–874.
107. Patel R, Uhl JR, Kohner P, Hopkins MK, Cockerill FR, 3rd. Multiplex PCR detection of *vanA*, *vanB*, *vanC-1*, and *vanC-2/3* genes in enterococci. J Clin Microbiol 1997; 35:703–707.
108. Dutka-Malen S, Molinas C, Arthur M, Courvalin P. Sequence of the *vanC*

gene of *Enterococcus gallinarum* BM4174 encoding a D-alanine:D-alanine ligase-related protein necessary for vancomycin resistance. Gene 1992; 112: 53–58.
109. Reynolds PE, Snaith HA, Maguire AJ, Dutka-Malen S, Courvalin P. Analysis of peptidoglycan precursors in vancomycin-resistant *Enterococcus gallinarum* BM4174. Biochem J 1994; 301:5–8.
110. Fines M, Perichon B, Reynolds P, Sahm DF, Courvalin P. VanE, a new type of acquired glycopeptide resistance in *Enterococcus faecalis* BM4405. Antimicrob Agents Chemother 1999; 43:2161–2164.
111. Arias CA, Martin-Martinez M, Blundell TL, Arthur M, Courvalin P, Reynolds PE. Characterization and modelling of VanT: a novel, membrane-bound, serine racemase from vancomycin-resistant *Enterococcus gallinarum* BM4174. Mol Microbiol 1999; 31:1653–1664.
112. Evers S, Casadewall B, Charles M, Dutka-Malen S, Galimand M, Courvalin P. Evolution of structure and substrate specificity in D-alanine:D-alanine ligases and related enzymes. J Mol Evol 1996; 42:706–712.
113. Sahm DF, Free L, Handwerger S. Inducible and constitutive expression of *vanC-1*–encoded resistance to vancomycin in *Enterococcus gallinarum*. Antimicrob Agents Chemother 1995; 39:1480–1484.
114. Navarro F, Courvalin P. Analysis of genes encoding D-alanine-D-alanine ligase-related enzymes in *Enterococcus casseliflavus* and *Enterococcus flavescens*. Antimicrob Agents Chemother 1994; 38:1788–1793.
115. Park IS, Lin CH, Walsh CT. Bacterial resistance to vancomycin: overproduction, purification, and characterization of VanC2 from *Enterococcus casseliflavus* as a D-Ala-D-Ser ligase. Proc Natl Acad Sci USA 1997; 94:10040–10044.
116. Patel R, Uhl JR, Kohner P, Hopkins MK, Steckelberg JM, Kline B, Cockerill FR, 3rd. DNA sequence variation with *vanA*, *vanB*, *vanC-1*, and *vanC-2/3* genes of clinical Enterococcus isolates. Antimicrob Agents Chemother 1998; 42:202–205.
117. Perichon B, Reynolds P, Courvalin P. VanD-type glycopeptide-resistant *Enterococcus faecium* BM4339. Antimicrob Agents Chemother 1997; 41:2016–2018.
118. Casadewall B, Courvalin P. Characterization of the *vanD* glycopeptide resistance gene cluster from *Enterococcus faecium* BM4339. J Bacteriol 1999; 181: 3644–3648.
119. Ostrowsky BE, Clark NC, Thauvin-Eliopoulos C, Venkataraman L, Samore MH, Tenover FC, Eliopoulos GM, Moellering RC, Jr, Gold HS. A cluster of VanD vancomycin-resistant *Enterococcus faecium*: molecular characterization and clinical epidemiology. J Infect Dis 1999; 180:1177–1185.
120. Bell JM, Paton JC, Turnidge J. Emergence of vancomycin-resistant enterococci in Australia: phenotypic and genotypic characteristics of isolates. J Clin Microbiol 1998; 36:2187–2190.
121. Marshall CG, Lessard IA, Park I, Wright GD. Glycopeptide antibiotic resistance genes in glycopeptide-producing organisms. Antimicrob Agents Chemother 1998; 42:2215–2220.
122. Marshall CG, Wright GD. DdlN from vancomycin-producing *Amycolatopsis*

orientalis C329.2 is a VanA homologue with D-alanyl-D-lactate ligase activity. J Bacteriol 1998; 180:5792–5795.
123. Sahm DF, Marsilio MK, Piazza G. Antimicrobial resistance in key bloodstream bacterial isolates: electronic surveillance with The Surveillance Network database—USA. Clin Infect Dis 1999; 29:259–263.
124. Fridkin SK, Yokoe DS, Whitney CG, Onderdonk A, Hooper DC. Epidemiology of a dominant clonal strain of vancomycin-resistant *Enterococcus faecium* at separate hospitals in Boston, Massachusetts. J Clin Microbiol 1998; 36: 965–970.
125. Fridkin SK, Steward CD, Edwards JR, Pryor ER, McGowan JE, Jr, Archibald LK, Gaynes RP, Tenover FC. Surveillance of antimicrobial use and antimicrobial resistance in United States hospitals: Project ICARE Phase 2. Clin Infect Dis 1999 29:245–252.
126. van den Braak N, van Belkum A, van Keulen M, Vliegenthart J, Verbrugh HA, Endtz HP. Molecular characterization of vancomycin-resistant enterococci from hospitalized patients and poultry products in The Netherlands. J Clin Microbiol 1998; 36:1927–1932.
127. Berchieri A, Jr. Intestinal colonization of a human subject by vancomycin-resistant *Enterococcus faecium*. Clin Microbiol Infect 1999; 5:97–100.
128. Jensen LB, Ahrens P, Dons L, Jones RN, Hammerum AM, Aarestrup FM. Molecular analysis of Tn1546 in *Enterococcus faecium* isolated from animals and humans. J Clin Microbiol 1998; 36:437–442.
129. Willems RJ, Top J, van den Braak N, van Belkum A, Mevius DJ, Hendriks G, van Santen-Verheuvel M, van Embden JD. Molecular diversity and evolutionary relationships of Tn1546-like elements in enterococci from humans and animals. Antimicrob Agents Chemother 1999; 43:483–491.
130. Roghmann MC, Qaiyumi S, Schwalbe R, Morris JG, Jr. Natural history of colonization with vancomycin-resistant *Enterococcus faecium* [letter] Infect Control Hosp Epidemiol 1997; 18:679–680.
131. Toye B, Shymanski J, Bobrowska M, Woods W, Ramotar K. Clinical and epidemiological significance of enterococci intrinsically resistant to vancomycin (possessing the vanC genotype) [published erratum appears in J Clin Microbiol 1998 May;36(5):1469]. J Clin Microbiol 1997; 35:3166–3170.
132. Kaplan AH, Gilligan PH, Facklam RR. Recovery of resistant enterococci during vancomycin prophylaxis. J Clin Microbiol 1988; 26:1216–1218.
133. Leclercq R, Derlot E, Weber M, Duval J, Courvalin P. Transferable vancomycin and teicoplanin resistance in *Enterococcus faecium*. Antimicrob Agents Chemother 1989; 33:10–15.
134. Noble WC, Virani Z, Cree RG. Co-transfer of vancomycin and other resistance genes from *Enterococcus faecalis* NCTC 12201 to *Staphylococcus aureus*. FEMS Microbiol Lett 1992; 72:195–198.
135. Poyart C, Pierre C, Quesne G, Pron B, Berche P, Trieu-Cuot P. Emergence of vancomycin resistance in the genus Streptococcus: characterization of a *vanB* transferable determinant in *Streptococcus bovis*. Antimicrob Agents Chemother 1997; 41:24–29.
136. Power EG, Abdulla YH, Talsania HG, Spice W, Aathithan S, French GL. *vanA*

genes in vancomycin-resistant clinical isolates of *Oerskovia turbata* and *Arcanobacterium (Corynebacterium) haemolyticum*. J Antimicrob Chemother 1995; 36:595–606.
137. Ligozzi M, Lo Cascio G, Fontana R. *vanA* gene cluster in a vancomycin-resistant clinical isolate of *Bacillus circulans*. Antimicrob Agents Chemother 1998; 42:2055–2059.
138. Rippere K, Patel R, Uhl JR, Piper KE, Steckelberg JM, Kline BC, Cockerill FR, 3rd, Yousten AA. DNA sequence resembling vanA and vanB in the vancomycin-resistant biopesticide *Bacillus popilliae*. J Infect Dis 1998; 178:584–588.
139. Schülin T, Wennersten CB, Moellering RC, Jr, Eliopoulos GM. In vitro activity of RU 64004, a new ketolide antibiotic, against gram-positive bacteria. Antimicrob Agents Chemother 1997; 41:1196–1202.
140. Schmitz FJ, Verhoef J, Fluit AC. Prevalence of resistance to MLS antibiotics in 20 European university hospitals participating in the European SENTRY surveillance programme. Sentry Participants Group. J Antimicrob Chemother 1999; 43:783–792.
141. Atkinson BA, Abu-Al-Jaibat A, LeBlanc DJ. Antibiotic resistance among enterococci isolated from clinical specimens between 1953 and 1954. Antimicrob Agents Chemother 1997; 41:1598–1600.
142. Martin B, Alloing G, Mejean V, Claverys JP. Constitutive expression of erythromycin resistance mediated by the *ermAM* determinant of plasmid pAM beta 1 results from deletion of 5′ leader peptide sequences. Plasmid 1987; 18:250–253.
143. Rosato A, Vicarini H, Leclercq R. Inducible or constitutive expression of resistance in clinical isolates of streptococci and enterococci cross-resistant to erythromycin and lincomycin. J Antimicrob Chemother 1999; 43:559–562.
144. Shaw JH, Clewell DB. Complete nucleotide sequence of macrolide-lincosamide-streptogramin B–resistance transposon Tn*917* in *Streptococcus faecalis*. J Bacteriol 1985; 164:782–796.
145. Leclercq R, Courvalin P. Bacterial resistance to macrolide, lincosamide, and streptogramin antibiotics by target modification [published erratum appears in Antimicrob Agents Chemother 1991 Oct;35(10):2165]. Antimicrob Agents Chemother 1991; 35:1267–1272.
146. Pepper K, Horaud T, Le Bouguenec C, de Cespedes G. Location of antibiotic resistance markers in clinical isolates of *Enterococcus faecalis* with similar antibiotypes. Antimicrob Agents Chemother 1987; 31:1394–1402.
147. LeBlanc DJ, Inamine JM, Lee LN. Broad geographical distribution of homologous erythromycin, kanamycin, and streptomycin resistance determinants among group D streptococci of human and animal origin. Antimicrob Agents Chemother 1986; 29:549–555.
148. Portillo A, Alonso A, Ruiz-Larrea E, Zarazaga M, Martinez JL, Torres C. Erythromycin resistance in *Enterococcus faecium* by an active efflux pump mechanism. 38th Interscience Conference on Antimicrobial Agents and Chemotherapy, American Society for Microbiology, San Diego, CA, 1998.
149. Malathum K, Coque TM, Singh KV, Murray BE. In vitro activities of two

ketolides, HMR 3647 and HMR 3004, against gram-positive bacteria. Antimicrob Agents Chemother 1999; 43:930–936.
150. Bozdogan B, Berrezouga L, Kuo MS, Yurek DA, Farley KA, Stockman BJ, Leclercq R. A new resistance gene, *linB*, conferring resistance to lincosamides by nucleotidylation in *Enterococcus faecium* HM1025. Antimicrob Agents Chemother 1999; 43:925–929.
151. Jones RN, Ballow CH, Biedenbach DJ, Deinhart JA, Schentag JJ. Antimicrobial activity of quinupristin-dalfopristin (RP 59500, Synercid) tested against over 28,000 recent clinical isolates from 200 medical centers in the United States and Canada. Diagn Microbiol Infect Dis 1998; 31:437–451.
152. Barriere JC, Berthaud N, Beyer D, Dutka-Malen S, Paris JM, Desnottes JF. Recent developments in streptogramin research. Curr Pharm Des 1998; 4: 155–180.
153. Caron F, Gold HS, Wennersten CB, Farris MG, Moellering RC, Jr, Eliopoulos GM. Influence of erythromycin resistance, inoculum growth phase, and incubation time on assessment of the bactericidal activity of RP 59500 (quinupristin-dalfopristin) against vancomycin-resistant *Enterococcus faecium*. Antimicrob Agents Chemother 1997; 41:2749–2753.
154. Allignet J, Loncle V, Mazodier P, el Solh N. Nucleotide sequence of a staphylococcal plasmid gene, *vgb*, encoding a hydrolase inactivating the B components of virginiamycin-like antibiotics. Plasmid 1988; 20:271–275.
155. Jensen LB, Hammerum AM, Aerestrup FM, van den Bogaard AE, Stobberingh EE. Occurrence of *satA* and *vgb* genes in streptogramin-resistant *Enterococcus faecium* isolates of animal and human origins in the Netherlands [letter]. Antimicrob Agents Chemother 1998; 42:3330–3331.
156. Bozdogan B, Leclercq R, Lozniewski A, Weber M. Plasmid-mediated coresistance to streptogramins and vancomycin in *Enterococcus faecium* HM1032. Antimicrob Agents Chemother 1999; 43:2097–2098.
157. Rende-Fournier R, Leclercq R, Galimand M, Duval J, Courvalin P. Identification of the *satA* gene encoding a streptogramin A acetyltransferase in *Enterococcus faecium* BM4145. Antimicrob Agents Chemother 1993; 37:2119–2125.
158. Hammerum AM, Jensen LB, Aarestrup FM. Detection of the *satA* gene and transferability of virginiamycin resistance in *Enterococcus faecium* from food-animals. FEMS Microbiol Lett 1998; 168:145–151.
159. Werner G, Klare I, Witte W. Association between quinupristin/dalfopristin resistance in glycopeptide-resistant *Enterococcus faecium* and the use of additives in animal feed. Eur J Clin Microbiol Infect Dis 1998; 17:401–402.
160. Werner G, Witte W. Characterization of a new enterococcal gene, *satG*, encoding a putative acetyltransferase conferring resistance to streptogramin A compounds. Antimicrob Agents Chemother 1990; 43:1813–1814.
161. Lautenbach E, Schuster MG, Bilker WB, Brennan PJ. The role of chloramphenicol in the treatment of bloodstream infection due to vancomycin-resistant Enterococcus. Clin Infect Dis 1998; 27:1259–1265.
162. Norris AH, Reilly JP, Edelstein PH, Brennan PJ, Schuster MG. Chloramphen-

icol for the treatment of vancomycin-resistant enterococcal infections. Clin Infect Dis 1995; 20:1137–1144.
163. Courvalin PM, Shaw WV, Jacob AE. Plasmid-mediated mechanisms of resistance to aminoglycoside-aminocyclitol antibiotics and to chloramphenicol in group D streptococci. Antimicrob Agents Chemother 1978; 13:716–725.
164. Trieu-Cuot P, de Cespedes G, Bentorcha F, Delbos F, Gaspar E, Horaud T. Study of heterogeneity of chloramphenicol acetyltransferase (CAT) genes in streptococci and enterococci by polymerase chain reaction: characterization of a new CAT determinant. Antimicrob Agents Chemother 1993; 37:2593–2598.
165. Lynch C, Courvalin P, Nikaido H. Active efflux of antimicrobial agents in wild-type strains of enterococci. Antimicrob Agents Chemother 1997; 41:869–871.
166. Charpentier E, Gerbaud G, Courvalin P. Presence of the Listeria tetracycline resistance gene *tet(S)* in *Enterococcus faecalis*. Antimicrob Agents Chemother 1994; 38:2330–2335.
167. Flannagan SE, Zitzow LA, Su YA, Clewell DB. Nucleotide sequence of the 18-kb conjugative transposon Tn916 from *Enterococcus faecalis*. Plasmid 1994; 32:350–354.
168. Franke AE, Clewell DB. Evidence for a chromosome-borne resistance transposon (Tn916) in *Streptococcus faecalis* that is capable of "conjugal" transfer in the absence of conjugative plasmid. J Bacteriol 1981; 145:494–502.
169. Bentorcha F, Clermont D, de Cespedes G, Horaud T. Natural occurrence of structures in oral streptococci and enterococci with DNA homology to Tn916. Antimicrob Agents Chemother 1992; 36:59–63.
170. Su YA, He P, Clewell DB. Characterization of the tet(M) determinant of Tn916: evidence for regulation by transcription attenuation. Antimicrob Agents Chemother 1992; 36:769–778.
171. McMurry LM, Park BH, Burdett V, Levy SB. Energy-dependent efflux mediated by class L (*tetL*) tetracycline resistance determinant from streptococci. Antimicrob Agents Chemother 1987; 31:1648–1650.
172. Bentorcha F, De Cespedes G, Horaud T. Tetracycline resistance heterogeneity in *Enterococcus faecium*. Antimicrob Agents Chemother 1991; 35:808–812.
173. Barry AL, Jones RN, Thornsberry C, Ayers LW, Gerlach EH, Sommers HM. Antibacterial activities of ciprofloxacin, norfloxacin, oxolinic acid, cinoxacin, and nalidixic acid. Antimicrob Agents Chemother 1984; 25:633–637.
174. Eliopoulos GM, Gardella A, Moellering RC, Jr. In vitro activity of ciprofloxacin, a new carboxyquinoline antimicrobial agent. Antimicrob Agents Chemother 1984; 25:331–335.
175. Tankovic J, Majhoubi F, Courvalin P, Duval J, Leclercq R. Development of fluoroquinolone resistance in *Enterococcus faecalis* and role of mutations in the DNA gyrase *gyrA* gene. Antimicrob Agents Chemother 1996; 40:2558–2561.
176. Eliopoulos GM. Activity of newer fluoroguinolones *in vitro* against gram-positive bacteria. Drugs 1999 58(Suppl 2):23–28.
177. Korten V, Huang WM, Murray BE. Analysis by PCR and direct DNA se-

quencing of *gyrA* mutations associated with fluoroquinolone resistance in *Enterococcus faecalis*. Antimicrob Agents Chemother 1994; 38:2091–2094.

178. Kanematsu E, Deguchi T, Yasuda M, Kawamura T, Nishino Y, Kawada Y. Alterations in the GyrA subunit of DNA gyrase and the ParC subunit of DNA topoisomerase IV associated with quinolone resistance in *Enterococcus faecalis*. Antimicrob Agents Chemother 1998; 42:433–435.
179. el Amin NA, Jalal S, Wretlind B. Alterations in GyrA and ParC associated with fluoroquinolone resistance in *Enterococcus faecium*. Antimicrob Agents Chemother 1999; 43:947–949.
180. Patel R, Rouse MS, Piper KE, Steckelberg JM. *In vitro* activity of linezolid against vancomycin-resistant enterococci, methicillin-resistant *Staphylococcus aureus* and penicillin-resistant *Streptococcus pneumoniae*. Diagn Microbiol Infect Dis 1999; 34:119–122.
181. Urban C, Mariano N, Mosinka-Snipas K, Wadee C, Chahrour T, Rahal JJ. Comparative *in-vitro* activity of SCH 27899, a novel everninomicin, and vancomycin. J Antimicrob Chemother 1996; 37:361–364.
182. Nicas TI, Mullen DL, Flokowitsch JE, Preston DA, Snyder NJ, Zweifel MJ, Wilkie SC, Rodriguez MJ, Thompson RC, Cooper RD. Semisynthetic glycopeptide antibiotics derived from LY264826 active against vancomycin-resistant enterococci. Antimicrob Agents Chemother 1996; 40:2194–2199.
183. Patel R, Rouse MS, Piper KE, Cockerill FR, 3rd, Steckelberg JM. In vitro activity of LY333328 against vancomycin-resistant enterococci, methicillin-resistant *Staphylococcus aureus*, and penicillin-resistant *Streptococcus pneumoniae*. Diagn Microbiol Infect Dis 1998; 30:89–92.
184. Bingen E, Lambert-Zechovsky N, Leclercq R, Doit C, Mariani-Kurkdjian P. Bactericidal activity of vancomycin, daptomycin, ampicillin and aminoglycosides against vancomycin-resistant *Enterococcus faecium*. J Antimicrob Chemother 1990; 26:619–626.

14

Methicillin Resistance in *Staphylococcus aureus*

Keeta S. Gilmore
and Michael S. Gilmore
University of Oklahoma Health Sciences Center, Oklahoma City, Oklahoma

Daniel F. Sahm
Focus Technologies, Inc., Herndon, Virginia

Methicillin-resistant strains of *Staphylococcus aureus* are those that have acquired the ability to grow in the presence of methylpenicillins and derivatives, including methicillin, oxacillin, and nafcillin. This resistance is caused by expression of an altered penicillin-binding protein, PBP2a (PBP2′) (7,8). Methicillin-susceptible strains are inhibited by oxacillin at concentrations of 4 μg/mL or methicillin at 8 μg/mL. However, *S. aureus* possessing the methicillin resistance trait (MRSA) grow in the presence of 16 μg/mL to over 2000 μg/mL of methicillin.

The targets of the antibiotic methicillin in sensitive strains of *S. aureus* are the penicillin-binding proteins (PBPs), essential enzymes that catalyze transpeptidation cross linking of peptidoglycan in the bacterial cell wall. Inhibition of this reaction with methicillin results in the arrest of cell wall biosynthesis, triggering death of the organism through induction of the autolytic response (9). Methicillin-resistant strains of *S. aureus* possess a 25-

to-50-kb DNA sequence that encodes, among other things, PBP2a and genes for regulation of its expression. As compared to the methicillin-inhibited counterpart on the surface of sensitive strains, PBP2a exhibits a decreased affinity for β-lactam antibiotics (7).

1 INTRODUCTION

The problems of antibiotic resistance among staphylococci, and the pathogenesis of staphylococcal infection, are well treated within this volume and elsewhere (1–6). The purpose of this chapter is to review the present and future challenge to health care specifically posed by methicillin-resistant strains of *S. aureus*. In particular, the subjects of this chapter are the nature of methicillin resistance and its origins, the epidemiology of methicillin resistance among nosocomial isolates and now also among community-acquired strains, the consequences of this resistance in limiting therapeutic options and its impact on health care costs, and the further evolution of increased glycopeptide resistance among methicillin-resistant strains.

Methicillin resistance in staphylococci is mediated by the production of a novel penicillin-binding protein, PBP2a, with decreased binding affinity to β-lactams (7,8). PBP2a is chromosomally encoded within an externally acquired section of DNA called *mec* DNA. Expression of PBP2a is controlled by two regulator genes on *mec* DNA, *mecI* and *mecR1*.

Methicillin-resistant *S. aureus* (MRSA) was initially detected in Europe in the 1960s shortly after the introduction of methicillin. Today, MRSA is present in the hospitals of most countries and is usually resistant to several antibiotics. Clinical infections are most common in patients in hospital intensive care units, nursing homes, and other chronic care facilities. All MRSA are currently susceptible to the glycopeptides such as vancomycin and teicoplanin; however, if resistance to these agents emerges, some staphylococcal infections could be untreatable.

2 EMERGENCE OF MRSA

Since the introduction of antibiotics into clinical use in the mid-1940s, microorganisms have shown a remarkable ability to protect themselves by developing and acquiring antibiotic resistance. By 1942, penicillin resistance was reported in *S. aureus* after only months of limited clinical trials (2). By 1953, 64–80% of *S. aureus* isolates were resistant to penicillin, with development of resistance to tetracycline, erythromycin, and other classes of antibiotics beginning to emerge (2). By 1960, using aggressive infection-control measures, antibiotic-resistant staphylococci had become the most common cause of hospital-acquired infection worldwide (2,10).

Methicillin, a β-lactam effective against penicillin-resistant *S. aureus* strains, became widely available in 1960. However, within a year of its introduction, methicillin-resistant strains of *S. aureus* (MRSA) were reported in the United Kingdom (11,12). Sporadic reports of clinical isolates of MRSA were soon observed in the United States (13), but the first well-documented outbreak in the United States did not occur until 1968 (14).

2.1 Nosocomial Infection

The MRSA problem arose initially in large tertiary care hospitals (20,21) with patients in burn (15–17), postoperative (15,16,18), and intensive care wards (18,19). Increased risk of MRSA infection was associated with use of multiple broad spectrum antibiotics (2,15,18,19,22), indwelling devices (18,22), ventilatory support (18), severity of underlying disease (2,15,22,23), and length of hospital stay (15,22,23). The National Nosocomial Infections Surveillance (NNIS) system reported that the percentage of MRSA among nosocomial isolates in the United States increased from 2.4% in 1975 (24) to 35% of staphylococcal isolates in 1996 (25). Data available from The Surveillance Network Databases—USA (TSN) (MRL, Herndon, VA) for the first 6 months of 1999 showed MRSA to represent 38.6% of staphylococcal isolates.

2.2 Community-Acquired Infection

The epidemiology of MRSA has shifted from that of an almost exclusively nosocomial problem to now being transmitted within the community with increasing frequency (26,27). Studies published in the early 1980s on community-acquired MRSA in the United States noted infection in intravenous drug abusers (28,29) and individuals with recognized predisposing risk factors such as persistent carriage, recently discharged patients with serious underlying diseases, previous antibiotic therapy, or residence in a nursing home (2,26). By the mid-1990s, community-acquired MRSA infections were described in individuals without identifiable risk factors (27). However, although nosocomially acquired MRSA isolates tended to be multidrug resistant, community-acquired MRSA strains obtained from patients without identified risk tended to be resistant only to methicillin (27). This more restricted set of antibiotic resistances has also been observed in studies of community-acquired MRSA strains among intravenous drug abusers compared with nosocomially acquired MRSA isolates (27,32,33).

2.3 Costs Attributable to MRSA

Methicillin resistance has created four decades of major therapeutic, management, and epidemiological problems throughout the world leading to increased hospital costs (15,34). Nosocomial bloodstream infection with MRSA was found to prolong hospitalization an average of 8 days over similar infections caused by methicillin-sensitive *S. aureus* (MSSA), resulting in an approximately threefold increase in direct costs (35). Studies have shown that treating an MRSA infection can cost 6–10% more than treating a methicillin-sensitive infection ($2500–$3700 per case) (36). This difference does not reflect greater virulence of MRSA; rather it reflects the increased cost of vancomycin treatment, longer hospital stay, and the cost of patient isolation and infection-control measures. In addition to increasing costs, the mortality rate attributable to MRSA infections has been observed in some studies to be more than 2.5 times higher than that attributable to MSSA infections (21 vs. 8%) (36). Although it should be noted that some of the death rate difference may be related to the underlying condition of patients who become infected with MRSA, such as older patients and patients previously exposed to antibiotics, as well as the lack of effectiveness of vancomycin to cure MRSA (36).

3 CURRENT EPIDEMIOLOGY OF MRSA

The anterior nares are the natural human reservoir for *S. aureus* where it can be isolated from 10–40% of healthy adults (2,37). From the nares, spread to the skin (especially eczematous lesions) and then to surgical wounds, foreign bodies (e.g., indwelling devices), burns, and the upper respiratory tract (21,37,38) is common, with the hands being the major mode of transmission (37,39). That a common cause of frequently severe infections is carried asymptomatically by a large proportion of the population in an accessible site, such as the anterior nares, challenges current paradigms of what constitutes a pathogen. Between 20 and 35% of the population are persistent *S. aureus* carriers, and 30–70% are intermittent carriers (40,41). Identification of carriers is an important key to containment, because strains associated with nasal colonization have been observed to account for 40–100% of staphylococcal sepsis, and surgical infection is 2–17 times more common among carriers than noncarriers (40).

The first reports of an endemic methicillin resistance problem emerged from large hospitals (>500 beds) in the mid-1980s (24,42), with subsequent occurrences in smaller hospitals and nonteaching hospitals (24,43). As a result, based on antibiotic susceptibility data for *S. aureus* isolates from 182 hospitals covering the period from 1975 through 1991, the rate of increase in MRSA differed significantly among hospitals categorized by bed capacity (24). The occurrence of MRSA in each category crossed the 5% threshold in 1983 for large hospitals (≥500 beds), 1985 for medium-size hospitals (200–499 beds), and for smaller hospitals (<200 beds) in 1987 (24).

S. aureus is now the leading cause of nosocomial pneumonia and surgical site infections (44), and behind coagulase-negative staphylococci (CoNS), is the second leading cause of nosocomial bloodstream infections (36,44). Infections and outbreaks are common throughout the world in nursing homes (45,46) and among outpatient populations (38,47), in addition to those reported in hospitals (21,37). Infection with MRSA is especially prominent in intensive care units (ICUs) (37).

Currently, approximately two million hospitalizations annually result in nosocomial infection (36). Surveillance databases, such as TSN, electronically collect and compile data daily from more than 200 clinical laboratories, identify potential laboratory testing errors, and detect emergence of resistance profiles and mechanisms that pose a public health threat (e.g., vancomycin-resistant staphylococci). It was noted in 1991, using data from the NNIS System, that the percentage of MRSA was greatest from hospitals reporting from the southeastern region of the United States (24). Using data collected from July 1998 through June 1999 by the TSN Database, this trend continues with the Southeast reporting 45.5% of *S. aureus* isolates to be MRSA compared to a national average of 35.7%.

MRSA is introduced into an institution primarily by admission of an infected or colonized patient who serves as a reservoir (18,19,38). Less frequently, MRSA can be brought in by colonized or infected health care workers who disseminate the organism directly to patients (23,38). The principal mode of transmission of MRSA within the hospital is via transiently colonized hands of health care workers, who acquire the organism after close contact with colonized patients, contaminated equipment, or their own flora (16,19,21,37,38,48). More rarely, patients can acquire MRSA via airborne transmission, as has been observed in burn units (16,37,38, 49–51).

Several risk factors for the acquisition of MRSA have been identified. These include prior hospitalization, admission to an ICU or burn unit, invasive procedures, skin lesions, age, and previous antimicrobial treatment (52–56). Current guidelines for MRSA control in hospitals focus on measures to control MRSA cross contamination and colonization (52). These guidelines include measures such as handwashing, the identification of human reservoirs, decontamination of the environment, patient isolation, and notification of known carriers when transferred to another institution

(52). Despite these procedures, MRSA continues to spread in most institutions, and has become endemic rather than epidemic (43).

Within the community, resistance to standard modes of therapy is emerging among common pathogens, including *S. aureus*, and reports of outpatient MRSA infections in both children and adults are increasingly common (57). In 1994, a survey from Canadian hospitals documented a substantial proportion of MRSA infection identified on admission to the hospital (26,58). In 1995, a retrospective review of the epidemiology of MRSA was undertaken by Moreno et al. to determine the occurrence of community MRSA infection (26). This study (26), as well as surveys from Canadian hospitals (58) and from a Connecticut hospital (53), revealed that community MRSA infection was more common than expected, and that the majority of isolates in the former study represented distinct strains rather than recent descendants of a single strain (26). Although previous reports of community-acquired methicillin-resistant *S. aureus* infections were generally limited to infections among intravenous drug users, individuals with serious underlying illness, individuals with a recent history of antibiotic therapy, or individuals confined to a nursing home (28,29,38, 58), these studies revealed cases of MRSA colonization and infection acquired in the community by individuals lacking predisposing factors (26, 53,58). Moreno et al. (26) were able to demonstrate no differences in risk for community acquisition of MRSA compared to MSSA.

4 MOLECULAR NATURE OF METHICILLIN RESISTANCE

The early introduction of β-lactam antibiotics quickly selected for the outgrowth of *S. aureus* strains possessing, or having acquired, the ability to express β-lactamases, achieving a resistance rate of 75% as early as 1952 (59). The outgrowth of β-lactamase–producing *S. aureus* prompted the commercial development of β-lactamase–resistant derivatives of penicillin such as methicillin, which possess an acyl side chain that prevents hydrolysis of the β-lactam ring by β-lactamases. The narrow-spectrum staphylococcal β-lactamases exhibit little activity against semisynthetic penicillins such as methicillin (60).

4.1 Mechanisms of Resistance to Methicillin

Under increased selective pressure, *S. aureus* developed multiple mechanisms of resistance to modified penicillins, including methicillin. Although methicillin is resistant to hydrolysis by small quantities of staphylococcal β-lactamase, strains of *S. aureus* have been isolated that are capable of producing large amount of β-lactamase (61). These hyperproducers of β-lactamase tend to resist methicillin through limited hydrolysis of the antibiotic, resulting in a phenotype that with respect to methicillin is intermediate between susceptible and resistant (61).

A second mechanism for achieving low-level resistance to methicillin by *S. aureus* strains involves the production of altered forms of native PBPs. *S. aureus* express at least four different PBPs, designated PBP1, 2, 3, and 4, that are the targets of β-lactam antibiotics (5). The β-lactam antibiotics serve as substrate analogues that covalently bind penicillin-binding proteins (PBPs), inactivating them at concentrations close to the MIC. PBPs are essential proteins that are anchored to the cytoplasmic membrane and catalyze the transpeptidation reaction that cross links the peptidoglycan of the bacterial cell wall; therefore the binding of β-lactam antibiotics to PBPs leads to a lethal event (60). Low level resistance to β-lactam antibiotics can be due to either a decrease in the binding affinities of PBPs for penicillins, or an increase in the production of PBPs, or both (5,62). However, the most prevalent means for achieving methicillin resistance is the acquisition of *mec* DNA, including a gene encoding a novel PBP, designated PBP2a or PBP2′ (5).

4.2 *mec* DNA

mec DNA is a large (approximately 30–50 kb) DNA fragment that does not occur in MSSA, and is always located at a fixed site in the *S. aureus* chromosome, specifically near the pur-nov-his gene cluster (60,63). *mec* DNA contains *mecA*, the structural gene for PBP2a; *mecI* and *mecR1*, regulatory elements controlling *mecA* transcription; and 20–45 kb of *mec*-associated DNA. The *mec*-associated DNA has been found to contain transposons and insertion elements providing a mechanism for the considerable variability found within the *mec* region. IS*431* is a common insertion sequence in the staphylococcal chromosome and plasmids, and is present within the *mec* DNA region. IS*431* serves as a trap for resistance determinants with similar IS elements, accounting for the multiple drug resistance phenotype common in MRSA (60). The transposon Tn*554* that contains *ermA*, the gene encoding for inducible erythromycin resistance, is located upstream from *mecA* in over 90% of MRSA (64).

4.2.1 *mecA*

Most MRSA (>90%) harbor *mecA*, the gene encoding the alternate PBP, PBP2a (5). *mecA* is inducible and encodes the 76-kD PBP2a polypeptide. The *mecA* gene occurs in both MRSA and methicillin-resistant coagulase-negative staphylococci (CoNS), and is highly conserved (65–68). Analysis of the nucleotide sequence of *mecA* and its operator region revealed that sequences contained within the 5′ end and operator were similar to sequences within the β-lactamase gene, *blaZ*, of *S. aureus*. The remainder of the structural gene exhibits sequence similarity to the PBP 2 and 3 genes of *Escherichia coli* (5,69). PBP2a shows characteristics typical of other membrane-bound PBPs, with a transglycosylase domain and conserved signatures of a transpeptidase (1).

The native PBPs in *S. aureus*, PBP1, 2, and 3, are essential for cell growth and survival of susceptible strains. These PBPs have a high affinity for most β-lactam antibiotics, and the binding of β-lactams by these PBPs triggers a lethal cascade (70,71). PBP2a binds β-lactams with much lower affinity than the native PBPs (7,72–74). In resistant strains of *S. aureus*, PBP2a can substitute for the essential functions of PBP1, 2, and 3 at otherwise lethal concentrations of antibiotic (7,60).

4.2.2 PBP2a

The four native PBPs become fully acylated by methicillin at concentrations of 5–10 mg/L. Under these conditions, the low-affinity PBP2a assumes the task of transpeptidation (4). In antibiotic-free medium, a highly resistant MRSA strain, strain COL (MIC = 1600 mg/L), produces a cell wall composed of a diverse family of over 35 muropeptide components (4). The majority of muropeptides (>60%) are trimers or higher oligomers. When methicillin is added to the medium at concentrations ranging from 5 mg/L to 750 mg/L, this complex wall is replaced by a simpler structure where the peptidoglycan is made of essentially two components: the pentaglycyl monomer and its dimer, with only a very small amount of trimers and traces of higher oligomers (4). From these observations, it would appear that PBP2a is limited in activity to linking two monomers, and is incapable of generating highly cross-linked oligomers that are typical products of the normal cell wall synthetic machinery (4).

4.3 Regulation of *mecA*

Expression of PBP2a is controlled by two regulator genes on the *mec* DNA, *mecI* and *mecR1*. Both genes are located immediately upstream of *mecA* (60) (Fig. 1), separated from *mecA* by its promotor and operator (75,76), and are divergently transcribed. Downstream from *mecA* is a variable segment of DNA that ends with an insertion-like element, IS*431* (77) that serves as a target for homologous recombination for other resistance determinants flanked by similar IS elements (1,78). Therefore, *mecA* and its associated DNA act as a trap for integration of other determinants, including genes for resistance to fluoroquinolones, aminoglycosides, tetracyclines, macrolides, and trimethoprim-sulfamethoxazole.

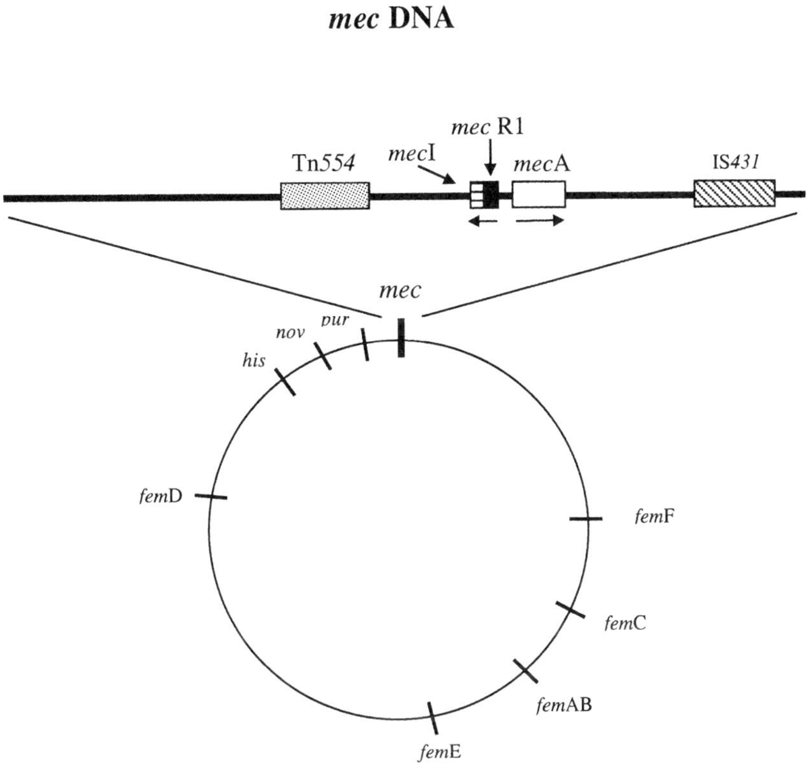

Figure 1 Organization of the *mec* region of DNA and chromosomal location. *mec* DNA is 30–50 kb containing the PBP2a structural gene, *mecA* and its upstream regulatory elements, *mecI-mecR1*. The regulatory genes are divergently transcribed from *mecA* as indicated by the arrows. Further upstream from *mecA* is Tn*554* and downstream from *mecA* is a variable region ending with IS*431*. (Adapted from Refs. 5 and 60.)

The MecR1 protein is a slow inducer of methicillin resistance and is similar to BlaR1, which is involved in the induction of staphylococcal β-lactamase. Both genes, *mecR1* and *blaR1*, encode signal-transducing PBPs that result in *mecA* and β-lactamase gene transcription in the presence of β-lactam antibiotic (60). Like BlaR1, MecR1 consists of two regions, a membrane-spanning domain and a penicillin-binding domain.

The MecI protein is a strong repressor of *mecA* transcription (76) and is highly related in primary structure to BlaI, the repressor of staphylococcal β-lactamase (1). Because of the extensive similarity to MecR1 and MecI,

BlaR1 and BlaI can also regulate *mecA* transcription (1,79). However, repression by BlaI is weaker than by MecI, and as a result, some PBP2a is produced in uninduced strains; induction of BlaI-repressed *mecA* by methicillin is as rapid as induction of BlaI-repressed β-lactamase synthesis (1,80). In the absence of β-lactamase regulatory elements and *mecI-mecRI* regulatory elements, PBP2a is produced constitutively (1).

Strains that contain functional *mecI-mecR1* regulatory elements are strongly repressed and produce PBP2a only after induction (1,80). However, induction is slow, and methicillin seems to be a relatively weak inducer (81). As a result, methicillin resistance is established slowly and may only appear after 48 hr on methicillin-containing plates, making these strains appear initially falsely susceptible at 24 hr (1).

Although *mecA* is present in all MRSA, there is considerable variation in the presence of the other genes (83). *mecR1-mecI* is present in 60–95% of *mecA*-positive *S. aureus* (82,84,85). Because *mecI* is such a strong repressor, it has been concluded that phenotypically resistant *mecA*-positive *S. aureus* strains either do not possess *mecI*, or have mutations within *mecI* which prevent it from functioning (5,82,84,86), or have mutations within the *mecA* promoter region corresponding to a presumptive operator of *mecA*, the binding site of the repressor protein (5,84). Inactivation of *mecI*, by either deletion or mutation, is an essential step in the production of PBP2a and expression of methicillin resistance (43,87). Two point mutations are frequently detected in the *mecI* gene—a substitution at nucleotide position 202 (C to T) or at position 260 (T to A)—both resulting in an in-frame stop codon in the middle of the *mecI* gene (5,82,84,86). In these stains, a functional repressor protein is not produced, allowing maximal expression of methicillin resistance (84). Point mutations in the operator region of the *mecA* promoter have also been identified (5,69).

A small number of *S. aureus* strains have been isolated that carry intact *mecI* and *mecR1*, together with *mecA*, and these strains have been termed pre-MRSA, as typified by prototype *S. aureus* strain N315 (5,84,86). Pre-MRSA are phenotypically susceptible to methicillin using routine tests (81–84). In these strains, the expression of methicillin resistance if fully repressed by *mecI* and is not induced by the presence of methicillin. However, when grown on selective media, resistant cells arise at a high frequency (10^{-5} to 10^{-6}) resulting from point mutations in the *mecI* gene (82,84), circumventing the *mecI*-mediated repression of *mecA* (84).

4.4 *fem* Genes

Although PBP2a mediates methicillin resistance in staphylococci, increased amounts of the protein do not correlate with an increased proportion of cells expressing resistance, leading to the recognition that other factors influence expression of methicillin resistance (88). Using insertional inactivation with *Tn551*, additional chromosomally located genes that are esstential for expression of methicillin resistance have been identified (88–91), leading to a current understanding that methicillin resistance in *S. aureus* is complex and requires not only expression of the *mecA* gene (63,92) but also the cooperation of auxiliary genes or *fem* genes (factor essential for the expression of methicillin resistance) (93,94). The *fem* genes are located throughout the staphylococcal genome, are physically distinct from *mec* DNA and are essential for maximum resistance (1,3,91,95). Eight *fem* genes, *femA–F, glmM,* and *mrp* have been described so far (1,60,96,97).

The *fem* genes (93,96–102) occur in both MRSA and MSSA and encode or regulate the activity of enzymes catalyzing reactions at different stages in peptidoglycan biosynthesis. Transposon mutagenesis of *fem* genes leads to a reduction in methicillin resistance. However, the mechanism by which they contribute to expression of methicillin resistance in clinical isolates is not clear. Inactivation of *fem* genes in susceptible strains of *S. aureus* results in hypersusceptibility to β-lactams (1).

The *fem*AB operon encodes two functionally related proteins required for formation of the pentaglycine interpeptide bridge that cross links peptidoglycan (1,60,103). Mutants lacking *fem*B are able to attach only the first three glycines to the cross bridge (1,60,104), whereas *fem*A mutants do not incorporate the second and third glycines into the bridge (1,60). The level of resistance in *fem*AB mutants is reduced to nearly susceptible levels (60). Disruption of *fem*C reduces the basal level of methicillin resistance in MRSA but still allows formation of a highly resistant subpopulation (1,60). Mutation in *fem*C produces a metabolic block in glutamine production. This block affects peptidoglycan composition by reducing the amidation of isoglutamate in the peptidoglycan stem pentapeptide, resulting in a reduction in the extent of cross linking in the peptidoglycan. Addition of glutamine to the culture restores both isoglutamate amidation and methicillin resistance (1,60).

*fem*D inactivation results in the loss of unsubstituted disaccharide pentapeptide monomer from the cell wall (2,60). *fem*F mutants are impaired in peptidoglycan precursor synthesis at the lysine addition step (1, 60,102) and inactivation of *glmM* inhibits the conversion of glucosamine-6-phosphate to glucosamine-1-phosphate, a reaction key to cell wall biosynthesis (97). The function of *fem*E and *mrp* is unknown at this time. It is evident from the growing list of auxiliary factors, involved in methicillin resistance, that any disruption in biosynthesis of peptidoglycan or membrane composition has the potential to reduce the optimal function of PBP2a when native PBPs are inactivated (1).

Interestingly, a strain of *S. aureus* resistant to methicillin has been described that possesses an intact *mecI* gene and *mec* promoter region (84). The basis for phenotypic resistance in these isolates is unknown. Given that *mecI* is part of an operon, its control regions could be affected by alterations in genes or control elements upstream (84) and, as demonstrated by inactivation of the *fem* genes, resistance is more than an interaction between the products of the *mecR1*, *mecI*, and *mecA* genes. Some strains appear to have acquired additional mechanisms to evade repression of *mecA* (84). In all probability, it is the interaction of many metabolic functions of the host microorganism with other factors yet to be identified that determines the efficacy of PBP2a and, subsequently, methicillin resistance levels (1).

4.5 Heterogeneous Resistance

Subpopulations of cells of a methicillin-resistant strain producing PBP2a vary markedly in the phenotypic expression of resistance. With most clinical isolates of MRSA, the majority of the population is relatively susceptible to β-lactam antibiotics, and only a small proportion of cells express high levels of resistance (3,99). Although all cells in an MRSA population have the potential to express resistance to methicillin, the population does not behave in a homogeneous manner (38,105,106).

There is a degree of heterogeneity in the phenotypic expression of antibiotic resistance from strain to strain and within the progeny of a single MRSA lineage (107). This characteristic of methicillin-resistant strains is called heterogeneous expression, with only one cell in 10^4 to 10^8 expressing detectable resistance (81,108). The proportion of cells expressing higher resistance levels is strain dependent (107), and the level of resistance in MRSA strains does not correlate to the quantity of PBP2a present (95, 99,109). In some strains, the highly resistant subpopulation will maintain the high level of resistance among descendants of this subpopulation (110). Among other clinical isolates, however, the highly resistant subclones return to their original resistance upon regrowth from a single colony in drug-free medium (1,110). Strains consistently producing populations of high-level resistant cells are termed homogeneous expression strains. Even though the subpopulation of highly resistant MRSA within a heterogeneous strain occurs at a low frequency, it can overgrow a culture under conditions of antibiotic pressure (107). The practical implication is that every MRSA strain, irrespective of whether expression is heterogeneous or homogeneous, may cause treatment failure in vivo (107).

This phenomenon of heterogeneous resistance makes it necessary for clinical laboratories to use special methods to ensure detection of

MRSA. Heterogeneous resistance can be overcome by low incubation temperature (30–35°C) and the incorporation of higher salt concentrations (2–4% NaCl) in the medium, which are conditions that favor enhanced expression of resistance (38,111).

5 ORIGINS OF METHICILLIN RESISTANCE

As noted above, nucleotide sequencing revealed that the *mecA* gene is composed of separate domains exhibiting sequence similarity to two distinct genes; the 5′ region of the *mecA* gene being similar to the penicillinase gene (*blaZ*) of *S. aureus* and the rest of the gene being related to *E. coli* PBP2 and PBP3 (5,69). Several theories on the origins of this gene have been proposed: that *mecA* emerged by homologous recombination between PBP and β-lactamase gene in an unknown organism (5,69) or *mecA* originated in a coagulase-negative staphylococcal species, perhaps a close evolutionary relative of *S. sciuri* (60,83). When bacterial isolates belonging to over 15 species of staphylococci were examined for reactivity with a DNA probe internal to the *mecA* of an MRSA strain, only one species, *S. sciuri*, positively hybridized in every one of 150^{+} independent isolates (112).

The *mecA* homologue in *S. sciuri* appears to be silent, as most *S. sciuri* isolates express no detectable resistance to either methicillin or penicillin (112,113). It has been suggested that due to the presence of the *mecA* homologue in *S. sciuri* that carries the structural motifs of PBPs, and the absence of a methicillin-resistant phenotype in the majority of *S. sciuri* strains, that the *mecA* homologue is native in this bacterium and performs some physiological function, such as cell wall biosynthesis (112). The product of *S. sciuri mecA* possesses a putative transglycosylase (TGase) domain with an N-terminal membrane anchor sequence and a transpeptidase (TPase) domain, similar to other high molecular weight PBPs. The *mecA* of *S. sciuri* exhibits an overall inferred amino acid sequence similarity of 88% and identity of 80% when compared to *mecA* of the MRSA (112). Comparison of the transpeptidase domain showed similarity of 96% and identity of 91%. However, comparison of the putative transglycosylase domains of the *S. sciuri* and the MRSA *mecA* showed a similarity of 80% and identity of only 68%, suggesting the potential for functional divergence in the TGase domain (112). With an overall similarity index of approximately 80%, the level is viewed as too low to implicate the *mecA* homologue of *S. sciuri* as a direct evolutionary precursor of the *mecA* of *S. aureus* (112). Then again, as the *S. sciuri mecA* homologue is by far the most closely related of known genes to *mecA* of MRSA, it appears to be certain that both genes share a common evolutionary ancestry, with intermediates most likely occurring elsewhere within the genus (112).

The earliest MRSA isolates appear to have descended from a single methicillin-resistant clone (114) and then entered other phylogenetic lineages of *S. aureus*. Alternately, it remains possible that the *mec* determinant was acquired at different times by different strains (1). Clonal analysis of MRSA (1,115) and of *mec* determinants stemming from *Staphylococcus* spp. other than *S. aureus* (mainly from *S. haemolyticus* and *S. epidermidis*) support the prospect that the *mec* determinant was disseminated by horizontal transfer, with CoNS possibly serving as the intermediary of the *mec* determinant for *S. aureus* (1,83). But whereas β-lactamases were rapidly and widely disseminated and are now present in about 80% of all staphylococci, the *mec* determinant is still largely restricted to discrete clonal lineages and seems to favor clonal over horizontal spread (1).

6 EXPECTATIONS FOR THE FUTURE

Currently, more than 95% of patients with *S. aureus* infections worldwide do not respond to first-line antibiotics such as penicillin or ampicillin (36,116). Moreover, MRSA are now found in the community, including in individuals who have never been hospitalized (52,53,117). Many multiresistant MRSA strains are presently only susceptible to a single class of clinically available bactericidal antibiotic, the glycopeptides (vancomycin and teicoplanin) and the acquisition of the *vanA* or *vanB* determinants from enterococci would be a potential public health disaster. Consequently, there is concern for the development of vancomycin resistance in multi–drug-resistant strains of MRSA, especially since the demonstration of transfer of the *vanA* gene from enterococci to *S. aureus* in vitro (118).

The enterococcal vancomycin-resistance mechanism has not yet been observed among clinical isolates of *S. aureus*, however, in 1996, reduced susceptibility (MIC = 8 μg/mL) to vancomycin was reported from Japan and the United States both in association with the failure of vancomycin treatment of MRSA infection (120–122). The emergence of MRSA strains with reduced susceptibility to vancomycin poses a potentially serious threat to public health. The mechanism for the reduced susceptibility is unknown, however, it appears to have developed de novo after antibiotic exposure (11). All MRSA strains with reduced susceptibility to vancomycin identified have had different patterns of antibiotic susceptibility (122,123), suggesting that these strains are developing independently (11). Current efforts in drug development target the enterococcal mechanism of vancomycin resistance, and it is, therefore, crucial that the mechanism of MRSA resistance to vancomycin be elucidated (118).

Several studies suggest that reduction of antibiotic use within the hospital could decrease nosocomial acquisition of multiresistant bacteria

(37,124–126) and scheduled rotation of antibiotic use may also reduce levels of antibiotic resistance (37,127). In addition to prudent use of antibiotics, strict compliance with infection control policies can aid in the reduction of nosocomial spread of multi–drug-resistant MRSA. This may, however, be harder to effect than decreasing antibiotic use, since studies have shown that compliance with simple handwashing in ICUs varies from only 20 to 40% (37,128–131).

Characterization of the interactions between PBP2a and β-lactams may elucidate the basis for the extremely low affinity for β-lactam antibiotics and contribute to the rational to design better PBP2a inhibitors, leading to more effective antibacterial agents for MRSA and other bacteria (132). PBP2a has already been utilized as a screening target for discovery of new β-lactam antibiotics with enhanced affinity and improved activity against MRSA (132,133).

There is an obvious need for more effective antibiotic therapy for infections with MRSA. Reports describing treatment failure of vancomycin for multi–drug-resistant MRSA infections have raised concern for the emergence of strains of MRSA for which there will be no effective therapy. However, new therapeutic agents alone will not provide a long-term solution, and our attention to prevention must remain constant. Strict adherence to hospital infection-control practices, as well as appropriate use of antibiotics and improved surveillance systems to track the emergence of resistance patterns, are of primary importance as we look to the future usefulness of antibiotic therapy against this extremely adaptive organism.

REFERENCES

1. Berger-Bachi B. Resistance not mediated by β-lactamase (methicillin resistance). In: Crossley KB, Archer GL, eds. The Staphylococci in Human Disease. New York: Churchill Livingstone, 1997:158–167.
2. Bradley SF. Methicillin-resistant *Staphylococcus aureus* infection. Clin Geriatr Med 1992; 8:853–868.
3. Chambers HF. Methicillin resistance in Staphylococci: molecular and biochemical basis and clinical implications. Clin Microbiol Rev 1997; 10:781–791.
4. de Lencastre H, de Jonge BLM, Matthews PR, Tomasz A. Molecular aspects of methicillin resistance in *Staphylococcus aureus*. J Antimicrob Chemother 1994; 33:7–24.
5. Hiramatsu K. Molecular evolution of MRSA. Microbiol Immunol 1995; 39(8):531–543.
6. Skurray RA, Firth N. Molecular evolution of multiple-antibiotic-resistant staphylococci. Ciba Found Symp 1997; 207:167–183.
7. Hartman BJ, Tomasz A. Low-affinity penicillin-binding protein associated

with β-lactam resistance in *Staphylococcus aureus*. J Bacteriol 1984; 158: 513–516.
8. Reynolds PE, Fuller C. Methicillin-resistant strains of *Staphylococcus aureus*; presence of identical additional penicillin-binding protein in all strains examined. FEMS Microbiol Lett 1986; 33:251–254.
9. Wise EM, Park JT. Penicillin: its basic site of action as an inhibitor of a peptide cross-linking reaction in cell wall mucopeptide synthesis. Proc Natl Acad Sci USA 1965; 54:75–81.
10. Wise RI, Ossman EA, Littlefield DR. Personal reflections on nosocomial staphylococcal infections and the development of hospital surveillance. Rev Infect Dis 1989; 11(6):1005–1019.
11. Hiramatsu K. The emergence of *Staphylococcus aureus* with reduced susceptibility to vancomycin in Japan. Am J Med 1998; 104:7S–10S.
12. Barbar M. Methicillin-resistant staphylococci. J Clin Pathol 1961; 14:385–393.
13. Bulger RJ. A methicillin-resistant strain of *Staphylococcus aureus*: clinical and laboratory experience. Ann Intern Med 1967; 67:81–89.
14. Barrett FF, McGehee RF, Finland M. Methicillin-resistant *Staphylococcus aureus* at Boston City Hospital. Bacteriologic and epidemiologic observations. N Engl J Med 1968; 279:441–448.
15. Boyce JM, Landry M. Deetz TR, DuPont HL. Epidemiologic studies of an outbreak of nosocomial methicillin-resistant *Staphylococcus aureus* infections. Infect Control 1981; 2:110–116.
16. Crossley K, Landesman B, Zaske D. An outbreak of infections caused by strains of *Staphylococcus aureus* resistant to methicillin and aminoglycosides. II. Epidemiologic Studies. J Infect Dis 1979; 139: 280–287.
17. Locksley RM, Cohen ML, Quinn TC, Thompkins LS, Coyle MB, Kirihara JM, Counts GW. Multiply antibiotic-resistant *Staphylococcus aureus*: introduction, transmission, and evolution of nosocomial infection. Ann Intern Med 1982; 97:317–324.
18. Craven DE, Reed C, Kollisch N, DeMaria A, Lichtenberg D, Shen K, McCabe WR. A large outbreak of infections caused by a strain of *Staphylococcus aureus* resistant to oxacillin and aminoglycosides. Am J Med 1981; 71:53–58.
19. Peacock JE Jr, Marsik FJ, Wenzel RP. Methicillin-resistant *Staphylococcus aureus*: introduction and spread within a hospital. Ann Intern Med 1980; 93: 526–532.
20. Boyce JM, Causey WA. Increasing occurrence of methicillin-resistant *Staphylococcus aureus* in the United States. Infect Control 1982; 3:377–383.
21. Haley RW, Hightower AW, Khabbaz RF, Thornsberry C, Martone JW, Allen JR, Hughes JM. The emergence of methicillin-resistant *Staphylococcus aureus* infections in United States hospitals. Possible role of the house staff-patient transfer circuit. Ann Intern Med 1982; 97:297–308.
22. Rimland D. Nosocomial infections with methicillin- and tobramycin-resistant *Staphylococcus aureus*—implication of physiotherapy in hospital-wide dissemination. Am J Med Sci 1985; 290:91–97.
23. Ward TT, Winn RE, Hartstein AL, Sewell DL. Observations relating to an interhospital outbreak of methicillin-resistant *Staphylococcus aureus*: role of antimicrobial therapy in infection control. Infect Control 1981; 2:453–459.

24. Panlilio AL, Culver DH, Gaynes RP, Banerjee S, Henderson TS, Tolson JS, Martone WJ. Methicillin-resistant *Staphylococcus aureus* in U.S. hospitals, 1975–1991. Infect Control Hosp Epidemiol 1992; 13:582–586.
25. Gaynes RP, Culver DH. National Nosocomial Infection Surveillance (NNIS) System. Nosocomial methicillin-resistant *Staphylococcus aureus* (MRSA) in the United States, 1975–1996. In: Proceedings of the Annual Meeting of the Infectious Disease Society of America (San Francisco). Alexandria, Virginia: IDSA, 1997:206.
26. Moreno F, Crisp C, Jorgensen JH, Patterson JE. Methicillin-resistant *Staphylococcus aureus* as a community organism. Clin Infect Dis 1995; 21:1308–1312.
27. Herold BC, Immergluck LC, Maranan MC, Lauderdale DS, Gaskin RE, Boyle-Vavra S, Leitch CD, Daum RS. Community-acquired methicillin-resistant *Staphylococcus aureus* in children with no identified predisposing risk. JAMA 1998; 279:593–598.
28. Saravolatz LD, Pohlod DJ, Arking LM. Community-acquired MRSA infections: a new source of nosocomial outbreaks. Ann Intern Med 1982; 97:325–329.
29. Saravolatz LD, Markowitz N, Arking L, Pohlod D, Fisher E. MRSA. Epidemiologic observations during a community-acquired outbreak. Ann Intern Med 1982; 96:11–16.
30. Pate KR, Nolan RL, Bannerman TL, Feldman S. Methicillin-resistant *Staphylococcus aureus* in the community. Lancet 1995; 346:132–133.
31. Berman DS, Eisner W, Kreiswirth B. Community-acquired methicillin-resistant *Staphylococcus aureus* infection. N Engl J Med 1993; 329:1896.
32. Craven DE, Rixinger AI, Goularte TA, McCabe WR. Methicillin-resistant *Staphylococcus aureus* bacteremia linked to intravenous drug abusers using a "shooting gallery." Am J Med 1986; 80:770–776.
33. Berman DS, Schafler S, Simberkoff MS, Rahal JJ. *Staphylococcus aureus* colonization in intravenous drug abusers, dialysis patients, and diabetics. J Infect Dis 1987; 155:829–831.
34. Pittet D, Tarara D, Wenzel RP. Nosocomial bloodstream infection in critically ill patients. Excess length of stay, extra costs, and attributable mortality. JAMA 1994; 271:1598–1601.
35. Abramsom MA, Sexton DJ. Nosocomial methicillin-resistant and methicillin-susceptible *Staphylococcus aureus* primary bacteremia: at what costs? Infect Control Hosp Epidemiol 1999; 20:408–411.
36. Rubin RJ, Harrington CA, Poon A, Dietrich K, Greene JA, Moiduddin A. The economic impact of *Staphylococcus aureus* infection in New York City hospitals. Emerg Infect Dis 1999; 5:9–17.
37. Dennesen PJW, Bonten MJM, Weinstein RA. Multiresistant bacteria as a hospital epidemic problem. Ann Med 1998; 30:176–185.
38. Mulligan ME, Murray-Leisure KA, Ribner BS, Standiford HC, John JF, Korvick JA, Kaufman CA, Yu VL. Methicillin-resistant *Staphylococcus aureus*: a consensus review of the microbiology, pathogenesis, and epidemiology with implications for prevention and management. Am J Med 1993; 94:313–328.
39. Edmond MB, Wenzel RP, Pasculle AW. Vancomycin-resistant *Staphylococcus aureus*: perspectives on measures needed for control. Ann Intern Med 1996; 124:329–334.

40. Casewell MW. The nose: an underestimated source of *Staphylococcus aureus* causing wound infection. J Hosp Infect 1998; 40:S4–S11.
41. Williams REO. Healthy carriage of *Staphylococcus aureus*: its prevalence and importance. Bacteriol Rev 1963; 27:56–71.
42. Gaynes RP, Culver DH, Horan TC, Henderson TS, Tolson JS, Martone WJ. Trends in MRSA in United States hospitals. Infect Dis Clin Practice 1994; 2: 452–455.
43. Schentag JJ, Hyatt JM, Carr JR, Paladino JA, Birmingham MC, Zimmer GS, Cumbo TJ. Genesis of methicillin-resistant *Staphylococcus aureus* (MRSA), how treatment of MRSA infections has selected for vancomycin-resistant Enterococcus faecium and the importance of antibiotic management and infection control. Clin Infect Dis 1998; 26:1204–1214.
44. Centers for Disease Control and Prevention. National Nosocomial Infection Surveillance System report: data summary from October 1986–April 1996. Atlanta: US Department of Health and Human Services, 1996.
45. Storch GA, Radcliff JL, Meyer PL, Hinrichs JH. Methicillin-resistant *Staphylococcus aureus* in a nursing home. Infect Control 1987; 8:24–29.
46. Kaufmann CA, Bradley SF, Terpenning MS. Methicillin-resistant *Staphylococcus aureus* in long-term care facilities. Infect Control Hosp Epidemiol 1990; 11:600–603.
47. Levine DP, Cushing RD, Jui J, Brown WJ. Community-acquired methicillin-resistant *Staphylococcus aureus* endocarditis in the Detroit Medical Center. Ann Intern Med 1982; 97:330–338.
48. Boyce JM, Opal SM, Potter-Bynoe G, Medeiros AA. Spread of methicillin-resistant *Staphylococcus aureus* in a hospital after exposure to a health care worker with chronic sinusitis. Clin Infect Dis 1993; 17:496–504.
49. Boyce JM, White RL, Causey WA, Lockwood WR. Burn units as a source of methicillin-resistant *Staphylococcus aureus* infections. JAMA 1983; 249:2803–2807.
50. Rutala WA, Katz EB, Sherertz RJ, Sarubbi FA Jr. Environmental study of a methicillin-resistant *Staphylococcus aureus* epidemic in a burn unit. J Clin Microbiol 1983; 18:683–688.
51. Farrington M, Ling J, Ling T, French GL. Outbreaks of infection with methicillin-resistant *Staphylococcus aureus* on neonatal and burn units of a new hospital. Epidemiol Infect 1990; 105:215–228.
52. Monnet DL. Methicillin-resistant *Staphylococcus aureus* and its relationship to antimicrobial use: possible implications for control. Infect Control Hosp Epidemiol 1998; 19:552–559.
53. Layton MC, Hierholzer WJ Jr, Patterson JE. The evolving epidemiology of methicillin-resistant *Staphylococcus aureus* at a university hospital. Infect Control Hosp Epidemiol 1995; 16:12–17.
54. Asensio A, Guerrero A, Quereda C, Lizan M, Martinez-Ferrer M. Colonization and infection with methicillin-resistant *Staphylococcus aureus*: associated factors and eradication. Infect Control Hosp Epidemiol 1996; 17:20–28.
55. Humphreys H, Duckworth G. Methicillin-resistant *Staphylococcus aureus* (MRSA)—a re-appraisal of control measures in the light of changing circumstances. J Hosp Infect 1997; 36:167–170.

56. Thompson RL, Cabezudo I, Wenzel RP. Epidemiology of nosocomial infections caused by methicillin-resistant *Staphylococcus aureus*. Ann Intern Med 1982; 97:309–317.
57. File TM Jr. Overview of resistance in the 1990s. Chest 1999; 115:3s–8s.
58. Embil J, Ramotar K, Romance L, Alfa M, Conly J, Cronk S, Taylor G, Sutherland B, Louie T, Henderson E, et al. Methicillin-resistant *Staphylococcus aureus* in tertiary care institutions on the Canadian prairies 1990–1992. Infect Control Hosp Epidemiol 1994; 15:646–651.
59. Finland M. Changing patterns of resistance of certain common pathogenic bacteria to antimicrobial agents. N Engl J Med 1955; 252:570–580.
60. Chambers HF. Penicillin-binding protein-mediated resistance in pneumococci and staphylococci. J Infect Dis 1999; 179:S353–S359.
61. McDougal LK, Thornsberry C. The role of beta-lactamase in staphylococcal resistance to penicillinase-resistant penicillins and cephalosporins. J Clin Microbiol 1986; 23:832–839.
62. Tomasz A, Drugeon HB, de Lencastre HM, Jabes D, McDougal L, Bille J. New mechanism for methicillin resistance in *Staphylococcus aureus*: clinical isolates that lack the PBP 2a gene and contain normal penicillin-binding proteins with modified penicillin-binding capacity. Antimicrob Agents Chemother 1989; 33:1869–1874.
63. Kuhl SA, Pattee PA, Baldwin JN. Chromosomal map location of the methicillin resistance determinant in *Staphylococcus aureus*. J Bacteriol 1978; 135: 460–465.
64. Kreiswirth B, Kornblum J, Arbeit RD, Eisner W, Maslow JN, McGeer A, Low DE, Novick RP. Evidence for a clonal origin of methicillin resistance in *Staphylococcus aureus*. Science 1993; 259:227–230.
65. Kobayashi N, Wu H, Kojima K, Taniguchi K, Urasawa S, Uehara N, Omizu Y, Kishi Y, Yagihashi A, Kurokawa I. Detection of mecA, femA, and femB genes in clinical strains of staphylococci using polymerase chain reaction. Epidemiol Infect 1994; 113:259–266.
66. Ryffel C, Tesch W, Birch-Machin I, Reynolds PE, Barberis-Maino L, Kayser FH, Berger-Bachi B. Sequence comparison of mecA genes isolated from methicillin-resistant *Staphylococcus aureus* and *Staphylococcus epidermidis*. Gene 1990; 94:137–138.
67. Suzuki E, Hiramatsu K, Yokota T. Survey of methicillin-resistant clinical strains of coagulase-negative staphylococci for mecA gene distribution. Antimicrob Agents Chemother 1992; 36:429–434.
68. Ubukata K, Nonoguchi R, Song MD, Matsuhashi M, Konno M. Homology of mecA gene in methicillin-resistant *Staphylococcus haemolyticus* and *Staphylococcus simulans* to that of *Staphylococcus aureus*. Antimicrob Agents Chemother 1990; 34:170–172.
69. Song MD, Wachi M, Doi M, Ishino F, Matsuhashi M. Evolution of an inducible penicillin-target protein in methicillin-resistant *Staphylococcus aureus* by gene fusion. FEBS Lett 1987; 221:167–171.
70. Georgopapadakou NH, Dix BA, Mauriz YR. Possible physiological functions of penicillin-binding proteins in *Staphylococcus aureus*. Antimicrob Agents Chemother 1986; 29:333–336.

71. Reynolds PE. The essential nature of staphylococcal penicillin-binding proteins. In: Actor P, Daneo-Moore L, Higgins ML, Salton MR, Shockman GD, eds. Antibiotic Inhibition of Bacterial Cell Surface Assembly and Function. Washington, DC: American Society for Microbiology, 1988:343–351.
72. Chambers HF. Coagulase-negative staphylococci resistant to beta-lactam antibiotics in vivo produce penicillin-binding protein 2a. Antimicrob Agents Chemother 1987; 31:1919–1924.
73. Georgopapadakou NH, Smith SA, Bonner DP. Penicillin-binding proteins in a *Staphylococcus aureus* strain resistant to specific beta-lactam antibiotics. Antimicrob Agents Chemother 1982; 22:172–175.
74. Brown DF, Reynolds PE. Intrinsic resistance to beta-lactam antibiotics in *Staphylococcus aureus*. FEBS Lett 1980; 122:275–278.
75. Tesch W, Ryffel C, Strassle A, Kayser FH, Berger-Bachi B. Evidence of a novel staphylococcal mec-encoded element (mecR) controlling expression of penicillin-binding protein 2′. Antimicrob Agents Chemother 1990; 34:1703–1706.
76. Hiramatsu K, Asada K, Suzuki E, Okonogi K, Yokota T. Molecular cloning and nucleotide sequence determination of the regulator region of mecA gene in methicillin-resistant *Staphylococcus aureus* (MRSA). FEBS Lett 1992; 298: 133–136.
77. Archer GL, Niemeyer DM. Origin and evolution of DNA associated with resistance to methicillin in staphylococci. Trends Microbiol 1994; 2:343–347.
78. Stewart PR, Dubin DT, Chikramane SG, Inglis B, Matthews PR, Poston SM. IS257 and small plasmid insertions in the mec region of the chromosome of *Staphylococcus aureus*. Plasmid 1994; 31:12–20.
79. Hackbarth CJ, Chambers HF. blaI and blaR1 regulate beta-lactamase and PBP 2a production in methicillin-resistant *Staphylococcus aureus*. Antimicrob Agents Chemother 1993; 37:1144–1149.
80. Ryffel C, Kayser FH, Berger-Bachi B. Correlation between regulation of mecA transcription and expression of methicillin resistance in staphylococci. Antimicrob Agents Chemother 1992; 36:25–31.
81. Boyce JM, Medeiros AA, Papa EF, O'Gara CJ. Induction of beta-lactamase and methicillin resistance in unusual strains of methicillin-resistant *Staphylococcus aureus*. J Antimicrob Chemother 1990; 25:73–81.
82. Suzuki E, Kuwahara-Arai K, Richardson JF, Hiramatsu K. Distribution of mec regulator genes in methicillin-resistant Staphylococcus clinical strains. Antimicrob Agents Chemother 1993; 37:1219–1226.
83. Archer GL, Niemeyer DM, Thanassi JA, Pucco MJ. Dissemination among staphylococci of DNA sequences associated with methicillin resistance. Antimicrob Agents Chemother 1994; 38:447–454.
84. Weller TMA. The distribution of mecA, mecR1 and mecI and sequence analysis of mecI and the mec promoter region in staphylococci expressing resistance to methicillin. J Antimicrob Chemother 1999; 43:15–22.
85. Kobayashi N, Taniguchi K, Kojima K, Urasawa S, Uehara N, Omizu Y, Kishi Y, Yagihashi A, Kurokawa I, Watanabe N. Genomic diversity of mec regulator genes in methicillin-resistant *Staphylococcus aureus* and *Staphylococcus epidermidis*. Epidemiol Infect 1996; 117:289–295.

86. Kobayashi N, Taniguchi K, Urasawa S. Analysis of diversity of mutations in the mecI gene and mecA promoter/operator region of methicillin-resistant *Staphylococcus aureus* and *Staphylococcus epidermidis*. Antimicrob Agents Chemother 1998; 42:717–720.
87. Kuwahara-Arai K, Kondo N, Hori S, Tateda-Suzuki E, Hiramatsu K. Suppression of methicillin resistance in a mecA-containing pre-methicillin-resistant *Staphylococcus aureus* strain is caused by the mecI-mediated repression of PBP 2′ production. Antimicrob Agents Chemother 1996; 40:2680–2685.
88. Hackbarth CJ, Miick C, Chambers HF. Altered production of penicillin-binding protein 2a can affect phenotypic expression of methicillin resistance in *Staphylococcus aureus*. Antimicrob Agents Chemother 1994; 38:2568–2571.
89. Berger-Bachi B, Strassle A, Kayser FH. Characterization of an isogenic set of methicillin-resistant and susceptible mutants of *Staphylococcus aureus*. Eur J Clin Microbiol 1986; 5:697–701.
90. Kornblum J, Hartman BJ, Novick RP, Tomasz A. Conversion of a homogeneously methicillin-resistant strain of *Staphylococcus aureus* to heterogeneous resistance by Tn551-mediated insertional inactivation. Eur J Clin Microbiol 1986; 5:714–718.
91. Murakami K, Tomasz A. Involvement of multiple genetic determinants in high-level methicillin resistance in *Staphylococcus aureus*. J Bacteriol 1989; 171:874–879.
92. Sjostrom JE, Lofdahl S, Philipson L. Transformation reveals a chromosomal locus of the gene(s) for methicillin resistance in *Staphylococcus aureus*. J Bacteriol 1975; 123:905–915.
93. Berger-Bachi B, Barberis-Maino L, Strassle A, Kayser FH. FemA, a host-mediated factor essential for methicillin resistance in *Staphylococcus aureus*: molecular cloning and characterization. Mol Gen Genet 1989; 219:263–269.
94. Tomasz A. Auxiliary genes assisting in the expression of methicillin resistance in *Staphylococcus aureus*. In: Novick RP, ed. Molecular Biology of the Staphylococci. New York: VCH, 1990:565–583.
95. Berger-Bachi B. Insertional inactivation of staphylococcal methicillin resistance by Tn551. J Baceteriol 1983; 154:479–487.
96. de Lencastre H, Tomasz A. Reassessment of the number of auxiliary genes essential for expression of high-level methicillin resistance in *Staphylococcus aureus*. Antimicrob Agents Chemother 1994; 38:2590–2598.
97. Wu SW, de Lencastre H. Mrp-A new auxiliary gene essential for optimal expression of methicillin resistance in *Staphylococcus aureus*. Microbial Drug Resist 1999; 5:9–18.
98. Berger-Bachi B. Expression of resistance to methicillin. Trends Microbiol 1994; 2:389–393.
99. Berger-Bachi B, Strassle A, Gustafson JE, Kayser FH. Mapping and characterization of multiple chromosomal factors involved in methicillin resistance in *Staphylococcus aureus*. Antimicrob Agents Chemother 1992; 36:1367–1373.
100. Gustafson J, Strassle A, Hachler H, Kayser FH, Berger-Bachi B. The femC locus of *Staphylococcus aureus* required for methicillin resistance includes the glutamine synthetase operon. J Bacteriol 1994; 176:1460–1467.

101. Ludovice AM, Wu SW, de Lencastre H. Molecular cloning and DNA sequencing of the *Staphylococcus aureus* UDP-N-acetylmuramyl tripeptide synthetase (murE) gene, essential for the optimal expression of methicillin resistance. Microb Drug Resist 1998; 4:85–90.
102. Ornelas-Soares A, de Lencastre H, de Jonge BL, Tomasz A. Reduced methicillin resistance in a new *Staphylococcus aureus* transposon mutant that incorporates muramyl dipeptides into the cell wall peptidoglycan. J Biol Chem 1994; 269:27246–27250.
103. de Jonge BL, Sidow T, Chang YS, Labischinski H, Berger-Bachi B, Gage DA, Tomasz A. Altered muropeptide composition in *Staphylococcus aureus* strains with an inactivated femA locus. J Bacteriol 1993; 175:2779–2782.
104. Henze U, Sidow T, Wecke J, Labischinski H, Berger-Bachi B. Influence of femB on methicillin resistance and peptidoglycan metabolism in *Staphylococcus aureus*. J Bacteriol 1993; 175:1612–1620.
105. Jorgensen JH. Laboratory and epidemiologic experience with methicillin-resistant *Staphylococcus aureus* in the USA. Eur J Clin Microbiol 1986; 5: 693–696.
106. Sabeth LD. Mechanisms of resistance to beta-lactam antibiotics in strains of *Staphylococcus aureus*. Ann Intern Med 1982; 97:339–344.
107. Tomasz A, Nachman S, Leaf H. Stable classes of phenotypic expression in methicillin-resistant clinical isolates of staphylococci. Antimicrob Agents Chemother 1991; 35:124–129.
108. Hackbarth CJ, Chambers HF. Methicillin-resistant staphylococci: detection methods and treatment of infections. Antimicrob Agents Chemother 1989; 33:995–999.
109. Chambers HF, Hackbarth CJ. Effect of NaCl and nafcillin on penicillin-binding protein 2a and heterogeneous expression of methicillin resistance in *Staphylococcus aureus*. Antimicrob Agents Chemother 1987; 31:1982–1988.
110. de Lencastre H, Figueiredo AM, Tomasz A. Genetic control of population structure in heterogeneous strains of methicillin resistant *Staphylococcus aureus*. Eur J Clin Microbiol Infect Dis 1993; 12:S13–S18.
111. Jorgensen JH, Redding JS, Maher LA, Ramirez PE. Salt-supplemented medium for testing methicillin-resistant staphylococci with newer beta-lactams. J Clin Microbiol 1988; 26:1675–1678.
112. Wu S, Piscitelli C, de Lencastre H, Tomasz A. Tracking the evolutionary origin of the methicillin resistance gene: cloning and sequencing of a homologue of mecA from a methicillin susceptible strain of *Staphylococcus sciuri*. Microb Drug Resist 1996; 2:435–441.
113. Couto I, de Lencastre H, Severina E, Kloos W, Webster JA, Hubner RJ, Sanches IS, Tomasz A. Ubiquitous presence of a mecA homologue in natural isolates of *Staphylococcus sciuri*. Microb Drug Resist 1996; 2:377–391.
114. Lacey RW, Grinsted J. Genetic analysis of methicillin-resistant strains of *Staphylococcus aureus*: evidence for their evolution from a single clone. J Med Microbiol 1973; 6:511–526.
115. Musser JM, Kapur V. Clonal analysis of methicillin-resistant *Staphylococcus aureus* strains from intercontinental sources: association of the mec gene with

divergent phylogenetic lineages implies dissemination by horizontal transfer and recombination. J Clin Microbiol 1992; 30:2058–2063.
116. Neu HC. The crisis in antibiotic resistance. Science 1992; 257:1064–1073.
117. Lugeon C, Blanc DS, Wenger A, Francioli P. Molecular epidemiology of methicillin-resistant *Staphylococcus aureus* at a low-incidence hospital over a 4-year period. Infect Control Hosp Epidemiol 1995; 16:260–267.
118. Krzysztof S, Roberts RB, Haber SW, Tomasz A. The development of vancomycin resistance in a patient with methicillin-resistant *Staphylococcus aureus* infection. N Engl J Med 1999; 340:517–523.
119. Noble WC, Virani Z, Cree RG. Co-transfer of vancomycin and other resistance genes from *Enterococcus faecalis* NCTC 12201 to *Staphylococcus aureus*. FEMS Microbiol Lett 1992; 72:195–198.
120. Hiramatsu K, Aritaka N, Hanaki H, Kawasaki S, Hosoda Y, Hori S, Fukuchi Y, Kobayashi I. Dissemination in Japanese hospitals of strains of *Staphylococcus aureus* heterogeneously resistant to vancomycin. Lancet 1997; 350: 1670–1673.
121. Centers for Disease Control and Prevention. *Staphylococcus aureus* with reduced susceptibility to vancomycin—United States. MMWR 1997; 46: 765–766.
122. Centers for Disease Control and Prevention. Update: *Staphylococcus aureus* with reduced susceptibility to vancomycin—United States. MMWR 1997; 46:813–815.
123. Centers for Disease Control and Prevention. Interim guidelines for prevention and control of staphylococcal infection associated with reduced susceptibility to vancomycin. MMWR 1997; 46:626–628,635.
124. Quale J, Landman D, Saurina G, Atwood E, DiTore V, Patel K. Manipulation of a hospital antimicrobial formulary to control an outbreak of vancomycin-resistant enterococci. Clin Infect Dis 1996; 23:1020–1025.
125. Rice LB, Wiley SH, Papanicolaou GA, Medeiros AA, Eliopoulos GM, Moellering RC Jr, Jacoby GA. Outbreak of ceftazidime resistance caused by extended-spectrum beta-lactamases at a Massachusetts chronic-care facility. Antimicrob Agents Chemother 1990; 34:2193–2199.
126. Meyer KS, Urban C, Eagan JA, Berger BJ, Rahal JJ. Nosocomial outbreak of Klebsiella infection resistant to late-generation cephalosporins. Ann Intern Med 1993; 119:353–358.
127. Kollef MH, Vlasnik J, Sharpless L, Pasque C, Murphy D, Fraser V. Scheduled change of antibiotic classes: a strategy to decrease the incidence of ventilator-associated pneumonia. Am J Respir Crit Care Med 1997; 156:1040–1048.
128. Gould D. Nurses' hand decontamination practice: results of a local study. J Hosp Infect 1994; 28:15–30.
129. Doebbeling BN, Stanley GL, Sheetz CT, Pfaller MA, Houston AK, Annis L, Li N, Wenzel RP. Comparative efficacy of alternative hand-washing agents in reducing nosocomial infections in intensive care units. N Engl J Med 1992; 327:88–93.
130. Albert RK, Condie F. Hand-washing patterns in medical intensive-care units. N Engl J Med 1981; 304:1465–1466.

131. Simmons B, Bryant J, Neiman K, Spencer L, Arheart K. The role of hand washing in prevention of endemic intensive care unit infections. Infect Control Hosp Epidemiol 1990; 11:589–594.
132. Lu WP, Sun Y, Bauer MD, Paule S, Koenigs PM, Kraft WG. Penicillin-binding protein 2a from methicillin-resistant *Staphylococcus aureus*: kinetic characterization of its interactions with β-lactams using electrospray mass spectrometry. Biochem 1999; 38:6537–6546.
133. Hecker SJ, Cho IS, Glinka TW, Zhang ZJ, Price ME, Lee VJ, Christensen BG, Boggs A, Chamberland S, Malouin F, Parr TR, Annamalai T, Blais J, Bond EL, Case L, Chan C, Crase J, Frith R, Griffith D, Harford L, Liu N, Ludwikow M, Mathias K, Rea D, Williams R. Discovery of MC-02,331, a new cephalosporin exhibiting potent activity against methicillin-resistant *Staphylococcus aureus*. J Antibiot 1998; 51:722–734.

15

Drug Resistance and Tuberculosis Chemotherapy—From Concept to Genomics

Alexander S. Pym
Liverpool University, Liverpool, England, and Institut Pasteur, Paris, France

Stewart T. Cole
Institut Pasteur, Paris, France

Despite the development of effective treatment over four decades ago, tuberculosis is still one of the most prominent infectious causes of morbidity and mortality. Since the mid-1980s rates of tuberculosis have been increasing even in some industrialized countries such as the United States, and this has been attributed to the breakdown of tuberculosis control programs, declining standards of living, and particularly the emergence of the human immunodeficiency virus pandemic. This worldwide increase in tuberculosis has been accompanied by the widespread appearance of strains of multi–drug-resistant *Mycobacterium tuberculosis*, the bacterium responsible for the vast majority of cases of human tuberculosis. These strains now pose a significant threat to world tuberculosis control. Individuals infected with these multi–drug-resistant strains are potentially untreatable or require prolonged multidrug therapy to be cured. Until recently, the mechanisms of resistance to antituberculosis drugs remained elusive. However, over the last decade, advances in mycobacterial genetics have enabled researchers to determine the principal mechanisms of resistance to the key antituberculosis drugs rifampin, isoniazid, pyrazinamide, ethambutol, and streptomycin. In addition, the availability of the complete genome sequence of *M. tuberculosis* has provided novel insights into the natural drug resistance of this unique organism as well as identifying a range of potential new drug targets. These scientific advances form the basis for developing new drugs and therapeutic strategies, as well as molecular methods for diagnosing drug resistance, urgently required to tackle the global problem of multi–drug-resistant tuberculosis.

1 INTRODUCTION

Over the last decade, tuberculosis has regrettably established itself as a global health emergency (1). It remains top of the list of infectious diseases with respect to global mortality and seventh in the list of all causes of mortality (2). Predictions for the future are grim, with an estimated 225 million new cases and 79 million deaths projected for the first three decades of our new century (3). Rates of tuberculosis have fallen steadily in the industrialized world throughout the 1900s, largely as a consequence of general improvements in health and living conditions. This led to the complacent view that tuberculosis would quietly disappear, and resulted in the failure to sustain or implement effective treatment and control measures. Since 1985 the annual rates of tuberculosis have risen (4), even in the United States, and this has been attributed to the breakdown of tuberculosis control programs, declining standards of living, and particularly the emergence of the human immunodeficiency virus (HIV) pandemic (4,5). This global "reemergence" of tuberculosis has been accompanied by the widespread appearance of strains of multi–drug-resistant *M. tuberculosis* (6,7) (the bacterium responsible for the vast majority of cases of human tuberculosis). These multi–drug-resistant epidemics now pose a significant threat to global tuberculosis control. Although other mycobacteria are human pathogens, notably *M. leprae* and the *M. avium* complex (MAC) (generally in immunodeficient individuals), this chapter will focus primarily on the recent rapid advances made in understanding the mechanisms of drug resistance in *M. tuberculosis* that have been prompted by the need to develop new drugs and therapeutic strategies to combat multi–drug-resistant tuberculosis. These advances have enabled the development of molecular techniques for the rapid diagnosis of drug resistance in *M. tuberculosis*, which are discussed more extensively in Chapter 11.

2 DEVELOPMENT OF TUBERCULOSIS CHEMOTHERAPY

The problem of resistance to antimycobacterial drugs is an old one. Within a decade of the development of the first effective agents against tuberculosis, drug resistance had been described and treatment strategies required to prevent it worked out. The first two drugs to enter formal clinical trials were developed in the early 1940s: streptomycin (SM), isolated from *Streptomyces griseus* (8), and *para*-aminosalicylic (PAS) acid (9), a synthetic derivative of salicylic acid. These compounds were rapidly shown to be effective in animal models, and early reports of their clinical use suggested they were effective against human tuberculosis (9,10). However, it was soon noted that patients with advanced forms of disease had less chance of responding and that early response to treatment in others was rapidly followed by deterioration and the emergence of drug-resistant strains. For example, in 1947, the MRC trial of SM versus bedrest in patients with acute, progressive, pulmonary tuberculosis showed that after 6 months therapy there were significantly less deaths in the SM-treated patients, and that this group was more likely to have had a bacteriological or radiological improvement (11). Unfortunately, 35 of the 41 SM-treated patients were found to be excreting drug-resistant bacilli, and after 5 years of follow-up, the mortality in the streptomycin group was only slightly better than in the controls (53 vs 63%) (12).

The priority of investigators then switched rapidly to investigating ways of preventing the emergence of resistance. It was soon shown that by combining SM with PAS the emergence of resistance to SM could be reduced from 70 to 9% (13). The discovery of a new more potent antimycobacterial agent, isoniazid (14) (INH), soon followed, and regimens combining this agent with SM and PAS were also found to be highly effective in preventing the emergence of drug resistance (15). Thus, in the space of little more than a decade, the first principle of modern tuberculosis chemotherapy had been established; namely, the necessity of combination drug therapy to combat the emergence of resistance. The biological basis for the need for combination therapy is thought to be due to the heavy pulmonary bacillary burden that exists prior to therapy, sufficiently large to contain spontaneous mutants resistant to a single antituberculosis drug, which will be rapidly selected for if treatment commences with only a single agent. Canetti, for example, quantified the number of bacteria found in surgically resected cavities from patients failing to respond to therapy and found this to be at least 10^8 (16). Subsequent estimates of the spontaneous mutation rates for drug resistance to an individual drug have been of the order of 1 in 10^6 for INH and SM (17,18).

Once the principle of combination therapy had been established, the research agenda switched to defining the minimal duration of therapy required to ensure an adequate cure rate. Using various combinations of INH, SM, and PAS, treatment periods for up to 18 months were required to obtain adequate results (19). However, the observation that pyrazinamide (PZA) (a derivative of nicotinamide first used clinically in 1952 [20] but subsequently reserved for use as a second-line agent because of fears about its toxicity [21]) and the newer agent rifampin (RMP) (22) were uniquely capable of sterilizing organs in animal models of tuberculosis lead to trials of shorter courses of therapy. In a series of painstaking and meticulous studies carried out through the 1970s (19), it was established that treatment regimens that contained either PZA or RMP could be reduced to 6 months (short-course therapy). These regimens were capable of curing (defined as patients free of tuberculosis after 2 years of follow-up) in excess of 95% of patients infected with fully sensitive organisms and are the basis for the current World Health Organization (WHO) guidelines, which recommend initial intensive treatment for 2 months with INH, RMP, and PZA (ethambutol or SM is added if there is a suspected high incidence of primary drug resistance) followed by a continuation phase of 4 months with RMP and INH (23).

3 DOTS

These studies also clearly demonstrated that drug resistance is not an intrinsic problem with the drugs themselves. Resistance to antimycobacterial agents should, therefore, be seen as a problem of ensuring that patients infected with tuberculosis are prescribed adequate treatment and adhere to these prescribed regimens. If patients adhere to appropriate therapy, drug resistance will not occur. Recent studies have thus focused on defining the most efficient way of drug delivery and of ensuring patient adherence to therapy. This has led to the recommendation that DOTS (directly observed therapy short course) should be the standard treatment protocol (23,24). Trials carried out from the 1960s (19) onward also established that treatment given intermittently two or three times a week could be as effective as daily regimens, making it possible for patients to be supervised (directly observed) either in a clinic or in the community (25) taking each dose of a 6-month (short course) regimen, thereby ensuring compliance (26).

DOTS has now been adopted by the WHO as their strategy for controlling tuberculosis, and a campaign has been launched to see that DOTS is introduced on a worldwide basis. In some areas where DOTS has been successfully implemented, it has been highly effective, although worldwide implementation has been slow (27). For example, in China, cure rates for tuberculosis were only 50% before the introduction of a DOTS program but rose in the space of a few years to 90% in a program that treated over 100,000 patients (28). A recent study from Botswana though showed that despite an exemplary DOTS program introduced in 1986, rates of tuberculosis have doubled in the last decade (29). This has been largely due to the HIV epidemic. On the basis of skin testing, it has been estimated that in excess of 75% of some African populations are latently infected with tuberculosis. In the absence of HIV infection, these individuals will have a 5–10% chance of developing active tuberculosis in their lifetime. Coinfection with HIV (rates in excess of 40% of young adults have been reported from some areas of southern Africa) increases this risk to 10% per year (30), as well as increasing susceptibility to acquiring tuberculosis and causing rapid progression to disease after infection (31,32). However, the study from Botswana (29) also demonstrated that, despite this catastrophic tuberculosis epidemic, DOTS prevented the emergence of drug-resistant tuberculosis, and reemphasizes that correctly administered treatment even under the most adverse conditions will prevent the appearance of drug-resistant tuberculosis.

4 DRUG-RESISTANT TUBERCULOSIS

4.1 Definition

Drug resistance is conventionally classified into two types: primary drug resistance occurs in individuals who are infected de novo with a drug-resistant strain and secondary (acquired) resistance which arises in an individual initially infected with a drug-sensitive strain from which resistant mutants emerge as a result of inadequate therapy. Multi–drug-resistant tuberculosis (MDR-TB) is conventionally described as resistance to two or more drugs, particularly RMP and INH. There is a certain redundancy built into the currently advocated short-course regimens which ensures that they will be effective in individuals infected with tuberculosis resistant to a single drug and probably to two drugs except for the combination of RMP and INH resistance (33). This is the basis for the current definition of MDR-TB, as individuals infected with INH/RMP–resistant strains will not respond to short-course therapy and will become rapidly resistant to the other front-line drugs (Table 1).

4.2 Epidemiology

Until the 1980s, MDR-TB was not perceived as a threat to tuberculosis control, and surveys from this period suggested that MDR strains were rare (34) and tended to occur only in the context of multiple courses of

TABLE 1 Front-Line Drugs Recommended for the Treatment of Tuberculosis

Drug	MIC[a] (γ/mL)	Peak serum levels (μg/mL)	Antituberculosis activity	Daily dose (mg)
Isoniazid	0.01–0.25	3–5	++++	300
Rifampin	0.06–0.25	8–20	++++	600
Pyrazinamide	6–50	20–60	+++	2000
Ethambutol	0.5–2.0	3–5	++	2500
Streptomycin	0.25–2.0	35–45	+++	1000

[a]Minimum inhibitory concentration.
Source: Ref. 53.

inadequate treatment. However, a new phenomenon appeared in the form of microepidemics of MDR-TB, which were first seen in health care settings in the United States (31,35–38) but also documented elsewhere (39–42). These were the result of infectious cases excreting and transmitting MDR strains of tuberculosis to numerous contacts, and occurred particularly among groups of HIV-infected individuals. Subsequent surveys found that MDR-TB was widespread in several cities in the United States. A report from New York revealed that 19% of all isolates were resistant to RMP and INH and 33% were resistant to at least one drug (43). Analysis of risk factors for MDR-TB found that history of previous treatment for tuberculosis was the strongest predictor; indicating that acquired drug resistance was a major factor in these epidemics. However, molecular epidemiology also demonstrated that 22% of all MDR-TB strains isolated in New York that year represented a single clone (strain W), suggesting that ongoing transmission leading to primary resistance was amplifying the epidemic (44). Isolates resembling strain W have been identified in Europe, South Africa, and Puerto Rico, as well as throughout the United States (45,46), revealing the potential for global dissemination of drug-resistant strains.

There are only scanty data on the levels of drug resistance found in the developing world; largely due to the absence in these countries of culture facilities necessary for determining drug susceptibility. A recent WHO-sponsored report from 35 countries though found that in certain regions of the world, where tuberculosis control programs were poor, there were already well-established MDR-TB epidemics, with levels of combined resistance (primary and secondary) to RMP and INH of over 20% (47).

4.3 Treatment of Drug-Resistant Tuberculosis

The first outbreaks of MDR-TB were associated with very high short-term case fatalities (greater than 80% in HIV-infected individuals (48), but following optimization of treatment, more recent studies have reported a better prognosis for MDR-TB–infected individuals (49–51). Treatment though is expensive (estimated at $180,000 per case in the United States) (52) and complex, involving at least four and possibly as many as six or seven antituberculosis drugs depending on the results of susceptibility testing (53,54). These second-line treatments will include agents such as cycloserine, PAS, and thiacetazone which are poorly tolerated. The minimum duration of treatment has not yet been determined, but for dual RMP/INH resistance, 2 years of therapy and resectional surgery have been advocated (53,54).

Although massive investment in tuberculosis control and treatment infrastructure appears to have brought the tuberculosis and MDR-TB epidemics under control in the United States (55,56), it remains a formidable challenge in the developing world. With the growing awareness of the scale of the global MDR-TB problem, the WHO has launched a "DOTS plus" campaign to incorporate treatment of drug-resistant tuberculosis into national tuberculosis control programs. To implement and financially support such a program will be a herculean task given the continuing paucity of existing tuberculosis treatment facilities in many areas of the world and the cost of additional treatment for countries that can barely afford RMP.

5 MOLECULAR BASIS OF DRUG RESISTANCE

5.1 Aminoglycosides

5.1.1 Streptomycin

Streptomycin (SM) is an aminoglycosidic aminocyclitol, the first of this class of antibiotics to be identified. Its initial clinical introduction marked the beginning of the chemotherapeutic era of tuberculosis control, and most of the clinical trials that defined the principles of antituberculosis chemotherapy used SM. Its adverse toxicity profile and the need for parenteral administration have resulted in it being replaced by other agents in the standard short-course chemotherapy advocated by the WHO. However, the emergence of drug resistance to other front-line antibiotics has meant that SM is still an important antimycobacterial agent. Other aminoglycoside antibiotics are also being used to treat multi–drug-resistant cases of tuberculosis, and therefore an understanding of the mechanism of resistance to SM has found a new relevance.

Aminoglycosides are broad-spectrum antibiotics and their mode of action, particularly that of SM, has been extensively studied in other organisms, which greatly facilitated the investigation of resistance mechanisms in *M. tuberculosis.* SM binds to the 30S ribosomal subunit leading to inhibition of translational initiation and misreading of messenger RNA (57). Resistance to SM and other aminoglycosides in gram-negative organisms is principally due to the acquisition of aminoglycoside-modifying enzymes. However, no plasmids or tranposons bearing the drug resistance genes have been detected in *M. tuberculosis,* and the aminoglycoside 2′-*N*-acetyltransferase, encoded by *aac(2′)-Ic,* is apparently unable to acetylate aminoglycosides (58). SM-resistant strains of *Escherichia coli* though have been isolated with mutations in two ribosomal components, the S12 protein and the 16S ribosomal RNA, that protect ribosomes from the disruptive action of SM. Analysis of these genes in *M. tuberculosis* was therefore a logical first approach to investigating SM resistance, particularly as resistance-conferring mutations in these genes would produce a dominant phenotype, as *M. tuberculosis* possesses only a single copy of these genes (59).

Various groups in parallel established that mutations in *rpsL,* the gene coding for the S12 ribosomal protein, were associated with resistance (60–68). These mutations were found to occur at codon 43 (Lys43Arg or rarely Lys43Thr) and less frequently at codon 88 (Lys88Arg or Lys88GLn) and to occur in isolates exhibiting high-level resistance to SM (MIC of greater than 500 μg/mL). The same mutations were selected under in vitro conditions in *M. smegmatis* (69), and substitutions at equivalent positions within the S12 protein of other organisms also confer resistance.

Sequence analysis of the *rrs* gene, which codes for the 16S rRNA, from SM-resistant clinical isolates also identified a series of single-nucleotide substitutions that were in general associated with an intermediate level of resistance. These mapped to two restricted regions which again corresponded to sites associated with SM resistance in *E. coli* and other organisms. One group—C to T substitution at positions 491, 512, or 516 and A to C/T at 513—mapped to the 530 loop region of the *E. coli* 16S rRNA and the other group—C to T at 798, G to C at 865, G to A at 877, A to G at 904, and A to C at position 906—mapped to the *E. coli* 912 region. A SM-dependent phenotype has also been associated with an insertional mutation in the 530 loop region (70). The 530 loop region is one of the most highly conserved regions of the 16S ribosomal RNA both in sequence and secondary structure, reflecting its importance for some translational function. The 912 region has been implicated in translational fidelity and is located at the junction of the three major domains of the 16S rRNA. Chemical footprinting experiments using *E. coli* ribosomes have demonstrated that this region is protected by the binding of SM (71), and mutations here reduce drug binding (72,73), suggesting it may be the primary site of action of SM.

The correspondence of *rpsL* and *rrs* mutations in *M. tuberculosis* with those in other model systems is compelling evidence that these are the principal SM-resistance–conferring mechanisms. This is supported by a number of studies that have documented the frequency of *rpsL* and *rrs* mutations in collections of SM-resistant clinical isolates of *M. tuberculosis*. Combining the results of these studies (Table 2) shows that *rpsL* and *rrs* mutations were found to occur in approximately 51 and 12% of isolates tested. However, in 37% of these SM-resistant strains, no mutations were found in either the *rpsL* or *rrs* genes. This substantial proportion of strains with no detectable ribosomal subunit mutations suggests that there is a third mechanism of resistance, although these strains are in general only resistant to low levels of SM (MIC < 50 μg/mL) and become susceptible in the presence of detergents (65). This is consistent with resistance resulting from a change in cellular permeability to SM, and compatible with the observation that certain species of mycobacteria, such as *M. avium*, are relatively drug impermeable and are naturally resistant to SM despite drug-susceptible *rrs* and *rpsL* alleles (60).

5.1.2 Other Aminoglycosides

The emergence of strains resistant to SM and other drugs has necessitated the use of other aminoglycosides to treat individuals infected with such strains. These aminoglycosides have a range of MICs for *M. tuberculosis*, and unlike SM, are made up of a 2-deoxystreptoamine ring rather than a streptidine ring. Given this structural difference, it is not surprising that

TABLE 2 Frequency of *rpsL* and *rrs* Gene Mutations in Collections of Streptomycin-Resistant Clinical Isolates of *Mycobacterium tuberculosis*

Reference	No. of SM-strains analyzed	No. with *rpsL* mutations	No. with *rrs* mutations	No. with no *rpsL/rrs* mutations
68	78	42 (54%)	8 (10%)	28 (36%)
92	25	13 (48%)	2 (8%)	10 (40%)
67	44	25 (47%)	5 (11%)	14 (32%)
66	44	24 (53%)	1 (2%)	19 (43%)
[a]	50	18 (36%)	8 (16%)	24 (48%)
63	38	20 (52%)	9 (24%)	9 (24%)
Total	279	142 (51%)	33 (12%)	104 (37%)

[a]Dobner P, Bretzel G, Rusch-Gerdes S, et al. Mol Cell Probes 1997; 2:123–126.

these 2-deoxystreptoamine aminoglycosides bind to different sites on the ribosome and appear to be fully active against *M. tuberculosis* strains harboring SM resistance–associated mutations in their *rrs* and *rpsL* genes (65). A structural study of a paromomycin-rRNA complex (74) indicates that these antibiotics bind to a region encompassing the 30S subunit A site, which includes position 1408, which has been demonstrated to be important for resistance to 2-deoxystreptoamine aminoglycosides in *E. coli* (75). In *M. tuberculosis*, position 1400 of the *rrs* gene (the equivalent of position 1408 of the *E. coli rrs*) was also found to be important in mediating resistance to 2-deoxystreptoamine aminoglycosides. Three studies found an A to G substitution at position 1408 of the *rrs* gene in 60 (76), 75 (77), and 76% (78) of kanamycin-resistant clinical isolates of *M. tuberculosis* analyzed by polymerase chain reaction (PCR) and sequencing. One of these studies (76) also described two other *rrs* mutations at positions 1401 and 1483, although these only occurred in 3 of the 43 resistant strains analyzed, indicating that 1400 is the principal site involved in mediating resistance to kanamycin. This is compatible with the observation that substitutions at positions equivalent to 1400 of the *M. tuberculosis rrs* gene can also confer amikacin/kanamycin resistance in *M. smegmatis* (77,79) and have been identified in kanamycin-resistant clinical isolates of *M. abscessus* and *M. chelonae* (80). As with SM resistance, low-level resistance to kanamycin was not found to be accompanied by mutations in the *rrs* gene, suggesting other pathways to resistance may be involved.

The degree of cross-resistance to other aminoglycosides conferred by the *rrs* substitution at position 1400 has not yet been extensively studied in *M. tuberculosis*, but it appears to confer at least resistance to amikacin and viomycin (76–79). In various fast-growing mycobacteria, a mutation at this position has been showed to convey resistance to five different 2-deoxystreptoamine aminoglycosides (80), and given that *M. tuberculosis* clinical isolates resistant to kanamycin are usually resistant to multiple aminoglycosides (81), it is likely that an A to G substitution at position 1400 is also a pan-deoxystreptoamine aminoglycoside–resistance conferring mutation in *M. tuberculosis*. Many kanamycin-resistant clinical isolates of *M. tuberculosis* appear to be fully sensitive to capreomycin, suggesting that this mutation does not convey resistance to this structurally different antibiotic, but this needs to be confirmed using isogenic strains.

5.2 Rifamycins

5.2.1 Development of Rifamycins

During a systematic search for new antibiotic compounds in the 1950s, workers at the Dow-Lepetit Research Laboratories in Milan observed that crude extracts from the fermentation broths of *Nocardia mediterranei* contained a mixture of microbiologically active species. These were subsequently characterized and found to be a group of closely related compounds with an ansa structure, an aromatic nucleus spanned by an aliphatic bridge, and were named rifamycins (22). Chemical modification of the rifamycins led to the isolation of rifampin or rifampicin (RMP) (82), a compound highly active against gram-positive and some gram-negative organisms with the appropriate chemotherapeutic properties. It was found to be a potent antituberculous agent whose clinical introduction enabled the duration of chemotherapy for tuberculosis to be reduced to 6 months (83). Its use was initially restricted to treating tuberculosis because of fears that more general clinical use would lead to the emergence of drug resistance. However, it is now employed for a wide range of other infections, such as chemoprophylaxis against *Neisseria meningitidis*. More recently, two other rifamycin derivatives, rifapentine (RPE) and rifabutin (RBU), have been licensed for the treatment of mycobacterial infections, and others such as KRM-1648 are being evaluated.

5.2.2 Rifampin

As was the case with SM, the mode of action and resistance mechanism of RMP had been well characterized in *E. coli*, making elucidation of the mechanism of RMP resistance in mycobacteria relatively straightforward. These studies established that, in *E. coli*, RMP inhibits transcription by targeting DNA-dependent RNA polymerase (84), and that mutations in several restricted and highly conserved regions of the β-subunit, coded for by the *rpoB* gene, lead to drug resistance (85–88). The availability of the *M. leprae rpoB* sequence (89) made it possible for two groups to isolate and characterize the *rpoB* gene from RMP-resistant strains of *M. tuberculosis* (90) and *M. leprae* (91). These studies demonstrated that, in both these organisms, missense mutations and short in-frame deletions, exclusively associated with RMP resistance, occurred in a central region of the *rpoB* gene, which corresponded to the region most commonly altered in RMP-resistant *E. coli* strains. Numerous subsequent studies of RMP-resistant strains of *M. tuberculosis* from globally dispersed sources (67,91–104) have confirmed these findings, and in the vast majority of strains analyzed (105), a point mutation or in-frame deletion/insertion can be found within an 81-bp region corresponding to codons 507–533 of the *rpoB* gene. A large number of mutations have been described, although substitutions at two positions, Ser531 and His526, were found to occur in the majority of strains analyzed. Mutations at these two positions (Ser531Leu, His526Tyr) and an Asp516Val mutation have been shown to confer resistance when episomal vectors carrying an appropriately mutated *rpoB* gene were transformed into *M. tuberculosis* (106,107). Although the complete array of *rpoB* mutations has not yet been genetically verified, the strict correspondence between RMP resistance–conferring mutations in *M. tuberculosis* and those in the genetically well-characterized *E. coli*, as well as in other organisms (108), is convincing evidence that *rpoB* mutations are the principal RMP resistance mechanism in *M. tuberculosis*. The realization that RMP resistance–conferring mutations are confined to a small genetic region has meant that molecular techniques for diagnosing RMP resistance have been relatively straightforward to develop, and numerous different molecular strategies have been successfully employed for detecting them (109,110).

Many of these surveys did not sequence the whole 3516-bp *rpoB* gene, so in the approximately 5% of RMP-resistant *M. tuberculosis* strains, which have a wild-type 81-bp RMP resistance determining region (RRDR), it is not clear to what extent mutations in other regions of this gene are also involved in mediating RMP resistance. Although several mutations have been reported outside of the RRDR (101), RMP-resistant strains with a wild-type *rpoB* gene have also been identified (111), indicating that RMP resistance can also arise through an *rpoB* gene–independent mechanism. *M. smegmatis* and many strains of the *M. avium* complex (MAC) are innately resistant to RMP despite a drug-sensitive RRDR sequence (112,113). In *M. avium*, this has been attributed to the impermeability of the cell wall (114), and in *M. smegmatis*, to ribosylative inactivation of the drug (113). There is no evidence that either of these mechanisms occur in *M. tuberculosis*.

5.2.3 Cross-Resistance Among Rifamycins

The emergence of *M. tuberculosis* strains resistant to RMP prompted various studies to evaluate if other rifamycins might have activity against these strains and therefore be of clinical use in the treatment of MDR tuberculosis. Cross-resistance among the rifamycins is also of importance, as the increasing use of RBU and RPE to treat other mycobacterial infections has the potential to select for RMP-resistant *M. tuberculosis*. RBU, derived from rifamycin-S, has greater activity against atypical mycobacteria (115), and has been extensively used for treatment of and prophylaxis against MAC infections in immunodeficient individuals, and there have been reports of the emergence of RMP-resistant strains of *M. tuberculosis* in HIV-infected individuals treated with this drug (116,117). RPE is a cyclopentyl-substituted rifamycin with a serum half-life five times that of rifamycin enabling it to be administered once weekly (118). One report though has indicated that these experimental once-weekly regimens could be associated with higher rates of acquired rifamycin resistance in HIV-infected individuals (119). It has also been suggested that HIV-infected individuals may be at risk of acquired RMP resistance, because poor drug absorption and drug interactions can lead to suboptimal serum levels of RMP likely to select for resistant strains (120).

Earlier studies suggested that RBU might be active against some strains of RMP-resistant tuberculosis (121–123). More recently, investigators have focused on systematically correlating specific *rpoB* mutations with rifamycin resistance profiles (107,124,125). Bodmer and others (124) correlated MICs of 26 RMP-resistant strains of *M. tuberculosis* to their *rpoB* gene sequences and found that substitutions at amino acid positions 513, 526, and 531 were all associated with high-level cross-resistance to RBU and RPE. However, several of the alleles studied (Leu511Pro, Asp516Tyr, Ser522Leu, and an in-frame deletion at position 518) exhibited only low-level resistance or remained moderately susceptible to RBU, although these strains were resistant to RPE and RMP (MIC > 8 μg/mL). Some of these observations hav been confirmed genetically. Transformation of *M. tuberculosis* with an *rpoB* gene harboring either a Ser531Leu or His526Tyr mutation led to resistance to RMP, RBU, and RPE, as well as the experimental rifamycin KRM-1648, whereas transformation with the Asp516Val mutant led to resistance only to RMP and RPE but not to RBU or KRM-1648 (107). Some other alleles, notably 514(Phe insertion), Asn519Lys, Ser512Thr, Ser531Trp, and Leu533Pro, have also been associated with resistance to RMP and RPE but not to RBU or KRM-1648 (107,125,126). Thus, the majority of RMP-resistant strains will possess pan-rifamycin resistance because of the high frequency of mutations at positions Ser531 and His526, but a small proportion of RMP-resistant strains will retain sensitivity to some other rifamycins, raising the possibility that individuals infected with these strains could be advantageously treated with a rifamycin as a second-line agent, although this needs to be tested experimentally.

5.3 Isoniazid

5.3.1 Historical Studies

Unlike SM and RMP, isonicotinic acid hydrazide (isoniazid, or INH) is a highly specific antimycobacterial agent, being exquisitely potent against *M. tuberculosis* (and the other members of the *M. tuberculosis* complex: *M. bovis*, *M. microti*, and *M. africanum*), but possessing no or little activity against *M. leprae*, atypical mycobacteria, or other organisms. Investigations into its mode of action were therefore restricted to the inherently difficult to manipulate members of the *M. tuberculosis* complex. Despite this, early studies were able to establish that INH-resistant strains were commonly catalase negative (127), with reduced virulence in animal models (128) and that the principal mode of action was likely to be through disruption of the cell wall, probably through inhibition of mycolic acid synthesis (a major cell wall component) (129). Unifying these observations had to wait until the development of the necessary tools to manipulate *M. tuberculosis* genetically.

5.3.2 Drug Activation

Although the role of catalase in the activation of INH to its active form was first proposed in 1958 (130), it was not until 1992 that the first genetic evidence for this hypothesis became available with the cloning of *katG*, the gene encoding the catalase-peroxidase enzyme of *M. tuberculosis* (131). It was shown that overexpression of *katG* in catalase-negative strains of *E. coli* or in an INH-resistant strain of *M. smegmatis* could render these organisms relatively sensitive to INH. It was further demonstrated that two clinical isolates of *M. tuberculosis* with high-level resistance to INH (MIC $> 50\ \mu g/mL^{-1}$) had a chromosomal deletion spanning the *katG* gene (131), and that transformation of these strains with *katG* could restore their INH sensitivity (132). Further characterization of *katG* has demonstrated that it encodes a dimeric, heme-containing enzyme with catalase and peroxidase activity, in keeping with its structural similarity to other eubacterial type I hydroperoxidases (HPI) (131–138).

Confirmation that INH is a prodrug requiring activation by KatG has been provided by demonstrating that InhA (a target for INH discussed below) is only rapidly inactivated by INH in the presence of KatG (138). A mechanism for the oxidation of INH to its bioactive form has also been proposed in which the drug is converted into a number of highly reactive species capable of either oxidizing or acylating groups in proteins (135).

A plethora of studies using different molecular methodologies to analyze collections of *M. tuberculosis* strains from diverse geographical locations have confirmed that the majority of clinical isolates resistant to INH have alterations to their *katG* gene (67,92,110,136,139–152). Although large-scale deletions of *katG* have been detected infrequently, missense mutations and small intragenic deletions are the commonest genetic modification associated with INH resistance. A large number of these mutations have been described (105), although the serine to threonine mutation at codon 315 is the commonest, occurring in over 40% of tested isolates (Table 3). The possible explanation for the apparent bias in selection by INH for this mutation over other resistance-conferring mutations in *katG* has been provided by several studies which have characterized the enzymatic properties of a recombinant KatG protein harboring a Ser315Thr mutation (153,154). In contrast to other INH resistance–conferring mutations, such as the Thr275Pro mutation which results in concomitant loss of peroxidatic activity and capacity to activate INH, the Ser315Thr mutation

TABLE 3 Proportion of Isoniazid-Resistant Clinical Isolates of *Mycobacterium tuberculosis* Harboring a Ser315Thr Mutation in the *katG* Gene or a Complete *katG* Gene Deletion

Methodology used	Number of INH-resistant strains analyzed	No. with a Ser315Thr mutation in *katG*	No. with complete *katG* gene deletion	Reference
Complete sequencing	34	16	0	145
Partial sequencing	51	29	3	145
SSCP/partial sequencing[a]	25	13	0	144
Polymerase chain reaction	53	na	4	143
SSCP/partial sequencing	36	5	2	136
SSCP/partial sequencing	26	1	0	142
Partial sequencing	24	22	0	148
Partial sequencing	54	6	3	147
SSCP/partial sequencing	42	1	0	67
SSCP/partial sequencing	17	3	0	152
Partial sequencing	87	52	0	149
Partial sequencing	50	26	2	150
Partial sequencing	25	19	0	151
Partial sequencing	29	12	2	99
Total	553	205 (41%)	16 (2.9%)	

[a]Single-stranded conformational polymorphism analysis.

results in an enzyme that appears to have a dissociative loss of its ability to activate INH, retaining at least 50% of its peroxidase and catalase activities. Catalase and peroxidase activities are thought to be important for protecting *M. tuberculosis* against reactive oxygen species encountered within macrophages and are essential for a fully virulent phenotype, as strains with reduced or absent activities are less virulent when assayed in animal models (128,155–157) and more susceptible to H_2O_2 intracellularly (158). The Ser315Thr mutation can therefore be seen as an adaptive response to the dual selective pressures of INH toxicity and the host immune system.

Although numerous missense mutations have been described in INH-resistant strains, only a small number have been characterized enzymatically (138,154,159,160), and only a single study has evaluated their effect directly on susceptibility to INH through complementation of a *katG*-negative strain of the *M. tuberculosis* complex (161). As other genes are involved in mediating INH resistance (discussed below), it is therefore unknown what the exact clinical significance of the majority of these mutations is. Some mutations map to the N-terminal region of KatG where the active site residues characteristic of peroxidases are located, and these are highly likely to interfere with enzymatic function (136). Others map to the C-terminal domain whose functional properties are obscure, and therefore on a structural basis, it is difficult to predict what effects they might have, although some (e.g., Leu587Pro) have been shown to diminish the stability of KatG (154). One N-terminal mutation, a replacement of arginine by leucine at codon 463, originally described to occur with high frequency in INH-resistant clinical isolates (146) is now thought to be a frequently occurring polymorphism, as it exists in some naturally INH-sensitive members of the *M. tuberculosis* complex such as *M. bovis*. This has been confirmed by the characterization of the recombinant Arg463Leu mutant protein (138) and by the demonstration that expression of this mutant in a catalase-negative strain could restore INH susceptibility (161). The apparent association of this polymorphism with INH resistance can be explained by its nonrandom worldwide distribution (162,163) and the sampling of INH-sensitive and resistant strains from different localities in some of the above studies.

5.3.3 Drug Targets

InhA. The studies summarized in Table 3 also revealed that there were a number of INH-resistant strains which possessed a wild-type *katG* gene, demonstrating that there were mechanisms of resistance independent of INH activation by KatG and suggesting that the intracellular targets of INH might be involved in mediating resistance. Various strategies have been used to identify these targets. The first successful approach involved the expression of genomic libraries, from two INH-resistant strains of mycobacteria isolated in vitro (from *M. smegmatis* and *M. bovis*) in the fast-growing mycobacterial species *M. smegmatis* to identify clones that could confer resistance to INH (164). Subsequent analysis of the INH resistance–conferring clones delineated a two-gene operon with homology to proteins involved in fatty acid biosynthesis from other organisms. Characterization of this operon revealed that only the second gene, *inhA* (165), was required for conferring INH resistance, and sequence analysis of the *inhA* gene from the parental strains revealed a serine to alanine substitution at position 94 relative to the wild-type gene. Confirmation that this mutation could confer a resistant phenotype was obtained by demonstrating that allelic exchange of the wild-type gene with the Ser94Ala mutant gene resulted in an INH-resistant transformant of *M. smegmatis*, although the significance of this experiment has been contested by some workers (166).

Biochemical studies revealed that *inhA* codes for a fatty acid enoyl-acyl carrier protein reductase. This enzymic activity is nicotinamide adenine dinucleotide (NADH) dependent and reduces the double bond at position two of a growing fatty acid chain linked to an acyl carrier protein (ACP), an activity common to all known fatty acid biosynthetic pathways. InhA has a marked preference for long-chain substrates (those containing 16 or more carbons) which are the precursors of very long alpha-branched fatty acids (C_{40}–C_{60}) known as mycolic acids, a major structural element of the mycobacterial cell wall (167).

Structural analysis of the InhA protein has characterized the nature of the interaction between INH and InhA (168). The observation that InhA inhibition by INH requires the presence of NADH and that the Ser94Ala mutant protein has a lower affinity for NADH and requires higher concentrations of this cofactor before inhibition occurs suggested that INH may interact with NADH rather than directly with InhA. This was elegantly demonstrated by cocrystallization of InhA with NADH and INH (169), since the structure showed that the activated form of INH was covalently linked to NADH within the active site of the enzyme. The modified NADH (isonicotinic acyl-NADH), consisting of an isonicotinic-acyl group derived from INH covalently attached to carbon four of the nicotinamide ring, was only formed in the presence of InhA, suggesting that the activated form of INH only interacts with bound NADH. Kinetic isotope analysis of InhA has suggested that the binding sequence of NADH and the long-chain acyl-ACP substrate is not strictly ordered but there is a preference for binding of NADH. It therefore has been proposed that binding of NADH renders InhA susceptible to attack by INH. The resulting in situ formation of isonicotinic acyl-NADH would then prevent the binding of substrate and ultimate elongation of the fatty acid chain required for mycolic acid synthesis. The relative resistance of the InhA Ser94Ala mutant to attack by INH can therefore be explained by its reduced affinity for NADH, which would result in an altered binding sequence with the substrate preferentially attaching first, thereby bypassing the INH-vulnerable state of enzyme-NADH complex. In addition, it is structurally plausible that the lowered NADH affinity of the mutant protein is also reflected in a lowered affinity for isonicotinic acyl-NADH, which would result in release of this compound from the active site after its formation, allowing a second chance for substrate binding and a normal cycle of substrate catalysis (168–171). Recently, this model has been challenged by Wilming and Johnsson (172), who were able to form isonicotinic acyl-NADH from INH and NAD^+ in the absence of InhA. This led them to propose an alternative model in which the isonicotinoyl radical, formed by oxidation of INH by KatG, reacts directly with NAD^+ outside the active site of InhA. This model is consistent with the observation that NADH-dehydrogenase defects in *M. smegmatis*, leading to a higher than normal NADH/NAD$^+$ ratio, are associated with a degree of INH resistance (173). Further experimental evidence is required to clarify the mechanism of activation of INH, a clearer understanding of which could facilitate the design of INH analogues which are active independently of KatG, thereby circumventing the principal mechanism of resistance.

Since the original description of the Ser94Ala mutation five different structural InhA gene substitutions have been identified through sequence analysis of INH-resistant clinical isolates and these are all situated in the NADH-binding pocket of InhA (94,174). Four of these mutant proteins have been characterized enzymatically and were found to have reduced affinity for NADH with a normal affinity for a fatty acyl-CoA substrate when compared to wild-type enzyme (175), confirming that InhA substitutions could also be selected for in vivo and that they have biochemical properties similar to the in vitro selected mutant Ser94Ala (164) (which has yet to be identified in a clinical isolate). Mutations in the upstream region of *mabA*, the putative promoter region for *inhA* expression, have also been described in INH-resistant clinical isolates (92,145,176), and several of these have been shown, using a gene fusion reporter construct, to confer expression levels from four- to eightfold greater than wild-type sequences (166), suggesting that upregulation of InhA may also be a resistance mechanism.

Although there are a limited amount of data, these InhA operon mutations appear to occur relatively frequently (promoter more commonly than structural mutations) and exclusively in INH-resistant mutants, and they have been described both in strains with mutated and wild-type *katG* genes. Strains with a wild-type *katG* and a mutation in the *inhA* operon have been found to possess a variable degree of resistance to INH, making it difficult to be certain what the exact clinical significance of these mutations is, particularly as it is likely that other unidentified resistance loci also exist and could occur in conjunction with substitutions in the *inhA* operon. Unfortunately, studies using isogenic strains have to date failed to clarify this issue. Although replacement by allelic exchange of *inhA* with the Ser94Ala mutant gene or multicopy vector expression of wild-type *inhA* resulted in a resistant phenotype when carried out in *M. smegmatis* (164), the published results for members of the *M. tuberculosis* complex are less convincing. The definitive gene replacement experiment has not yet been carried out, and of the two studies reporting the results of expression on a multicopy plasmid (estimated to be 4–10 copies) of wild-type *inhA* or the Ser94Ala mutant gene (166,177), only one was able to demonstrate a significant increase in the MIC for INH (177). Furthermore, chromosomal integration of a single copy of the entire *mabA-inhA* coding region from several INH-resistant clinical isolates, including the promoter region harboring an upregulating single-base substitution, only produced a modest effect on INH sensitivity (166), and failed to confer a resistance to 1 μg/mL of INH (as seen with the parent strains of the sequences tested). A possible explanation for these apparent discrepancies is that the mechanisms of INH resistance in *M. smegmatis* are different to those in *M. tuberculosis*. This is supported by the observation that treatment with INH leads to accumulation of shorter chain fatty acids in *M. smegmatis* compatible with an inhibition of a relatively short-chain fatty acid biosynthetic pathway, such as InhA, but to accumulation of 24- and 26-carbon fatty acids in *M. tuberculosis*, suggesting possible inhibition at a more distal point in mycolic acid synthesis (166). This illustrates the problems associated with extrapolating findings made in the relatively easy to manipulate fast-growing mycobacteria to slow-growing *M. tuberculosis*, and also emphasizes the importance of rigorous laboratory-based studies on isogenic strains for the interpretation of population-based genetic data gathered from analysis of drug-resistant clinical isolates.

KasA. A second successful approach used to determine the targets of activated INH has been to examine the early adaptive response of *M. tuberculosis* following a challenge with INH. Inhibition at a particular point in a metabolic pathway will result in upstream accumulation of substrate and subsequent compensatory alterations in protein expression to deal with such a bottleneck. By examining two-dimensional gel electrophoretic protein profiles after treatment with 1 μg/mL of INH, Mdluli and others (178) identified two upregulated proteins of 12 and 80 kD. Subsequent analysis of these protein species revealed the 12-kD species to be an acyl carrier protein (AcpM) and the 80-kD species to be a covalent complex of AcpM and KasA (β-ketoacyl-ACP synthase), which was labeled following treatment with [^{14}C]INH, indicating covalent attachment of INH to this protein complex. These two proteins form part of a fatty acid synthase (FAS) type II system which produces the meromycolate branch of full-length mycolic acids. Purified AcpM from INH-treated cells was found to be saponified with predominantly hexacosanoic acid, the chemical species found to accumulate after treatment with INH, supporting the view that the AcpM-KasA complex is a target for INH.

The original study of Mdluli and others (178) reported the sequences of *kasA* from 43 INH-susceptible and 23 INH-resistant clinical isolates. Four mutations were found within the *kasA* gene (Asp66Asn, Gly269Ser, Gly312Ser, and Phe413Leu) exclusively within INH-resistant strains; two of which occurred in strains with no other INH-resistant mutations, suggesting that *kasA* was not only a target for INH but also a gene involved in mediating resistance. The crystal structure of the *E. coli* homologue of *kasA* suggests that three of these mutations occur at residues within the active site of the enzyme (178). However, a subsequent study of 144 resistant clinical isolates found three mutations (Arg121Lys, Gly269Ser, Gly387Asp) in *kasA* that were unique to resistant strains, which occurred in only 3.2% of resistant strains tested, most of which had mutations in other genes associated with INH resistance (179). The Gly312Ser mutation was also described, but it was found to occur in 19% of the INH-sensitive strains tested. A more recent study using a molecular beacon assay to detect mutations at *kasA* codons 66, 269, 312, and 413 detected mutations only at codon 269, which occurred with equal frequency in both resistant and sensitive isolates (110). These two studies suggest that *kasA* mutations, like *inhA* mutations, are not a clinically important mechanism of INH resistance.

Changes in gene expression following treatment with INH have also been examined using a number of other techniques. Wilson and others (180), employing a DNA microarray representing 97% of the predicted open reading frames (ORF) of *M. tuberculosis*, were able to demonstrate a rapid induction (within 20 min of a challenge of INH) of a five-gene operon that includes *acpM* and *kasA*, thereby confirming that the interruption of the FAS-II cycle is rapidly sensed and responded to at the transcriptional level. No changes in *inhA* expression were seen. *fbpC*, which encodes a major exported antigen known to have trehalose-dimycolyl transferase activity (181), was also found to be induced, which is in keeping with a previous study (182). This enzyme is thought to esterify mycolates to specific carbohydrates in the cell wall and may be upregulated in response to a diminution in the pool of mature mycolates. However, no role has yet been established for *fbpC* in resistance to INH.

Studies that have analyzed the response to an INH challenge at the level of the transcriptome or proteome have identified a number of other genes induced by INH: a three-gene operon of unknown function (Rv0341, Rv0342, Rv0343) (182), a two-gene operon (Rv3139, Rv3140) coding for two fatty acyl-CoA dehydrogenases (FadE23, FadE24) (180), and *efpA* predicted to code for a proton motive force energized transporter (180). FadE23 and FadE24 are likely to be involved in fatty acid degradation, and their induction is probably a mechanism for recycling accumulated fatty acid precursors. The induction of *efpA* is more intriguing, as the deduced protein secondary structure is similar to members of the QacA transporter family which mediate antibiotic and chemical resistance in bacteria and yeast (184), and it was found to be most closely related to *pur8* of *Streptomyces alboniger*, a gene responsible for puromycin resistance. However, since the molecules it translocates are still unknown, and no *efpA* mutations have yet been identified in drug-resistant strains (184), it remains speculative to attribute a protective drug efflux function to this protein.

5.3.4 Compensatory Mutations

One of the problems associated with defining the molecular basis of INH resistance has been to distinguish between the primary drug resistance–conferring mutations and secondary compensatory mutations that arise to counteract the potentially detrimental effects on bacterial physiology of the primary mutation. This is exemplified by the difficulties associated with establishing the exact function of a group of mutations that occur in some INH-resistant strains, which result in the upregulation of alkyl hydroperoxide reductase. Initial interest in this gene derived from the observation that an *oxyR* null mutant of *E. coli* (the wild type of this organism is highly resistant to INH) was susceptible to INH (185). *oxyR* is a member of the LysR family of transcriptional regulators and controls the expression of *katG* and *ahpC-ahpF* (the two subunits of alkyl hydrogen peroxide reductase) in *E. coli*. Transformation of an *E. coli oxyR* mutant with *katG* or the *ahp* genes found that only the latter were capable of restoring resistance, indicating that alkyl hydrogen peroxide reductase was involved in resistance to INH in *E. coli* (185). This led to characterization of the *oxyR-ahpC* locus from different mycobacterial species (186,187). Although *M. leprae* organisms were found to have a functional *oxyR* (188), which was linked to and divergently transcribed from *ahpC*, members of the *M. tuberculosis* complex were found to possess an *oxyR* pseudogene in conjunction with a functional but feebly expressed *ahpC* gene. Transformation of the *M. leprae oxyR-ahpC* locus into *M. tuberculosis* rendered this organism resistant to INH (186), and suggested that the lack of a functional *oxyR*, leading to a low level of *ahpC* expression, could at least partially explain the hypersensitivity of *M. tuberculosis* to INH (186,189), and that possible mutations at this locus could be involved in INH resistance.

This was supported by the observation that some catalase-deficient clinical isolates, notably those with *katG* deletions, had point mutations in the upstream region of *ahpC* (109,190,191) which were associated with enhanced promoter activity when assayed using *luxAB* (187) or *lacZ* (192) as a reporter gene. However, overexpression of AhpC, using a number of different constructs, to levels higher than those seen in the INH-resistant isolates, had little or no effect on INH sensitivity in either *M. tuberculosis* or *M. bovis* BCG (187) but did confer resistance to both hydrogen and cumene peroxide (an organic hydrogen peroxide). Studies using *M. smegmatis* suggested that in this organism (resistant to higher levels of INH than *M. tuberculosis*) (193) the situation is different, as a more significant effect of

AhpC overexpression on INH sensitivity has been reported (192), and an *ahpC* knockout mutant was found to be hypersensitive to INH (194). This led to the view that, at least in *M. tuberculosis*, AhpC upregulation acts only as a compensatory mechanism to protect *M. tuberculosis* against the additional burden of reactive oxygen species due to the absence or reduction in catalase-peroxidase activity associated with mutations in *katG*. These compensatory mutations would be particularly important in vivo, as *katG* is a virulence factor that is thought to protect *M. tuberculosis* from the reactive oxygen species encountered within the macrophage. This hypothesis has been tested in vivo by comparing the virulence of isogenic strains expressing different levels of AhpC. In a subcutaneous guinea pig model, in which AhpC levels of *M. bovis* were manipulated with various antisense constructs (195), upregulation did appear to compensate for the loss of *katG*, whereas in an intravenous mouse model using in vitro selected strains of *M. tuberculosis* with differing levels of AhpC expression, AhpC had no detectable effect on virulence (156).

The picture is further complicated by the observation that INH itself induces AhpC within several hours of treatment (180), and that studies of clinical isolates have found that *ahpC* promoter mutations do occur in the absence of *katG* mutations (109). In addition, some INH-resistant clinical isolates have been identified with higher than normal levels of AhpC (189) but with no apparent promoter mutations, indicating that there are probably other mutations that can alter the expression levels of this protein. It is, therefore, still unclear whether AhpC is involved directly in the detoxification of activated INH and its reactive oxygen intermediates or whether it is important for restoring the bacterial defense mechanisms against oxidative stress. It is likely that the dual selective pressures of the host immune system and INH toxicity encountered in an INH-treated individual lead to the selection of a complex array of mutations that adapt *M. tuberculosis* to this particular environment, and the pattern of these mutations may differ from strain to strain.

5.3.5 Detection of Isoniazid Resistance by Molecular Methods

Conventional methods for detecting resistance have been based on subculturing primary isolates in the presence of differing concentrations of antibiotic, resulting in delays of over 2 months from the time of the initial culture for the diagnosis of drug-resistant tuberculosis. Although newer culture-based techniques, exploiting, for example, radiometric (196), colorimetric (197), or light emission measurements (198) to assay mycobacterial growth, have considerably shortened this delay, the molecular determination of resistance has the potential to give "real-time" results. This is particularly important in the case of tuberculosis to allow swift initiation of an appropriate combination of drugs to prevent the emergence of secondary resistance and to implement appropriate control measures. *M. tuberculosis* appears to be an organism particularly suited to molecular diagnostic techniques, as a global survey of strains has revealed an extremely low frequency of polymorphisms in structural genes (estimated to be over 1000-fold less than in *E. coli*) (199). This limited structural allelic variation means that usually only a single wild-type sequence needs to be used as the reference with which to compare a sequence from a test strain, effectively removing the "noise" caused by synonymous mutations from any molecular diagnostic strategy.

Despite this, the detection of INH resistance by molecular techniques is a particular challenge owing to the number of potential resistance-conferring genetic loci that need to be interrogated and the plethora of mutations that have been described in *katG*, the principal mechanism of resistance. However, various studies have demonstrated that targeting a limited number of mutations at selected loci can produce a diagnostic sensitivity sufficiently high to be of practical utility. Telenti and others analyzed 38 INH-resistant strains from a single reference laboratory by single-strand conformation polymorphism mutation analysis of PCR products (PCR-SSCP) generated from selected regions of *katG*, and the promoter regions of *inhA* and *ahpC*. Using this strategy, they were able to identify INH-resistant strains with a sensitivity of 87% and specificity of 100% (109). A similar result was obtained using molecular beacons targeted to detect a more specific range of mutations (110). For primers designed to detect mutations between codon 313 and 318 of *katG* over the *inhA* ribosomal binding site and from two selected regions in the *ahpC-oxyR* intergenic region generated a specificity of 85% for the 100 isolates tested. Interrogating selected codons within *kasA* did not improve the discriminatory power of their approach. The specificity was significantly higher for MDR strains when compared to single drug-resistant strains. One explanation for this difference is that INH resistance–conferring mutations are accumulated sequentially, and conditions likely to select for MDR strains such as prolonged suboptimal therapy or prior treatment for tuberculosis are also likely to favor the accumulation of multiple mutations.

5.4 Pyrazinamide

The introduction to the antituberculous pharmacopeia of the nicotinamide analogue pyrazinamide (PZA) had far-reaching consequences, since it enabled the duration of treatment to be shortened from more than a year to 6 months. Used during the initial intensive phase of short-course chemotherapy, PZA, which is more potent at acidic pH, is believed to be particularly active on intracellular *M. tuberculosis*. For instance, the MIC of the drug for a strain grown in vitro at pH 7 is > 250 μg/mL but can be reduced to 15 μg/mL by lowering the pH to 5. It was initially thought that this potentiation effect could be explained by tubercle bacilli residing in acidified phagosomes that concentrated the drug (200). However, it was subsequently shown that the pH within these vesicles was neutral (201). Considerable insight into PZA uptake and resistance mechanisms was provided recently by Zhang et al. in an elegant series of publications addressing the issues of the remarkable specificity of the drug for *M. tuberculosis* and its relationship with a broad-spectrum amidase (202–208).

PZA resistance has long been associated with the loss of activity of PZase, a cytosolic enzyme of 186 amino acids that hydrolyzes both PZA and nicotinamide, and which may play a role in pyridine nucleotide metabolism. In a seminal study (202), Scorpio and Zhang cloned the *pncA* gene encoding PZase from *M. tuberculosis* and demonstrated restoration of drug susceptibility upon its expression in the naturally resistant organism BCG. On further examination, *pncA* mutations that lowered or abolished PZase activity were found in PZA-resistant isolates of *M. tuberculosis*, and also in *M. bovis* and BCG (203). Subsequently, other workers surveyed their strain collections for altered *pncA* genes and found a near-perfect association between the presence of mutant alleles and PZA resistance (205,209–211). Indeed, detection of resistance by molecular techniques is now preferable to susceptibility testing by microbiological methods, as these are notoriously error prone owing to the pH effects discussed above.

Although many mycobacteria contain *pncA* genes and elicit PZase, PZA is most active against *M. tuberculosis* and its close relatives *M. africanum* and *M. microti*. The natural PZA resistance of *M. bovis*, the other member of the *M. tuberculosis* complex, and its descendants stems from the amino acid substitution His57Asp in the PncA protein. The *M. tuberculosis* PZase shows 69.9 and 67.7% identity to those of *M. kansasii* and *M. avium*, respectively, and 35.5% identity with the nicotinamidase of *E. coli*. Both these mycobacterial PZase restored drug susceptibility to levels similar to those conferred by the *M. tuberculosis* enzyme when expressed in a resistant host such as BCG, although the *M. tuberculosis* protein probably has higher nicotinamidase and PZase activities (206,207). Nevertheless, both *M. kansasii* and *M. avium* are naturally resistant to the drug, and the likely reason for this will be explained below. *M. smegmatis* also produces a highly active PZase, PzaA, with an apparent molecular weight of 50 kD, that is quite distinct from the other enzymes yet hydrolyzes both PZA and nicotinamide (212). PzaA probably corresponds to a broad-spectrum amidase capable of hydrolyzing a wide range of substrates (212). Furthermore,

M. smegmatis also contains a PncA homologue (Y. Zhang, personal communication). The potential contribution of other amidases to PZA and nicotinamide metabolism has been recognized and discussed by others (213).

The mutations present in *pncA* from a large number of drug-resistant isolates of *M. tuberculosis* are known. The majority of these (69%) correspond to missense mutations, although frameshifts, insertions, deletions, and nonsense mutations (31%) also occur (203,205,209–211). The higher frequency of missense mutants suggests that the corresponding proteins may retain some residual activity that could confer a competitive advantage. In the one study where such details are presented, it is clear that missense mutations abolished PZase activity, but nicotinamidase levels were not measured (203). It is of some interest to examine the distribution of the amino acid substitutions resulting from missense mutations. These occur throughout the PncA protein but are generally located in positions that are conserved in all four enzymes. This suggests that these amino acid residues may play critical roles in substrate binding and catalysis, but confirmation by biochemical characterization of well-defined PZase variants is now required. PZase is distantly related to the *N*-carbamoyl-sarcosine aminohydrolase from *Arthrobacter* spp., a functionally related enzyme whose crystal structure is known (211). From multiple alignments and knowledge of the catalytic mechanism of the *Arthrobacter* enzyme, Lemaitre et al. tentatively predicted that Asp-8, Trp-68, Ala-134, Thr-135, and Cys-138 of PZase should be key residues for the hydrolysis of PZA. However, as PZase is monomeric whereas *N*-carbamoyl-sarcosine aminohydrolase is a tetramer, and their primary sequences are only weakly related, structural information for the mycobacterial enzyme is desirable in order truly to understand its catalytic properties.

The toxicity of PZA results from its conversion to pyrazinoic acid (214), and the PZase enzyme from resistant organisms such as BCG is unable to catalyze this reaction which occurs at alkaline, natural, and acidic pH. However, pyrazinoic acid only accumulates in susceptible tubercle bacilli when the external pH is acidic. In naturally resistant mycobacteria such as *M. smegmatis* and *M. vaccae*, efficient conversion of PZA occurs, but pyrazinoic acid is rapidly excreted by a highly active efflux system that can be inhibited by reserpine or valinomycin (208). *M. tuberculosis* also appears to have a pyrazinoic acid efflux system, but as this is many orders of magnitude less efficient, copious amounts of the acid build up in the cytoplasm. The natural PZA resistance of *M. kansasii* is attributable to the much lower activity of its PZase rather than to efflux mechanisms, since the introduction of the *M. avium* gene into *M. kansasii* results in hugely increased PZA susceptibility and pH-dependent pyrazinoic acid accumulation (207). By contrast, in *M. avium*, which shows lower levels of PZA resistance than *M. kansasii*, an efflux mechanism has been reported whose efficiency is intermediate between those of *M. smegmatis* and *M. tuberculosis* (207).

Prior to its interaction with PZase, PZA must cross the cell wall and enter the cytoplasm. Daffé and coworkers demonstrated that the radiolabeled drug diffused passively through the outer envelope of *M. tuberculosis* and was then actively transported by an ATP-dependent uptake system. This transporter was inhibited by arsenate albeit at very high concentrations and also appears to transport nicotinamide (213). Similar transport systems were also detected in *M. avium* and *M. kansasii* but not in *M. bovis* BCG. The latter observation was also made by Zhang et al. (208), who found that the drug did not accumulate in PZA-resistant strains belonging to the *M. tuberculosis* complex. However, upon production were restored. This could suggest that the transport system is only expressed when PZase is present; similar observations regarding nicotinamide uptake and the presence of nicotinamidase have been made in *E. coli* mutants lacking the amidase (215) and might reflect the existence of regulatory pathways for pyridine nucleotide metabolism. Alternatively, it is conceivable that PZA also diffuses across the cytoplasmic membrane of mycobacteria in a passive manner and is then converted to pyrazinoic acid by PZase which is trapped in the cell at low pH in *M. tuberculosis* but excreted by the naturally resistant species. The inhibition of PZA uptake observed in the presence of nicotinamide (213) may simply reflect the fact that as nicotinamide is the preferred substrate, PZase hydrolyses the drug at much lower level.

To summarize, three components appear to be involved in mediating PZA susceptibility in pathogenic mycobacteria: the putative uptake system, PZase, and an efflux pump. The relative contributions of these three factors determine the level of drug susceptibility. Clinically relevant mutations have been described in the *M. tuberculosis* complex that result in loss of PZase activity and concomitantly the absence of pyrazinoic acid. To date, no genes or mutations affecting the PZA transporter (if this exists) or the efflux system have been reported in tubercle bacilli. Overexpression of the putative efflux system would lead to increased excretion of pyrazinoic acid. Such mutants might exist and could explain the resistance observed in a small number of clinical isolates with wild-type *pncA* genes.

Since pyrazinoic acid is the active agent, it was logical to test this compound directly, but its bactericidal activity for tubercle bacilli in infected mice was found to be much lower than that of the acid generated endogenously from PZA (214). To identify the target for the bioactive form of PZA, attempts were made at isolating laboratory mutants of *M. tuber-*

culosis that show increased resistance to pyrazinoic acid. These have been repeatedly unsuccessful (203). Moreover, in well-characterized clinical isolates that display reproducible PZA resistance, most strains harbored defective *pncA* genes (203). These observations suggest that the drug target must be an essential enzyme, and current thinking is centered on fatty acid synthase I.

5.5 Ethambutol

Since its introduction in the early 1960s, ethambutol (ETH) was thought to act on the mycobacterial envelope, as treatment perturbed the biogenesis of several cell wall components. Through biochemical studies, performed mainly with *M. smegmatis*, it was found that the primary site of action was arabinan synthesis (216), which in turn impacts on arabinogalactan and lipoarabinomannan production (217). It is now clear from work with ETH-resistant mutants and molecular genetics that the main drug target is the arabinosyltransferase(s) involved in the polymerization of arabinan. Using complementary approaches with *M. tuberculosis* (218) and *M. avium* (219), these enzymes were shown to be encoded by linked genes that have evolved by a gene-duplication mechanism probably controlled by the regulatory gene *embR*. In *M. avium*, *embR* is transcribed divergently from the adjacent *embAB* genes (219), whereas in *M. tuberculosis*, the *embCAB* operon (218) is situated 1.87 Mb distal to *embR* (59).

The arabinosyltransferases are membrane-bound enzymes (219) with 12 predicted membrane-spanning segments and a large extracytoplasmic domain at the COOH-terminus (218), although experimental support for the topological model is missing. Missense mutations located in a tetrapeptide at positions 303–306 of a putative cytoplasmic loop of EmbB have been shown to be responsible for acquired drug resistance in the majority of clinical isolates of *M. tuberculosis* (220) and in laboratory mutants of *M. smegmatis* (221). Mutations affecting Met306 are predominant, and replacement by Leu or Val is associated with higher resistance levels (40 μg/mL) than the substitution Met306Ile (220). High-level resistance has also been described for *M. tuberculosis* strains harboring *embB* genes with Thr630Ile or Phe330Val mutations, although causality has not yet been demonstrated. The natural ETH resistance of a variety of nontuberculous mycobacteria such as *M. leprae*, *M. fortuitum*, and *M. abscessus* results from the presence of one or more alterations to the otherwise well-conserved motif at positions 303–306 (222). Yet again, as in the case of INH and PZA, serendipity has played a major role in the susceptibility of the tubercle bacillus to a front-line drug.

In roughly 70% of ETH-resistant clinical isolates of *M. tuberculosis*, drug resistance can be explained by alterations to *embB*. However, nothing is known of the mechanism operational in the remaining 30%, and these generally display lower levels of resistance (15 μg/mL) (220). Telenti has proposed that high-level resistance results from a stepwise mutational process in *M. smegmatis* (218), although other workers provide evidence in favor of a single event (221). There is a clear indication from heterologous expression of *emb* genes in this host that increased gene dosage confers higher resistance levels (218), and it is conceivable that unlinked mutations leading to overexpression of the *M. tuberculosis embCAB* operon may occur in clinical isolates displaying low-level ETH resistance. A strong candidate locus is *embR*, which is required for transcription of the operon (219), although increased efflux of the drug cannot be excluded.

5.6 Fluoroquinolones

Fortunately, the search for new antituberculosis drugs, prompted by the increase in MDR tuberculosis, has produced one new class of compounds with useful antimycobacterial properties. The discovery that the broad-spectrum bacteriocidal activity of fluoroquinolones extended to mycobacteria led to their rapid clinical deployment (223,224). Although there is still a lack of large-scale controlled clinical trials using fluoroquinolones, they have now been used extensively for over a decade, and they have established themselves as a key element of therapy for cases of MDR-TB (53). Fluoroquinolones have also been used successfully to treat MAC and other atypical mycobacterial infections. Their tolerability and oral route of administration make them particularly useful drugs. Most studies have been conducted with ofloxacin and ciprofloxacin, but recent reports have suggested that some newer compounds are considerably more potent (225,226).

The main mechanism of resistance to fluoroquinolones in *M. tuberculosis* appears to be similar to that in *E. coli*. Fluoroquinolones have been shown to target DNA gyrase, a type II DNA topoisomerase composed of two A and two B subunits encoded by *gyrA* and *gyrB*, that catalyses negative supercoiling of DNA (227). Sequencing of the equivalent of the *E. coli* quinolone resistance–determining region (QRDR) of *gyrA* (228) in fluoroquinolone-resistant clinical isolates showed that a number of distinct amino acid substitutions were associated with resistance: Ala90Val, Ser91Pro, Asp94 to His, Tyr, Asn, Gly, or Ala (229, 230–233). A naturally occurring polymorphism unrelated to resistance, Ser95Thr, was also identified (231). There are some discrepancies in the levels of fluoroquinolone resistance that these individual mutations confer which have yet to be resolved by genetic studies. But combining the results of the above studies conducted on resistant clinical isolates with those conducted exclusively on in vitro selected strains (234,235) suggests that the Ala90Val and Asp91Pro substitutions result in lower levels of resistance (MIC for ciprofloxacin of approximately 4 μg/mL) than substitutions occurring at position 94 (MIC of 16 μg/mL), with the Asp94Gly producing the highest level of resistance. The degree of cross-resistance conferred by these mutations has also been studied in vitro. Although Alangaden and others (236) found that the Asp94Asn, Asp94Ala, and Ala90Val substitutions were associated with cross-resistance to ciprofloxacin, ofloxacin, sparfloxacin, and five experimental fluoroquinolones, other reports seem to suggest that, although cross-resistance to ciprofloxacin and ofloxacin is invariable, some substitutions such as Ala94Tyr have less of an effect on sparfloxacin susceptibility (230,234).

Undoubtedly there are pathways to fluoroquinolone resistance other than mutations in the QRDR, as fluoroquinolone-resistant clinical isolates (230,232) wild-type at this locus have been identified. In addition, in vitro selection of ciprofloxacin-resistant mutants of *M. tuberculosis* appears to be a two-step process requiring an initial selection for low-level resistance before high-level resistance can be selected for (231,234). Analysis of these first-step low-level resistant mutants (234) found that some of them were wild type at the QRDR locus, and sequencing of *gyrB* from the high-level resistant mutants identified substitutions at a position in this gene that have also been described in quinolone-resistant *E. coli* (237). Other fluoroquinolone resistance mechanisms have been described in different organisms including alterations in DNA topoisomerase IV, decreased cell wall permeability, and drug efflux mechanisms (238,239). A putative membrane efflux pump, LfrA, has been identified in the nonpathogenic *M. smegmatis* (240) which can confer low-level fluoroquinolone resistance when overexpressed on a multicopy vector, but there are no published data on whether any of the putative transporters identified in the genomic sequence of *M. tuberculosis* can mediate fluoroquinolone resistance.

6 USING GENOMICS TO UNDERSTAND NATURAL DRUG RESISTANCE

The availability of the complete genome sequence of *M. tuberculosis* provides a powerful resource for understanding the basis of the natural resistance of the organism to many of the leading antimicrobial agents (59) and may even enable existing drugs to be used more efficiently. Genes encoding antibiotic-modifying enzymes such as β-lactamases and aminoglycoside acetyl transferases are present and may account for the intrinsic resistance to certain antibiotics belonging to the β-lactam and aminoglycoside classes (58,241). Knowledge of these potential resistance mechanisms allows us to consider the use of specific inhibitors, and it is of some interest to note that ampicillin shows modest efficacy against the tubercle bacillus when given in association with sulbactam, a β-lactamase inhibitor (242). Knowledge of the genome sequence also allows the rapid development of molecular diagnostic reagents for monitoring drug resistance.

Another potential mechanism that could contribute to natural antimicrobial resistance is drug efflux, and in many pathogenic bacteria, this is known to be mediated by transmembrane proteins belonging to the major facilitator superfamily (MFS), the small multidrug resistance family (SMR) and the resistance/nodulation/cell division family (RND) (243–245). At least six subfamilies comprise the MFS, and these differ in their substrate specificity and the number of transmembrane (TM) segments; 12 or 14 (243, 245). *M. tuberculosis* appears to have 14 MFS proteins with 14 TM segments that could serve as proton motive force–dependent drug pumps, and it has already been demonstrated that one such protein from *M. tuberculosis*, Rv1258c, also known as Tap, probably acts in drug efflux, as its expression in *M. smegmatis* leads to increased resistance to several aminoglycosides and tetracycline (246). Another MFS protein, EfpA, was identified on the basis of its similarity to the multidrug resistance pump, QacA, but has not bene shown to confer drug resistance (184). There are also about a dozen members of the MFS with 12 TM segments, but it is unknown whether any of these are involved in the transport of drugs. The LfrA protein of *M. smegmatis* is an MFS member that has been shown to act as a fluoroquinolone pump (240), although the tubercle bacillus has no direct LfrA counterpart, there are at least six proteins that show significant similarity.

M. tuberculosis has only one member of the SMR family, the 107n residue Mmr protein (also known as Rv3065, or EmrE) and like its relatives that has four TM stretches (59,247). When expressed in *M. smegmatis*, the gene *mmr* confers resistance to a variety of compounds including criflavine, erythromycin, ethidium bromide, safranin O, and pyronin Y. *M. tuberculosis* is somewhat unusual, as its genome (59) contains 15 genes encoding members of the RND superfamily (248), an exceptionally high number compared to other fully sequenced bacterial genomes (245). In the gram-negative pathogens *Pseudomonas aeruginosa* and *E. coli*, RND proteins, such as MexAB or AcrAB (~1000 amino acids), act as proton motive force–dependent efflux systems and confer high levels of resistance to fluoroquinolones and other antimicrobial agents (249,250). The genetic context of the *mmpL* genes, encoding the *M. tuberculosis* RND proteins, suggested an involvement in the export of lipids or glycolipids (251), and experimental evidence in support of this proposal is accumulating. It is possible that the MmpL proteins can also act in drug efflux. However, at present, there are no compelling data associating any of these MFS, SMR, or RND members with clinical drug resistance in the tubercle bacillus.

7 IDENTIFYING NEW DRUG TARGETS FROM THE GENOME SEQUENCE

In recent years there has been growing concern that the future of short-course chemotherapy was menaced by the steady increase in the incidence of drug-resistant strains of *M. tuberculosis* and that the specter of untreatable disease could soon become a reality. Although the standard four-drug regimen is excellent in treating drug-susceptible disease, it could certainly be improved. One of the reasons for the emergence of drug resistance is poor compliance and here the length of treatment is a major factor. It is incontestable that better adherence to treatment could be obtained by reducing the duration of therapy, provided complete sterilization could be ensured, and this would also lower the risk of side effects such as hepatotoxicity. The drugs currently used to treat tuberculosis are active at relatively high concentrations, and again this can lead to nausea and other undesirable consequences. If a new treatment were to be devised, this should involve drugs that are considerably more active than those currently available so that the above failings can be alleviated.

With detailed knowledge of the ~4000 genes of *M. tuberculosis* at our disposal (see http://bioweb.pasteur.fr/GenoList/TubercuList/ for further details), various strategies for the identification of new drug targets may be envisaged. The genes can be divided into three broad categories according to the level of functional information available from bioinformatic and experimental approaches (252). Roughly 20% of the genes belong to the orphan class (group I), as they show no similarity to genes described in other organisms. Some level of information is available for a further 40% of the genes (group 2), whereas precise functions can be attributed to the remaining 40% on the strength of their very strong similarity to genes of known function (group 3). The latter group encodes many of the classic targets of chemotherapy such as ribosomal components, RNA polymerase, or the central metabolic pathways, and the genomic information will facilitate drug development in these areas. Present in group 2 are genes that are often referred to as conserved hypotheticals, as they occur in most sequenced bacterial genomes but escaped detection by classic genetic or biochemical approaches. Many of these genes probably encode essential functions and thus represent attractive targets for new broad-spectrum antibiotics. Their essentiality can be established by a battery of approaches such as allelic exchange (253) or signature-tagged mutagenesis (254).

Armed with some biological knowledge, a variety of potential new drug targets can be readily identified. Prominent in the proteome of the tubercle bacillus are several large protein families that have evolved via gene-duplication events (59,251), and foremost among these are a variety of enzymes involved in the biosynthesis of the lipid, glycolipid, and polyketide components of the cell envelope, a fully validated drug target. As mentioned above, the MmpL proteins belonging to the RND superfamily appear to be required for export of certain complex lipids. Inactivation of RND efflux pumps renders naturally resistant bacteria more susceptible to certain antimicrobial agents (255), so targeting the MmpL proteins of *M. tuberculosis* with a novel inhibitor could have a dual effect. Also prominent in the proteome are the histidine kinases comprising the two component regulatory systems and 10 eukaryotic-like serine threonine protein kinases that probably also respond to environmental stimuli. The pharmaceutical industry has batteries of lead compounds for both of these protein kinase families, since there has been intensive research into the histidine kinases as novel drug targets in recent years (256,257). Numerous inhibitors of serine threonine protein kinases have been developed for use in cancer treatment.

Another remarkable feature of the *M. tuberculosis* proteome is the abundance of cytochrome P-450–dependent monoxygenases. Twenty different cytochrome P-450 genes are present, whereas none are found in *E. coli*, suggesting that the corresponding enzymes should play important roles in mycobacteria. The antifungal agent fluoconazole targets the cytochrome P-450–containing enzyme 14α-sterol demethylase and blocks ergosterol biosynthesis (258). There is growing evidence that mycobacterial cell walls contain sterols and, together with the abundance of the cytochrome P-450 species, this raises the possibility that azole or imidazole derivatives would represent novel inhibitors if any of the cytochrome P-450–containing enzymes of the tubercle bacillus proved to be essential.

Finally, it is our opinion that one of the reasons for the success of the current antituberculosis chemotherapy is the fact that it associates three highly specific drugs with a broad-spectrum antibiotic. The great specificity of ETH, INH, and PZA has probably restricted the emergence of the transferable resistance mechanisms that have bedeviled treatment of bacterial infections with aminoglycoside and β-lactam antibiotics. It would be prudent, therefore, to develop new drugs that target functions confined to mycobacteria, and their genes can be identified by comparative genomics (259). These will probably include representatives of gene classes 1 and 2; namely, those for which little or no functional information is currently available.

ACKNOWLEDGMENTS

The authors wish to acknowledge the financial support of the Institut Pasteur, the Association Française Raoul Follereau, the Wellcome Trust, and the Biomed Programme of the European Union (grants; BMH4-CT96-1241 and BMH4-CT97-2277). A.S.P. received a Wellcome Trust Research Fellowship in Clinical Tropical Medicine.

REFERENCES

1. WHO Tuberculosis Control Programme. TB: A global emergency-report on the TB epidemic. WHO Report 1994; WHO/TB/94.177.
2. Murray CJ, Lopez AD. Global mortality, disability, and the contribution of risk factors: Global Burden of Disease Study. Lancet 1997; 349:1436–1442.
3. Murray CJ, Salomon JA. Modeling the impact of global tuberculosis control strategies. Proc Natl Acad Sci USA 1998; 95:13881–13886.
4. Raviglione MC, Snider DE, Kochi A. Global epidemiology of tuberculosis. Morbidity and mortality of a worldwide epidemic. JAMA 1995; 273:220–226.
5. Dye C, Scheele S, Dolin P, Pathania V, Ravigliione MC. Consensus statement. Global burden of tuberculosis: estimated incidence, prevalence, and mortality by country. WHO Global Surveillance and Monitoring Project. JAMA 1999; 282:677–686.
6. Pablos MA, Raviglione MC, Laszlo A, et al. Global surveillance for antituberculosis drug resistance, 1994–1997. WHO/IUATLD Working Group on antituberculosis drug resistance surveillance. N Engl J Med 1998; 338:1641–1649.
7. Cohn DL, Bustreo F, Raviglione MC. Drug-resistant tuberculosis: review of the worldwide situation and the WHO/IUATLD Global Surveillance Project. Clin Infect Dis 1997; 24(Suppl):S121–S130.
8. Schatz A, Bugie E, Waksman SA. Streptomycin. Substance exhibiting activity against Gram-positive and Gram-negative bacteria. Proc Soc Exp Biol Med 1944; 55:66–69.
9. Lehman J. Para-aminosalicylic acid in the treatment of tuberculosis. Lancet 1946; 250:15–16.
10. Pfeutze KH, Pyle MM, Hinshaw HC. The first clinical trial of streptomycin in human tuberculosis. Am Rev Tuberc 1955; 71:752–754.
11. Medical Research Council. Streptomycin treatment of pulmonary tuberculosis. Br Med J 1948; 2:769–782.
12. Fox W, Sutherland I, Daniels M. A five year assessment of patients in a controlled trial of streptomycin in pulmonary tuberculosis. Q J Med 1954; 23: 347–366.
13. Medical Research Council. Treatment of pulmonary tuberculosis with streptomycin and para-aminosalicylic acid. Br Med J 1950; 2:1073–1085.
14. Robitzek EH, Selikoff IJ. Hydrazine derivatives of isonicotinic acid (Rimifon, Marsilid) in the treatment of active progressive caseous-pneumonic tuberculosis. A preliminary report. Am Rev Tuberc 1952; 65:402–428.

15. Council MR. Various combinations of isoniazid with streptomycin or with PAS in the treatment of pulmonary tuberculosis. BMJ 1955; 1:435–445.
16. Canetti G. Modifications des populations des foyers tuberculeux au cours de la chimiothérapie antibacillaire. Ann Inst Pasteur 1960; 53–79.
17. Canetti G, Grosset J. Teneur des souches sauvages de *Mycobacterium tuberculosis* en variants résitants à la streptomycine sur milieu de Lowenstein-Jensen. Ann Inst Pasteur 1961; 101:28–46.
18. David HL. Probability distribution of drug-resistant mutants in unselected populations of *Mycobacterium tuberculosis*. Appl Microbiol 1970; 20:810–814.
19. Fox W, Ellard GA, Mitchison DA. Studies on the treatment of tuberculosis undertaken by the British Medical Research Council tuberculosis units, 1946–1986, with relevant subsequent publications. Int J Tuberc Lung Dis 1999; 3:5231–5279.
20. Yeager R, Monroe WE, Dessau FI. Pyrazinamide in the treatment of pulmonary tuberculosis. Am Rev Tuberc 1952; 65:523–546.
21. American Thoracic Society. A statement on pyrazinamide by the committee on therapy. Am Rev Tuberc 1957; 75:1012–1015.
22. Sensi P, Greco AM, Ballotta R. Rifomycin. I. Isolation and properties of rifomycin B and rifomycin complex. Antibiot Ann 1960; 1959:262–270.
23. WHO Global Tuberculosis Control Programme. Treatment of tuberculosis: Guidelines for National Programmes. WHO report 1996; WHO/TB/97.220.
24. Kochi A. Tuberculosis control—is DOTS the health breakthrough of the 1990s? World Health Forum 1997; 18:225–232.
25. Wilkinson D, High-compliance tuberculosis treatment programme in a rural community. Lancet 1994; 343:647–648.
26. bayer R, Wilkinson D. Directly observed therapy for tuberculosis: history of an idea. Lancet 1995; 345:1545–1548.
27. WHO Global Tuberculosis Control Programme. Global Tuberculosis Control. WHO report 1998; WHO/TB/98-237.
28. China Tuberculosis Collaboration. Results of directly observed short-course chemotherapy in 112,842 Chinese patients with smear-positive tuberculosis. Lancet 1996; 347:358–362.
29. Kenyon TA, Mwasekaga MJ, Huebner R, Rumisha D, Binkin N, Maganu E. Low levels of drug resistance amidst rapidly increasing tuberculosis and human immunodeficiency virus co-epidemics in Botswana. Int J Tuberc Lung Dis 1999; 3:4–11.
30. Narain JP, Raviglione MC, Kochi A. HIV-associated tuberculosis in developing countries: epidemiology and strategies for prevention. Tuberc Lung Dis 1992; 73:311–321.
31. Small PM, Shafer RW, Hopewell PC, et al. Exogenous reinfection with multidrug-resistant *Mycobacterium tuberculosis* in patients with advanced HIV infection. N Engl J Med 1993; 328:1137–1144.
32. Daley CL, Small PM, Schecter GF, et al. An outbreak of tuberculosis with accelerated progression among persons infected with the human immunodeficiency virus. An analysis using restriction-fragment-length polymorphisms. N Engl J Med 1992; 326:231–235.

33. Mitchison DA, Nunn AJ, Influence of initial drug resistance on the response to short-course chemotherapy of pulmonary tuberculosis. Am Rev Respir Dis 1986; 133:423–430.
34. Kopanoff DE, Kilburn JO, Glassroth JL, Snider DJ, Farer LS, Good RC. A continuing survey of tuberculosis primary drug resistance in the United States: March 1975 to November 1977. A United States Public Health Service cooperative study. Am Rev Respir Dis 1978; 118:835–842.
35. Edlin BR, Tokars JI, Grieco MH, et al. An outbreak of multidrug-resistant tuberculosis among hospitalized patients with the acquired immunodeficiency syndrome. N Engl J Med 1992; 326:1514–1521.
36. Alland D, Kalkut GE, Moss AR, et al. Transmission of tuberculosis in New York City. An analysis by DNA fingerprinting and conventional epidemiologic methods. N Engl J Med 1994; 330:1710–1716.
37. Valway SE, Greifinger RB, Papania M, et al. Multidrug-resistant tuberculosis in the New York State prison system, 1990–1991. J Infect Dis 1994; 170:151–156.
38. Beck SC, Dooley SW, Hutton MD, et al. Hospital outbreak of multidrug-resistant *Mycobacterium tuberculosis* infections. Factors in transmission to staff and HIV-infected patients. JAMA 1992; 268:1280–1286.
39. Multidrug-resistant tuberculosis outbreak on an HIV ward—Madrid, Spain, 1991–1995. MMWR 1996; 45:330–333.
40. Ritacco V, Di LM, Reniero A, et al. Nosocomial spread of human immunodeficiency virus-related multidrug-resistant tuberculosis in Buenos Aires. J Infect Dis 1997; 176:637–642.
41. Portugal I, Covas MJ, Brum L, et al. Outbreak of multiple drug-resistant tuberculosis in Lisbon: detection by restriction fragment length polymorphism analysis. Int J Tuberc Lung Dis 1999; 3:207–213.
42. Sacks LV, Pendle S, Orlovic D, Blumberg L, Constantinou C. A comparison of outbreak- and non outbreak-related multidrug-resistant tuberculosis among human immunodeficiency virus-infected patients in a South African hospital. Clin Infect Dis 1999; 29:96–101.
43. Frieden TR, Sterling T, Pablos MA, Kilburn JO, Cauthen GM, Dooley SW. The emergence of drug-resistant tuberculosis in New York City. N Engl J Med 1993; 328:521–526.
44. Moss AR, Alland D, Telzak E, et al. A city-wide outbreak of a multiple-drug-resistant strain of *Mycobacterium tuberculosis* in New York. Int J Tuberc Lung Dis 1997; 1:115–121.
45. Fang Z, Doig C, Rayner A, Kenna DT, Watt B, Forbes KJ. Molecular evidence for heterogeneity of the multiple-drug-resistant *Mycobacterium tuberculosis* population in Scotland (1990 to 1997). J Clin Microbiol 1999; 37:998–1003.
46. Agerton TB, Valway SE, Blinkhorn RJ, et al. Spread of strain W, a highly drug-resistant strain of *Mycobacterium tuberculosis*, across the United States. Clin Infect Dis 1999; 29:85–92.
47. WHO Global Tuberculosis Control Programme. Anti-Tuberculosis Drug Resistance in the World. The WHO/IUATLD Global Project on Anti-Tuberculosis Drug Resistance Surveillance 1997; WHO/TB?97.229.
48. Nosocmial transmission of multidrug-resistant tuberculosis among HIV-

infected persons—Florida and New York, 1988–1991. MMWR 1991; 40: 585–591.
49. Park MM, Davis AL, Schluger NW, Cohen H, Rom WN. Outcome of MDR-TB patients, 1983–1993. Prolonged survival with appropriate therapy. Am J Respir Crit Care Med 1996; 153:317–324.
50. Turett GS, Telzak EE, Torian LV, et al. Improved outcomes for patients with multidrug-resistant tuberculosis. Clin Infect Dis 1995; 21:1238–1244.
51. Mannheimer SB, Sepkowitz KA, Stoeckle M, Friedman CR, Hafner A, Riley LW. Risk factors and outcome of human immunodeficiency virus-infected patients with sporadic multidrug-resistant tuberculosis in New York City. Int J Tuberc Lung Dis 1997; 1:319–325.
52. Mahmoudi A, Iseman MD. Pitfalls in the care of patients with tuberculosis. Common errors and their association with the acquisition of drug resistance. JAMA 1993; 270:65–68.
53. Iseman MD. Treatment of multidrug-resistant tuberculosis. N Engl J Med 1993; 329:784–791.
54. Iseman MD. Management of multidrug-resistant tuberculosis. Chemotherapy 1999; 2:3–11.
55. Liu Z, Shilkret KL, Finelli L. Epidemiology of drug-resistant tuberculosis in New Jersey from 1991 to 1995. Int J Epidemiol 1998; 27:121–126.
56. Frieden TR, Fujiwara PI, Washko RM, Hamburg MA. Tuberculosis in New York City—turning the tide. N Engl J Med 1995; 333:229–233.
57. Noller HF. Structure of ribosomal RNA. Annu Rev Biochem 1984; 53:119–162.
58. Ainsa JA, Perez E, Pelicic V, Berthet FX, Gicquel B, Martin C. Aminoglycoside 2′-N-acetyltransferase genes are universally present in mycobacteria: characterization of the *aac(2′)-Ic* gene from *Mycobacterium tuberculosis* and the *aac(2′)-Id* gene from *Mycobacterium smegmatis*. Mol Microbiol 1997; 24: 431–441.
59. Cole ST, Brosch R, Parkhill J, et al. Deciphering the biology of *Mycobacterium tuberculosis* from the complete genome sequence. Nature 1998; 393:537–544.
60. Honoré N, Cole ST. Streptomycin resistance in mycobacteria. Antimicrob Agents Chemother 1994; 38:238–242.
61. Meier A, Kirschner P, Bange FC, Vogel U, Bottger EC. Genetic alterations in streptomycin-resistant *Mycobacterium tuberculosis*: mapping of mutations conferring resistance. Antimicrob Agents Chemother 1994; 38:228–233.
62. Nair J, Rouse DA, Bai GH, Morris SL. The *rpsL* gene and streptomycin resistance in single and multiple drug-resistant strains of *Mycobacterium tuberculosis*. Mol Microbiol 1993; 10:521–527.
63. Finken M, Kirschner P, Meier A, Wrede A, Bottger EC. Molecular basis of streptomycin resistance in *Mycobacterium tuberculosis*, alterations of the ribosomal protein S12 gene and point mutations within a functional 16S ribosomal RNA pseudoknot. Mol Microbiol 1993; 9:1239–1246.
64. Douglass J, Steyn LM. A ribosomal gene mutation in streptomycin-resistant *Mycobacterium tuberculosis* isolates. J Infect Dis 1993; 167:1505–1506.
65. Meier A, Sander P, Schaper KJ, Scholz M, Bottger EC. Correlation of molecular resistance mechanisms and phenotypic resistance levels in streptomycin-

resistant *Mycobacterium tuberculosis*. Antimicrob Agents Chemother 1996; 40:2452–2454.
66. Cooksey RC, Morlock GP, McQueen A, Glickman SE, Crawford JT. Characterization of streptomycin resistance mechanisms among *Mycobacterium tuberculosis* isolates from patients in New York City. Antimicrob Agents Chemother 1996; 40:1186–1188.
67. Morris S, Bai GH, Suffys P, Portillo GL, Fairchok M, Rouse D. Molecular mechanisms of multiple drug resistance in clinical isolates of *Mycobacterium tuberculosis*. J Infect Dis 1995; 171:954–960.
68. Sreevatsan S, Pan X, Stockbauer KE, William DL, Kreiswirth BN, Musser JM. Characterization of *rpsL* and *rrs* mutations in streptomycin-resistant *Mycobacterium tuberculosis* isolates from diverse geographic localities. Antimicrob Agents Chemother 1996; 40:1024–1026.
69. Kenney TJ, Churchward G. Genetic analysis of the *Mycobacterium smegmatis rpsL* promoter. J Bacteriol 1996; 178:3564–3571.
70. Honoré N, Marchal G, Cole ST. Novel mutation in 16S rRNA associated with streptomycin dependence inn *Mycobacterium tuberculosis*. Antimicrob Agents Chemother 1995; 39:769–770.
71. Moazed D, Noller HF. Interaction of antibiotics with functional sites in 16S ribosomal RNA. Nature 1987; 327:389–394.
72. Powers T, Noller HF. A functional pseudoknot in 16S ribosomal RNA. EMBO J 1991; 10:2203–2214.
73. Leclerc D, Melancon P, Brakier GL. Mutations in the 915 region of *Escherichia coli* 16S ribosomal RNA reduce the binding of streptomycin to the ribosome. Nucleic Acids Res 1991; 19:3973–3977.
74. Fourmy D, Recht MI, Blanchard SC, Puglisi JD. Structure of the A site of Escherichia coli 16S ribosomal RNA complexed with an aminoglycoside antibiotic. Science 1996; 274:1367–1371.
75. Beauclerk AA, Cundliffe E. Sites of action of two ribosomal DNA methylases responsible for resistance to aminoglycosides. J Mol Biol 1987; 193:661–671.
76. Suzuki Y, Katsukawa C, Tamaru A, et al. Deletion of kanamycin-resistant *Mycobacterium tuberculosis* by identifying mutations in the 16S rRNA gene. J Clin Microbiol 1998; 36:1220–1225.
77. Taniguchi H, Chang B, Abe C, Nikaido Y, Mizuguchi Y, Yoshida SI. Molecular analysis of kanamycin and viomycin resistance in *Mycobacterium smegmatis* by use of the conjugation system. J Bacteriol 1997; 179:4795–4801.
78. Alangaden GJ, Kreiswirth BN, Aouad A, et al. Mechanism of resistance to amikacin and kanamycin in *Mycobacterium tuberculosis*. Antimicrob Agents Chemother 1998; 42:1295–1297.
79. Sander P, Prammananan T, Bottger EC. Introducing mutations into a chromosomal rRNA gene using a genetically modified eubacterial host with a single rRNA operon. Mol Microbiol 1996; 22:841–848.
80. Prammananan T, Sander P, Brown BA, et al. A single 16S ribosomal RNA substitution is responsible for resistance to amikacin and other 2-deoxystreptamine aminoglycosides in *Mycobacterium abscessus* and *Mycobacterium chelonae*. J Infect Dis 1998; 177:1573–1581.

81. Ho YI, Chan CY, Cheng AF. In-vitro activities of aminoglycoside-aminocyclitols against mycobacteria. J Antimicrob Chemother 1997; 40:27–32.
82. Maggi N, Pasqualucci CR, Ballota R, Sensi P. Rifampicin: a new orally active rifamycin. Chemotherapia 1966; 11:285–292.
83. East Africa/British Medical Research Council. Controlled clinical trial of short course (6 months) regimens of chemotherapy for treatment of pulmonary tuberculosis. Lancet 1972; 1:1072–1085.
84. Hartmann G, Honikel KO, Knusel F, Nuesch J. The specific inhibition of the DNA-directed RNA synthesis by rifamycin. Biochim Biophys Acta 1967; 145: 843–844.
85. Ovchinnikov YA, Monastyrskaya GS, Guriev SO, et al. RNA polymerase rifampicin resistance mutations in *Escherichia coli*: sequence changes and dominance. Mol Gen Genet 1982; 190:344–348.
86. Ovchinnikov Y, Monastyrskaya GS, Gubanov VV, et al. Primary structure of *Escherichia coli* RNA polymerase nucleotide substitution in the beta subunit gene of the rifampicin resistant *rpoB* 255 mutant. Mol Gen Genet 1981; 184: 536–538.
87. Jin DJ, Gross CA. Mapping and sequencing of mutations in the *Escherichia coli rpoB* gene that lead to rifampicin resistance. J Mol Biol 1988; 202:45–58.
88. Lisitsyn NA, Sverdlov ED, Moiseyeva EP, Danilevskaya ON, Nikiforov VG. Mutation to rifampicin resistance at the beginning of the RNA polymerase beta subunit gene in *Escherichia coli*. Mol Gen Genet 1984; 196:173–174.
89. Honoré N, Bergh S, Chanteau S, et al. Nucleotide sequence of the first cosmid from the *Mycobacterium leprae* genome project: structure and function of the Rif-Str regions. Mol Microbiol 1993; 7:207–214.
90. Telenti A, Imboden P, Marchesi F, et al. Detection of rifampicin-resistance mutations in *Mycobacterium tuberculosis*. Lancet 1993; 341:647–650.
91. Honoré N, Cole ST. Molecular basis of rifampin resistance in *Mycobacterium leprae*. Antimicrob Agents Chemother 1993; 37:414–418.
92. Heym B, Honoré N, Truffot PC, et al. Implications of multidrug resistance for the future of short-course chemotherapy of tuberculosis: a molecular study. Lancet 1994; 344:293–298.
93. Kapur V, Li LL, Iordanescu S, et al. Characterization by automated DNA sequencing of mutations in the gene (*rpoB*) encoding the RNA polymerase beta subunit in rifampin-resistant *Mycobacterium tuberculosis* strains from New York City and Texas. J Clin Microbiol 1994; 32:1095–1098.
94. Kapur V, Li LL, Hamrick MR, et al. Rapid Mycobacterium species assignment and unambiguous identification of mutations associated with antimicrobial resistance in *Mycobacterium tuberculosis* by automated DNA sequencing. Arch Pathol Lab Med 1995; 119:131–138.
95. Matsiota BP, Vrioni G, Marinis E. Characterization of *rpoB* mutations in rifampin-resistant clinical *Mycobacterium tuberculosis* isolates from Greece. J Clin Microbiol 1998; 36:20–23.
96. Williams DL, Waguespack C, Eisenach K, et al. Characterization of rifampin-resistance in pathogenic mycobacteria. Antimicrob Agents Chemother 1994; 38:2380–2386.

97. Donnabella V, Martiniuk F, Kinney D, et al. Isolation of the gene for the beta subunit of RNA polymerase from rifampicin-resistant *Mycobacterium tuberculosis* and identification of new mutations. Am J Respir Cell Mol Biol 1994; 11:639–643.
98. Rinder H, Dobner P, Feldmann K, et al. Disequilibria in the distribution of *rpoB* alleles in rifampicin-resistant *M. tuberculosis* isolates from Germany and Sierra Leone. Microb Drug Resist 1997; 3:195–197.
99. Gonzalez N, Torres MJ, Aznar J, Palomares JC. Molecular analysis of rifampin and isoniazid resistance of *Mycobacterium tuberculosis* clinical isolates in Seville, Spain. Tuberc Lung Dis 1999; 79:187–190.
100. Yuen LK, Leslie D, Coloe PJ. Bacteriological and molecular analysis of rifampin-resistant *Mycobacterium tuberculosis* strains isolated in Australia. J Clin Microbiol 1999; 37:3844–3850.
101. Schilke K, Weyer K, Bretzel G, et al. Universal pattern of *rpoB* gene mutations among multidrug-resistant isolates of *Mycobacterium tuberculosis* complex from Africa. Int J Tuberc Lung Dis 1999; 3:620–626.
102. Hirano K, Abe C, Takahashi M. Mutations in the *rpoB* gene of rifampin-resistant *Mycobacterium tuberculosis* strains isolated mostly in Asian countries and their rapid detection by line probe assay. J Clin Microbiol 1999; 37:2663–2666.
103. Pozzi G, Meloni M, Iona E, et al. *rpoB* mutations in multidrug-resistant strains of *Mycobacterium tuberculosis* isolated in Italy. J Clin Microbiol 1999; 37:1197–1199.
104. Kiepiela P, Bishop K, Kormuth E, Roux L, York DF. Comparison of PCR-heteroduplex characterization by automated DNA sequencing and line probe assay for the detection of rifampicin resistance in *Mycobacterium tuberculosis* isolates from KwaZulu-Natal, South Africa. Microb Drug Resist 1998; 4:263–269.
105. Ramaswamy S, Musser JM. Molecular genetic basis of antimicrobial agent resistance in *Mycobacterium tuberculosis*: 1998 update. Tuberc Lung Dis 1998; 79:3–29.
106. Miller LP, Crawford JT, Shinnick TM. The *rpoB* gene of *Mycobacterium tuberculosis*. Antimicrob Agents Chemother 1994; 38:805–811.
107. Williams DL, Spring L, Collins L, et al. Contribution of *rpoB* mutations to development of rifamycin cross-resistance in *Mycobacterium tuberculosis*. Antimicrob Agents Chemother 1998; 42:1853–1857.
108. Heep M, Beck D, Bayerdorffer E, Lehn N. Rifampin and rifabutin resistance mechanism in *Helicobacter pylori*. Antimicrob Agents Chemother 1999; 43: 1497–1499.
109. Telenti A, Honoré N, Bernasconi C, et al. Genotypic assessment of isoniazid and rifampin resistance in *Mycobacterium tuberculosis*: a blind study at reference laboratory level. J Clin Microbiol 1997; 35:719–723.
110. Piatek AS, Telenti A, Murray MR, et al. Genotypic analysis of *Mycobacterium tuberculosis* in two distinct populations using molecular beacons: implications for rapid susceptibility testing. Antimicrob Agents Chemother 2000; 44:103–110.

111. Musser JM. Antimicrobial agent resistance in mycobacteria: molecular genetic insights. Clin Microbiol Rev 1995; 8:496–514.
112. Guerrero C, Stockman L, Marchesi F, Bodmer T, Roberts GD, Telenti A. Evaluation of the *rpoB* gene in rifampicin-susceptible and -resistant *Mycobacterium avium* and *Mycobacterium intracellulare*. J Antimicrob Chemother 1994; 33:661–663.
113. Quan S, Venter H, Dabbs ER. Ribosylative inactivation of rifampin by *Mycobacterium smegmatis* is a principal contributor to its low susceptibility to this antibiotic. Antimicrob Agents Chemother 1997; 41:2456–2460.
114. Hui J, Gordon N, Kajioka R. Permeability barrier to rifampin in mycobacteria. Antimicrob Agents Chemother 1977; 11:773–779.
115. Kunin CM. Antimicrobial activity of rifabutin. Clin Infect Dis 1996; 11(Suppl): S3–S13.
116. Bishai WR, Graham NM, Harrington S, et al. Brief report: rifampin-resistant tuberculosis in a patient receiving rifabutin prophylaxis. N Engl J Med 1996; 334:1573–1576.
117. Weltman AC, Righi SP, DiFerdinando GJ, Jovell RJ, Driscoll JR. Rifampicin-resistant *Mycobacterium tuberculosis*. Lancet 1995; 345:1513.
118. Temple ME, Nahata MC. Rifapentine: its role in the treatment of tuberculosis. Ann Pharmacother 1999; 33:1203–1210.
119. Vernon A, Burman W, Benator D, Khan A, Bozeman L. Acquired rifamycin monoresistance in patients with HIV-related tuberculosis treated with once-weekly rifapentine and isoniazid. Tuberculosis Trials Consortium. Lancet 1999; 353:1843–1847.
120. March F, Garriga X, Rodriguez P, et al. Acquired drug resistance in *Mycobacterium tuberculosis* isolates recovered from compliant patients with human immunodeficiency virus-associated tuberculosis. Clin Infect Dis 1997; 25: 1044–1047.
121. Hong Kong Chest Service/British Medical Research Council. A controlled study of rifabutin and an uncontrolled study of ofloxacin in the retreatment of patients with pulmonary tuberculosis resistant to isoniazid, streptomycin and rifampicin. Tuberc Lung Dis 1992; 73:59–67.
122. Dickinson JM, Mitchison DA. In vitro activity of new rifamycins against rifampicin-resistant *M. tuberculosis* and MAIS-complex mycobacteria. Tubercle 1987; 68:177–182.
123. Woodley CL, Kilburn JO. In vitro susceptibility of *Mycobacterium avium* complex and *Mycobacterium tuberculosis* strains to a spiro-piperidyl rifamycin. Am Rev Respir Dis 1982; 126:586–587.
124. Bodmer T, Zurcher G, Imboden P, Telenti A. Mutation position and type of substitution in the beta-subunit of the RNA polymerase influence in-vitro activity of rifamycins in rifampicin-resistant *Mycobacterium tuberculosis*. J Antimicrob Chemother 1995; 35:345–348.
125. Moghazeh SL, Pan X, Arain T, Stover CK, Musser JM, Kreiswirth BN. Comparative antimycobacterial activities of rifampin, rifapentine, and KRM-1648 against a collection of rifampin-resistant *Mycobacterium tuberculosis* isolates with known *rpoB* mutations. Antimicrob Agents Chemother 1996; 40:2655–2657.

126. Yang B, Koga H, Ohno H, et al. Relationship between antimycobacterial activities of rifampicin, rifabutin and KRM-1648 and *rpoB* mutations of *Mycobacterium tuberculosis*. J Antimicrob Chemother 1998; 42:621–628.
127. Middlebrook G. Isoniazid-resistance and catalase activities of tubercle bacilli. A preliminary report. Am Rev Tuberc 1954; 471–472.
128. Middlebrook G, Cohn ML. Some observations on the pathogenicity of isoniazid-resistant variants of tubercle bacilli. Science 1953; 118:297–299.
129. Winder FG. Mode of action of the antimycobacterial agents and associated aspects of the molecular biology of mycobacteria. In: Ratledge C, Stanford J, eds. The Biology of Mycobacteria. New York, Academic Press 1982; 353–438.
130. Kruger-Thiemer E. Isonicotinic acid hypothesis of the antituberculous action of isoniazid. Am Rev Tuberc 77:364–367.
131. Zhang Y, Heym B, Allen B, Young D, Cole S. The catalase-peroxidase gene and isoniazid resistance of *Mycobacterium tuberculosis*. Nature 1992; 358: 591–593.
132. Zhang Y, Garbe T, Young D. Transformation with *katG* restores isoniazid-sensitivity in *Mycobacterium tuberculosis* isolates resistant to a range of drug concentrations. Mol Microbiol 1993; 8:521–524.
133. Gayathri Devi B, Shaila MS, Ramakrishnan T. The purification and properties of peroxidase in *Mycobacterium tuberculosis* H37Rv and its possible role in the mechanism of action of isonicotinic acid hydrazide. Biochem J 1975; 149:187–197.
134. Heym B, Zhang Y, Poulet S, Young D, Cole ST. Characterization of the *katG* gene encoding a catalase-peroxidase required for the isoniazid susceptibility of *Mycobacterium tuberculosis*. J Bacteriol 1993; 175:4255–4259.
135. Johnsson K, Schultz PG. Mechanistic studies of the oxidation of isoniazid by the catalase peroxidase from *Mycobacterium tuberculosis*. J Am Chem Soc 1994; 116:7425–7426.
136. Heym B, Alzari PM, Honoré N, Cole ST. Missense mutations in the catalase-peroxidase gene, *katG*, are associated with isoniazid resistance in *Mycobacterium tuberculosis*. Mol Microbiol 1995; 15:235–245.
137. Nagy JM, Cass AE, Brown KA. Purification and characterization of recombinant catalase-peroxidase, which confers isoniazid sensitivity in *Mycobacterium tuberculosis*. J Biol Chem 1997; 272:31265–31271.
138. Johnsson K, Froland WA, Schultz PG. Overexpression, purification, and characterization of the catalase-peroxidase KatG from *Mycobacterium tuberculosis*. J Biol Chem 1997; 272:2834–2840.
139. Stoeckle MY, Guan L, Riegler N, et al. Catalase-peroxidase gene sequences in isoniazid-sensitive and -resistant strains of *Mycobacterium tuberculosis* from New York City. J Infect Dis 1993; 168:1063–1065.
140. Altamirano M, Marostenmaki J, Wong A, FitzGerald M, Black WA, Smith JA. Mutations in the catalase-peroxidase gene from isoniazid-resistant *Mycobacterium tuberculosis* isolates. J Infect Dis 1994; 169:1162–1165.
141. Rouse DA, Morris SL. Molecular mechanisms of isoniazid resistance in *Mycobacterium tuberculosis* and *Mycobacterium bovis*. Infect Immun 1995; 63:1427–1433.
142. Rouse DA, Li Z, Bai GH, Morris SL. Characterization of the *katG* and *inhA*

genes of isoniazid-resistant clinical isolates of *Mycobacterium tuberculosis*. Antimicrob Agents Chemother 1995; 39:2472–2477.

143. Ferrazoli L, Palaci M, Telles MA, et al. Catalase expression, *katG*, and MIC of isoniazid for *Mycobacterium tuberculosis* isolates from Sao Paulo, Brazil. J Infect Dis 1995; 171:237–240.
144. Victor TC, Pretorius GS, Felix JV, Jordaan AM, van HP, Eisenach KD. *katG* mutations in isoniazid-resistant strains of *Mycobacterium tuberculosis* are not infrequent. Antimicrob Agents Chemother 1996; 40:1572.
145. Musser JM, Kapur V, Williams DL, Kreiswirth BN, van SD, van EJ. Characterization of the catalase-peroxidase gene (*katG*) and *inhA* locus in isoniazid-resistant and -susceptible strains of *Mycobacterium tuberculosis* by automated DNA sequencing: restricted array of mutations associated with drug resistance. J Infect Dis 1996; 173:196–202.
146. Cockerill FR, Uhl JR, Temesgen Z, et al. Rapid identification of a point mutation of the *Mycobacterium tuberculosis* catalase-peroxidase (*katG*) gene associated with isoniazid resistance. J Infect Dis 1995; 171:240–245.
147. Marttila HJ, Soini H, Huovinen P, Viljanen MK. *katG* mutations in isoniazid-resistant *Mycobacterium tuberculosis* isolates recovered from Finnish patients. Antimicrob Agents Chemother 1996; 40:2187–2189.
148. Marttila HJ, Soini H, Eerola E, et al. A Ser315 Thr substitution in KatG is predominant in genetically heterogenous multidrug-resistant *Mycobacterium tuberculosis* isolates originating from the St. Petersburg area in Russia. Antimicrob Agents Chemother 1998; 42:2443–2445.
149. Haas WH, Schilke K, Brand J, et al. Molecular analysis of *katG* gene mutations in strains of *Mycobacterium tuberculosis* complex from Africa. Antimicrob Agents Chemother 1997; 41:1601–1603.
150. Dobner P, Rusch GS, Bretzel G, et al. Usefulness of *Mycobacterium tuberculosis* genomic mutations in the genes *katG* and *inhA* for the prediction of isoniazid resistance. Int J Tuberc Lung Dis 1997; 1:365–369.
151. Escalante P, Ramaswamy S, Sanabria H, et al. Genotypic characterization of drug-resistant *Mycobacterium tuberculosis* isolates from Peru. Tuberc Lung Dis 1998; 79:111–118.
152. Cingolani A, Antinori A, Sanguinetti M, et al. Application of molecular methods for detection and transmission analysis of *Mycobacterium tuberculosis* drug resistance in patients attending a reference hospital in Italy. J Infect Dis 1999; 179:1025–1029.
153. Wengenack NL, Todorovic S, Yu L, Rusnak F. Evidence for differential binding of isoniazid by *Mycobacterium tuberculosis* KatG and the isoniazid-resistant mutant KatG(S315T). Biochemistry 1998; 37:15825–15834.
154. Saint Joanis B, Souchon H, Wilming M, Johnsson K, Alzari PM, Cole ST. Use of site-directed mutagenesis to probe the structure, function and isoniazid activation of the catalase/peroxidase, KatG, from *Mycobacterium tuberculosis*. Biochem J 1999; 338:753–760.
155. Jackett PS, Aber VR, Lowrie DB. Virulence and resistance to superoxide, low pH and hydrogen peroxide among strains of *Mycobacterium tuberculosis*. J Gen Microbiol 1978; 104:37–45.

156. Heym B, Stavropoulos E, Honoré N, et al. Effects of overexpression of the alkyl hydroperoxide reductase AhpC on the virulence and isoniazid resistance of *Mycobacterium tuberculosis*. Infect Immun 1997; 65:1395–1401.
157. Li Z, Kelley C, Collins F, Rouse D, Morris S. Expression of *katG* in *Mycobacterium tuberculosis* is associated with its growth and persistence in mice and guinea pigs. J Infect Dis 1998; 177:1030–1035.
158. Manca C, Paul S, Barry CR, Freedman VH, Kaplan G. *Mycobacterium tuberculosis* catalase and peroxidase activities and resistance to oxidative killing in human monocytes in vitro. Infect Immun 67:74–79.
159. Wengenack NL, Uhl JR, St AA, et al. Recombinant *Mycobacterium tuberculosis* KatG(S315T) is a competent catalase-peroxidase with reduced activity toward isoniazid. J Infect Dis 1997; 176:722–727.
160. Nagy JM, Jesmin J, Servos S, Cass AE, Brown KA. Site-directed mutants of the catalase-peroxidase from *Mycobacterium tuberculosis*. Biochem Soc Trans 1998; 26:5281.
161. Rouse DA, DeVito JA, Li Z, Byer H, Morris SL. Site-directed mutagenesis of the *katG* gene of *Mycobacterium tuberculosis*: effects on catalase-peroxidase activities and isoniazid resistance. Mol Microbiol 1996; 22:583–592.
162. Lee AS, Tang LL, Lim IH, Ling ML, Tay L, Wong SY. Lack of clinical significance for the common arginine-to-leucine substitution at codon 463 of the *katG* gene in isoniazid-resistant *Mycobacterium tuberculosis* in Singapore. J Infect Dis 1997; 176:1125–1127.
163. Musser JM. Letter. J Infect Dis 1997; 176:1126–1127.
164. Banerjee A, Dubnau E, Quemard A, et al. *inhA*, a gene encoding a target for isoniazid and ethionamide in *Mycobacterium tuberculosis*. Science 1994; 263: 227–230.
165. Banerjee A, Sugantino M, Sacchettini JC, Jacobs WJ. The mabA gene from the *inhA* operon of *Mycobacterium tuberculosis* encodes a 3-ketoacyl reductase that fails to confer isoniazid resistance. Microbiology 1998; 144:2697–2704.
166. Mdluli K, Sherman DR, Hickey MJ, et al. Biochemical and genetic data suggest that InhA is not the primary target for activated isoniazid in *Mycobacterium tuberculosis*. J Infect Dis 1996; 174:1085–1090.
167. Quemard A, Sacchettini JC, Dessen A, et al. Enzymatic characterization of the target for isoniazid in *Mycobacterium tuberculosis*. Biochemistry 1995; 34: 8235–8241.
168. Dessen A, Quemard A, Blanchard JS, Jacobs WJ, Sacchettini JC. Crystal structure and function of the isoniazid target of *Mycobacterium tuberculosis*. Science 1995; 267:1638–1641.
169. Rozwarski DA, Grant GA, Barton D, Jacobs WJ, Sacchettini JC. Modification of the NADH of the isoniazid target (InhA) from *Mycobacterium tuberculosis*. Science 1998; 279:98–102.
170. Basso LA, Zheng R, Blanchard JS. Kinetics of inactivation of WT and C243S mutant of *Mycobacterium tuberculosis* enoyl reductase by activated isoniazid. J Am Chem Soc 1996; 118:11301–11302.
171. Rozwarski DA, Vilcheze C, Sugantino M, Bittman R, Sacchettini JC. Crystal structure of the *Mycobacterium tuberculosis* enoyl-ACP reductase, InhA, in

complex with NAD+ and a C16 fatty acyl substrate. J Biol Chem 1999; 274: 15582–15589.

172. Wilming M, Johnsson K. Spontaneous formation of the bioactive form of the tuberculosis drug isoniazid. Angew Chem Int Ed Engl 1999; 38:2588–2590.
173. Miesel L, Weisbrod TR, Marcinkeviciene JA, Bittman R, Jacobs WJ. NADH dehydrogenase defects confer isoniazid resistance and conditional lethality in *Mycobacterium smegmatis*. J Bacteriol 1998; 180:2459–2467.
174. Ristow M, Mohlig M, Rifai M, Schatz H, Feldmann K, Pfeiffer A. New isoniazid/ethionamide resistance gene mutation and screening for multidrug-resistant *Mycobacterium tuberculosis* strains. Lancet 1995; 346:502–503.
175. Basso LA, Zheng R, Musser JM, Jacobs WJ, Blanchard JS. Mechanisms of isoniazid resistance in *Mycobacterium tuberculosis*: enzymatic characterization of enoyl reductase mutants identified in isoniazid-resistant clinical isolates. J Infect Dis 1998; 178:769–775.
176. Heym B, Philipp W, Cole ST. Mechanisms of drug resistance in *Mycobacterium tuberculosis*. Curr Top Microbiol Immunol 1996; 215:49–69.
177. Wilson TM, de LG, Collins DM. Effect of *inhA* and *katG* on isoniazid resistance and virulence of *Mycobacterium bovis*. Mol Microbiol 1995; 15:1009–1015.
178. Mdluli K, Slayden RA, Zhu Y, et al. Inhibition of a *Mycobacterium tuberculosis* beta-ketoacyl ACP synthase by isoniazid. Science 1998; 280:1607–1610.
179. Lee AS, Lim IH, Tang LL, Telenti A, Wong SY. Contribution of *kasA* analysis to detection of isoniazid-resistant *Mycobacterium tuberculosis* in Singapore. Antimicrob Agents Chemother 1999; 43:2087–2089.
180. Wilson M, DeRisi J, Kristensen HH, et al. Exploring drug-induced alterations in gene expression in *Mycobacterium tuberculosis* by microarray hybridization. Proc Natl Acad Sci USA 1999; 96:12833–12838.
181. Belisle JT, Vissa VD, Sievert T, Takayama K, Brennan PJ, Besra GS. Role of the major antigen of *Mycobacterium tuberculosis* in cell wall biogenesis. Science 1997; 276:1420–1422.
182. Garbe TR, Hibler NS, Deretic V. Isoniazid induces expression of the antigen 85 complex in *Mycobacterium tuberculosis*. Antimicrob Agents Chemother 1996; 40:1754–1756.
183. Alland D, Kramnik I, Weisbrod TR, et al. Identification of differentially expressed mRNA in prokaryotic organisms by customized amplification libraries (DECAL): the effect of isoniazid on gene expression in *Mycobacterium tuberculosis*. Proc Natl Acad Sci USA 1998; 95:13227–13232.
184. Doran JL, Pang Y, Mdluli KE, et al. *Mycobacterium tuberculosis efpA* encodes an efflux protein of the QacA transporter family. Clin Diagn Lab Immunol 1997; 4:23–32.
185. Rosner JL. Susceptibilities of *oxyR* regulon mutants of *Escherichia coli* and *Salmonella typhimurium* to isoniazid. Antimicrob Agents Chemother 1993; 37: 2251–2253.
186. Deretic V, Philipp W, Dhandayuthapani S, et al. *Mycobacterium tuberculosis* is a natural mutant with an inactivated oxidative-stress regulatory gene: implications for sensitivity to isoniazid. Mol Microbiol 1995; 17:889–900.

187. Sherman DR, Mdluli K, Hickey MJ, et al. Compensatory *ahpC* gene expression in isoniazid-resistant *Mycobacterium tuberculosis*. Science 1996; 272:1641–1643.
188. Dhandayuthapani S, Mudd M, Deretic V. Interactions of *OxyR* with the promoter region of the *oxyR* and *ahpC* genes from *Mycobacterium leprae* and *Mycobacterium tuberculosis*. J Bacteriol 1997; 179:2401–2409.
189. Zhang Y, Dhandayuthapani S, Deretic V. Molecular basis for the exquisite sensitivity of *Mycobacterium tuberculosis* to isoniazid. Proc Natl Acad Sci USA 1996; 93:13212–13216.
190. Sreevatsan S, Pan X, Zhang Y, Deretic V, Musser JM. Analysis of the *osyR-aphC* region in isoniazid-resistant and -susceptible *Mycobacterium tuberculosis* complex organisms recovered from diseased humans and animals in diverse localities. Antimicrob Agents Chemother 1997; 41:600–606.
191. Kelley CL, Rouse DA, Morris SL. Analysis of *aphC* gene mutations in isoniazid-resistant clinical isolates of *Mycobacterium tuberculosis*. Antimicrob Agents Chemother 1997; 41:2057–2058.
192. Wilson TM, Collins DM. *ahpC*, a gene involved in isoniazid resistance of the *Mycobacterium tuberculosis* complex. Mol Microbiol 1996; 19:1025–1034.
193. Sherman DR, Sabo PJ, Hickey MJ, et al. Disparate responses to oxidative stress in saprophytic and pathogenic mycobacteria. Proc Natl Acad Sci USA 1995; 92:6625–6629.
194. Deretic V, Pagan RE, Zhang Y, Dhandayuthapani S, Via LE. The extreme sensitivity of *Mycobacterium tuberculosis* to the front-line antituberculosis drug isoniazid. Nat Biotechnol 1996; 14:1557–1561.
195. Wilson T, de LG, Marcinkeviciene JA, Blanchard JS, Collins DM. Antisense RNA to *ahpC*, an oxidative stress defence gene involved in isoniazid resistance, indicates that AhpC of *Mycobacterium bovis* has virulence properties. Microbiology 1998; 144:2687–2695.
196. Siddiqi DE, Libonati JP, Middlebrook G. Evaluation of a rapid radiometric method for drug susceptibility testing of *Mycobacterium tuberculosis*. J Clin Microbiol 1981; 13:908–912.
197. Yajko DM, Madej JJ, Lancaster MV, et al. Colorimetric method for determining MICs of antimicrobial agents for *Mycobacterium tuberculosis*. J Clin Microbiol 1995; 33:2324–2327.
198. Jacobs WJ, Barletta RG, Udani R, et al. Rapid assessment of drug susceptibilities of *Mycobacterium tuberculosis* by means of luciferase reporter phages. Science 1993; 260:819–822.
199. Sreevatsan S, Pan X, Stockbauer KE, et al. Restricted structural gene polymorphism in the *Mycobacterium tuberculosis* complex indicates evolutionarily recent global dissemination. Proc Natl Acad Sci USA 1997; 94:9869–9874.
200. Mackaness GB. The intracellular activation of pyrazinamide and nicotinamide. Am Rev Tuberc Pulmon Dis 1956; 74:718–728.
201. Sturgill KS, Schlesinger PH, Chakraborty P, et al. Lack of acidification in Mycobacterium phagosomes produced by exclusion of the vesicular proton-ATPase. Science 1994; 263:678–681.

202. Scorpio A, Zhang Y. Mutations in *pncA*, a gene encoding pyrazinamidase/nicotinamidase, cause resistance to the antituberculous drug pyrazinamide in tubercle bacillus. Nat Med 1996; 2:662–667.
203. Scorpio A, Lindholm LP, Heifets L, et al. Characterization of *pncA* mutations in pyrazinamide-resistant *Mycobacterium tuberculosis*. Antimicrob Agents Chemother 1997; 41:540–543.
204. Scorpio A, Collins D, Whipple D, Cave D, Bates J, Zhang Y. Rapid differentiation of bovine and human tubercle bacilli based on a characteristic mutation in the bovine pyrazinamidase gene. J Clin Microbiol 1997; 35:106–110.
205. Sreevatsan S, Pan X, Zhang Y, Kreiswirth BN, Musser JM. Mutations associated with pyrazinamide resistance in *pncA* of *Mycobacterium tuberculosis* complex organisms. Antimicrob Agents Chemother 1997; 41:636–640.
206. Sun Z, Scorpio A, Zhang Y. The *pncA* gene from naturally pyrazinamide-resistant *Mycobacterium avium* encodes pyrazinamidase and confers pyrazinamide susceptibility to resistant *M. tuberculosis* complex organisms. Microbiology 1997; 143:3367–3373.
207. Sun Z, Zhang Y. Reduced pyrazinamidase activity and the natural resistance of *Mycobacterium kansasii* to the antituberculosis drug pyrazinamide. Antimicrob Agents Chemother 1999; 43:537–542.
208. Zhang Y, Scorpio A, Nikaido H, Sun Z. Role of acid pH and deficient efflux of pyrazinoic acid in unique susceptibility of *Mycobacterium tuberculosis* to pyrazinamide. J Bacteriol 1999; 181:2044–2049.
209. Hirano K, Takahashi M, Kazumi Y, Rukasawa Y, Abe C. Mutation in *pncA* is a major mechanism of pyrazinamide resistance in *Mycobacterium tuberculosis*. Tuberc Lung Dis 1998; 78:117–122.
210. Marttila HJ, Marjamäki M, Vyshnevskaya E, et al. *pncA* mutations in pyrazinamide-resistant *Mycobacterium tuberculosis* isolates from Northwestern Russia. Antimicrob Agents Chemother 1999; 43:1764–1766.
211. Lemaitre N, Sougakoff W, Truffot-Pernot C, Jarlier V. Characterization of new mutations in pyrazinamide-resistant strains of *Mycobacterium tuberculosis* and identification of conserved regions important for the catalytic activity of the pyrazinamidase PncA. Antimicrob Agents Chemother 1999; 43:1761–1763.
212. Boshoff HI, Mizrahi V. Purification, gene cloning, targeted knockout, overexpression, and biochemical characterization of the major pyrazinamidase from *Mycobacterium smegmatis*. J Bacteriol 1998; 180:5809–5814.
213. Raynaud C, Lanéelle M-A, Senaratne RH, Draper P, Lanéelle G, Daffé M. Mechanisms of pyrazinamide resistance in mycobacteria: importance of lack of uptake in addition to lack of pyrazinamidase activity. Microbiology 1999; 145:1359–1367.
214. Konno K, Feldman FM, McDermott W. Pyrazinamide susceptibility and amidase activity of tubercle bacilli. Am Rev Respir Dis 1967; 95:461–469.
215. McLaren J, Ngo DTC, Olivera M. Pyridine nucleotide metabolism in *Escherichia coli*. J Biol Chem 1973; 248:5144–5149.
216. Takayama K, Kilburn JO. Inhibition of synthesis of arabinogalactan by eth-

ambutol in *Mycobacterium smegmatis*. Antimicrob Agents Chemother 1989; 33:1493–1499.

217. Khoo KH, Douglas E, Azadi P, et al. Truncated structural variants of lipoarabinomannan in ethambutol drug-resistant strains of *Mycobacterium smegmatis*. Inhibition of arabinan biosynthesis by ethambutol. J Biol Chem 1996; 271:28682– 28696.
218. Telenti A, Philipp WJ, Sreevatsan S, et al. The emg operon, a gene cluster of *Mycobacterium tuberculosis* involved in resistance to ethambutol. Nat Med 1997; 3:567–570.
219. Belanger AE, Besra GS, Ford ME, et al. The *ambAB* genes of *Mycobacterium avium* encode an arabinosyl transferase involved in cell wall arabinan biosynthesis that is the target for the antimycobacterial drug ethambutol. Proc Natl Acad Sci USA 1996; 93:11919–11924.
220. Sreevatsan S, Stockbauer KE, Pan X, et al. Ethambutol resistance in *Mycobacterium tuberculosis*: critical role of *embB* mutations. Antimicrob Agents Chemother 1997; 41:1677–1681.
221. Lety MA, Nair S, Berche P, Escuyer V. A single point mutation in the *embB* gene is responsible for resistance to ethambutol in *Mycobacterium smegmatis*. Antimicrob Agents Chemother 1997; 41:2629–2633.
222. Alcaide F, Pfyffer GE, Telenti A. Role of *embB* in natural and acquired resistance to ethambutol in mycobacteria. Antimicrob Agents Chemother 1997; 41:2270–2273.
223. Tasukamura M. In vitro antituberculosis activity of a new antibacterial substance ofloxacin (DL8280). Am Rev Respir Dis 1985; 131:348–351.
224. Tsukamura M, Nakamura E, Yoshii S, Amano H. Therapeutic effect of a new antibacterial substance ofloxacin (DL8280) on pulmonary tuberculosis. Am Rev Respir Dis 1985; 131:352–356.
225. Woodcock JM, Andrews JM, Boswell FJ, Brenwald NP, Wise R. In vitro activity of BAY 12-8039, a new fluoroquinolone. Antimicrob Agents Chemother 1997; 41:101–106.
226. Takahata M, Mitsuyama J, Yamashiro Y, et al. In vitro and in vivo antimicrobial activities of T-3811ME, a novel des- F(6)-quinolone. Antimicrob Agents Chemother 1999; 43:1077–1084.
227. Wang JC. DNA topoisomerases. Ann Rev Biochem 1985; 54:665–697.
228. Yoshida H, Bogaki M, Nakamura M, Nakamura S. Quinolone resistance-determining region in the DNA gyrase *gyrA* gene of *Escherichia coli*. Antimicrob Agents Chemother 1990; 34:1271–1272.
229. Cambau E, Sougakoff W, Besson M, Truffot PC, Grosset J, Jarlier V. Selection of a *gyrA* mutant of *Mycobacterium tuberculosis* resistant to fluoroquinolones during treatment with ofloxacin. J Infect Dis 1994; 170:479–483.
230. Xu C, Kreiswirth BN, Sreevatsan S, Musser JM, Drlica K. Fluoroquinolone resistance associated with specific gyrase mutations in clinical isolates of multidrug-resistant *Mycobacterium tuberculosis*. J Infect Dis 1996; 174:1127–1130.
231. Takiff HE, Salazar L, Guerrero C, et al. Cloning and nucleotide sequence of

Mycobacterium tuberculosis gyrA and *gyrB* genes and detection of quinolone resistance mutations. Antimicrob Agents Chemother 1994; 38:773–780.
232. Sullivan EA, Kreiswirth BN, Palumbo L, et al. Emergence of fluoroquinolone-resistant tuberculosis in New York City. Lancet 1995; 345:1148–1150.
233. Williams KJ, Chan R, Piddock LJ, *gyrA* of ofloxacin-resistant clinical isolates of *Mycobacterium tuberculosis* from Hong Kong. J Antimicrob Chemother 1996; 37:1032–1034.
234. Kocagoz T, Hackbarth CJ, Unsal I, Rosenberg EY, Nikaido H, Chambers HF. Gyrase mutations in laboratory-selected, fluoroquinolone-resistant mutants of *Mycobacterium tuberculosis* H37Ra. Antimicrob Agents Chemother 1996; 40:1768–1774.
235. Revel VV, Truong QC, Moreau N, Jarlier V, Sougakoff W. Sequence analysis, purification, and study of inhibition by 4-quinolones of the DNA gyrase from *Mycobacterium smegmatis*. Antimicrob Agents Chemother 1996; 40: 2054–2061.
236. Alangaden GJ, Manavathu EK, Vakulenko SB, Zvonok NM, Lerner SA. Characterization of fluoroquinolone-resistant mutant strains of *Mycobacterium tuberculosis* selected in the laboratory and isolated from patients. Antimicrob Agents Chemother 1995; 39:1700–1703.
237. Yoshida H, Bogaki M, Nakamura M, Yamanaka LM, Nakamura S. Quinolone resistance-determining region in the DNA gyrase *gyrB* gene of *Escherichia coli*. Antimicrob Agents Chemother 1991; 35:1647–1650.
238. Ferrero L, Cameron B, Manse B, et al. Cloning and primary structure of *Staphylococcus aureus* DNA topoisomerase IV: a primary target of fluoroquinolones. Mol Microbiol 1994; 13:641–653.
239. Ng EY, Trucksis M, Hooper DC. Quinolone resistance mediated by *norA*: physiologic characterization and relationship to *flqB*, a quinolone resistance locus on the *Staphylococcus aureus* chromosome. Antimicrob Agents Chemother 1994; 38:1345–1355.
240. Takiff HE, Cimino M, Musso MC, et al. Efflux pump of the proton aniporter family confers low-level fluoroquinolone resistance in *Mycobacterium smegmatis*. Proc Natl Acad Sci USA 1996; 93:362–366.
241. Quinting B, Reyrat JM, Monnaie D, et al. Contribution of beta-lactamase production to the resistance of mycobacteria to beta-lactam antibiotics. FEBS Lett 1997; 406:275–278.
242. Herbert D, Paramasivan CN, Venkatesan P, Kubendiran G, Prabhakar R, Mitchison DA. Bactericidal action of ofloxacin, sulbactam-ampicillin, rifampin, and isoniazid on logarithmic- and stationary-phase cultures of *Mycobacterium tuberculosis*. Antimicrob Agents Chemother 1996; 40:2296–2299.
243. Pao SS, Paulsen TT, Saier MJ. Major facilitator superfamily. Microbiol Mol Biol Rev 1998; 62:1–34.
244. Paulsen IT, Brown MH, Skurray RA. Proton-dependent multidrug efflux systems. Microbiol Rev 1996; 60:575–608.
245. Paulsen IT, Sliwinski MK, Saier MJ. Microbial genome analyses: global comparisons of transport capabilities based on phylocenies, bioenergetics and substrate specificities. J Mol Biol 1998; 277:573–592.

246. Ainsa JA, Blokpoel MCJ, Otal I, Young DB, de Smet KAL, Martin C. Molecular cloning and characterization of Tap, a putative multidrug efflux pump present in *Mycobacterium fortuitum* and *Mycobacterium tuberculosis*. J Bacteriol 1998; 180:5836–5843.
247. de Rossi E, Branzoni M, Cantoni R, Milano A, Riccardi G, Ciferri O. *mmr*, a *Mycobacterium tuberculosis* gene conferring resistance to small cationic dyes and inhibitors. J Bacteriol 1998; 180:6068–6071.
248. Tseng T-T, Gratwick KS, Kollmann J, et al. The RND permease superfamily: an ancient, ubiquitous and diverse family that includes human disease and development proteins. J Mol Microbiol Biotechnol 1999; 1:107–125.
249. Li XZ, Nikaido H, Poole K. Role of *mexA-mexB-oprM* an antibiotic efflux in *Pseudomonas aeruginosa*. Antimicrob Agents Chemother 1995; 39:1948–1953.
250. Ma D, Cook DN, Alberti M, Pon NG, Nikaido H, Hearst JE. Genes *arcA* and *acrB* encode a stress-induced efflux system of *Escherichia coli*. Mol Microbiol 1995; 16:45–55.
251. Tekaia F, Gordon SV, Garnier T, Brosch R, Barrell BG, Cole ST. Analysis of the proteome of *Mycobacterium tuberculosis in silico*. Tuberc Lung Dis 1999; 79: 329–342.
252. Cole ST. Learning from the genome sequence of *Mycobacterium tuberculosis* H37Rv. FEBS Lett 1999; 452:7–10.
253. Pelicic V, Jackson M, Reyrat JM, Jacobs WJ, Gicquel B, Guilhot C. Efficient allelic exchange and transposon mutagenesis in *Mycobacterium tuberculosis*. Proc Natl Acad Sci USA 1997; 94:10955–10960.
254. Hensel M, Shea JE, Gleeson C, Jones MD, Dlton E, Holden D. Simultaneous identification of bacterial virulence genes by negative selection. Science 1995; 269:400–403.
255. Lomovskaya O, Lee A, Hoshino K, et al. Use of a genetic approach to evaluate the consequences of inhibition of efflux pumps in *Pseudomonas aeruginosa*. Antimicrob Agents Chemother 1999; 43:1340–1346.
256. Barrett JF, Goldschmidt RM, Lawrence LE, et al. Antibacterial agents that inhibit two-component signal transduction systems. Proc Natl Acad Sci USA 1998; 95:5317–5322.
257. Hlasta DJ, Demers JP, Foleno BD, et al. Novel inhibitors of bacterial two-component systems with gram positive antibacterial activity: pharmacophore identification based on the screening hit closantel. Bioorgan Med Chem Lett 1998; 8:1923–1928.
258. Georgopapadakou NH. Antifungals: mechanism of action and resistance, established and novel drugs. Curr Opin Microbiol 1998; 1:547–557.
259. Cole ST. Comparative mycobacterial genomics. Curr Opin Microbiol 1998; 1: 567–571.

16

Antibiotic Resistance in Enterobacteria

Nafsika H. Georgopapadakou
Newbiotics, Inc., San Diego, California

Enterobacteria cause a variety of nosocomial and community-acquired (foodborne) infections and include some of the most deadly pathogens. As a result, their resistance to antibiotics has profound clinical implications. The major antibiotic classes currently in use for enterobacterial infections are the β-lactams, quinolones, aminoglycosides, tetracyclines, and sulfonamides. Resistance to β-lactams is relatively common and involves β-lactamases: inducible, chromosomal (class C) as well as constitutive, plasmid-mediated, extended spectrum (classes A and D). Resistance to quinolones, relatively uncommon in enterobacteria, is primarily associated with the target DNA gyrase and affects quinolones in use as well as in clinical development. Reduced accumulation in the cell, due to active efflux through the cytoplasmic membrane and decreased influx through the outer membrane, may facilitate the emergence of resistance. Resistance to aminoglycosides is predominantly due to enzymatic inactivation in the periplasmic space, the exact nature of the modification depending on the particular aminoglycoside. The major mechanisms for tetracycline resistance involves an active efflux system; ribosomal protection is not a clinically important mechanism in enterobacteria. Sulfonamide resistance is due to an additional, plasmid-mediated, sulfonamide-resistant, dihydropteroate synthase target. Overall, the biggest clinical concern is the gradual erosion of the effectiveness of β-lactams and quinolones, two bactericidal and generally safe antibacterial classes.

1 INTRODUCTION

The family Enterobacteriaceae is the widest and most heterogeneous group of medically important gram-negative bacteria (1). It includes many species found in the gastrointestinal tract of humans and animals, as well as in soil, water, and plants. About a third of the 30 genera known are human pathogens, causing a variety of diseases ranging from mild intestinal infections to urinary tract infections, nosocomial respiratory tract infections, wound infections, and septic shock (Table 1). Individual species, famously *Yersinia pestis* (responsible for the Black Death of the Middle Ages) (2–4), more modestly *Escherichia coli* O157:H7 (responsible for 70,000 cases of infection and 60 deaths in the United States yearly (http://www.cdc.gov/ncidod/dbmd/diseaseinfo/escherichiacoli_g.htm) (5) have been associated with specific epidemics that continue to the present.

The pathogenicity of enterobacteria is associated with clusters of genes (plasmidborne or chromosomeborne) that play critical roles in bacterial colonization and virulence (6–9). Pathogenic strains often live in a sea of avirulent strains, the latter representing a reservoir of potential hosts for the mobile genetic elements that encode virulence factors.

Antibiotic resistance is a direct consequence of antibiotic use both in humans and in animals (10–12). For example, quinolone resistance in *Salmonella*, a foodborne pathogen causing perhaps a million cases of (mostly self-limited) infection and a thousand deaths in the United States yearly, almost certainly originated from animals (13–17). The overreliance on antibiotics, to the exclusion of infection-control measures and improved hygiene, has eroded the effectiveness of older, inexpensive agents and threatens the effectiveness of recently introduced ones (18).

Antibiotic resistance is commonly detected by susceptibility testing, which provides the resistance phenotype of an organism and has practical implications for patient treatment. For epidemiological/surveillance purposes, strain typing is often performed by serological methods or, more precisely, by pulsed-field gel electrophoresis (PFGE) of digested DNA (19–21). Resistance is further characterized by biochemical and molecular biology techniques. The former include function assays; the latter DNA restriction analysis, DNA probes, and nucleic acid amplification by the polymerase chain reaction (PCR) (22). Resistance mechanisms may operate synergistically—for example, transport-associated resistance and antibiotic inactivation—and the contribution of each to the overall resistance may be difficult to assess (23).

Table 1 Pathogenic Enterobacteria and the Diseases They Produce

Genus	Species	Infection/disease	Comments
Citrobacter	*freundii*	UT, RT, wound, blood	Nosocomial
Enterobacter	*aerogenes*	UT, RT, wound, blood	Nosocomial
	cloacae	UT, RT, wound, blood	Nosocomial
Escherichia	*coli*	GIT,[a] UT,[b], RT, wound, blood	Foodborne, nosocomial
Klebsiella	*oxytoca*	UT, RT, wound, blood	Nosocomial
	pneumonia	UT, RT, wound, blood	Nosocomial
Morganella	*morganii*	UT, RT, wound, blood	Nosocomial
Proteus	*mirabilis*	UT, RT, wound, blood	Nosocomial
	vulgaris	UT, RT, wound, blood	Nosocomial
Providencia	*rettgeri*	UT, wound, pneumonia, blood	Nosocomial
	stuartii	UT, wound, pneumonia, blood	
Salmonella	*enteritidis*	GIT	Foodborne
	typhi	GIT, typhoid fever	Foodborne
	typhimurium	GIT	Foodborne
Serratia	*marcescens*	UT, wound, pneumonia, blood	Nosocomial
Shigella[c]	*dysenteriae*	GIT (shigellosis)	Foodborne
	flexneri	GIT (shigellosis)	Foodborne
	sonnei	GIT (shigellosis)	Foodborne
Yersinia	*enterocolitica*	GIT	Foodborne/ waterborne
	pseudotuberculosis	GIT	Zoonotic, foodborne
	pestis	Lymph nodes (bubonic plague)	Zoonotic (rodents/fleas)

Abbreviations: GIT, gastrointestinal tract; UT, urinary tract; RT, respiratory tract.
[a]Infections caused by particularly virulent *E. coli* (EC) strains: enterotoxigenic (ETEC), diarrhea; enteropathogenic (EPEC), infantile diarrhea; enteroinvasive (EIEC), dysentery; enterohemorrhagic (EHEC), such as serotype O157:H7, hemorrhagic colitis.
[b]Most common cause of UT infection.
[c]In the United States, *Shigella sonnei* (group D *Shigella*) accounts for over two thirds of the shigellosis, while *S. flexneri* (group B *Shigella*) accounts for most of the rest. *S. dysenteriae* type 1 is found in the developing world, where it causes deadly epidemics.

2 RESISTANCE MECHANISMS IN ENTEROBACTERIA

2.1 Genetic Mechanisms

Resistance in enterobacteria can result from gene mutations or transfer of resistance determinants (R-determinants) between strains or species. Clinically, gene transfer is the most common mechanism of transferring resistance (24). R-determinants are typically on plasmids, but they may also be part of mobile genetic elements (transposons, integrons, gene cassettes) which can move between plasmids or chromosomes in the same organism or to a new organism (25–27).

Plasmids (4–400 kb) are self-replicating, extrachromosomal elements that contain genes for resistance, virulence, and other functions and are dispensable under certain conditions. Some larger plasmids are conjugative (R-plasmids) and can transfer between organisms, spreading along resistance genes. For example, conjugative plasmids were responsible for the spread of sulfonamide resistance to *Shigella dysenteriae* in the 1950s. Resistance genes can thus disseminate independently of the host organism (horizontal transfer) in addition to disseminating along with the host (clonal spread).

Transposons (2–20 kb) are mobile genetic elements that contain insertion sequences (0.2–6.0 kb) and one or more resistance genes. They are not capable of autonomous self-replication, but can move (transpose) from one site on the chromosome to another site on the same or different chromosome or plasmid and replicate along with it. Transposition is made possible by short inverted repeats of DNA.

Integrons are mobile genetic elements of specific structure that consist of two conserved segments flanking a central region in which resistance gene cassettes are inserted (25). On the 5′-conserved segment is an *int* gene that encodes a site-specific recombinase capable of capturing DNA, including resistance genes. Although the probability of capture of a resistance gene is low, it can confer a selective advantage to its host. Adjacent to it are a suitably oriented promoter for expression of the cassette genes and the receptor site for the gene cassettes (*attI* site). On the 3′-conserved segment, which is of variable length, is typically the *sul1* gene that encodes a sulfonamide-resistant dihydropteroate synthase (28). Additional resistance genes can also be present, their distance from the promoter determining their level of expression.

Alarmingly, as resistance genes move to other plasmids or to the chromosome, they sometimes link with other resistance genes in resistance clusters, whose transfer can then result in simultaneous acquisition of resistance to several unrelated drugs (multidrug resistance) (24,25).

2.2 Biochemical Mechanisms

Biochemical mechanisms of antibiotic resistance include altered transport (influx or efflux) and thereby reduced intracellular accumulation (29); enzymatic inactivation (hydrolysis or derivatization); altered or additional resistant target; bypassed target; and compensatory changes downstream of target. Table 2 summarizes resistance mechanisms for specific antibacterial classes used against enterobacteria.

Table 2 Biochemical Resistance Mechanisms in Enterobacteria

Antibiotic class	New/altered enzyme protein/gene	Gene location	Comments	Ref.
Antibiotic inactivation				
Hydrolysis				
β-Lactams	β-Lactamase	Ch, P	Most common resistance mechanism	37,38
Modification				
Aminoglycosides	*N*-Acetyltransferase, *O*-adenylyltransferase, *O*-phosphotransferase	Ch, P		79,80, 85,90
Chloramphenicol	*O*-Acetyltransferase			124
Altered/additional target				
Decreased binding				
Quinolones	DNA gyrase, topo IV	Ch, P	Most common resistance mechanism	69,70
Aminoglycosides	RNA, ribosomal protein S12			79,80
Tetracyclines	Ribosomal protection (*tetM, O*)			111
Sulfonamides	Dihydropteroate synthetase	P		28
Trimethoprim	Dihydrofolate reductase (DHFR)	P		121
Overproduced target				
Trimethoprim	DHFR (type IV)	P		121
Decreased intracellular accumulation				
Decreased uptake				
β-Lactams	Porins	Ch		58–62
Quinolones	Porins	Ch		69,70
Aminoglycosides	Altered active transport			96
Increased efflux				
β-Lactams				
Quinolones	AcrAB-TolC, EmrAB	Ch		69,70
Tetracyclines	tet A,B,C,D,E,K,L (I)	Ch, P	Most common resistance mechanism	106

Abbreviations: Ch, chromosomal; I, inducible; P, plasmid-mediated.

3 RESISTANCE TO MAJOR ANTIBIOTIC CLASSES IN ENTEROBACTERIA

3.1 β-Lactams

β-Lactam antibiotics constitute the most enduring and widely used class of antibacterials, encompassing a large number of mostly semisynthetic compounds. The clinically useful β-lactams are divided on the basis of structure into penams, penems, and cephems (30) (Fig. 1). Their targets are peptidoglycan transpeptidases, cell wall–synthesizing enzymes located on the outer face of the cytoplasmic membrane (31). These enzymes are ubiquitous in bacteria and are commonly detected by their ability to bind covalently and specifically penicillin and other β-lactam antibiotics (hence the name, penicillin-binding proteins, or PBPs). Not all PBPs are peptidoglycan transpeptidases or essential; in enterobacteria, only three of the eight PBPs are.

Clinically, the most important mechanism of resistance in enterobacteria is hydrolysis by β-lactamases, common bacterial enzymes related to the cell wall targets (32). β-Lactamases are divided into four groups based on amino acid sequence homology (Ambler classification) (33,34). In enterobacteria, all four classes are represented (Table 3), but classes A and C are of the greatest clinical significance. Both penams and cephems are affected, although rarely are penems affected. Surprisingly, *Y. pestis* appears to have decreased susceptibility to penems but not to penams or cephems (35,36).

In class A, the most important enzymes are TEM-1 and SHV-1. TEM-1, which originated in *E. coli*, is now very common in *Klebsiella* and other enterobacteria, whereas SHV-1 is commonly found in *K. pneumoniae* and can be plasmid-mediated or chromosomal (32,37,38). Of great clinical concern is the emergence of extended-spectrum variants of TEM and SHV enzymes in the 1980s that continues to the present. To date, over 70 TEM variants and over 20 SHV enzymes have been identified (32,39–41). These extended-spectrum β-lactamases (ESBLs) are particularly problematic, because they can hydrolyze oxyimino β-lactams (cefotaxime, ceftriaxone, ceftazidime) and can easily spread to other species. They are generally sensitive to inhibition by clavulanic acid (42), although resistant variants have been reported (43). Clinical isolates that produce ESBLs are frequently associated with nosocomial outbreaks (44,46).

Many clinical laboratories lack the necessary technology, and thus ESBL detection in the clinical microbiology laboratory is often problematic. For example, in a survey by Tenover et al. (47), the percentage of laboratories in Connecticut that failed to detect resistance in the ESBL or ampC-producing isolates ranged from 24 to 32%. A 1998 survey of 369

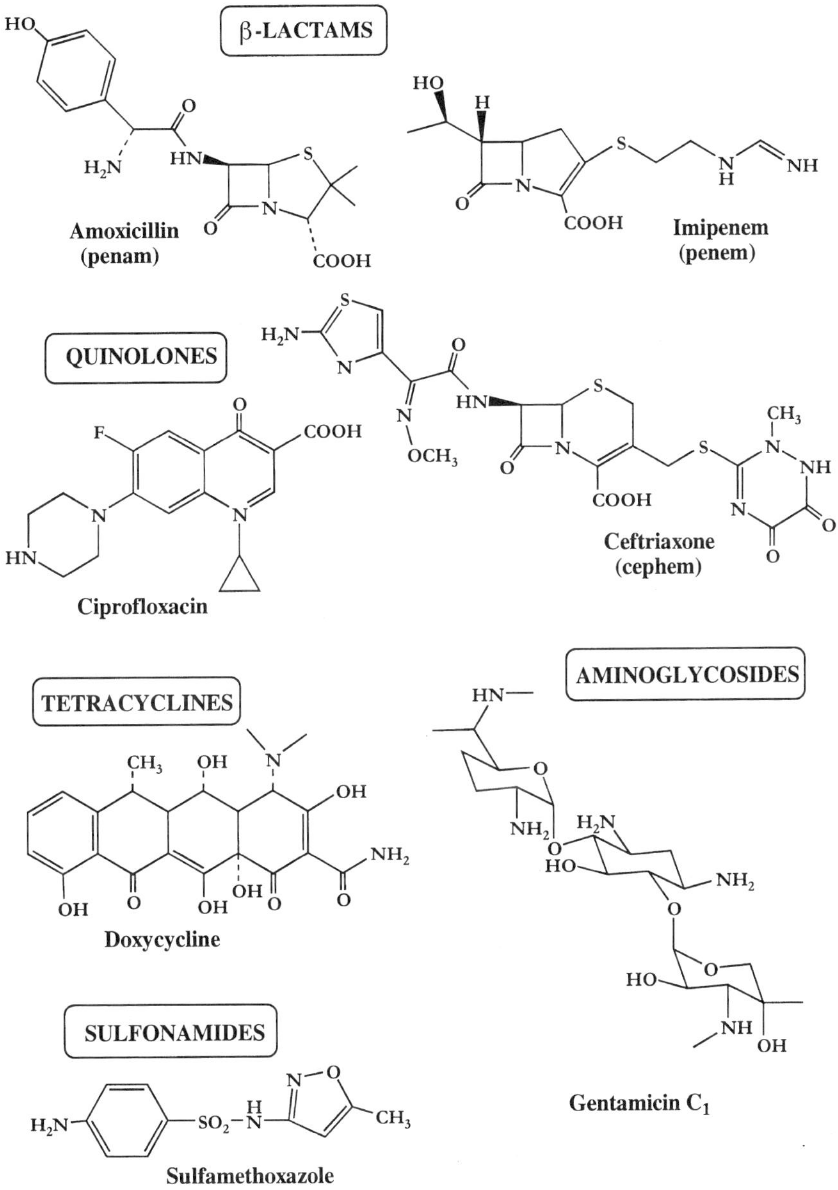

Figure 1 Structures of antibiotic classes used against enterobacteria.

Table 3 Common β-Lactamases in Enterobacteria

Enzyme	Original host	Substrate profile	Inhibitor profile	Comments
Class A: serine enzymes (~30 kDa), mostly plasmid-mediated, constitutively expressed				
TEM-1	*E. coli*	pen/ceph	clox/clav/sulb/taz	Most common type
TEM-3	*E. coli*	pen/ceph	clox/clav/sulb/taz	Differs from TEM-1 by one amino acid
TEM-3 to TEM-70	*K. pneumoniae*	pen/ceph	clox/clav/sulb/taz	ESBL variants of TEM-1, in nosocomial outbreaks (see www.lahey.org/lcinternet/studies/webt/htm)
SHI-1	*K. pneumoniae*	pen/ceph	clav/sulb/taz	a.k.a. PIT-2; ESBL, often chromosomal; 24 variants to-date (www.lahey.org/lcinternet/studies/webt.htm)
K-1	*K. oxytoca*	ceph		Chromosomal, extended spectrum
Class B: metalloenzymes (~22 kDa), mostly chromosomal				
IMP-1	*Stentrophomonas maltophila*	pen/ceph/cpen		Uncommon; reported in *S. marcescens*, *S. flexneri* (integronborne)
Class C: serine enzymes (~40 kDa), mostly chromosomal, inducible				
AmpC	*E. coli*	ceph	taz	
P99	*E. cloacae*	ceph	taz	
MIR-1	*K. pneumoniae*	ceph		Plasmid-mediated
CMY-1 to CMY-5	*K. pneumoniae*	ceph		Plasmid-mediated
Class D: serine enzymes (~12 kDa), plasmid-mediated				
OXA-1	*E. coli*	pen/ceph		Related, less common enzymes: OXA-3 to OXA-7

Abbreviations: ceph, cephems; clav, clavulanic acid; clox, cloxacillin; cpen, carbapenems; ESBL, extended-spectrum β-lactamases; pen, penams; taz, tazobactam.

microbiology laboratories participating in the Centers for Disease Control and Prevention (CDC) Emerging Infections Network found that only 32% tested for ESBL producers, and of that subset only 17% used adequate methods to confirm ESBL presence (48).

Ambler class C β-lactamases are produced by most enterobacteria but are particularly important in clinical isolates of *Enterobacter cloacae* and *E. aerogenes*, *Citrobacter freundii*, and *Serratia marcescens* (49–52). They hydrolyze both penicillins and cephalosporins, including cephamycins such as cefoxitin, and are resistant to the classic β-lactamase inhibitor clavulanic acid. They are normally inducible, with regulation of expression being linked to the cell wall synthesis and recycling (53,54).

A major factor contributing to β-lactam resistance is decreased outer membrane permeability (55,56). Because they live in the gut, enterobacteria have developed a particularly "finicky" outer membrane. This is a protein-rich asymmetrical lipid bilayer that contains liposaccharide in the outer leaflet and envelops the peptidoglycan. It functions as a molecular sieve, having water-filled channels (pores) formed by 35- to 40-kD protein trimers (porins). It is through these channels that nonspecific transport of small hydrophilic molecules, such as β-lactams, occurs. There are at least two porin species in *E. coli*, OmpC and OmpF, which form channels of 11 and 12A diameter, respectively, with an exclusion limit of 600–800 D. Other enterobacteria also have two to three porins, homologous to those of *E. coli* (57–59). Porin-deficient mutants of enterobacteria have reduced β-lactam susceptibility relative to isogenic wild-type strains (60–62).

A key aspect to the susceptibility of enterobacteria to β-lactams is the interplay of outer membrane permeability, affinity/turnover for β-lactamases in the periplasmic space, and affinity for target PBPs. Although β-lactamases constitute the major form of β-lactam resistance in enterobacteria, it is the combined presence of β-lactamases and reduced outer membrane permeability that affects resistance. This cooperative action effectively reduces the concentration of β-lactams in the periplasm. Decreased target affinity has not been reported in clinical isolates of enterobacteria, perhaps because of the fitness price it entails (63).

3.2 Quinolones

Quinolones are broad-spectrum, bactericidal antibiotics whose potent activity, including activity even against intracellular pathogens, and ease of administration (oral, parenteral), have firmly established them both in the hospital and the community. Quinolones enter bacterial cells through the porins in the outer membrane and by diffusion through the cytoplasmic membrane (64,65). They then complex immediately, selectively, and reversibly with DNA gyrase and the related topoisomerase IV, bacterial enzymes essential for transcription, replication, and chromosome decatenation. They trap a covalent enzyme-DNA complex (cleavable complex) in which the enzyme has broken the phosphodiester backbone of the DNA, and thereby inhibit the subsequent relegation of DNA (66). The result is inhibition of supercoiling (DNA gyrase), chromosome decatenation (topoisomerase IV), and, most importantly, the induction of DNA lesions which trigger the SOS response (66a) and ultimately lead to cell death.

With a single exception (67), quinolone resistance is exclusively chromosomal, spreading along with the resistant organism. The most common mechanism of clinical resistance in enterobacteria is associated with mutations in *gyrA* which encodes subunit A of DNA gyrase (68–70). Resistance mutations tend to cluster between residues 67 and 106 (quinolone resistance–determining region, QRDR) (71). Mutations in *parC*, which encodes the homologous subunit A of topoisomerase IV, are also associated with resistance (72). Reduced accumulation in the cell, due to active efflux through the cytoplasmic membrane combined with decreased influx through the outer membrane, appears to cause only low levels of resistance but can facilitate the emergence of fluoroquinolone-resistant strains (73–76).

Clinical resistance is relatively uncommon in enterobacteria (77) despite the widespread use of quinolones. The only exception is the foodborne pathogen *Salmonella*, where resistance in some specific phage types has been found in Europe, most likely due to the extensive use of quinolones in food animals (13). Nevertheless, because resistance affects not only quinolones in use but also in clinical development, it is a cause for continuous vigilance.

3.3 Aminoglycosides

Aminoglycosides are bactericidal, broad-spectrum antibiotics discovered in the 1940s. They are still widely used (gentamicin, amikacin), usually in combination with β-lactam agents, against problem pathogens despite their ototoxicity and nephrotoxicity (78). Structurally, aminoglycosides are polycationic amino sugars, the amino groups being protonated in biological media. A number of subclasses have been identified, and semisynthetic derivatives less prone to enzymatic inactivation have been developed (79,80).

Aminoglycosides enter through the outer membrane via a porin-independent, "self-promoted" pathway, and pass through the cytoplasmic membrane via an energy-dependent pathway (83,84). The subsequent binding to the A site of the 16S ribosomal RNA results in inhibition of protein synthesis (81,82).

Clinically, the most significant mechanism of aminoglycoside resistance is enzymatic modification (85), the exact profile depending on the aminoglycoside being used (86–88). The modifying enzymes, *N*-acetyltransferases, *O*-phosphoryltransferases, *O*-adenyltransferases, have broad substrate specificity and can catalyze more than one reaction (89–93). Their origin is diverse (94). Impaired uptake may also contribute to resistance (95–97).

3.4 Tetracyclines

Tetracyclines are broad-spectrum, bacteriostatic agents that also act by inhibiting protein synthesis (98). They bind reversibly to a single, high-affinity site on the 30S ribosomal subunit and disrupt the codon-anticodon interaction between aminoacyl-tRNA and mRNA, thereby inhibiting the binding of aminoacyl-tRNA to the acceptor site on the ribosome. Their selective antibacterial toxicity may be due, at least in part, to selective, concentrative uptake by bacteria (99,100).

The major mechanism for tetracycline resistance involves an inducible active efflux system whereby the intracellular concentration of these compounds is reduced (101–103). Several genes encoding for components of this system (*tetA-E* in gram-negative bacteria) (104–109), located mostly on plasmids, have been identified. Different tetracyclines are not equally recognized by transport proteins; for example, TetA does not recognize minocycline and doxycycline. Ribosomal protection by a soluble, usually plasmid-mediated, 72-kD protein homologous to the elongation factor G, involved in protein synthesis (110,112), is not a significant resistance mechanism in enterobacteria. A notable development in the field is the glycylcyclines, minocycline derivatives, one of which (GAR 936) is currently in clinical development (112–114). They are active against most tetracycline-resistant strains. Another recent development is the potentiation of the antibacterial activity of tetracyclines by inhibitors of Tet efflux proteins (115).

3.5 Antifolates: Sulfonamides and Trimethoprim

Sulfonamides, the oldest, totally synthetic antibacterial agents, are competitive inhibitors of dihydropteroate synthetase by virtue of their active (sulfone) form being a structural analogue of the *para*-aminobenzoic acid substrate. Clinically, the most common and important mechanism of resistance to these bacteriostatic agents is altered, usually plasmid-mediated, target enzyme (28,116). Two distinct types of altered dihydropteroate synthetase have been characterized in gram-negative bacteria, I and II, encoded by *sulI* and *sulII*, respectively (117–119). They have reduced binding to inhibitors but remarkably maintain normal binding to the *para*-aminobenzoic acid substrate. The *sulI* gene is often located in transposons related to Tn*21* or on large R-plasmids with a resistance region similar to Tn*21* (28, 119). The *sulII* gene is carried mainly on small nonconjugative plasmids.

Trimethoprim, also totally synthetic and commonly used in combination with sulfonamides, is a bactericidal agent. It is a selective, potent, competitive inhibitor of the bacterial dihydrofolate reductase (DHFR). The resulting tetrahydrofolate depletion affects methyl transfer reactions, particularly the one involved in thymine biosynthesis, thereby causing thymineless death. The most common mechanism of trimethoprim resistance is altered, usually plasmid-mediated, target enzyme (120–122). Mutant forms of the normal, chromosomal DHFR are far less common in clinical isolates. Seven major types of plasmid-encoded, trimethoprim-resistant DHFRs (I–VII) have been found in gram-negative bacteria. They share variable homology with each other and with the normal, chromosomal enzyme, suggesting both divergent and convergent evolution.

3.6 Chloramphenicol

Chloramphenicol, still an important bacteriostatic agent, owes its selective antibacterial activity to inhibition of the peptidyltransferase reaction of protein synthesis via binding to the 50S ribosomal subunit. The major mechanism of clinical resistance to chloramphenicol is its inactivation by acetylation (123). Three genetically distinct groups of chloramphenicol acetyltransferases have been found so far, some inducible and others constitutive, but all sharing sequence homology (124). As previously stated, multi–drug-resistant *S. typhimurium* (DT104), which represents approximately 10% of the *Salmonella* isolates in the United States, is resistant to chloramphenicol (125).

4 FUTURE DIRECTIONS

Enterobacteria cause a variety of nosocomial and community-acquired (foodborne) infections, and thus their resistance to antibiotics has profound clinical implications. Of the five major antibiotic classes currently used for enterobacterial infections (β-lactams, quinolones, aminoglycosides, tetracyclines, sulfonamides), quinolones (ciprofloxacin) and β-lactams (amoxycilin/clavulanic acid combination and third-generation cephalosporins) are the least affected by resistance. The biggest threat for the future is the gradual erosion of the effectiveness of these two bactericidal and generally safe antibacterial classes (126,127).

The fact that enterobacteria infections are treatable with existing antibiotics has kept them out of the limelight. *Enterobacter* and *Klebsiella,* although often resistant to several antibiotics, are simply not in the *Pseudomonas/ Enterococcus* resistance league. Neither are as spectacularly invasive as some *Streptococcus* strains. Yet, they are very common pathogens, and their emerging resistance in institutional settings should be a cause for concern (128–132).

Since drug development is a long, tortuous process, we need to be more proactive and start targeting enterobacteria now with new agents that do not cross react with existing ones. The possibility of also covering *Pseudomonas* would be an added bonus. In this context, it would be valuable to draw on the recent advances in our understanding of efflux mechanisms, with the goal of perhaps avoiding them rather than targeting them. The hydra-like nature of transport proteins, whereby suppression of one unmasks another, argues for caution. Nevertheless, recent work has shown that it is possible to potentiate antibacterial activity by inhibiting drug efflux (133,134), just as earlier work had shown that it was possible to potentiate antibacterial action by promoting drug influx through the outer membrane (135,136).

REFERENCES

1. Holt JG, Krieg NR, Sneath PHA, Staley JT, William ST, eds. Bergey's Manual of Determinative Bacteriology. 9th ed. Baltimore: Williams & Wilkins, 1993.
2. Achtman M, Zurth K, Morelli G, Torrea G, Guiyoule A, Carniel E. *Yersinia pestis*, the cause of plague, is a recently emerged clone of *Yersinia pseudotuberculosis*. Proc Natl Acad Sci USA 1999; 96:14043–14048.
3. Skurnik M, Peippo A, Ervela E. Characterization of the O-antigen gene clusters of *Yersinia pseudotuberculosis* and the cryptic O-antigen gene cluster of *Yersinia pestis* shows that the plague bacillus is most closely related to and has evolved from *Y. pseudotuberculosis* serotype O:1b. Mol Microbiol 2000; 37: 316–330.
4. Dykhuizen DE. *Yersinia pestis*: an instant species? Trends Microbiol 2000; 8: 296–298.
5. Armstrong GL, Hollingsworth J, Morris JG Jr. Emerging foodborne pathogens: *Escherichia coli* O157:H7 as a model of entry of a new pathogen into the food supply of the developed world. Epidemiol Rev 1996; 18:29–51.
6. Hacker J, Blum-Oehler G, Muhldorfer I, Tschape H. Pathogenicity islands of virulent bacteria: structure, function and impact on microbial evolution. Mol Microbiol 1997; 23:1089–1097.
7. Wood MW, Jones MA, Watson PR, Hedges S, Wallis TS, Galyov EE. Identi-

fication of a pathogenicity island required for *Salmonella* enteropathogenicity. Mol Microbiol 1998; 29:883–891.

8. Hensel M, Nikolaus T, Egelseer C. Molecular and functional analysis indicates a mosaic structure of *Salmonella* pathogenicity island 2. Mol Microbiol 1999; 31:489–498.
9. Vokees SA, Reeves SA, Torres AG, Payne SM. The aerobactin iron transport system genes in *Shigella flexneri* are present within a pathogenicity island. Mol Microbiol 1999; 33:63–73.
10. Tenover FC, McGowan JE Jr. Reasons for the emergence of antibiotic resistance. Am J Med Sci 1996; 311:9–16.
11. Barbosa TM, Levy SB. The impact of antibiotic use on resistance development and persistence. Drug Resist Updates 2000; 3:303–311.
12. Stohr K, Wegener HC. Animal use of antimicrobials: impact on resistance. Drug Resist Updates 2000; 3:207–209.
13. Angulo FJ, Johnson KR, Tauxe RV, Cohen ML. Origins and consequences of antimicrobial-resistant nontyphoidal *Salmonella*: implications for the use of fluoroquinolones in food animals. Microb Drug Resist 2000; 6:77–83.
14. Glynn MK, Bopp C, Dewitt W, Dabney P, Mokhtar M, Angulo FJ. Emergence of multidrug-resistant *Salmonella enterica* serotype Typhimurium DT104 infections in the United States. N Engl J Med 1998; 338:1333–1338.
15. Akkina JE, Hogue AT, Angulo FJ, Johnson R, Petersen KE, Saini PK, Fedorka-Cray PJ, Schlosser WD. Epidemiologic aspects, control, and importance of multiple-drug resistant *Salmonella typhimurium* DT104 in the United States. J Am Vet Med Assoc 1999; 214:790–798.
16. Briggs CE, Fratamico PM. Molecular characterization of an antibiotic resistance gene cluster of *Salmonella typhimurium* DT104. Antimicrob Agents Chemother 1999; 43:846–849.
17. Angulo FJ, Griffin PM. Changes in antimicrobial resistance in *Salmonella enterica* serovar Typhimurium [letter]. Emerg Infect Dis 2000; 6:436–438.
18. Hughes JM, Tenover FC. Approaches to limiting emergence of antimicrobial resistance in bacteria in human populations. Clin Infect Dis 1997; 24(Suppl 1): S131–S135.
19. Podzorski RP, Persing DH. Molecular detection and identification of microorganisms. In: Murray PR, Baron JE, Pfaller MA, Tenover FC, Yolken RH, eds. Manual of Clinical Microbiology. 6th ed. Washington, DC: ASM Press, 1995:130–157.
20. Tenover FC, Arbeit RD, Goering RV, Mickelsen PA, Murray BE, Persing DH, Swaminathan B. Interpreting chromosomal DNA restriction patterns produced by pulsed-field gel electrophoresis: criteria for bacterial strain typing. J Clin Microbiol 1995; 33:2233–2239.
21. Tenover FC, Arbeit RD, Goering RV. How to select and interpret molecular strain typing methods for epidemiological studies of bacterial infections: a review for healthcare epidemiologists. Molecular Typing Working Group of the Society for Healthcare Epidemiology of America. Infect Control Hosp Epidemiol 1997; 18:6426–6439.

22. Courvalin P. Impact of molecular biology on antibiotic susceptibility: testing and therapy. Am J Med 1995; 99(6A):21S–25S.
23. Hiraoka M, Okamoto R, Inoue M, Mitsuhashi S. Effects of β-lactamases and *omp* mutation on susceptibility to β-lactam antibiotics in *Escherichia coli.* Antimicrob Agents Chemother 1989; 33:382–386.
24. Courvalin P. Transfer of antibiotic resistance genes between gram-positive and gram-negative bacteria. Antimicrob Agents Chemother 1994; 38:1447–1451.
25. Hall RM, Collis CM. Antibiotic resistance in gram-negative bacteria: the role of gene cassettes and integrons. Drug Resist Updates 1998; 1:109–119.
26. Hall RM. Mobile gene cassettes and integrons: moving antibiotic resistance genes in gram-negative bacteria. Ciba Found Symp 1997; 207:192–202; discussion 202–205.
27. Liebert CA, Hall RM, Summers AO. Transposon Tn*21*, flagship of the floating genome. Microbiol Mol Biol Rev 1999; 63:507–522.
28. Skold O. Sulfonamide resistance: mechanisms and trends. Drug Resist Updates 2000; 3:155–160.
29. Bauernfeind A, Georgopapadakou NH. Clinical significance of antibacterial transport. In: Georgopapadakou NH, ed. Drug Transport in Antimicrobial and Anticancer Chemotherapy. New York: Marcel Dekker, 1995:1–19.
30. Neuhaus FC, Georgopapadakou NH. Strategies in β-lactam design. In: Sutcliffe JA, Georgopapadakou NH, eds. Emerging Targets in Antibacterial and Antifungal Chemotherapy. New York: Chapman and Hall, 1992:204–273.
31. Georgopapadakou NH. Penicillin-binding proteins and bacterial resistance to beta-lactams. Antimicrob Agents Chemother 1993; 37:2045–2053.
32. Rice LB, Bonomo RA. β-Lactamases: which ones are clinically important? Drug Resist Updates 2000; 3:178–189.
33. Ambler RP. The structure of beta-lactamases. Philos Trans R Soc Lond B Biol Sci 1980; 289:321–331.
34. Ledent P, Raquet X, Joris B, Van Beeumen J, Frere JM. A comparative study of class-D beta-lactamases. Biochem J 1993; 292:555–562.
35. Wong JD, Barash JR, Sandfort RF, Janda JM. Susceptibilities of *Yersinia pestis* strains to 12 antimicrobial agents. Antimicrob Agents Chemother 2000; 44: 1995–1996.
36. Galimand M, Guiyoule A, Gerbaud G, Rasoamanana B, Chanteau S, Carniel E, Courvalin P. Multidrug resistance in *Yersinia pestis* mediated by a transferable plasmid. N Engl J Med 1997; 337:10,677–680.
37. Bush K, Jacoby GA, Medeiros AA. A functional classification scheme for β-lactamases and its correlation with molecular structure. Antimicrob Agents Chemother 1995; 39:1211–1233.
38. Jacoby GA. Extended-spectrum beta-lactamases and other enzymes providing resistance to oxyimino-beta-lactams. Infect Dis Clin North Am 1997; 11: 875–887.
39. Rice LB, Carias LL, Hujer AM, Bonafede M, et al. High-level expression of chromosomally encoded SHV-1 beta-lactamase and an outer membrane pro-

tein change confer resistance to ceftazidime and piperacillin/tazobactam in a clinical isolate of *Klebsiella pneumoniae*. Antimicrob Agents Chemother 2000; 44:362–367.
40. Rasheed JK, Jay C, Metchock B, Berkowitz F, Weigel L, Crellin J, Steward C, Hill B, Medeiros AA, Tenover FC. Evolution of extended-spectrum beta-lactam resistance (SHV-8) in a strain of *Escherichia coli* during multiple episodes of bacteremia. Antimicrob Agents Chemother 1997; 41:647–653.
41. Rasheed JK, Anderson GJ, Yigit H, Queenan AM, Domenech-Sanchez A, Swenson JM, Biddle JW, Ferraro MJ, Jacoby GA, Tenover FC. Characterization of the extended-spectrum beta-lactamase reference strain, *Klebsiella pneumoniae* K6 (ATCC 700603), which produces the novel enzyme SHV-18. Antimicrob Agents Chemother 2000; 44:2382–2388.
42. MGP Page. β-Lactamase inhibitors. Drug Resist Updates 2000; 3:109–125.
43. Stapleton PD, Shannon KP, French GL. Construction and characterization of mutants of the TEM beta-lactamase containing amino acid substitutions associated with both extended-spectrum resistance and resistance to beta-lactamase inhibitors. Antimicrob Agents Chemother 1999; 43:1881–1887.
44. Rahal JJ, Urban C, Horn D, et al. Class restriction of cephalosporin use to control total cephalosporin resistance in nosocomial *Klebsiella*. JAMA 1998; 280:1233–1237.
45. Lucet JC, Decre D, Fichelle A, et al. Control of a prolonged outbreak of extended-spectrum beta-lactamase-producing enterobacteriaceae in a university hospital. Clin Infect Dis 1999; 29:1411–1418.
46. Monnet DL, Biddle JW, Edwards JR, Culver DH, Tolson JS, Martone WJ, Tenover FC, Gaynes RP. Evidence of interhospital transmission of extended-spectrum beta-lactam–resistant *Klebsiella pneumoniae* in the United States, 1986 to 1993. The National Nosocomial Infections Surveillance System. Infect Control Hosp Epidemiol 1997; 18:492–498.
47. Tenover FC, Mohammed MJ, Gorton TS, Dembek ZE. Detection and reporting of organisms producing extended-spectrum beta-lactamases: survey of laboratories in Connecticut. J Clin Microbiol 1999; 37:4065–4070.
48. Laboratory capacity to detect antimicrobial resistance. MMWR 2000; 48: 1167–1171.
49. Martinez-Martinez L, Conejo MC, Pascual A, Hernandez-Alles S, Ballesta S, Ramirez De Arellano-Ramos E, Benedi VJ, Perea EJ. Activities of imipenem and cephalosporins against clonally related strains of *Escherichia coli* hyper-producing chromosomal beta-lactamase and showing altered porin profiles. Antimicrob Agents Chemother 2000; 44:2534–2536.
50. Joris B, De Meester F, Galleni M, Frere JM, Van Beeumen J. The K1 beta-lactamase of *Klebsiella pneumoniae*. Biochem J 1987; 243:561–567.
51. Martinez-Martinez L, Pascual A, Hernandez-Alles S, Alvarez-Diaz D, Suarez AI, Tran J, Benedi VJ, Jacoby GA. Roles of beta-lactamases and porins in activities of carbapenems and cephalosporins against *Klebsiella pneumoniae*. Antimicrob Agents Chemother 1999; 43:1669–1673.
52. Weindorf H, Schmidt H, Martin HH. Contribution of overproduced chromosomal beta-lactamase and defective outer membrane porins to resistance to

extended-spectrum beta-lactam antibiotics in *Serratia marcescens*. J Antimicrob Chemother 1998; 41:189–195.
53. Wiedemann B, Pfeiffle D, Wiegand I, Janas E. β-Lactamase induction and cell wall recycling in gram-negative bacteria. Drug Resist Updates 1998; 1: 223–226.
54. Jacobs C. Life in the balance: cell walls and antibiotic resistance. Science 1997; 278:1731–1732.
55. Nikaido H, Vaara M. Molecular basis of bacterial outer membrane permeability. Microbiol Rev 1985; 49:1–32.
56. Georgopapadakou NH. Antibiotic permeation through the bacterial outer membrane. J Chemother 1990; 2:275–279.
57. Alcantar-Curiel MD, Garcia-Latorre E, Santos JI. *Klebsiella pneumoniae* 35 and 36 kDa porins are common antigens in different serotypes and induce opsonizing antibodies. Arch Med Res 2000; 31:28–36.
58. Hernandez-Alles S, Alberti S, Alvarez D, Domenech-Sanchez A, Martinez-Martinez L, Gil J, Tomas JM, Benedi VJ. Porin expression in clinical isolates of *Klebsiella pneumoniae*. Microbiology 1999; 145:673–679.
59. Hutsul JA, Worobec E. Molecular characterization of the *Serratia marcescens* OmpF porin, and analysis of *S. marcescens* OmpF and OmpC osmoregulation. Microbiology 1997; 143:2797–2806.
60. Chevalier J, Pages JM, Mallea M. In vivo modification of porin activity conferring antibiotic resistance to *Enterobacter aerogenes*. Biochem Biophys Res Commun 1999; 266:248–251.
61. Mallea M, Chevalier J, Bornet C, Eyraud A, Davin-Regli A, Bollet C, Pages JM. Porin alteration and active efflux: two in vivo drug resistance strategies used by *Enterobacter aerogenes*. Microbiology 1998; 144:3003–3009.
62. Domenech-Sanchez A, Hernandez-Alles S, Martinez-Martinez L, Benedi VJ, Alberti S. Identification and characterization of a new porin gene of *Klebsiella pneumoniae*: its role in beta-lactam antibiotic resistance. J Bacteriol 1999; 181:2726–2732.
63. Bjorkman J, Andersson DI. The cost of antibiotic resistance from a bacterial perspective. Drug Resist Updates 2000; 3:237–245.
64. Chapman JS, Georgopapadakou NH. Routes of quinolone permeation in *Escherichia coli*. Antimicrob Agents Chemother 1988; 32:438–442.
65. McCaffrey C, Bertasso A, Pace J, Georgopapadakou NH. Quinolone accumulation in *Escherichia coli*, *Pseudomonas aeruginosa*, and *Staphylococcus aureus*. 36:1601–1605.
66. Hooper DC. Mode of action of fluoroquinolones. Drugs 1999; 58(Suppl 2): 6–10.
66a. Little JW, Mount DW. The SOS regulatory system of *Escherichia coli*. Cell 1982; 29:11–22.
67. Martinez-Martinez L, Pascual A, Jacoby GA. Quinolone resistance from a transferable plasmid. Lancet 1998; 351:797–799.
68. Weigel LM, Steward CD, Tenover FC. *gyrA* mutations associated with fluoroquinolone resistance in eight species of Enterobacteriaceae. Antimicrob Agents Chemother 1998; 42:2661–2667.

69. Hooper DC. Mechanisms of fluoroquinolone resistance. Drug Resist Updates 1999; 2:38–55.
70. Piddock LJ. Mechanisms of fluoroquinolone resistance: an update 1994–1998. Drugs 1999; 58(Suppl 2):11–18.
71. Yoshida H, Bogaki M, Nakamura M, et al. Quinolone-resistance-determining region in the DNA gyrase *gyrA* gene of *Escherichia coli*. Antimicrob Agents Chemother 1990; 34:1271–1272.
72. Breines DM, Ouabdesselam S, Ng EY, Tankovic J, Shah S, Soussy CJ, Hooper DC. Quinolone resistance locus *nfxD* of *Escherichia coli* is a mutant allele of the *parE* gene encoding a subunit of topoisomerase IV. Antimicrob Agents Chemother 1997; 41:175–179.
73. Georgopapadakou NH. Quinolone uptake and efflux. In: Georgopapadakou NH, ed. Drug Transport in Antimicrobial and Anticancer Chemotherapy. New York: Marcel Dekker, 1995:245–267.
74. Cohen SP, Hooper DC, Wolfson JS, et al. An endogenous active efflux of norfloxacin in *Escherichia coli*. Antimicrob Agents Chemother 1988; 32:1187–1191.
75. Chapman JS, Bertasso A, Georgopapadakou NH. Fleroxacin resistance in *Escherichia coli*. Antimicrob Agents Chemother 1989; 33:239–241.
76. Ishii H, Sato K, Hoshino K, et al. Active efflux of ofloxacin by a highly quinolone-resistant strain of *Proteus vulgaris*. J Antimicrob Chemother 1991; 28:827–836.
77. O'Hara CM, Steward CD, Wright JL, Tenover FC, Miller JM. Isolation of *Enterobacter intermedium* from the gallbladder of a patient with cholecystitis. J Clin Microbiol 1998; 36:3055–3056.
78. Dworkin RJ. Aminoglycosides for the treatment of gram-negative infections: therapeutic use, resistance and future outlook. Drug Resist Updates 1999; 2:173–179.
79. Wright GD, Berghuis AM, Mobashery S. Aminoglycoside antibiotics. Structures, functions, and resistance. Adv Exp Med Biol 1998; 456:27–69.
80. Kotra LP, Haddad J, Mobashery S. Aminoglycosides: perspectives on mechanisms of action and resistance and strategies to counter resistance. Antimicrob Agents Chemother 2000; 44:3249–3256.
81. Fourmy D, Recht MI, Blanchard SC, Puglisi JD. Structure of the A site of *Escherichia coli* 16S ribosomal RNA complexed with an aminoglycoside antibiotic. Science 1996; 274:1367–1371.
82. Moazed D, Noller HF. Interaction of antibiotics with functional sites in 16S ribosomal RNA. Nature 1987; 327:389–394.
83. Bryan LE. Mechanisms of action of aminoglycoside antibiotics. In: Root RK, Sande MA, eds. New Dimensions in Antimicrobial Therapy. New York: Churchill Livingstone, 1984:17–36.
84. Bryan LE, Kwan S. Roles of ribosomal binding, membrane potential, and electron transport in bacterial uptake of streptomycin and gentamicin. Antimicrob Agents Chemother 1983; 23:835–845.
85. Davies J, Wright GD. Bacterial resistance to aminoglycoside antibiotics. Trends Microbiol 1997; 5:234–240.
86. Gerding DN, Larson TA, Hughes RA, et al. Aminoglycoside resistance and

aminoglycoside usage: ten years of experience in one hospital. Antimicrob Agents Chemother 1991; 35:1284–1290.
87. Schmitz FJ, Verhoef J, Fluit AC. Prevalence of aminoglycoside resistance in 20 European university hospitals participating in the European SENTRY Antimicrobial Surveillance Programme. Eur J Clin Microbiol Infect Dis 1999; 18:414–421.
88. Busch-Sorensen C, Sonmezoglu M, Frimodt-Moller N, Hojbjerg T, Miller GH, Espersen F. Aminoglycoside resistance mechanisms in Enterobacteriaceae and *Pseudomonas* spp. from two Danish hospitals: correlation with type of aminoglycoside used. APMIS 1996; 104:763–768.
89. Shaw KJ, Rather PN, Hare RS, Miller GH. Molecular genetics of aminoglycoside resistance genes and familial relationships of the aminoglycoside-modifying enzymes. Microbiol Rev 1993; 57:138–163.
90. Thompson PR, Schwartzenhauer J, Hughes DW, Berghuis AM, Wright GD. The COOH terminus of aminoglycoside phosphotransferase (3′)-IIIa is critical for antibiotic recognition and resistance. J Biol Chem 1999; 274:30697–30706.
91. Sandvang D, Aarestrup FM. Characterization of aminoglycoside resistance genes and class I integrons in porcine and bovine gentamicin-resistant *Escherichia coli*. Microb Drug Resist 2000; 6:19–27.
92. Lambert T, Gerbaud G, Courvalin P. Characterization of transposon *Tn1528*, which confers amikacin resistance by synthesis of aminoglycoside 3′-O-phosphotransferase type VI. Antimicrob Agents Chemother 1994; 38:702–706.
93. Wu HY, Miller GH, Blanco MG, Hare RS, Shaw KJ. Cloning and characterization of an aminoglycoside 6′-N-acetyltransferase gene from *Citrobacter freundii* which confers an altered resistance profile. Antimicrob Agents Chemother 1997; 41:2439–2447.
94. Rather PN. Origins of the aminoglycoside-modifying enzymes. Drug Resist Updates 1998; 1:285–291.
95. Perlin MH, Lerner SA. High-level amikacin resistance in *E. coli* due to phosphorylation and impaired aminoglycoside uptake. Antimicrob Agents Chemother 1986; 29:216–224.
96. Acosta MB, Ferreira RC, Padilla G, Ferreira LC, Costa SO. Altered expression of oligopeptide-binding protein (OppA) and aminoglycoside resistance in laboratory and clinical *Escherichia coli* strains. J Med Microbiol 2000; 49: 409–413.
97. Kashiwagi K, Tsuhako MH, Sakata K, Saisho T, Igarashi A, da Costa SO, Igarashi K. Relationship between spontaneous aminoglycoside resistance in *Escherichia coli* and a decrease in oligopeptide binding protein. J Bacteriol 1998; 180:5484–5488.
98. Chopra I, Hawkey PM, Hinton M. Tetracyclines, molecular and clinical aspects. J Antimicrob Chemother 1992; 29:245–277.
99. Chopra I. Tetracycline uptake and efflux in bacteria. In: Georgopapadakou NH, ed. Drug Transport in Antimicrobial and Anticancer Chemotherapy. New York: Marcel Dekker, 1995:221–243.
100. Nikaido H, Thanassi DG. Penetration of lipophilic agents with multiple

protonation sites into bacterial cells: tetracyclines and fluoroquinolones as examples. Antimicrob Agents Chemother 1993; 37:1393–1399.
101. Speer BS, Shoemaker NB, Salyers AA. Bacterial resistance to tetracycline: mechanisms, transfer, and clinical implications. Clin Microbiol Rev 1992; 5:387–399.
102. Levy SB. Active efflux mechanisms for antimicrobial resistance. Antimicrob Agents Chemother 1992; 36:695–703.
103. Thanassi DG, Suh GS, Nikaido H. Role of outer membrane barrier in efflux-mediated tetracycline resistance of *Escherichia coli.* J Bacteriol 1995; 177:998–1007.
104. Levy SB, McMurry LM, Barbosa TM, Burdett V, Courvalin P, Hillen W, Roberts MC, Rood JI, Taylor DE. Nomenclature for new tetracycline resistance determinants. Antimicrob Agents Chemother 1999; 43:6,1523–1524.
105. McMurry LM, Petrucci RE, Levy SB. Active efflux of tetracycline encoded by four genetically different tetracycline resistant determinants in *Escherichia coli.* Proc Natl Acad Sci USA 1980; 77:3974–3977.
106. Robert MC. Tetracycline resistance determinants: mechanisms of action, regulation of expression, genetic mobility, and distribution. FEMS Microbiol Rev 1996; 19:1–24.
107. Roberts MC. Genetic mobility and distribution of tetracycline resistance determinants. Ciba Found Symp 1997; 207:206–218; discussion 219–222.
108. Makino S, Asakura H, Obayashi T, Shirahata T, Ikeda T, Takeshi K. Molecular epidemiological study on tetracycline resistance R plasmids in enterohaemorrhagic *Escherichia coli* O157:H7. Epidemiol Infect 1999; 123:25–30.
109. Magalhaes VD, Schuman W, Castilho BA. A new tetracycline resistance determinant cloned from *Proteus mirabilis.* Biochim Biophys Acta 1998; 1443:262–266.
110. Taylor DE, Chau A. Tetracycline resistance mediated by ribosomal protection. Antimicrob Agents Chemother 1996; 40:1–5.
111. Oliva B, Chopra I. *tet* Determinants provide poor protection against some tetracyclines: further evidence for division of tetracyclines into two classes. Antimicrob Agents Chemother 1992; 36:876–878.
112. Testa RT, Petersen PJ, Jacobus NV, et al. In vitro and in vivo antibacterial activities of the glycylglycines, a new class of semisynthetic tetracyclines. Antimicrob Agents Chemother 1993; 37:2270–2277.
113. Tally FT, Ellestad GA, Testa RT. Glycylcyclines: a new generation of tetracyclines. J Antimicrob Chemother 1995; 35:449–452.
114. Petersen PJ, Jacobus NV, Weiss WJ, Sum PE, Testa RT. In vitro and in vivo antibacterial activities of a novel glycylcycline, the 9-t-butylglycylamido derivative of minocycline (GAR-936). Antimicrob Agents Chemother 1999; 43:738–744.
115. Nelson ML, Levy SB. Reversal of tetracycline resistance mediated by different bacterial tetracycline resistance determinants by an inhibitor of the Tet(B) antiport protein. Antimicrob Agents Chemother 1999; 43:1719–1724.
116. Huovinen P, Sundstrom L, Swedberg G, Skold O. Trimethoprim and sulfonamide resistance. Antimicrob Agents Chemother 1995; 39:279–289.

117. Swedberg G, Skold O. Characterization of different plasmid-borne dihydropteroate synthases mediating bacterial resistance to sulfonamides. J Bacteriol 19809; 142:1–7.
118. Radstrom P, Swedberg G. RSF1010 and a conjugative plasmid contain *sulII*, one of the two known genes for plasmid-borne sulfonamide resistant dihydropteroate synthase. Antimicrob Agents Chemother 1988; 32:1684–1692.
119. Sundstrom L, Radstrom P, Swedberg G, Skold O. Site-specific recombination promotes linkage between trimethoprim- and sulfonamide resistance genes. Sequence characterization of *dhfrV* and *sulI* and a recombination active locus of *Tn21*. Mol Gen Genet 1988; 213:191–201.
120. Amyes SGB, Towner KJ. Trimethoprim resistance: epidemiology and molecular aspects. J Med Microbiol 1990; 31:1–9.
121. Grey D, Hamilton-Miller JMT, Brumfitt W. Incidence and mechanisms of resistance to trimethoprim in clinically isolated gram-negative bacteria. Chemotherapy 1979; 25:147–156.
122. Huovinen P. Increases in rates of resistance to trimethoprim. Clin Infect Dis 1997; 24(Suppl 10):S63–S66.
123. Shaw WV. Chloramphenicol acetyltransferase: enzymology and molecular biology. Crit Rev Biochem 1983; 14:1–46.
124. Parent R, Roy PH. The chloramphenicol acetyltransferase gene of Tn*2424*: a new breed of *cat*. J Bacteriol 1992; 174:2891–2897.
125. Areangioli MA, Leroy-Setrin S, Martel JL, Chaslus-Dancla E. Evolution of chloramphenicol resistance, with emergence of cross-resistance to florfenicol, in bovine *Salmonella typhimurium* strains implicates definitive phage type (DT) 104. J Med Microbiol 2000; 49:103–110.
126. Centers for Disease Control and Prevention/National Nosocomial Infections Surveillance System. Semiannual report. December 1999 (corrected March 2000) Atlanta: CDC/NNIS. 2000 [database online: http://www.cdc.gov/ncidod/hip/NNIS/dec99sar.pdf].
127. Jacoby GA. Antimicrobial-resistant pathogens in the 1990s. Annu Rev Med 1996; 47:169–179.
128. Struelens MJ, Byl B, Vincent J-L. Antibiotic policy: a tool for controlling resistance of hospital pathogens. Clin Microbiol Infect 1999; 5:S19–S24.
129. Flynn DM, Weinstein RA. Nathan C, et al. Patients' endogenous flora as the source of "nosocomial" *Enterobacter* in cardiac surgery. J Infect Dis 1987; 156:363–368.
130. Gupta K, Scholes D, Stamm WE. Increasing prevalence of antimicrobial resistance among uropathogens causing acute uncomplicated cystitis in women. JAMA 1999; 281:736–738.
131. Wiener J, Quinn JP, Bradford PA, et al. Multiple antibiotic-resistant *Klebsiella* and *Escherichia coli* in nursing homes. JAMA 1999; 281:517–523.
132. Guyot A, Barrett SP, Threlfall EJ, Hampton MD, Cheasty T. Molecular epidemiology of multi-resistant *Escherichia coli*. J Hosp Infect 1999; 43:39–48.
133. Nikaido H. The role of outer membrane and efflux pumps in the resistance of gram-negative bacteria. Can we improve access? Drug Resist Updates 1998; 1:93–98.

134. Lee A, Mao W, Warren MS, Mistry A, Hoshino K, Okumura R, Ishida H, Lomovskaya O. Interplay between efflux pumps may provide either additive or multiplicative effects on drug resistance. J Bacteriol 2000; 182:3142–3150.
135. Hancock REW. Alterations in outer membrane permeability. Annu Rev Microbiol 1984; 38:237.
136. Viljanen P, Kayhty H, Vaara M, Vaara T. Susceptibility of gram-negative bacteria to the synergistic bactericidal action of serum and polymyxin B nonapeptide. Can J Microbiol 1986; 32:66–69.

17

Public Health Responses to Antimicrobial Resistance in Outpatient and Inpatient Settings

Richard E. Besser, Julia Y. Morita, and Scott K. Fridkin
Centers for Disease Control and Prevention, Atlanta, Georgia

In 1998, Centers for Disease Control and Prevention (CDC) issued a report, Preventing Emerging Infectious Diseases, which outlined a plan designed to address key emerging infectious disease issues (1). Antimicrobial resistance was seen as one of the major infectious disease issues facing the world as we enter the new millennium. In this chapter, we will review some of the approaches that have been taken to combat this problem in outpatient and inpatient settings.

We will focus on issues related to pneumococcal resistance in the community: the importance of surveillance to measure the magnitude of the problem and to assess the impact of interventions designed to reduce resistance; the role that inappropriate antibiotic use has in promoting the development of resistant bacteria; and interventions that have been undertaken or are underway to promote the judicious use of antibiotics.

In the inpatient setting, we address the unique nature of the hospital environment, which makes this aspect of health care delivery a focus for the emergence and spread of many antimicrobial-resistant pathogens. We review surveillance data that have shown increasing rates of resistance for most pathogens associated with nosocomial infections. And we will describe many opportunities to prevent the emergence and spread of these resistant pathogens through a systematic review of surveillance data on antimicrobial resistance and antimicrobial prescribing and improved use of established infection control measures.

1 OVERCOMING PNEUMOCOCCAL RESISTANCE IN THE COMMUNITY

1.1 Background

1.1.1 Burden of Pneumococcal Disease

Streptococcus pneumoniae infections are among the leading causes of illness and death in young children, persons with underlying medical conditions, and the elderly. Each year in the United States, pneumococcal disease is estimated to cause 3000 cases of meningitis and 60,000 cases of bacteremia (2). The overall case-fatality rate for pneumococcal bacteremia is 15–20% among adults. The case-fatality rate for elderly patients has been estimated to be between 30 and 40% (3).

S. pneumoniae is the most common cause of community-acquired bacterial pneumonia and otitis media, each year causing as many as 500,000 cases of pneumonia and 7 million cases of otitis media (4–7). It has been estimated to account for approximately 25–35% of cases of community-acquired bacterial pneumonia in persons who require hospitalization. In the United States, acute otitis media cases—20–50% of which are caused by pneumococci—results in more than 24 million visits to pediatricians per year (3).

1.1.2 Emergence and Spread of Drug-Resistant *Streptococcus pneumoniae*

Therapy for invasive (e.g., bacteremia, meningitis) and noninvasive (e.g., otitis media) pneumococcal infections is frequently empiric, because isolates are usually unavailable from patients with pneumonia and otitis media and susceptibility patterns are not initially available for patients with meningitis and bacteremia. In the past, pneumococci were uniformly susceptible to penicillin, allowing most physicians to treat patients who had pneumococcal infections with penicillin alone without testing for resistance.

Since the 1960s, however, resistance to penicillin and other antimicrobial agents has emerged and spread. Penicillin resistance among *S.*

pneumoniae was first recognized in Australia and New Guinea in the 1960s, followed by South Africa during the 1970s, and then throughout countries in Africa, Asia, and Europe during the 1980s (8–11). In the United States, case reports of antimicrobial-resistant strains of *S. pneumoniae* began appearing during the late 1970s and early 1980s (12–14). Surveillance data from CDC verified that antimicrobial resistance among invasive isolates of *S. pneumoniae* remained at low levels from 1979 to 1987 (Fig. 1) (15). However, in the 1990s, antimicrobial-resistant strains, including those with reduced susceptibility to multiple antimicrobial drugs, became increasingly prevalent in many parts of the country (16–20).

The trend of increasing antimicrobial resistance among invasive *S. pneumoniae* isolates has been well documented by surveillance data from CDC. These data documented a >60-fold increase in invasive isolates resistant to penicillin in 1992 compared to 1979–1987 (16) and continued increases during 1993–1998 (see Fig. 1) (2,21). Penicillin and cephalosporin treatment failures have been reported in both adult and pediatric patients with meningitis (22–28). Antimicrobial resistance among *S. pneumoniae* isolates obtained from the middle ear of children with otitis media has been reported in the range of 17–42% (20,29,30). As with meningitis,

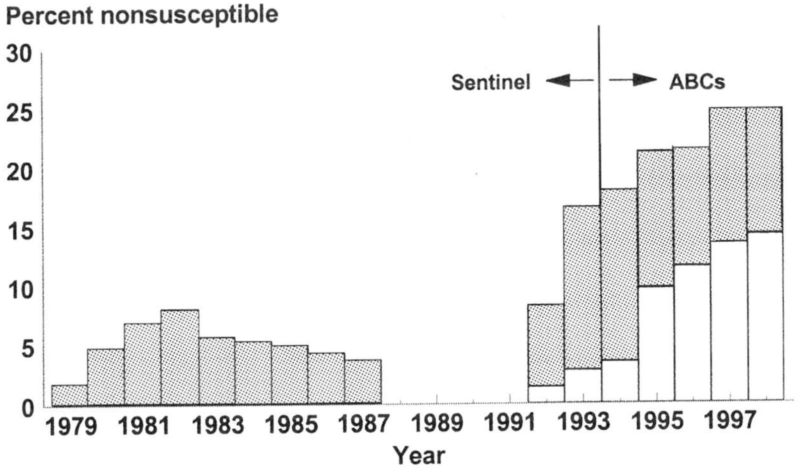

Figure 1 Proportion of pneumococcal isolates resistant to penicillin, United States, 1979–1998, as reported to the sentinel surveillance system (1979–1993) and the Active Bacterial Core Surveillance system (1994–1998). The dark portion of each bar indicates intermediate level resistance; the light portion indicates high-level resistance.

bacteriological and clinical failures in response to treatment with oral cephalosporins have been documented among children with otitis media caused by penicillin-resistant *S. pneumoniae* (31–34).

1.2 Public Health Response to Emergence of Drug-Resistant *Streptococcus pneumoniae*

1.2.1 Surveillance

In 1994, in response to the increasing prevalence of drug-resistant *Streptococcus pneumoniae* (DRSP), CDC convened a working group of public health practitioners, clinical laboratorians, health care providers, and representatives of key professional societies. To minimize the impact of DRSP, the working group developed a strategy which included three major goals: surveillance, epidemiological investigation, and prevention and control. The primary goals of establishing surveillance include monitoring the prevalence and the geographical distribution of DRSP and rapid recognition of new patterns of resistance. The primary goal of prevention and control of DRSP is to minimize complications of antimicrobial-resistant pneumococcal infections. Nationwide surveillance data would provide necessary information for fulfilling the objectives identified for prevention and control. Area-specific data could be used by clinicians to heighten their awareness and guide their selection of antimicrobial drugs for treatment of infections likely to be caused by *S. pneumoniae*. Public health officials could use this information to improve vaccination use, targeting areas most likely to benefit from intervention (e.g., regions with high levels of resistance to antimicrobial drugs or communities of persons at highest risk for infection), to promote the judicious use of antimicrobial agents in areas with high levels of antimicrobial resistance, and to publish national and regional trends in pneumococcal antimicrobial resistance (7,35).

An ideal nationwide, electronic, laboratory-based surveillance system for invasive pneumococcal infections and their antimicrobial susceptibility patterns was designed and proposed by the working group (35). CDC is in the early stages of developing a system modeled after the proposed system. Alternative sources of information have been and will continue to be used until this nationwide system is functioning.

Sentinel Surveillance System. In 1978, CDC established a sentinel surveillance system for the purpose of conducting pneumococcal vaccine efficacy studies (see Fig. 1). The 54 hospitals in 26 states that initially participated in this system were selected on the basis of their willingness to submit isolates. These hospitals were asked to submit all *S. pneumoniae* isolates from normally sterile sites and to provide demographical and clinical information. Pneumococcal isolates were sent to CDC for serotyping and, until recently, for antimicrobial susceptibility testing. Isolates were serotyped on the basis of capsular swelling with type-specific antisera prepared at CDC (36). Antimicrobial susceptibility testing was done by the broth microdilution procedure (36). Resistance was defined using National Committee for Clinical Laboratory Standards (NCCLS) criteria (37). Although this system was useful for documenting increases in penicillin resistance among invasive *S. pneumoniae* isolates during the 1990s (16,21), results were not representative of communities not included in this system. In fact, it was not possible to generalize to other hospitals near participating institutions given what we now know from population-based surveillance (38). The system, now with only 12 participating hospitals, no longer tests for antimicrobial susceptibility, but continues to focus on vaccine efficacy, where selection bias is not a problem.

DRSP Reporting Requirements. Before the Working Group convened, several state health departments and the New York City Department of Health had already instituted regulations requiring laboratories to report DRSP isolates from certain anatomical sites (e.g., cerebrospinal fluid, blood). From 1994 to 1996, the New York City Department of Health received reports of DRSP isolated from all anatomical sites. Reports were evaluated by telephone consultation with the reporting laboratory to determine anatomical site, the MIC testing methodology, and the quantitative MIC. The New York City Department of Health used these data to identify high-risk groups and to provide laboratory directors, hospital infection-control departments, and clinicians with regular updates about DRSP within New York City (18). A nationwide requirement for reporting DRSP could lead to better characterization of the epidemiology of DRSP infections at the local level. In response to a recommendation from the working group, the Council of State and Territorial Epidemiologists (CSTE) approved a proposal to make invasive infections caused by DRSP reportable. Although regulatory authority for reporting nationally notifiable diseases resides at the state level, approval by CSTE provides a basis for state health officials to encourage their state legislators to adopt the measure. Currently, 27 state health departments and the New York City Department of Health have regulations requiring laboratories or providers to report DRSP isolates.

Surveillance that is restricted to antimicrobial-resistant infections can provide useful information. However, it does not provide information about the proportion of all isolates in a community that are resistant to antimicrobial agents (18). To obtain such information, surveillance for all invasive pneumococcal infections is necessary. Of the 28 state and city health departments that require reporting of DRSP, only 16 require reporting of all invasive pneumococcal infections. The following section provides a description of a population-based surveillance system for invasive infections caused by antimicrobial-susceptible and antimicrobial-nonsusceptible *S. pneumoniae*.

Active Bacterial Core Surveillance. Since 1994, CDC has performed active, population-based, surveillance for invasive pneumococcal infections as part of the Emerging Infections Program of the National Center for Infectious Diseases (see Fig. 1). Currently, *S. pneumoniae* is one of five bacteria under surveillance by the Active Bacterial Core Surveillance (ABCs). Objectives of this system include determining the incidence and epidemiological characteristics of invasive disease caused by bacteria under surveillance; determining the molecular and epidemiological patterns and microbiological characteristics of public health relevance for isolates causing invasive disease; and providing an infrastructure for additional special studies of these bacteria. A case of invasive pneumococcal disease is defined as isolation of *S. pneumoniae* from a normally sterile site in a resident of one of the surveillance sites. In 1998, the total population under surveillance for pneumococcal disease was 17,070,953. The sites include the entire state of Connecticut, one county in California, 20 counties in Georgia, 6 counties in Maryland, 7 counties in Minnesota, 7 counties in New York, 3 counties in Oregon, and 5 counties in Tennessee (2).

Case finding is active and laboratory based. Surveillance personnel contact acute care hospitals' and reference microbiology laboratories every 2–4 weeks to identify new cases and request isolate submission. Isolates are submitted to either CDC's laboratory (Georgia isolates) or reference laboratories (all other isolates) for antimicrobial susceptibility testing and serotyping. Antimicrobial susceptibility testing is performed by the broth microdilution method (36). Nonsusceptibility and resistance are defined by using NCCLS criteria (37). Isolates are serotyped on the basis of capsular swelling with type-specific antisera prepared at CDC (36). Medical record review is performed by either infection control personnel, county health department personnel, or area surveillance personnel. Data from case report forms, isolate forms, and special study forms are entered into a computerized database at each surveillance site. The data are transmitted to CDC where they are verified and aggregated. Every month, summary tables and laboratory testing results are sent to the surveillance sites from CDC. Data are posted annually on the ABCs website (http://www.cdc.gov/abcs).

Population-based data have been useful in describing the epidemiology of invasive infections caused by DRSP. For 1997, the prevalence of penicillin-nonsusceptible *S. pneumoniae* varied geographically. Although overall 25% of invasive isolates were nonsusceptible to penicillin, the proportion of nonsusceptible isolates ranged from a low of 15.3% in Maryland to a high of 38.3% in Tennessee. Additionally, the prevalence of penicillin-nonsusceptible *S. pneumoniae* ranged widely within surveillance sites. In Connecticut, for example, only 7 (32%) of the 22 hospitals participating in ABCs had proportions within 5% of the overall proportion of nonsusceptible isolates for the entire state (18.1%) (38). Given the expense entailed in conducting active, population-based surveillance, CDC is evaluating the utility of alternative surveillance systems such as pooled hospital antibiograms, electronic laboratory surveillance, and sentinel networks that use many hospitals in selected regions of a state to see if these cheaper methods might provide equivalent information.

1.2.2 Epidemiological Investigations of DRSP

The ABCs system provides a framework for special epidemiological investigations of DRSP to be performed. A study using ABCs population-based surveillance data for the eight-county metropolitan Atlanta area in 1994 identified 27 and 24% of invasive *S. pneumoniae* isolates as being penicillin nonsusceptible among children and adults, respectively. Higher proportions of whites than blacks had infections caused by penicillin-resistant or multi–drug-resistant strains of *S. pneumoniae.* Higher proportions of white children less than 6 years of age had infections caused by penicillin-resistant, cefotaxime-resistant, or multi–drug-resistant strains of *S. pneumoniae* than black children of the same age. Suburban residence was also associated with an increased risk of infections with an antimicrobial-resistant organism (19). The high rate of antimicrobial-resistant infections found in adults suggested that antimicrobial resistance is not a problem limited to pediatric patients. Because of concern about antimicrobial-resistant infections in children, recommendations for empirical therapy for children with suspected life-threatening pneumococcal infections were developed (39). This study's results provided evidence for the necessity of recommendations for empirical therapy of pneumococcal infections in adults.

A case-control study to identify risk factors for invasive pediatric pneumococcal disease was performed within the ABCs system. Along with risk factors identified for invasive pediatric pneumococcal disease, recent antimicrobial use was associated with increased risk for invasive infection with penicillin-resistant *S. pneumoniae* (40). Another study performed using ABCs data evaluated the relationship of county-level macrolide and β-lactam antimicrobial sales and the proportion of invasive infections caused by erythromycin-resistant and penicillin-nonsusceptible *S.*

pneumoniae, respectively. The results of this study suggested that counties with high antimicrobial sales also had high proportions of DRSP (41). These results support the strategy of judicious use of antimicrobial agents to decrease the spread of antimicrobial resistance among *S. pneumoniae*.

These data were useful for identifying geographical areas, specific populations, and other risk factors associated with a higher prevalence of infections caused by antimicrobial-resistant organisms within the surveillance areas. Clinicians within the areas under surveillance can use these data to guide their selection of antimicrobial therapy for presumed *S. pneumoniae* infections. Additionally, intervention strategies including vaccination campaigns and judicious use of antimicrobial agent campaigns can be developed using these data to identify appropriate geographical areas and populations. Collecting similar data at a nationwide level would provide clinicians and public health officials throughout the United States with similar information, which could be used to prevent and control antimicrobial-resistant pneumococcal disease.

1.2.3 Treatment of Pneumococcal Infections in an Era of Increasing DRSP Prevalence

In 1996, to assist clinicians in treating patients with presumed pneumococcal infections in an era of increasing DRSP prevalence, CDC convened the DRSP Therapeutic Working Group. This group is composed of practicing pediatricians, family practitioners, internists, academicians, and public health practitioners. To date, guidelines for the treatment of otitis media and pneumonia have been published (42,43). These guidelines were written in response to concern about the increasing prevalence of antimicrobial resistance among *S. pneumoniae* and the growing body of literature documenting treatment failures of DRSP infections. Although appropriate treatment of DRSP infections must be encouraged, more impact on DRSP can be achieved through primary prevention of these infections with vaccines.

1.2.4 Promotion of Pneumococcal Vaccination

CDC is working with state health departments to promote the use of pneumococcal vaccines. In 1997, among persons aged ≥65 years, 46% reported having received pneumococcal vaccine during the preceding year. This percentage was higher than in 1995, when only 36% of persons in this age group recalled receiving pneumococcal vaccine (44,45). However, these vaccination rates are still far from the national health objective for year 2000, when pneumococcal and influenza vaccination levels were expected to be ≥60% among persons at high risk for severe disease (45).

The 23-valent pneumococcal vaccine contains 23 purified pneumococcal capsular polysaccharide antigens, which comprise at least 85–90% of the serotypes that cause invasive disease in the United States and include serotypes of drug-resistant strains that most frequently cause invasive disease (21,46,47). Underutilization of the vaccine may be partly due to some ongoing controversy regarding vaccine efficacy, duration of protection, side effects, and adverse reactions. Nonetheless, pneumococcal polysaccharide vaccine has been shown to be effective against bacteremic disease caused by organisms whose serotypes are contained in the vaccine (48). Effectiveness in case-control studies has ranged from 56 to 81%. Vaccine effectiveness of 65–84% has been demonstrated among immunocompetent persons aged 65 years or older and among persons with diabetes mellitus, coronary vascular disease, congestive heart failure, chronic obstructive pulmonary disease, and anatomical asplenia (3,49–52). In addition, a meta-analysis of nine randomized controlled trials concluded that pneumococcal vaccine is efficacious in reducing the frequency of bacteremic pneumococcal pneumonia among low-risk adults (53).

This vaccine is not recommended for children under 2 years old, because antibody response to most pneumococcal capsular types is generally poor or inconsistent in this age group (3). However, in February, 2000, a new pneumococcal protein conjugate vaccine was licensed for use in young children and has been recommended for use by the Advisory Committee on Immunization Practices. Preliminary results demonstrate high efficacy for the prevention of meningitis and bacteremia, with moderate efficacy against clinically defined otitis media and pneumonia due to *S. pneumoniae* (54). CDC is currently undertaking studies to assess the impact of this vaccine on prevalence of DRSP in immunized populations. Because conjugate vaccines appear to reduce carriage of pneumococci of vaccine serotypes, and approximately 80% of DRSP occurs within serotypes included in the heptavalent vaccine, their impact on transmission of DRSP may be considerable (55,56).

1.2.5 National Campaign to Promote the Judicious Use of Antibiotics

Background. Numerous studies have documented the association between recent antibiotic use and both carriage of and invasive disease by DRSP (57). Children who have recently received an antibiotic are two to six times more likely to be colonized with DRSP than are children without recent antibiotic use. In addition, children with invasive disease caused by DRSP are significantly more likely to report recent antibiotic use than are children with invasive disease due to sensitive pneumococci.

If antibiotics were being used exclusively for conditions for which they are known to be clinically effective, this increased risk would be viewed as one of the necessary but unavoidable consequences of therapy. However, this is not the case. In an analysis of data from the National Ambulatory Medical Care Survey, McCaig and Hughes documented that over 30% of all antibiotics are prescribed for colds, upper respiratory infections (URIs), and bronchitis (58). These conditions are largely viral in etiology and would not be expected to improve with antibiotic therapy (59). An analysis of the same data by Gonzales et al. demonstrated that antibiotics were prescribed to 51% of adults with colds, 52% with URIs, and 66% with bronchitis (59). In children, antibiotics were prescribed to 44% with colds, 46% with URIs, and 75% with bronchitis (60). These are all conditions for which prescribing could be reduced or eliminated without adversely affecting patient care.

Physicians prescribe unnecessary antibiotics for many reasons; however, studies indicate that physicians believe they overprescribe because of patient/parent demand. Focus groups conducted with pediatricians and family practitioners in Atlanta identified parental expectations for antibiotics as the primary reason for this inappropriate use (61). In a recent national survey of pediatricians, 48% reported that parents pressure them to prescribe antibiotics (62). Seventy-eight percent of surveyed pediatricians felt that the single most important thing that could be done to promote judicious antibiotic use would be to educate parents about the proper indications for antibiotic use. This is supported by a study addressing the relationship between parental expectations and pediatrician antimicrobial prescribing. Physician perception of parental expectation was the only significant predictor of prescribing for conditions of viral origin (63). When pediatricians believed that a parent wanted an antibiotic, they prescribed them in 62% of cases as compared with 7% of cases when they believed the parent did not want an antibiotic. Interestingly, there was no association between parents' true expectations and physician prescribing. Hamm found very similar findings in a study of antibiotic prescribing for respiratory tract infections in adults (64). Although prescribing was related to physician-perceived patient expectations, patient satisfaction was not related to receipt of an antibiotic. One could conclude from these studies that if physicians and parents were able to communicate more openly, inappropriate prescribing might be reduced.

Although more difficult to document, it is likely that physicians' lack of understanding of the wide variety of presentations and natural history of viral illnesses plays a role in overprescribing. For example, although green nasal discharge is normal in a child with a cold (65), many physicians use this finding as an indication for antibiotic prescribing (61). This lack of understanding, combined with the diagnostic uncertainty inherent in most clinical encounters and the time pressures of outpatient practice, contributes to the problem of inappropriate antibiotic prescribing.

Components of the National Campaign. CDC is undertaking a national campaign to promote judicious antibiotic use. This campaign is designed to address concerns over the rising prevalence of antibiotic-resistant organisms, driven in part by inappropriate antibiotic use for outpatient respiratory infections. Judicious antibiotic use is defined as: (1) prescribing antibiotic therapy only when it is likely to be beneficial; (2) using an appropriate (targeted) agent; and (3) using a drug only at its appropriate dose and duration.

The objectives of the campaign are to decrease inappropriate antimicrobial use in the outpatient setting and thereby reduce the spread of antibiotic resistance. The approach centers on (1) establishing partnerships; (2) developing educational materials for physicians and the public; (3) developing and implementing interventions; and (4) assessing the impact on antibiotic use, resistance, and physician/patient satisfaction.

To address the multifaceted nature of the problem of inappropriate antibiotic use, CDC has undertaken partnerships with state and local health departments, managed-care organizations, pharmacy benefit management companies, pharmaceutical companies, health care purchasers, professional associations, and medical schools. Each partner has a very important role to play and is able to target a different audience.

EDUCATIONAL MATERIALS FOR CLINICIANS. CDC has produced a variety of materials for use by clinicians. These materials are available electronically from CDC website (http://www.cdc.gov/antibioticresistance) and may be downloaded and copied freely. In 1998, CDC, in collaboration with the American Academy of Pediatrics and members of the American Academy of Family Physicians, published Principles of Judicious Use of Antimicrobial Agents for Pediatric Upper Respiratory Tract Infections (66). This supplement to the journal *Pediatrics* provides the scientific basis for judicious antibiotic use for patients with otitis media, pharyngitis, acute sinusitis, cough illness/bronchitis, and the common cold. From these principles, CDC has developed detailing sheets for use in physician education, day care notes to allow children with nonbacterial infections back to class without an antibiotic, and viral prescription pads so that physicians can provide tips for relief of viral symptoms. Principles for judicious use of antibiotics for adult URIs are currently being developed in collaboration with various professional associations and will be available in 2001.

A slide set has been developed for use in continuing medical education courses, grand rounds, and other educational forums (67). A medical school curriculum is currently under development and will be pilot tested in 2002. The course has three components: 1) didactic lectures; 2) small-group interactive sessions; and 3) on-line case-based learning. The materials developed for this course can form the foundation for the development of separate curricula for pediatric, family practice, and internal medicine residency programs and for use in continuing medical education courses. These curricula will be evaluated to assess their impact on knowledge, attitudes, and skills regarding antibiotic use. The curricula are designed to give clinicians the skills they need to withhold appropriately antibiotics in outpatient encounters.

Educational Materials for the Public. It is clear that for prescribing behavior to change, public knowledge and behavior must be addressed. CDC has developed informational materials that will make it easier for clinicians to discuss their decisions to withhold appropriately antibiotics from patients with viral illnesses. Pamphlets targeting pediatric and adult clinic populations are designed to draw clearly the distinction between viral and bacterial diseases. These materials are available from CDC website (67). Posters are available for use in clinics and offices featuring two messages: "Sometimes a mother's care is the best medicine," and "When your child is sick, antibiotics are not always the answer."

Interventions to Promote Judicious Use. In June 1999, CDC convened a panel of investigators to evaluate the impact of selected intervention studies designed to promote judicious antimicrobial use. The projects ranged from managed-care and community-level interventions to large-scale, statewide interventions, all focusing on educating medical care providers, parents, and patients about appropriate indications for and use of antibiotics (68,69). Projects used a variety of strategies and materials to improve communication between physicians and patients and to promote the use of symptomatic therapy as an alternative to antibiotics. Projects used a variety of strategies and materials to improve communication between health care providers and patients and to promote the use of symptomatic therapy as an alternative to antibiotics.

Although many of the projects that were presented are ongoing or have yet to be analyzed fully, a number of conclusions were reached. First, interventions to promote judicious use of antibiotics clearly may be effective in reducing inappropriate prescribing. Successful projects all combined physician and patient education, acknowledging the role that both groups play in the promotion of inappropriate prescribing. Second, to be successful, projects must present their messages via multiple vehicles to make use of the variety of means by which people learn. Third, the problem of inappropriate prescribing is not uniform across the country. For instance, some investigators documented significant antibiotic overprescribing for the common cold, whereas other investigators found that this was rarely occurring. In sites where this was not occurring, physicians were often offended by being told not to prescribe for this condition. Successful programs must tailor the prevention messages to local conditions.

Two successful projects highlighted these findings. Gonzales et al. were able to reduce antibiotic prescribing for adults with bronchitis by 40% through an intervention targeting clinicians and patients in a managed-care setting. This reduction was achieved without any increase in adverse events or decrease in patient satisfaction (69). This controlled intervention compared a full intervention that provided education to physicians and patients to either no intervention or an intervention directed only at physicians. These researchers demonstrated that an intervention directed solely at physicians was ineffective. Physicians will reduce their antibiotic prescribing only when they feel that their patients are receptive to this change. In Alaska, preliminary analysis indicates that Petersen et al. were able to achieve a 22% reduction in prescribing for upper respiratory infections in children in villages receiving an intervention directed at medical assistants and patients (70). No reductions were seen in control villages. This intervention focused on improving the diagnosis of otitis media and included extensive patient and physician education components. By addressing the local problem of overprescribing for otitis media, the investigators were able to have a major impact on overall antibiotic prescribing.

FUTURE DIRECTIONS. CDC has funded a 5-year statewide intervention to promote judicious antibiotic use in Wisconsin. This campaign will target the general public as well as health care providers with educational materials. Partnerships are being forged among industry, medical societies, and the public health community. In addition to looking at the impact of the intervention on prescribing practices and resistant invasive pneumococcal infections, the investigators will attempt to address issues of cost effectiveness and acceptability. Hopefully, the materials created in this project and the lessons learned will be shared nationally so that unfunded projects can move forward in directions that are likely to achieve results.

Currently, many local and state health departments are developing and implementing strategies to promote the judicious use of antibiotics. Many of these projects involve partnerships between clinicians, public health professionals, and private industry. Managed care organizations, pharmacy benefit management companies, and large health care purchasers are teaming up with public health practitioners to promote appropriate antibiotic use as a means of improving health and reducing health care costs. Lessons learned from the projects that have been critically evaluated are being used to guide the development of many of these projects. Through the use of surveillance for antimicrobial resistance, and the systematic analysis of prescriber databases we will be able to see whether these efforts are effective in slowing the rise or reversing the trends in antimicrobial resistance in the community.

2 OVERCOMING ANTIMICROBIAL RESISTANCE IN THE HOSPITAL SETTING

2.1 Overview

The unique nature of the hospital environment makes this aspect of health care delivery a focus for the emergence and spread of many antimicrobial-resistant pathogens. There are ample opportunities for the cross transmission of resistant bacteria from patient to patient, and patients are commonly exposed to broad-spectrum antimicrobial agents. Rates of resistance have increased for most pathogens associated with nosocomial infections, with highest rates being among patients in intensive care units (ICUs). However, there are many opportunities to prevent the emergence and spread of these resistant pathogens through implementation of a systematic review of surveillance data, both for antimicrobial resistance and antimicrobial use, and improved utilization of established infection control measures (patient isolation, handwashing, glove use, and appropriate gown use).

2.2 Hospital Environment

Several factors unique to hospitals contribute to cross transmission of antimicrobial-resistant pathogens. First, the urgent nature of critical care often does not allow for necessary aseptic technique or handwashing. Second, the large number and wide variety of health care workers attending to patients' needs leads to inconsistent training and compliance with handwashing, gloving, and gowning. Evidence suggests that antimicrobial-resistant pathogens are carried from patient to patient (exogenous flora) by way of the unwashed hands of health care workers (71). Third, specific agents used for handwashing, and the degree of asepsis used in maintaining invasive devices, may have an impact on the cross transmission of these pathogens as well (72). Finally, the introduction of antimicrobial-resistant bacteria into a hospital may occur upon transfer of critically ill patients unknowingly colonized or infected with such bacteria from other facilities. All of these factors contribute to make surveillance of antimicrobial resistance difficult. Rates of resistance may fluctuate monthly as these different factors become more or less prevalent in the hospital. To complicate things further, the selective pressure of antimicrobial use may have an impact on rates of resistance in particular hospital areas as well.

2.3 Antimicrobial Use

Perhaps no other factor is more important in the development of antimicrobial resistance than antimicrobial use (73–75). Of studies involving hospital-acquired pathogens, 22 reviewed by McGowan (73) have shown a fairly consistent association between use and resistance. Unfortunately, nearly all of these studies were reports from single hospitals, which may not be representative of other hospitals. However, a previous multicenter study in the 1970s demonstrated that changes in aminoglycoside use paralleled changes in aminoglycoside-resistant gram-negative bacilli (76). Also, one other multicenter study demonstrated this type of relationship among several antimicrobials and the corresponding resistant pathogens, including ceftazidime use and ceftazidime-resistant *Enterobacter cloacae* (77). These data begin to demonstrate the usefulness of surveillance of antimicrobial use in hospitals and of identifying scenarios in which extreme amounts of usage are leading to extreme amounts of resistance. Conversely, such surveillance may illustrate areas where rates of resistance are out of proportion with the amount of selective pressure (i.e., antimicrobial use), pointing to a problem of cross transmission in that hospital area.

2.4 Rates of Antimicrobial Resistance in Hospital-Acquired Infections

Although several systems currently exist in the United States and other countries that aggregate antimicrobial susceptibility data among multiple institutions, CDCs National Nosocomial Infections Surveillance System (NNIS) has received susceptibility reports on pathogens associated with nosocomial infections since the 1980s. Examination of the rates of antimicrobial resistance among these pathogens show that rates of methicillin-resistant *Staphylococcus aureus* (MRSA) and methicillin-resistant coagulase-negative staphylococci have increased steadily over the past decade (Fig. 2a–b). Perhaps in response to the increasing numbers of infections with MRSA, which requires treatment with vancomycin, the percentage of enterococcal isolates resistant to vancomycin has dramatically risen from 0.5% in 1989 to 22% in 1997 among ICU patients with nosocomial infection reported to the NNIS (Fig. 2c).

Other common antimicrobial-resistant pathogens encountered among ICU patients include *Pseudomonas aeruginosa* resistant to imipenem and *P. aeruginosa*, *Klebsiella pneumoniae*, or *Enterobacter* spp. resistant to third-generation cephalosporins such as cefotaxime, ceftriaxone, or ceftazi-

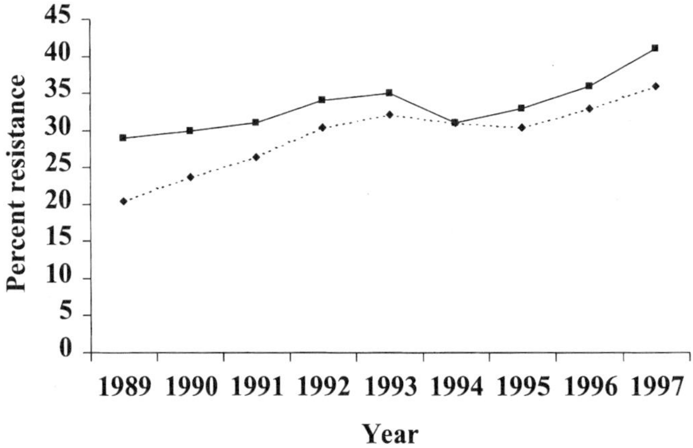

(a)

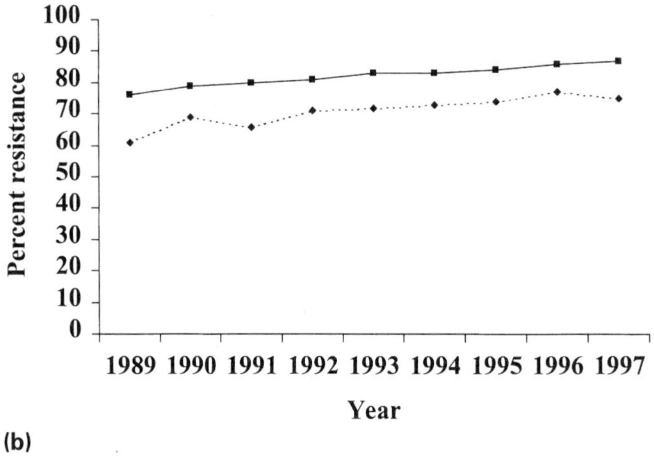

(b)

Figure 2 Proportion of isolates associated with a nosocomial infection among intensive care unit (*solid line*) or nonintensive care unit (*dotted line*) patients which were (a) *Staphylococcus aureus* resistant to methicillin, (b) coagulase-negative staphylococci resistant to methicillin, (c) enterococci resistant to vancomycin, (d) *Klebsiella pneumoniae* resistant to third-generation cephalosporins (i.e., ceftriaxone, cefotaxime, or ceftazidime), (3) *Enterobacter* spp. resistant to third-generation cephalosporins, (f) *P. aeruginosa* resistant to third-generation cephalosporins, (g) *P. aeruginosa* resistant to imipenem, and (h) *P. aeruginosa* resistant to ofloxacin or ciprofloxacin. (From National Nosocomial Infection Surveillance System, January 1989–June 1998.)

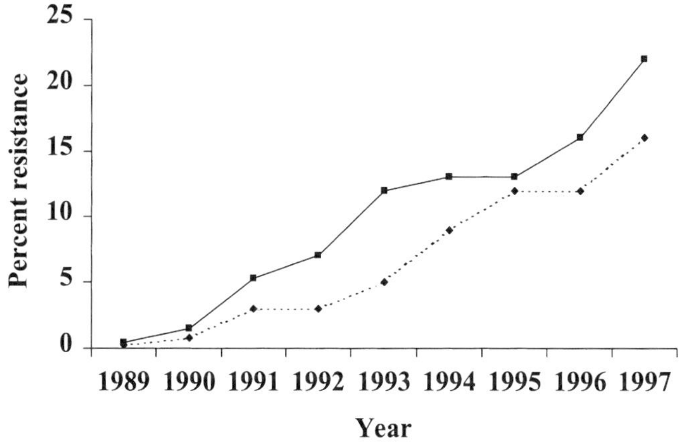
Percent resistance
25
20
15
10
5
0
1989 1990 1991 1992 1993 1994 1995 1996 1997
Year

(c)

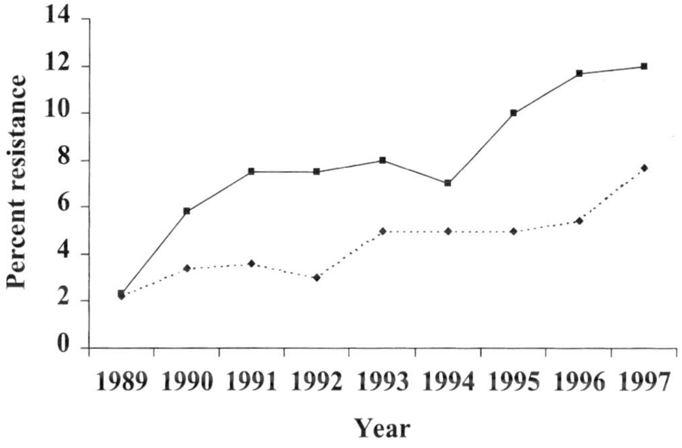
Percent resistance
14
12
10
8
6
4
2
0
1989 1990 1991 1992 1993 1994 1995 1996 1997
Year

(d)

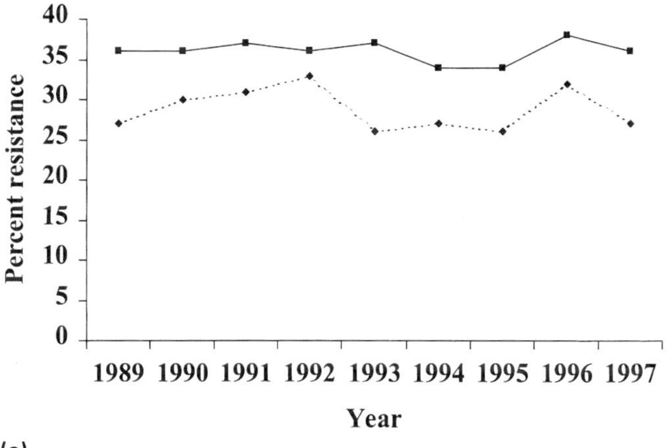
Percent resistance
40
35
30
25
20
15
10
5
0
1989 1990 1991 1992 1993 1994 1995 1996 1997
Year

(e)

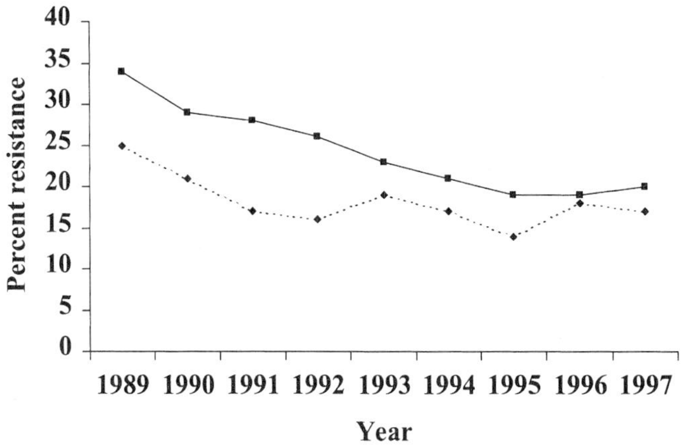

(f)

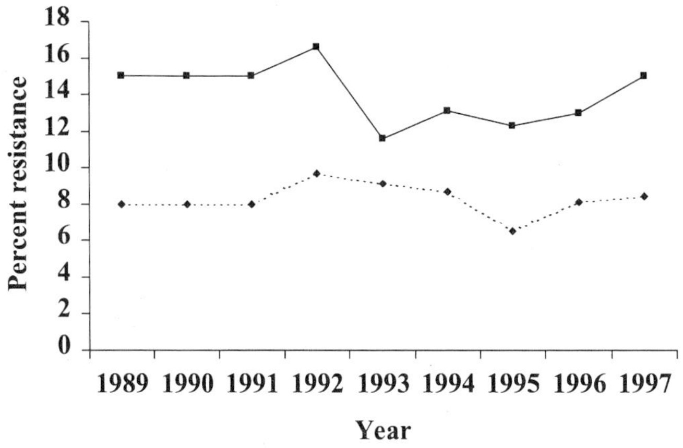

(g)

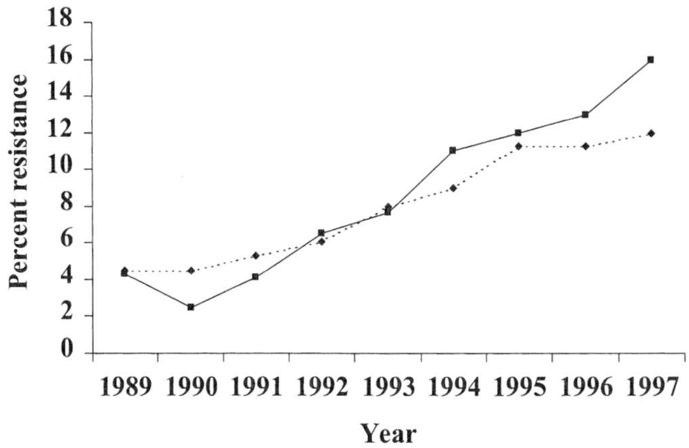

(h)

Figure 2 Continued

dime. Examination of data from NNIS hospitals shows that rates of resistance among these pathogens again appear higher among isolates from ICU patients compared to non-ICU patients (Fig. 2d–g) (Table 1). However, the rates of resistance have been relatively stable over the past decade. Finally, fluoroquinolone resistance (i.e., resistance of ofloxacin or ciprofloxacin) among *P. aeruginosa* isolates reported to the NNIS shows a rapid increase over the past decade (Fig. 2h). However, in contrast to all the other pathogens discussed so far, quinolone-resistant *P. aeruginosa* is not more prevalent among ICU patients compared to non-ICU patients (see Table 1). Probably many reasons account for this. Two contributing factors may be the large amounts of quinolones used by patients outside the ICU, or the development of fluoroquinolone resistance among *P. aeruginosa* unrelated to the ICU setting.

These data illustrate the importance of the ICU in examining rates of antimicrobial resistance. In general, for almost all organisms evaluated, the rates of resistance were significantly higher in patients cared for in the ICU than in non-ICU patients (see Table 1). This may be for a combination of factors and includes the parallel increased use of most antimicrobial agents

TABLE 1 Relative Risk of Isolating the Specific Antimicrobial-Resistant Pathogen from a Nosocomial Infection Occurring in an Intensive Care Unit Patient Compared to Other Patients, National Nosocomial Infection Surveillance (NNIS) System, January 1989 to July 1998

Pathogen	Antimicrobial resistance	Relative risk among ICU patients (95% CI)[a]
Coagulase-negative staphylococci	Methicillin	1.22 (1.21–1.24)
Staphylococcus aureus	Methicillin	1.09 (1.07–1.16)
Enterococci spp.	Vancomycin	1.16 (1.13–1.20)
Enterobacter spp.	Third-generation cephalosporins	1.11 (1.09–1.13)
Klebsiella pneumoniae	Third-generation cephalosporins	1.24 (1.20–1.30)
Pseudomonas aeruginosa	Imipenem	1.16 (1.13–1.21)
Pseudomonas aeruginosa	Third-generation cephalosporins	1.13 (1.11–1.16)
Pseudomonas aeruginosa	Ciprofloxacin/ ofloxacin	1.03 (1.00–1.05)

[a]Data from NNIS system, common relative risk and 95% confidence interval, by Mantel-Haenszel Statistic, controlling for year of infection.

in ICUs compared with other areas of the hospital (78). This has broad implications for interpreting and reporting antimicrobial resistance rates.

2.5 Aggregating Surveillance Data in the Hospital

A workshop sponsored by CDC and the National Foundation for Infectious Diseases made recommendations that hospitals monitor antimicrobial resistance and antimicrobial use in an attempt to reduce the emergence and spread of antimicrobial-resistant pathogens (78). Such monitoring also can aid the infection-control team in determining how to focus its efforts in reducing the emergence and spread of antimicrobial-resistant pathogens (73). Most importantly, controlling antimicrobial resistance (and use) is a multifaceted problem requiring a multidisciplinary approach (79).

2.5.1 Antimicrobial Resistance Data

One method to optimize use includes providing feedback data to ICU clinicians. Such data would help clinicians make wise empiric therapy choices and provide direction in altering antimicrobial choice in efforts to reduce specific problems with resistance. One study demonstrated that rates of antimicrobial resistance may differ between specific types of ICUs and that feedback to clinicians on ICU-specific rates of resistance leads to antimicrobial selection changes and subsequent reduction in the ICU-specific resistance rates (80).

Providing ICU-specific feedback has been utilized by CDC's Project Intensive Care Antimicrobial Resistance Epidemiology (ICARE) (78). Data from this multicenter study are reported for each hospital's ICU separately and from the non-ICU areas combined. Analyses have repeatedly shown that rates of resistance are highest in the ICU areas compared to other areas. Accordingly, CDC provides these data aggregated for each individual ICU back to the hospital's infection-control staff. The rates of resistance may be similar among all ICUs; in such cases, it may be easiest for infection-control staff to report to clinicians one resistance rate for all ICUs, one for all non-ICU areas, and one for all outpatient areas. These data will allow clinicians to target their empiric therapy for the patient population for whom they are providing care. In the scenario where specific ICUs may have very different rates of resistance, such as coronary care units versus general surgical units, several antibiograms may be generated. However, for most hospitals, such an approach will not be feasible. For example, if only nine isolates of enterococci are sent from patients in a coronary care unit over a 12-month period, calculating a resistance rate on nine isolates may not provide an accurate measure of the true frequency of resistance in that unit.

In project ICARE, hospitals are able to use the aggregated data from over 40 hospitals as benchmarks for their institution. However, only comparisons between similar hospital areas may be valid. Many hospitals participating in project ICARE have reported using these comparative data to make focused quality improvement efforts. Hospitals pharmacy and therapeutics committees may also use ICU-specific or hospitalwide susceptibility data to make changes to clinical practice guidelines based on local susceptibility patterns. Although this process may take a lot of effort, it often is the most rewarding, because significant input from local opinion leaders has a greater impact than supplying clinicians with resistance information without guidance on interpreting it.

Currently, there is no consensus on how to best provide feedback of antimicrobial resistance patterns to clinicians. NCCLS is currently developing guidelines for clinical microbiology laboratories to use when aggregating cumulative susceptibility results.

2.5.2 Antimicrobial Use Surveillance

Along with feedback of resistance rates to help ICU-based clinicians optimize antimicrobial therapy and reduce resistance rates, feedback of antimicrobial use data may be necessary as well. Infection-control, pharmacy and therapeutics committees, and critical care staff may gain insight into current prescribing practices by aggregating data on antimicrobial use by hospital area.

Unfortunately, these data may be very difficult to obtain. Currently, most pharmacy systems can report amounts of antimicrobial agents dispensed, but rarely can they track what is actually consumed. Therefore, these data must be examined carefully for accuracy. Regardless, obtaining such data can be very rewarding.

Hospitals may use comparative data of antimicrobial use, such as those provided by project ICARE, to determine if specific ICUs or the entire hospital is overusing antimicrobial agents. Caution must be used in making any comparisons of antimicrobial use data, as antimicrobial use will depend on the types of patients cared for in the ICU. Data from project ICARE illustrate that different ICU types use different amounts of specific antimicrobial drugs (78). Therefore, CDC reports use rates by specific type of ICU back to ICARE hospitals. Accounting for the different type of ICU, when an ICU is using a specific antimicrobial at a rate beyond the 90th percentile, and then evaluating how and why that usage is so extreme, may help optimize use.

2.5.3 Patterns of Antimicrobial Use

After an institution determines that it is overusing antimicrobial agents, a detailed examination of patterns is needed. Antimicrobial use can be divided into three categories: empiric therapy, definitive therapy, and prophylaxis. Each component may need to be addressed by different means to achieve any benefit. Surprisingly, only about 30% of all antimicrobial drugs in hospitals are used for definitive therapy in which the susceptibility patterns for the infection-associated pathogen are known. The problem behind a specific ICU's excessive use of an antimicrobial agent may result from misuse within any or all of the above three categories. Most data on reducing inappropriate use of antimicrobial agents have involved vancomycin (81,82). These studies documented that 30–80% of empiric and 20–25% of definitive vancomycin therapy is inappropriate. Implementing antimicrobial control programs tailored to the areas of most inappropriate use and/or greatest amount of use may be needed to optimize use. For instance, one study demonstrated that most vancomycin use occurred during the first 3 days of therapy, and that focusing efforts on improving initial empiric therapy reduced inappropriate use greatly (81).

2.6 Interventions

Using surveillance data will allow infection control teams to implement reasonable interventions in hopes of reducing rates of antimicrobial resistant pathogens that cause hospital-acquired infections. Some studies have documented decreased rates of colonization or infection with antimicrobial-resistant bacteria after interventions (83). These interventions usually include some restriction policy on specific antimicrobial agents with or without other mechanisms, such as automatic stop orders after 72 hr of empiric use (74). A recent study by White et al. demonstrated that preapproval for selected parenteral agents reduced rates of antimicrobial-resistant pathogens without compromising patient outcomes, with the greatest effect occurring within the ICUs (84). In general, the efficacy of specific aspects of programs to improve antimicrobial use remains unclear, and their effectiveness in reducing antimicrobial resistance has been difficult to assess (83). In addition, implementation of either criteria-based guidelines (i.e., appropriate vs inappropriate use) or diagnosis-based guidelines (e.g., community-acquired pneumonia) has been promulgated by professional societies, but their effectiveness at optimizing antimicrobial use has not yet been determined but should be promising.

2.7 On-line Source for Further Information

The Alliance for the Prudent Use of Antibiotics (APUA) is one of many groups working nationally and internationally to promote appropriate antibiotic use. This organization, founded in 1981, has taken a leadership role in the effort to combat antimicrobial resistance. It has materials and programs directed at clinicians and consumers that are available through its website (http://www.healthsci.tufts.edu/apua/apua.html). Futher information regarding antimicrobial resistance programs, materials, and guidelines are available at CDC's webiste (http://cdc.gov/antibioticresistance).

2.8 Conclusions

Several considerations must be kept in mind when evaluating antimicrobial resistance in the hospital setting. Interpretation of the magnitude of the problem is best made with knowledge of a hospital's (or the individual ICU's) pattern of antimicrobial use. Dramatic differences in antimicrobial resistance exist within individual hospitals and may depend on both antimicrobial use and infection control practices. Only by improving surveillance of antimicrobial resistance and antimicrobial use can hospitals begin to make rational decisions about allocating scarce resources toward improving patient care by reducing rates of infections with antimicrobial-resistant bacteria. No strategy for controlling resistance or optimizing antimicrobial use will be successful unless the entire healthcare delivery system views this problem as vital. A multidisciplinary, systems-oriented approach involving the hospital leadership will be required to succeed at combating the growing problem of antimicrobial resistance in ICUs (79).

REFERENCES

1. Centers for Disease Control and Prevention. Preventing Emerging Infectious Diseases: A Strategy for the 21st Century. Atlanta, Georgia: U.S. Department of Health and Human Services, 1998.
2. Centers for Disease Control and Prevention. Active bacterial core surveillance (ABCs) report, emerging infections program network, streptococcus pneumoniae 1998. http://www.cdc.gov/ncidod/dbmd/abcs/survreports/spneu98.pdf: Centers for Disease Control and Prevention, 1999.
3. Centers for Disease Control and Prevention. Prevention of pneumococcal disease. Recommendations of the Advisory Committee on Immunization Practices (ACIP). MMWR 1997; 46:1–19.
4. Williams W, Hickson M, Kane M, Kendal A, Spika J, Hinman A. Immunization policies and vaccine coverage among adults: the risk for missed opportunities. Ann Intern Med 1988; 108:616–625.
5. Stool S, Field M. The impact of otitis media. Pediatr Infect Dis J 1989; 8:s11–s14.
6. Centers for Disease Control and Prevention. Pneumococcal polysaccharide vaccine usage, United States. MMWR 1984; 33:273–276.
7. Jernigan D, Cetron M, Breiman R. Minimizing the impact of drug-resistant *Streptococcus pneumoniae* (DRSP). A strategy from the DRSP Working Group. JAMA 1996; 275:206–209.
8. Hansman D, Andrews G. A resistant pneumococcus. Lancet 1967; 2:264–265.
9. Caputo G, Applebaum P, Liu H. Infections due to penicillin-resistant pneu-

mococci: clinical, epidemiologic, and microbiologic features. Arch Intern Med 1993; 153:1301–1307.
10. Jacobs M, Koornhof H, Robins-Browne R. Emergence of multiply resistant pneumococci. N Engl J Med 1978; 299:735–740.
11. Applebaum P. World-wide development of antibiotic resistance in pneumococci. Eur J Clin Microbiol 1987; 6:367–377.
12. Cates K, Gerrard J, Giebink G. A penicillin-resistant pneumococcus. J Pediatr 1978; 93:624–626.
13. Thornsberry C, Swenson J, Facklam R. Antimicrobial resistance in pneumococci from the USA: are there trends? Antimicrob Newslett 1988; 5:22.
14. Istre G, Humphreys J, Albrecht K, Thornsberry C, Swenson J, Hopkins R. Chloramphenicol and penicillin resistance in pneumococci from blood and cerebrospinal fluid: a prevalence study in Denver. J Clin Microbiol 1983; 17:472–475.
15. Spika J, Facklam R, Plikaytis B, Oxtoby M, The Pneumococcal Surveillance Working Group. Antimicrobial resistance of *Streptococcus pneumoniae* in the United States, 1979–1987. J Infect Dis 1991; 163:1273–1278.
16. Breiman R, Butler J, Tenover F, Elliot J, Facklam R. Emergence of drug-resistant pneumococcal infections in the United States. JAMA 1994; 271:1831–1835.
17. Duchin J, Breiman R, Diamond A, Lipman H, Block S, Hedrick J, Finger R, Elliott J. High prevalence of multidrug-resistant *Streptococcus pneumoniae* among children in a rural Kentucky community. Pediatr Infect Dis J 1995; 14:745–750.
18. Centers for Disease Control and Prevention. Surveillance for penicillin-nonsusceptible *Streptococcus pneumoniae*—New York City, 1995. MMWR 1997; 46:297–299.
19. Hofmann J, Cetron M, Farley M, Hofmann J, Cetron M, Farley M, Baughman W, Facklam R, Elliott J, Deaver K, Breiman R. The prevalence of drug-resistant *Streptococcus pneumoniae* in Atlanta. N Engl J Med 1995; 333:481–486.
20. Centers for Disease Control and Prevention. Drug-resistant *Streptococcus pneumoniae*—Kentucky and Tennessee, 1993. MMWR 1994; 43:23–31.
21. Butler J, Hofmann J, Cetron M, Elliott J, Facklam R, Breiman R. The continued emergence of drug-resistant *Streptococcus pneumoniae* in the United States: an update from the Centers for Disease Control and Prevention's Pneumococcal Sentinel Surveillance System. J Infect Dis 1996; 174:986–993.
22. Naraqui S, Kirkpatrick G, Kabins S. Relapsing pneumococcal meningitis: isolation of an organism with decreased susceptibility to penicillin G. J Pediatr 1974; 85:671–673.
23. Paredes A, Taber L, Yow M, Clark D, Nathan W. Prolonged pneumococcal meningitis due to an organism with increased resistance to penicillin. Pediatrics 1976; 58:378–381.
24. Jackson M, Shelton S, Nelson J, McCracken G. Relatively penicillin-resistant pneumococcal infections in pediatric patients. Pediatr Infect Dis J 1984; 3:129–132.
25. Gartner J, Michaels R. Meningitis from a pneumococcus moderately resistant to penicillin. JAMA 1979; 241:1707–1709.

26. Vliadrich P, Gudiol F, Linares J, Rufi G, Ariza J, Pallares R. Characteristics and antibiotic therapy of adult meningitis due to penicillin-resistant pneumococci. Am J Med 1988; 84:839–846.
27. Weingarten RD, Markiewicz Z, Gilbert DN. Meningitis due to penicillin-resistant *Streptococcus pneumoniae* in adults. Rev Infect Dis 1990; 12:118–124.
28. Bradley J, Connor J. Ceftriaxone failure in meningitis caused by *Streptococcus pneumoniae* with reduced susceptibility to beta-lactam antibiotics. Pediatr Infect Dis J 1991; 10:871–873.
29. Doern G, Brueggemann A, Holley H, Rauch A. Antimicrobial resistance of *Streptococcus pneumoniae* recovered from outpatients in the United States during the winter months of 1994 to 1995: results of a 30-center national surveillance study. Antimicrob Agents Chemother 1996; 40:1208–1213.
30. Block S, Harrison C, Hedrick J. Penicillin resistant *Streptococcus pneumoniae* in acute otitis media: risk factors, susceptibility patterns and antimicrobial management. Pediatr Infect Dis J 1995; 14:751–759.
31. Jacobs M. Increasing importance of antibiotic-resistant *Streptococcus pneumoniae* in acute otitis media. Pediatr Infect Dis J 1996; 15:940–943.
32. Dagan R, Abramson O, Leibovitz E, Lang R, Goshen S, Greenberg D, Yagupsky P, Leiberman A, Fliss D. Impaired bacteriologic response to oral cephalosporins in acute otitis media caused by pneumococci with intermediate resistance to penicillin. Pediatr Infect Dis J 1996; 15:980–985.
33. Dagan R, Abramson O, Leibovitz E, Greenberg D, Lang R, Goshen S, Yagupsky P, Leiberman A, Fliss D. Bacteriologic response to oral cephalosporins: are established susceptibility breakpoints appropriate in the case of acute otitis media? J Infect Dis 1997; 176:1253–1259.
34. del Castillo F, Baquero-Artigao F, Garcia-Perea A. Influence of recent antibiotic therapy on antimicrobial resistance of *Streptococcus pneumoniae* in children with acute otitis media in Spain. Pediatr Infect Dis J 1998; 17:94–97.
35. Centers for Disease Control and Prevention. Defining the public health impact of drug-resistant *Streptococcus pneumoniae*: report of a working group. MMWR 1996; 45:1–20.
36. Facklam R, Washington JI. Streptococcus and related catalase-negative gram positive cocci. In: Balows A, Hausler WJ, Herrmann KL, Isenberg H, Shadomy H, eds. Manual of Clinical Microbiology. Washington, DC: American Society for Microbiology, 1991:238–257.
37. National Committee on Laboratory Standards, National Committee for Clinical Laboratory Standards. Performance standards for antimicrobial susceptibility testing. Vol. NCCLS document no. M100-s5, 1994.
38. Centers for Disease Control and Prevention. Geographic variation in penicillin resistance in *Streptococcus pneumoniae*—selected sites, United States, 1997. MMWR 1999; 48:656–661.
39. American Academy of Pediatrics Committee on Infectious Diseases. Therapy for children with invasive pneumococcal infections. Pediatrics 1997; 99:289–299.
40. Levine OS, Farley MM, Harrison LH, Lefkowitz L, McGeer A, Schwartz B. Risk factors for invasive pneumococcal disease in children: a population-based case-control study in North America. Pediatrics 1999; 103:E28.
41. Schwartz B, Kolczak M, Whitney C, Kool J, Schuchat A. U.S. counties with

higher rates of antibiotic use have significantly higher proportions of beta lactam and macrolide nonsusceptible *S. pneumoniae* antimicrobial resistance. Abstracts of the 38th Interscience Conference on Antimicrobial Agents and Chemotherapy. San Diego, CA, 1998:76.
42. Dowell SF, Butler JC, Giebink GS, Jacobs MR, Jernigan D, Musher D, Rakowsky A, Schwartz B. Acute otitis media: management and surveillance in an era of pneumococcal resistance—a report from the Drug-resistant *Streptococcus pneumoniae* Therapeutic Working Group. Pediatr Infect Dis J 1999; 18:1–9.
43. Heffelfinger J, Dowell S, Jorgensen J, et al. Management of community-acquired pneumonia in the era of pneumococcal resistance: a report from the Drug-resistant *Streptococcus pneumoniae* Therapeutic Working Group (DRSP-TWG). Arch Intern Med 2000; 160:1399–1408.
44. Centers for Disease Control and Prevention. Influenza and pneumococcal vaccination levels among adults aged 65 years—United States. MMWR 1997; 46:913–919.
45. Centers for Disease Control and Prevention. Influenza and pneumococcal vaccination levels among adults aged greater than or equal to 65 years—United States. MMWR 1998; 47:797–802.
46. Breiman R, Spika J, Navarro V, Darden P, Darby C. Pneumococcal bacteremia in Charleston County, South Carolina: a decade later. Arch Intern Med 1990; 150:1401–1405.
47. Istre G, Tarpay M, Anderson M, Pryor A, Welch D. Pneumococcus Study Group. Invasive disease due to *Streptococcus pneumoniae* in an area with a high rate of relative penicillin resistance. J Infect Dis 1987; 156:732–735.
48. Butler J, Breiman R, Campbell J, Lipman H, Broome C, Facklam R. Pneumococcal polysaccharide vaccine efficacy: an evaluation of current recommendations. JAMA 1993; 270:1826–1831.
49. Shapiro E, Clemens J. A controlled evaluation of the protective efficacy of pneumococcal vaccine for patients at high risk of serious pneumococcal infections. Ann Intern Med 1984; 101:325–330.
50. Sims R, Steinmann W, McConville J, King L, Zwick W, Schwartz J. The clinical effectiveness of pneumococcal vaccine in the elderly. Ann Intern Med 1988; 108:653–657.
51. Shapiro E, Berg A, Austrian R. The protective efficacy of polyvalent pneumococcal polysaccharide vaccine. N Engl J Med 1991; 325:1453–1460.
52. Farr BM, Johnston BL, Cobb DK, Fisch MJ, Germanson TP, Adal KA, Anglim AM. Preventing pneumococcal bacteremia in patients at risk: results of a matched case-control study. Arch Intern Med 1995; 155:2336–2340.
53. Fine MJ, Smith MA, Carson CA, Meffe F, Sankey SS, Weissfeld LA, Detsky AS, Kapoor WN. Efficacy of pneumococcal vaccination in adults: a meta-analysis of randomized controlled trials. Arch Intern Med 1994; 85(Suppl):S14–S30.
54. Black S, Shinefield H, Ray P, Lewis E, Fireman B. Efficacy of heptavalent conjugate pneumococcal vaccine (Wyeth Lederle) in 37,000 infants and children: impact on pneumonia, otitis media, and an update on invasive disease—results of the northern California Kaiser Permanente Efficacy Trial. 39th Interscience Conference on Antimicrobial Agents and Chemotherapy. San Francisco, ASM, 1999:379.

55. Butler J. Epidemiology of pneumococcal serotypes and conjugate vaccine formulations. Microbial Drug Resistance 1997; 3:125–129.
56. Dagan R, Melamed M, Muallem L, Piglansky D, Greenberg D, Abramson O, Mendelman P, Bohidar N, Yagupsky P. Reduction of nasopharyngeal carriage of pneumococci during the second year of life by a heptavalent conjugate pneumococcal vaccine. J Infect Dis 1996; 174:1271–1278.
57. Dowell S, Schwartz B. Resistant pneumococci: protecting patients through judicious use of antibiotics. Am Fam Physician 1997; 55:1647–1654.
58. McCaig L, Hughes J. Trends in antimicrobial drug prescribing among office-based physicians in the United States. JAMA 1996; 273:214–219.
59. Gonzales R, Steiner J, Sande M. Antibiotic prescribing for adults with colds, upper respiratory tract infections, and bronchitis by ambulatory care physicians. JAMA 1997; 278:901–904.
60. Nyquist A-C, Gonzales R, Steiner J, Sande M. Antibiotic prescribing for children with colds, upper respiratory tract infections, and bronchitis. JAMA 1998; 279:875–877.
61. Barden LS, Dowell SF, Schwartz B, Lackey C. Current attitudes regarding use of antimicrobial agents: results from physicians' and parents' focus group discussions. Clin Pediatr 1998; 37:665–672.
62. Bauchner H, Pelton S, Klein J. Parents, physicians, and antibiotic use. Pediatrics 1999; 103:395–401.
63. Mangione-Smith R, McGlynn E, Elliott M, Krogstad P, Brook R. The relationship between perceived parental expectations and pediatrician antimicrobial prescribing behavior. Pediatrics 1999; 103:711–718.
64. Hamm R, Hicks R, Bemben D. Antibiotics and respiratory infections: are patients more satisfied when expectations are met? J Fam Pract 1996; 43:56–62.
65. Todd J. Bacteriology and clinical relevance of nasopharyngeal and oropharyngeal cultures. Pediatr Infect Dis J 1984; 3:159–163.
66. Dowell SF. Principles of judicious use of antimicrobial agents for pediatric upper respiratory tract infections. Pediatrics 1998; 101:S163–S184.
67. Centers for Disease Control and Prevention. Antibiotic resistance: a new threat to you and your family's health. http://www.cdc.gov/ncidod/dbmd/antibioticresistance: Centers for Disease Control and Prevention, 1999.
68. Belongia E, Sullivan B, Chyou P, Madagame E, Reed K, Schwartz B. A community intervention trial to decrease unnecessary antibiotic use and reduce colonization with penicillin-nonsusceptible *Streptococcus pneumoniae* in children (abst). 39th Interscience Conference on Antimicrobial Agents and Chemotherapy. San Francisco: American Society for Microbiology, 1999.
69. Gonzales R, Steiner J, Lum A, Barrett P. Decreasing antibiotic use in ambulatory practice: impact of a multidimensional intervention on the treatment of uncomplicated acute bronchitis in adults. JAMA 1999; 281:1512–1519.
70. Petersen K, Hennessy T, Parkinson A, Bruden D, Chiou L, Kokesh J, Schwartz B, Butler J. Provider and community education decreases antimicrobial use and carriage of penicillin-resistant *Streptococcus pneumoniae* (PRSP) in rural Alaska communities (abst). Abstracts of the 1999 IDSA Conference. Philadelphia: Infectious Disease Society of America, 1999.
71. Centers for Disease Control and Prevention. Guidelines for handwashing and

hospital environmental control, 1985. MMWR 1988; 36.
72. Doebbeling BN, Stanley GL, Sheetz CT, Pfaller MA, Houston AK, Annis L, Li N, Wenzel RP. Comparative efficacy of alternative hand-washing agents in reducing nosocomial infections in intensive care units. N Engl J Med 1992; 327: 88–93.
73. McGowan JE, Jr. Antimicrobial resistance in hospital organisms and its relation to antibiotic use. Rev Infect Dis 1983; 5:1048.
74. Shlaes DM, Gerding DN, John JF, Craig WA, Bornstein DL, Duncan RA, Eckman MR, Farrer WE, Greene WH, Lorian V, Levy S, McGowan JEJ, Paul S, Ruskin J, Tenover FC, Watanakunakorn C. Society for Healthcare Epidemiology of American and Infectious Diseases Society of America Joint Committee on the Prevention of Antimicrobial Resistance: guidelines for the preventions of antimicrobial resistance in hospitals. Infect Control Hosp Epidemiol 1997; 18:275–291.
75. Gold HS, Moellering RC Jr. Antimicrobial-drug resistance. N Engl J Med 1996; 335:1445–1453.
76. Gerding DN, Larson TA. Resistance surveillance programs and the incidence of gram-negative bacillary resistance to amikacin from 1967–1985. Am J Med 1986; 80(6B):22–28.
77. Ballow CH, Schentag JJ. Trends in antibiotic utilization and bacterial resistance: Report of the National Nosocomial Resistance Surveillance Group. Diagn Microbiol Infect Dis 1992; 15:37S–42S.
78. Hospital Infections Program, National Center for Infectious Diseases, Centers for Disease Control and Prevention. Intensive Care Antimicrobial Resistance Epidemiology (ICARE) Surveillance Report. Data Summary from January 1996 through December 1997. A report from the National Nosocomial Infections Surveillance (NNIS) System. Am J Infect Control 1999; 27:279–284.
79. Goldmann DA, Weinstein RA, Wenzel RP, Tablan OC, Duma RJ, Gaynes RP, Schlosser J, Martone WJ. Strategies to prevent and control the emergence and spread of antimicrobial-resistant microorganisms in hospitals. JAMA 1996; 275:234–240.
80. Jarvis WR. Preventing the emergence of multidrug-resistant microorganisms through antimicrobial use controls: the complexity of the problem. Infect Control Hosp Epidemiol 1996; 17:490–495.
81. Singer MV, Haft R, Barlam T, Aronson M, Shafer A, Sands KE. Vancomycin control measures at a tertiary-care hospital: impact of interventions on volume and patterns of use. Infect Control Hosp Epidemiol 1998; 19:248–253.
82. Evans ME, Kortas KJ. Vancomycin use in a university medical center: comparison with Hospital Infection Control Practices Advisory Committee Guidelines. Infect Control Hosp Epidemiol 1996; 17:356–359.
83. McGowan JE Jr. Do intensive hospital antibiotic control programs prevent the spread of antibiotic resistance? Infect Control Hosp Epidemiol 1994; 15: 478–483.
84. White ACJ, Atmar RL, Wilson J, Cate TR, Stager CE, Greenberg SB. Effects of requiring prior authorization for selected antimicrobials: expenditures, susceptibilities, and clinical outcomes. Clin Infect Dis 1997; 25:230–239.

18

Approaches to New Antimicrobial Targets in the Age of Genomics

Philip J. Youngman
Millennium Pharmaceuticals, Cambridge, Massachusetts

The traditional targets exploited thus far by the pharmaceutical industry for discovery of new antibiotics have largely been limited to enzymes or structural proteins involved in the well-characterized major essential pathways of macromolecule biosynthesis, including protein synthesis, production or assembly of cell wall peptidoglycan, and DNA synthesis. The total number of specific molecular targets in these pathways acted upon by therapeutically relevant antibiotics in current use represents a small fraction of the hundreds of essential gene products conserved among eubacteria that might in principle be developed as screening targets. It is anticipated that screens for inhibitors of novel essential gene products with functions mechanistically different from those of traditional targets will yield inhibitory compounds with novel chemical structures that will not be substrates for known cellular mechanisms of antibiotic resistance. Genes essential for infectivity or virulence of pathogenic bacteria, but not strictly essential for bacterial viability in vitro, represent another kind of novel target of potential utility for discovery of new antibiotics unrelated chem-

Current affiliation: Elitra Pharmaceuticals, San Diego, California.

ically to drugs currently compromised by the emergence of antibiotic resistance. This chapter is devoted primarily to a review of approaches for using genomics information and technologies to identify potential new essential-gene targets or infectivity targets for drug-discovery screening, but it will also consider ways in which new information and technologies can enhance paradigms of discovery other than target-based screening.

1 INTRODUCTION

It is sobering to consider how unsuccessful the pharmaceutical industry has been in recent decades in its efforts to develop new antibiotics that act through novel mechanisms; and it is interesting to consider the development history of the most recent success story—that of the oxazolidinones, a compound class first described in 1987 (1). The oxazolidinones were discovered several years prior to 1987 at DuPont in a simple cell-based screen for synthetic compounds that inhibited bacterial growth. Development of the initial compound series was discontinued by DuPont when toxicity was observed in phase I trials. An incomplete understanding of the mechanism of action (MOA) of the oxazolidinones was probably a factor in the decision by DuPont not to pursue more aggressively the synthesis of analogues that might differentiate toxicity from potency. Although it had been demonstrated that oxazolidinones block some early step in the protein synthesis (2), the peptidyl transferase reaction is not inhibited and the precise MOA still remains obscure (3). In the target-based discovery paradigm that currently dominates the drug development process in the pharmaceutical industry, identification of the specific molecular target for a lead compound is usually a prerequisite for commitment of lead-improvement chemistry resources. Nevertheless, development of the oxazolidinones was subsequently pursued successfully by Pharmacia and Upjohn (4), culminating with linezolid, which recently completed phase III trials.

No antibacterial compound series derived directly from target-based screening has yet advanced to clinical trials. Nevertheless, target-based screening remains the preferred strategy for identifying antibacterial lead compounds. The failure of this discovery paradigm yet to produce a clinical development candidate is generally attributed to an insufficiency of screening targets. Because of its potential for bringing many new targets into play, the advent of genomics information and technology has raised hopes that the overall productivity of the drug-discovery process will be significantly enhanced. This chapter will therefore focus primarily on how bacterial genome sequence information can enhance the productivity of target-based screening. As a secondary goal, it will consider alternatives to the target-based discovery paradigm and alternative ways in which the application of genomics technologies might aid the search for novel antibiotics. In particular, it will explore ways in which genomics technologies might enhance the effectiveness of cell-based screens, perhaps enabling a productive return to the use of antimicrobial activity as the primary criterion for identifying lead compounds, the only antibiotic discovery paradigm with a track record of success.

2 THE PROMISE OF GENOMICS

2.1 Status of Genome Sequencing Efforts

Since publication of the first complete bacterial genome sequence (*Haemophilus influenzae*) in 1995 (5), additional complete or partial genomic sequences have become available at an ever-increasing rate. A web site maintained by the Institute for Genomic Research (TIGR) lists 27 completed and published genomic sequences for prokaryotic organisms as of April, 2000, and 118 additional bacterial sequencing projects in progress (http://www.tigr.org/tdb/mdb/mdb.html). Many additional bacterial genome sequences are available to the pharmaceutical industry by commercial subscription from companies such as Incyte and Genome Therapeutics. Complete genome sequences are already available for several eukaryotic microbial organisms, and many additional sequencing efforts are underway. The first essentially complete sequence of an animal genome was reported in late 1998 (6), and the sequencing of several vertebrate and plant genomes is nearing completion. Perhaps most impressive of all, the International Human Genome Project announced in February 1999 that a working draft of the complete human genome sequence would be available by mid-2000 (see http://www.ornl.gov/ghmis/home.html), which is more than 2 years ahead of schedule. We are rapidly approaching an era in which the genetic content of both bacterial and eukaryotic organisms is explicitly defined and in which the remaining goal will be to understand the physiological role of every gene product in those organisms.

Of equal importance to the rapid accumulation of genome sequence information is the rapid development of technologies that allow the monitoring of genomewide gene expression patterns at the level of transcription, translation, and posttranslational modifications (7,8) and the maturing of computational methods (bioinformatics) both for the effective analysis of the massive quantities of information generated in expression profiling experiments and for inference of structure and function from raw gene sequence data (9,10). In its current usage, the term *genomics* has come to encompass both the information and the technology. The question to be considered here is: How will the information and the technology be converted into targets for the discovery of novel antibiotics?

2.2 Sequence + Informatics = Targets?

Since the object of antibacterial therapy is to kill bacteria, or at least to prevent their proliferation, any bacterial gene product essential for viability or growth could in principle serve as a target for lead discovery. The contribution expected of genomics information and technologies is to define comprehensively the full complement of essential bacterial genes and to provide a means for ranking them in priority of attractiveness as lead-discovery targets. The ideal product profile for an antibacterial therapeutic is broad-spectrum and negligible toxicity. This implies that the ideal target would be a gene product conserved broadly among bacteria but one which lacks a highly related counterpart in eukaryotic cells. How many such targets exist? How many were already discovered through pregenomics molecular genetics analysis? What is the most effective way to identify those remaining to be discovered?

Although most efforts to bring genomics to bear on these questions are not conducted in the public domain, a good example of a genomics-based approach to antibacterial target discovery is offered by the recent work of Arigoni and colleagues (11). These investigators made the reasonable assumption that the genome of *Mycoplasma genitalium*, which is the smallest of any bacterial species known, would be highly enriched for essential genes. They reasoned further that because *M. genitalium* and the genetically tractable gram-negative enteric bacterium *Escherichia coli* are distant relatives phylogenetically (12), genes conserved in sequence between the two would be broadly conserved in the larger bacterial world. Because the genomic sequences of *M. genitalium* and *E. coli* are available in public databases (13,14), it was straightforward to apply standard computer algorithms to determine how many hypothetical *E. coli* genes of unknown function have related counterparts (presumptive orthologues) in *M. genitalium*. Excluding members of the ABC transporter superfamily, this process generated a list of 26 *E. coli* genes. To determine which of these genes actually encoded proteins essential for viability or growth, in-frame deletions (gene "knockouts") were obtained in each by the method of Link et al. (15). The surprising result of this analysis was that only 6 of the 26 proved to be essential. This was considered to be surprising, because about 12–15% of all genes in a bacterium like *E. coli* are essential regardless of their conservation in other species (16). Thus, demanding both broad conservation and the existence of a presumptive orthologue in the minimal genome of a mycoplasma resulted in a relatively modest enrichment for potential essential gene targets. Moreover, of the six novel essential *E. coli* genes identified in this manner, only five proved to be essential in the gram-positive species *Bacillus subtilis*. Still others among these five target candidates were found by other investigators not to be essential in another gram-positive species, *Streptococcus pneumoniae* (C. Fritz and P. Youngman, unpublished). One lesson to be learned from the study described above is that previously unidentified essential genes broadly conserved in bacteria are relatively rare. Another is that conservation across phylogenetic distance is not in itself a reliable predictor of whether a particular gene will prove to be essential for viability or growth.

2.3 Functional Genomics Approaches for the Discovery of New Essential Gene Targets

The realization that sequence information alone is not sufficient to identify targets of potential value for the discovery of novel antibacterial compounds has led to the development of several creative approaches for systematic functional assessment of bacterial genes.

2.3.1 Conditional Lethal Mutations

Genes whose products are essential for viability will of course result in lethality when they suffer a mutation that results in loss of function. Thus, a traditional means for identifying essential genes is the isolation of mutants in which a mutation results in loss of viability only under certain conditions, such as elevated temperature. The exhaustive isolation and characterization of such mutants will result in the identification of a large proportion of all potential essential gene targets. Moreover, as championed most notably by researchers at Microcide Pharmaceuticals (17), the mutant collection itself offers the possibility of a novel strategy for identifying antibacterial compounds. The basis for the Microcide strategy is the assumption that a conditional-lethal mutation in a particular gene will reduce the function and/or concentration of the mutant gene product at a semi-permissive temperature and will therefore increase sensitivity of the mutant strain to compounds that specifically inhibit function of the mutant protein. This could, in principle, facilitate the discovery of antibacterial compounds in several ways. First, it should detect inhibitory compounds at a concentration below that normally required for inhibition of growth. More importantly, this approach could distinguish target-specific mechanisms of growth inhibition from nonspecific mechanisms and might provide a way to identify rapidly the molecular targets of orphan antibacterials (i.e., synthetic compounds or natural products with known antimicrobial activity but whose targets or mechanisms of action remained unelucidated).

It should be noted, however, that this approach is not without pitfalls or unproven aspects. Compounds that inhibit function of a mutant protein under destabilizing conditions might not in many cases act as effective inhibitors of the native structure. Hypersensitivity of a mutant to a particular inhibitor could also occur through indirect mechanisms. Moreover, isolation and characterization of conditional lethal mutants is a laborious method for identification of essential genes.

2.3.2 Genetic Footprinting

An extremely elegant solution to the problem of essential gene identification was outlined in the work of Smith and colleagues (18). This approach, referred to as genetic footprinting, in essence allows all genes in a sequenced microbial organism to be surveyed simultaneously to determine whether disruption by transposon insertion can be tolerated. Failure of a gene to tolerate such disruption can then be taken as evidence that the gene is essential for viability, with the caveat that polar effects on expression of downstream genes in bacterial operons must be excluded.

Implementation of genetic footprinting in a particular organism of interest requires only the existence of a transposable element capable of relatively random and tightly regulated transposition. In the original description of the method, the yeast *Saccharomyces cerevisiae* was used as the model system and the transposable element was a modified form of the retrotransposon Ty*1* in which expression of transposition functions was placed under control of the galactose-inducible GAL1 promoter. After transposition was induced in the presence of galactose, cells were transferred to fresh culture broth lacking adenine, a nutritional supplement that would be required by certain auxotrophic mutants, in particular any mutant having suffered an insertion of a Ty*1* element within the ADE2 gene. In other words, under these conditions, it was expected that insertions of Ty*1* into ADE2 could not be tolerated. The actual presence of Ty*1* insertions within ADE2 and their relative abundance was monitored by carrying out polymerase (PCR) reactions primed in one direction by an oligonucleotide that annealed at different known locations within or near ADE2 and primed in the other direction by an oligonucleotide specific for Ty*1* (Fig. 1). The size-distribution of the products of amplification by each primer pair thus revealed the locations of Ty*1* insertions relative to the site of annealing for the ADE2-specific oligonucleotide when the products were fractionated by electrophoresis (Fig. 1). In this way, Ty*1* insertions could be detected at multiple locations within ADE2 at time 0 in the experiment; the time at which the cells were transferred to culture broth lacking adenine. After a period of growth in the new medium, however, the number of insertion sites and the abundance of insertions were dramatically reduced. Similarly, genes whose products are generally essential for viability under any condition would be unable to tolerate insertions at all, and their positions should be revealed as DNA intervals devoid of transposon-

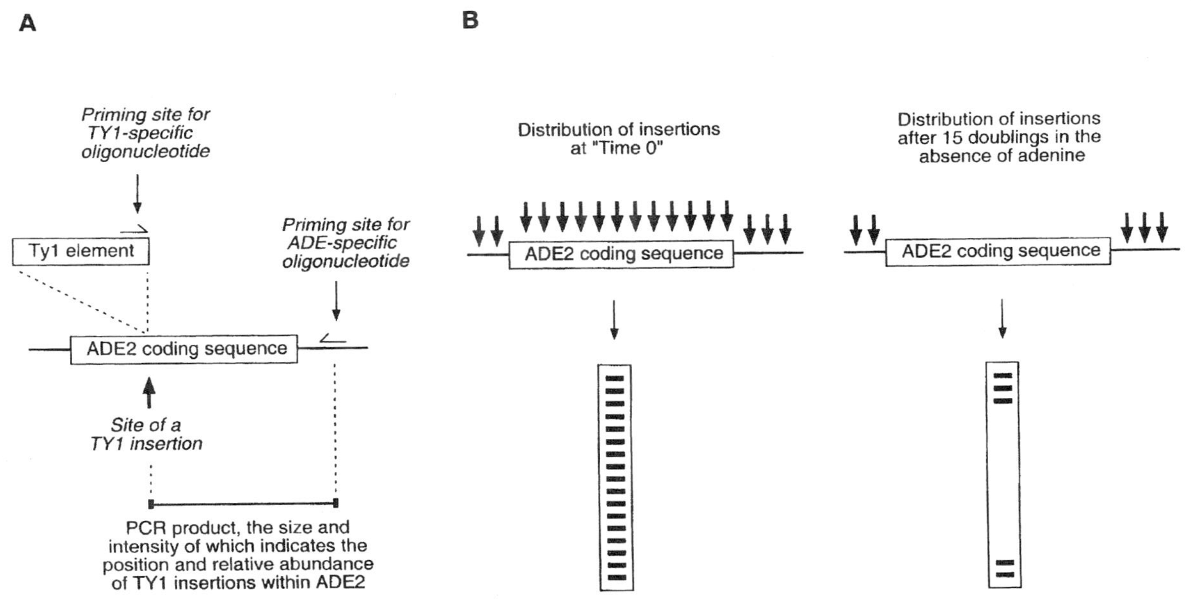

Figure 1 The principle of genetic footprinting (18). As indicated schematically in Panel A, following induction of Ty*1* transposition, PCR amplification is carried out, using one primer specific for the transposon (annealing within and near one end) and another primer complementary to a site within or near the gene of interest. Panel B illustrates an idealized result showing how the positional distribution of Ty*1* insertions within a segment of DNA can be inferred from the size distribution of PCR products generated as described in Panel A. "Time 0" is the time at which induction of transposition was discontinued and cells were transferred to a minimal medium lacking adenine. The absence of insertions in ADE2 after 15 doublings in minimal medium results in a discontinuity, or footprint, in the size distribution of PCR products.

primed PCR products. The method should therefore be broadly applicable to any bacterial or fungal organism in which the appropriate transposon tools are available and should be capable of identifying very rapidly any genomic segment in those organisms likely to contain an essential gene target.

2.3.3 Scanning for Essential Genes by Shuttle Mutagenesis

Bacteria that possess a system of natural "competence" for efficient transformation by exogenous DNA offer several advantages for essential gene discovery. One important advantage arises from the fact that genomic DNA from the naturally competent bacterium can be subjected to insertional mutagenesis externally (either in vitro or in another organism) and then returned to the source organism by transformation. As demonstrated in the recent work of Akerley and colleagues (19) with a method referred to as GAMBIT (genomic analysis and mapping by in vitro transposition), subjecting segments of a bacterial chromosome to external mutagenesis enables a particularly effective form of genetic footprinting.

In the GAMBIT approach, the genomic segment that serves as the mutagenesis target is generated by extended-length PCR (Fig. 2). This has the advantage over using cloned DNA of ensuring equal representation of all chromosomal regions, and also makes it possible to focus the analysis on a specified genomic interval. Mutagenesis is carried out with a mini-transposon derivative of the *mariner* transposable element engineered to carry a drug-resistance gene that provides a selection in bacteria. In principle, any of a number of similarly modified transposable elements might be used in such an application, but the *mariner* system for in vitro transposition offers a particularly attractive combination of biochemical simplicity and low insertion site specificity (20). Randomly mutagenized DNA, now tagged with a selectable drug marker, is then used to transform naturally competent bacteria, resulting in transfer of the inserted element to the chromosome by homologous recombination. Because this recombination event occurs by a marker-replacement mechanism, insertions within genes whose products are essential for growth lead to nonviable recombinants. Thus, insertions are distributed at relatively random locations throughout the mutagenized interval *except* within essential genes. When sites of insertion are visualized by element-primed PCR, essential genes are apparent as gaps in the distribution. The utility of GAMBIT was first demonstrated in *S. pneumoniae* and *H. influenzae*, but the method should be equally applicable to other naturally transformable species. A conceptually similar approach, referred to as genome scanning, was demonstrated by another group (21).

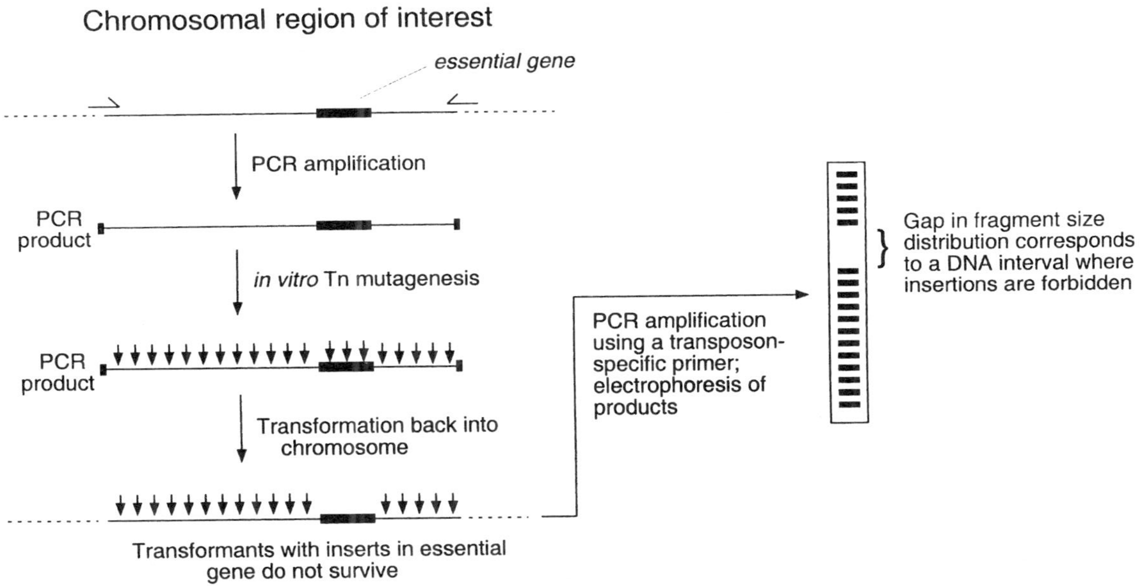

Figure 2 The principle of GAMBIT (19). The DNA segment (~10 kb) to be subjected to GAMBIT is first amplified by extended-length PCR. The PCR product is then subjected to mutagenesis in vitro with a customized derivative of the *mariner* transposable element engineered to carry a drug-resistance marker selectable in the bacterial species of interest (the distribution of insertions is indicated schematically with arrowheads). The *mariner*-marked insertions can then be transferred back into the bacterial chromosome by transformation, selecting for the transposon-associated drug-resistance gene. When an insertion falls within an essential gene (*solid black rectangle*), transformants will not survive. PCR amplification is then carried out on DNA isolated from the transformant population using one primer that anneals to the transposon and one primer that anneals to a fixed location near one end of the segment of interest, and PCR products are fractionated by electrophoresis. DNA intervals where insertions are forbidden (corresponding to essential genes) can be visualized as a gap, or footprint, in the size distribution of PCR products.

2.3.4 Directed Disruption of Genes with Integrational Vectors

An advantage of the genetic footprinting strategies outlined above is that they provide a rapid way to establish where insertional disruptions are forbidden; a disadvantage is that they depend upon assumptions that may not be fully satisfied in practice (e.g., randomness of insertion). Results must also be regarded as preliminary and must be confirmed by further genetic validation. If the genomic sequence of the target organism is known, why not simply disrupt each orf (open reading frame) of potential interest in a systematic way? Moreover, why not create these disruptions in a way that provides information about the regulation of the gene or operon suffering the disruption? A system for accomplishing precisely these objectives was recently described by Vagner and colleagues for *B. subtilis* (22). The system depends upon a family of integrational vectors such as pMUTIN (Fig. 3) into which DNA segments internal to *B. subtilis* orfs of interest are cloned. In the example given in Figure 3, the cloned fragment is an internal segment of an open reading frame, *orfY*, the middle gene in a hypothetical operon consisting of *orfX*, *orfY*, and *orfZ*, coordinately transcribed from the P_{xyz} promoter. In this case, the consequence of integrative recombination of the pMUTIN vector is disruption of *orfY* and placement of the *orfZ* gene under transcriptional control of the IPTG-inducible promoter P_{spac}. Another consequence of integrative recombination is to create a *lacZ* transcriptional fusion under control of P_{xyz}. Integrative recombination will be a forbidden event if *orfY* is an essential gene. If *orfZ*, but not *orfY* is essential, growth will become dependent upon the presence of IPTG.

In practice, however, the pMUTIN system suffers from several shortcomings. First, systematic construction of clones covering all orfs of potential interest (thousands) would be a very slow process. Second, the regulatory elements of such constructions are imperfect (e.g., the P_{spac} promoter is not tightly turned off when uninduced), leading to ambiguous results in certain instances. Moreover, the advantages derived from *lacZ* fusions are largely superseded by genomewide transcription profiling using DNA microarrays (see below). Nevertheless, a systematic effort is currently being undertaken by a consortium of approximately 20 laboratories in Europe, the United States, and Japan to disrupt many of the 4000+ orfs of *B. subtilis* using the pMUTIN technology, and this effort will undoubtedly contribute significantly to the functional genomics information database for this important gram-positive model system.

2.3.5 High-Throughput–Directed Gene Disruption

To those enamored of clever genetic strategies for essential gene discovery, it is sometimes disappointing to recognize that knowing the DNA se-

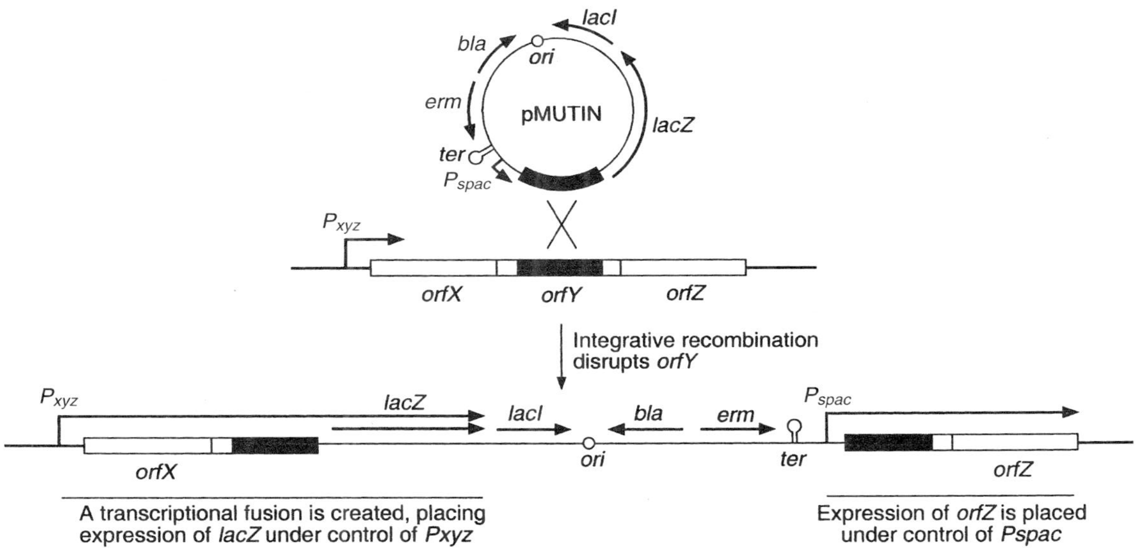

Figure 3 Gene disruption and simultaneous construction of *lacZ* transcriptional fusions using the pMUTIN integrational vector system: *ori*, a plasmid replication origin that functions in *E. coli* but not in *B. subtilis*; *bla*, a drug-resistance gene selective in *E. coli*; *erm*, a drug-resistance gene selectable in *B. subtilis*; P_{spac}, a promoter regulated by IPTG; *ter*, a factor-independent transcription termination sequence; *lacI*, a gene encoding the *lac* repressor of *E. coli*; *lacZ*, a promoterless copy of the *E. coli* β-galactosidase gene (see text for further details). (Adapted from Ref. 22.)

quence of an entire bacterial genome to start with can make the application of high-throughput "brute force" methods far more efficient than the use of elegantly conceived integrational vectors or the use of sophisticated genetic footprinting strategies. Among the best species for construction of high-throughput gene disruptions is *S. pneumoniae*, a gram-positive pathogen with an extremely efficient natural transformation system for uptake and recombinational integration of linear segments of DNA. For example, a segment of DNA consisting of a selectable drug-resistance marker flanked on either side by approximately 1 kb of DNA homologous to sequences present in the *S. pneumoniae* genome would generate several thousand transformants when added to 50 μL of *S. pneumoniae* bacteria in a condition of transformation competence. If the DNA sequence of an *S. pneumoniae* orf of interest is known, a segment of DNA spanning the orf but interrupted by a drug-resistance gene can be obtained by overlap-extension PCR (23) using four unique oligonucleotide primers per orf (Fig. 4). Thus, if the entire DNA sequence of the *S. pneumoniae* genome were known, PCR products suitable for disruption of all orfs of interest might be synthesized over a period of a few days by a research group with access to automated primer-picking software and willing to invest in the synthesis of the required number of oligonucleotide primers.

With the genomic sequence of *S. pneumoniae* in hand, the process of systematic PCR product-mediated gene disruption might begin by using bioinformatics tools to ask the question: Which orfs (genes) of *S. pneumoniae* are highly conserved (at least among gram-positive bacteria), have no very close homologues in eukaryotic genomes, and have not previously been tested for essentiality in other bacteria (through investigation of their presumptive orthologues)? Depending upon the stringency with which such a question was asked, the result might be a list of some 1000 orfs of potential interest. Orf-specific PCR products might then be synthesized as described above and used to transform small samples of *S. pneumoniae* bacteria. In cases where the targeted gene is essential, no transformants would arise. In cases where the targeted gene is nonessential, thousands of transformants would arise. Because the difference in possible outcomes is so extreme, transformation can be carried out and transformed cells plated to assess the results on a very small scale (e.g., agar plugs dispensed into 96-well microtiter plates). Thus, hundreds of disruption trials might be conducted in a single week. It is safe to assume that industry-based research groups with access to the *S. pneumoniae* genomic sequence, or the capability to generate their own sequence, and access to the required bioinformatics tools and primer-synthesis capabilities, have already completed the process of using such approaches to identify all broadly conserved essential gene targets in bacteria.

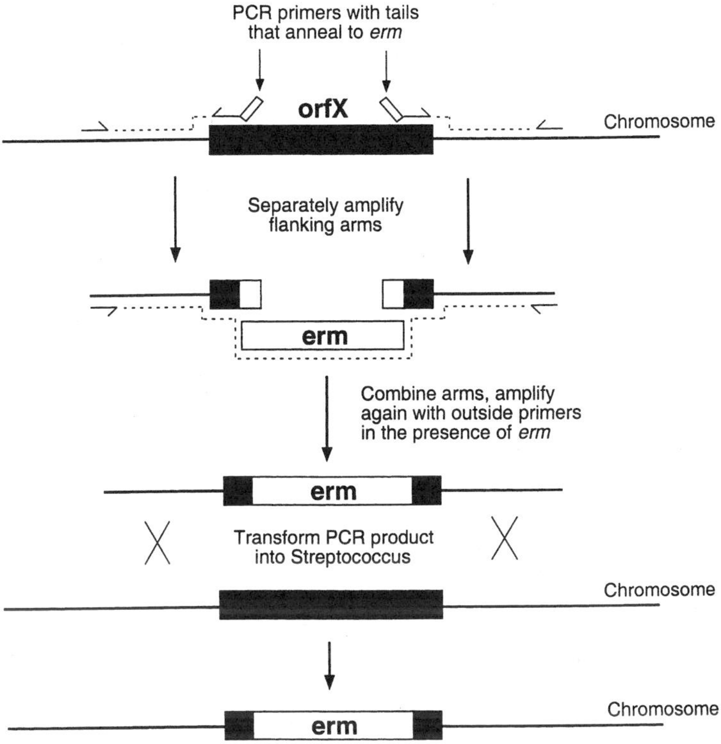

Figure 4 High-throughput gene disruption in *S. pneumoniae*. The target for gene disruption is *orfX* (*solid rectangle*). In the first step, flanking arms of chromosomal DNA on either side of *orfX* are amplified by PCR. Each PCR reaction uses a primer that anneals within *orfX* and which includes a non-annealing tail homologous to sequences in an *erm* cassette that includes the complete coding sequence of *erm* but which lacks its promoter and transcription terminator. The *erm* gene encodes resistance to the antibiotic erythromycin. In the second step, the flanking arm PCR products are combined with the *erm* cassette and amplified again using the two outside primers. The resulting PCR product is used to transform competent *S. pneumoniae* bacteria, selecting for erythromycin resistance. If *orfX* is essential for growth, no transformants will be obtained. If *orfX* is nonessential, many thousands of transformants will be produced from <50 μL of cell suspension.

3 VIRULENCE FACTORS AS POTENTIAL DRUG TARGETS

The products of genes that are essential for bacterial growth or viability under all conditions may not be the only targets of potential utility for the discovery of therapeutically useful antibacterial agents. To establish a productive infection, pathogenic bacteria require highly specialized functions to adhere to certain host cells, to evade humoral or cell-based immune responses, or in some cases to enter and proliferate within certain cells of the infected host. Properties of pathogens that influence the efficiency with which they carry out infections are referred to as virulence factors. It is reasonable to assume that many, or perhaps most, virulence factors are irrelevant or of marginal importance to the viability or growth of bacteria under laboratory conditions. Thus, virulence factors would not be discovered through approaches outlined in the previous sections. On the other hand, if virulence factors essential for infectivity of pathogenic bacteria could be identified, they might serve as effective targets for anti-infective agents.

3.1 In Vitro Expression Technology

Ground-breaking work in the identification of virulence factors that might include genes whose products are essential for certain stages of the infection process was carried out in the laboratory of J. Mekalanos and reported in a series of papers in the mid-1990s, starting with the influential publication of Mahan et al. in 1993 (24). The approach outlined in this work is referred to as IVET (in vivo expression technology). The fundamental rationale of IVET is that genes whose products are essential for infectivity are likely to be expressed much more strongly during infection than on ordinary laboratory media. Thus, if it were possible to select a set of genes strongly expressed during infection and then to screen them to identify ones that were weakly expressed in vitro, this set should be strongly enriched for genes of interest.

The IVET selection/screen is carried out using vectors such as pIVET1 (Fig. 5), which was designed for identifying virulence genes of *Salmonella typhimurium*. Central to the IVET strategy is the fact that the *purA* gene product is essential for infectivity of *S. typhimurium* in the mouse model of infection. Thus, a *purA* mutant of *S. typhimurium* that contains pIVET1 will not productively infect a mouse unless a promoter-containing DNA fragment is inserted into the cloning site of pIVET1 appropriately oriented to drive expression of its *purA* gene. Moreover, such a promoter would clearly need to be active in the in vivo environment of the infected mouse. The purpose of the *lacZY* genes is to provide a chromogenic indicator of expression in vitro: On a particular kind of agar

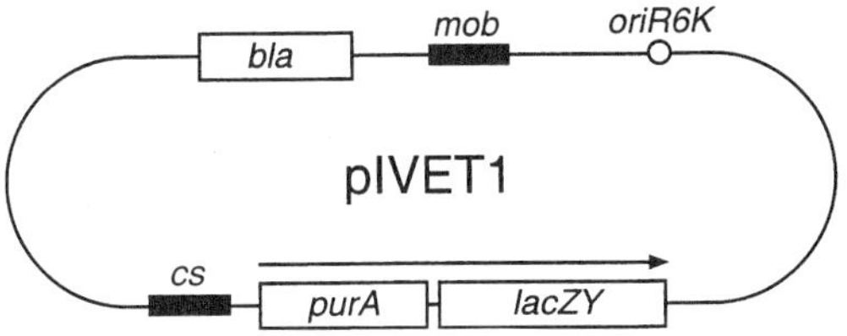

Figure 5 Relevant features of pIVET1, a vector used for identification of promoters whose expression is induced during infection of an animal model (24): *bla*, a drug-resistance marker selectable in either *E. coli* or *S. typhimurium*; *mob*, a mobilization signal that makes the plasmid transferable by mating from one strain of *E. coli* or *S. typhimurium* to another mediated by the RP4 conjugation system (29); *oriR6K*, a suicide plasmid replication origin dependent upon the Pi protein; *purA* a promoterless gene whose product is required for purine biosynthesis; *lacZY*, a promoterless operon fragment encoding β-galactosidase and lactose permease; *cs*, a cloning site immediately upstream from the promoterless *purA*, *lacZY* gene cluster suitable for insertion of small DNA fragments.

medium, bacteria expressing *lacZ* and *lacY* will form red colonies and bacteria not expressing these genes or expressing them poorly will form white colonies. The overall IVET selection strategy involves first creating a clone library of DNA fragments in pIVET1 in *E. coli* and then transferring the library into a *purA* mutant strain of *S. typhimurium* by conjugation. The transconjugant population is then used en masse to infect a mouse strain. Bacteria harvested from the mouse spleen after 3 days are plated on the indicator medium. A high percentage of the colonies yielded from such a regimen are red on the indicator plates, indicating the presence in pIVET1 of cloned fragments that are presumably active both in vivo and in vitro; the rare white colonies reflect the presence of a fragmentborne promoter that is relatively inactive in vitro but that is strongly induced in vivo. Targeted disruptions of genes associated with the latter kind of promoter reveal that approximately 10% of them contribute to virulence, although no novel genes absolutely essential for infectivity have yet to come forward from the original or from subsequent IVET-like studies.

An elegant variation on the IVET theme has been developed more recently by Valdivia and Falkow (25), taking advantage of the green fluorescent protein (GFP) of *Aequorea victoria*. These investigators created a promoter-trap vector in which promoters within cloned DNA fragments controlled expression of *gfp*. Thus, bacteria containing cloned fragments with promoters activated after engulfment by macrophages could be recognized by the fluorescence of the GFP protein. More importantly, a fluorescence-activated cell sorter (FACS) could automatically identify and isolate macrophages containing GFP-expressing bacteria. After release from the macrophage by lysis and passage on laboratory media in vitro, the bacteria themselves could be sorted by FACS according to the strength of the fluorescence signal they produced. In this way, differential fluorescence-induction (DFI) could be used to identify bacteria containing cloned fragments carrying promoters specifically activated after engulfment by host macrophages. As with the IVET approach, some genes identified by DFI could be demonstrated to contribute to virulence by testing for the ability of gene-disruption mutants to compete in mixed infections with wild-type bacteria. However, as with IVET, no virulence factors identified thus far by DFI have proved to be essential for infectivity. Thus, although IVET and DFI have clear value for identifying and understanding the biological roles of genes that contribute to bacterial virulence, their value as tools to identify therapeutic targets remains speculative.

3.2 Signature-Tag Mutagenesis

One limitation of both IVET and DFI is that they rely upon differential expression of genes in vivo and in vitro as the criterion to qualify for further functional characterization. Many virulence factors, indeed genes whose products are essential for viability or growth in vivo, but which are not essential in a laboratory environment, would not necessarily exhibit differential expression. Thus, without disputing the value of IVET or DFI as tools for basic research, it should be recognized that they may not enrich for targets of interest from the standpoint of developing antibacterial therapeutics; in fact, they may exclude the targets of greatest interest. The most direct way to identify genes whose products are essential for infectivity would of course be to test individual mutants or strains created by directed gene disruption for their ability to propagate in an animal infection model. Although it would be prohibitively expensive and time consuming to carry out such infectivity tests serially, the signature-tag mutagenesis (STM) strategy devised by Holden and colleagues (26) provides a way to test hundreds of mutant strains simultaneously.

The central concept in STM is to create disruption mutations using cassettes that contain short, unique sequences that can be distinguished from one another by hybridization. In the original description of the approach, mutations were constructed by transposon-mediated mutagenesis with mini-Tn5 elements containing 40-bp random-sequence tags flanked by invariable arms that provided annealing sites for PCR primers (Fig. 6). Because the variable-sequence portions of the tags were chosen such that they would exhibit no significant cross-hybridization under

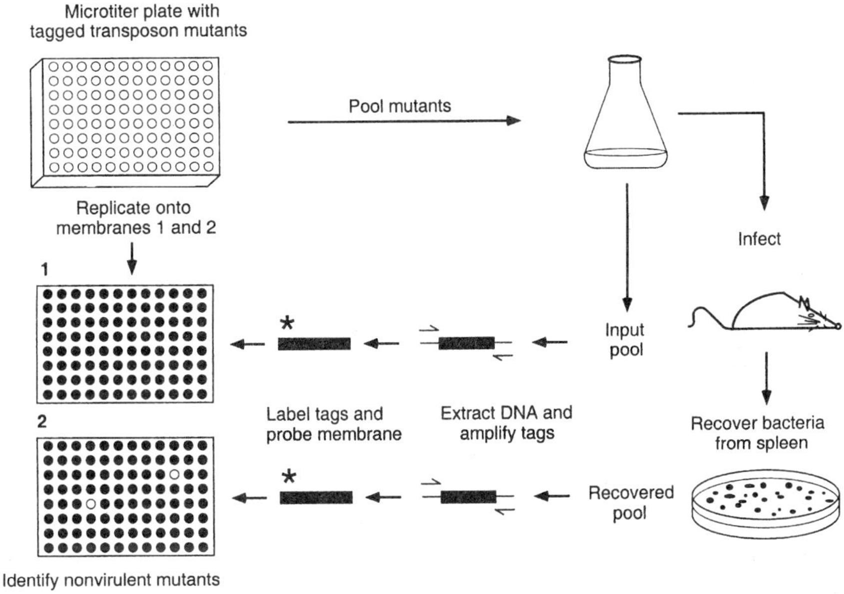

Figure 6 Negative selection to identify genes essential for propagation in an infected animal. Individual mutants generated with signature-tagged transposons are first grown separately in the wells of a 96-well microtiter plate. Replicate colony blots are prepared from the mutant array and mutants are pooled for infection of a mouse. A sample of the infection pool (input pool) is also used to prepare amplified and radiolabeled signature tags, anticipating that input pool tags will hybridize to all positions on the replicate colony blots. After the pooled mutants are passaged through the mouse and harvested from the spleen (recovered pool), labeled tags are again prepared and hybridized back to the replicate colony blots. The labeled tags prepared from the output pool will fail to hybridize only to positions corresponding to mutants unable to propagate in the mouse. (Adapted from Ref. 26.)

moderately stringent conditions, mixed probes prepared from pooled mutants would hybridize only to the tag of origin. Thus, if the tag regions from 100 distinctly tagged mutants were separately amplified by PCR and arrayed on a nylon membrane and a hybridization probe were prepared from a mixture of all 100 mutants, the probe would hybridize to all 100 positions in the array. But if one mutant were omitted from the mixture used to prepare the probe, the mixed probe would hybridize to all positions in the array except for the one corresponding to the omitted mutant. The ability to use hybridization in this manner to detect the absence of specific members of the mixed population enabled Holden and colleagues to pool mutants in 96-fold complexity (the number of wells in a standard microtiter plate), to passage the mixed population through a mouse, and then to detect mutants that failed to propagate in the animal (Fig. 6).

3.3 Are Virulence Factors Good Drug Targets?

Although approaches like STM would appear to be capable of identifying genes whose products are essential for infectivity but not for viability or growth in vitro, the question must be asked: What product profile could be anticipated from a drug-development campaign based on such a target? Would the drug be bactericidal? Would the drug have a broad spectrum of activity? Would the target organism be able to evade treatment by mutation to resistance?

Unfortunately, these important questions are impossible to answer in the abstract. However, given the high cost of a typical drug-development campaign, currently estimated to be in excess of $500 million, some consideration of the risk-reward equation might be appropriate even if it can only be speculative. *Cidality*: if a gene product is not essential for viability under all circumstances, it would seem unlikely that inhibiting its function would lead to rapid cidality in vivo. Although one could argue that certain pathogens that propagate in very specialized in vivo environments might not remain viable in the absence of a gene product that is dispensable in vitro, this would probably not generally be the case. *Spectrum*: a gene product that is specifically essential in vivo would seem likely to play a specialized role associated with the way a particular pathogen must colonize host tissues or evade host defenses, and therefore would be likely to be pathogen-specific in its essentiality. What is wrong with a pathogen-specific therapeutic? If we accept the premise that a pathogen-specific drug is no easier (no less expensive) to develop than a broad-spectrum drug, then the much larger market expected for the broad-spectrum drug would make it the clear favorite. *Mutation to resistance*: there would seem to be no clear basis for predicting whether infectivity targets would have a greater or lesser chance than generally essential gene products to evade treatment by mutation to resistance. However, it is clear that detecting the rate of mutation to resistance is far easier in the latter case and somewhat problematic in the former.

Perhaps the most problematic aspect of infectivity genes or virulence factors as targets for drug development relates to practical requirements of lead-optimization chemistry. Drugs are not discovered by target-based screening, only leads are discovered. Antibacterial lead compounds become clinical candidates through a laborious and expensive process in which many hundreds or thousands of analogue compounds are prepared and then tested for possible improved activity against the targets protein in vitro. These results are then correlated with their ability to affect bacterial growth through a target-specific mechanism; the latter being determined through the use of a specific in vivo secondary assay. Those with first-hand experience can attest to the challenge of pursuing such a lead-optimization effort to a successful conclusion even when the test for antibacterial potency is a simple matter of determining a minimal inhibitory concentration (MIC) in a standard nutrient broth, and the secondary assay is a simple biochemical manipulation of a cell extract. It would be difficult to justify investing the resources represented by a team of 10–15 medicinal chemists for a period of 12–18 months in a lead-development program where the biological activity of the analogues they produced could only be assessed in a whole-animal model of infection and where a secondary assay to confirm target-specificity of the biological activity may not be available at all.

4 BACK TO THE FUTURE

As noted earlier, despite the fact that target-based screening is the discovery paradigm that dominates the search for new antibacterial therapeutics, no drug candidate compound derived from target-based screening has yet to advance to clinical trials. Is there a flaw in this paradigm? Are there viable alternatives? Are there ways in which genomics information and technologies might facilitate discovery paradigms other than target-based screening?

In considering these questions, it may be helpful to start by reviewing the advantages and disadvantages of the older discovery paradigm of cell-based screens for compounds that inhibit bacterial growth. One advantage is in the simplicity of the screen itself. Screening compounds are added to a suspension of ordinary bacteria and the readout is simply their ability to achieve a threshold optical density after a period of incubation: no washing steps, no exotic reagents, no specialized instrumentation. Another advantage is that compounds detected as actives in the primary screen are guaranteed by the nature of the screen itself to have the ability to pass through the cell membrane (with the rare but important exception of compounds that act against a target on the exterior of the cell). Lead candidates detected in target-based screens often have impressive potency against the target protein but show no antibacterial activity because of poor membrane penetration. The tools of modern medicinal chemistry grow more powerful each year, but initiating a chemistry program on an antibacterial lead with no in vitro activity is still appropriately regarded as an act of desperation. The most important advantage of the cell-based screen for antibacterial activity, however, is in the predictive power of the paradigm with respect to ultimate success of the lead-discovery program. A lead compound with target-specific antibacterial activity in vitro can be tested rapidly and inexpensively for activity in an animal model of infection; and a lead compound with antibacterial activity in vitro and in vivo is an excellent prospect for investment of lead-optimization chemistry resources.

What are the disadvantages of cell-based screens for antibacterial activity? Many in the industry would say that they have "been there and done that." In fact, libraries of natural product extracts from actinomycete, fungal, plant, and marine sources have been extensively screened for antibacterial activity, and it is undoubtedly true that the low-hanging fruit is long gone. However, this may not be true of synthetic compound libraries. Proprietary collections of synthetic compounds have expanded enormously in the past decade, and many of them have not been screened exhaustively for antibacterial activity. A more thoughtful objection to cell-based screens for antibacterial activity is the argument that it is relatively easy to find a compound with antibacterial activity but more difficult to find a compound for which the antibacterial activity reflects a target-specific MOA rather than nonspecific toxicity, and it may not be easy to make the distinction. Moreover, chemistry optimization of a lead compound requires not only that its MOA is target-specific but also that the target itself has been identified. Finally, it might be argued that a primary screen for antibacterial activity is not sensitive enough to detect all hits of interest and utility from the standpoint of establishing a preliminary structure-activity relationship (SAR) among a related family of hit compounds.

Do genomic technologies provide any remedy for these important objections? A technology that may is transcriptional profiling with the use of DNA microarrays. Transcriptional profiling provides a means for separately quantifying changes in expression of every orf in the bacterial genome. Consider, for example, the task of establishing whether a compound with antibacterial activity is acting through a target-specific mechanism. Gross specificity of MOA is currently investigated simply by treating bacteria briefly with the compound in question in the presence of radiolabeled precursors for the major pathways of macromolecule biosynthesis (e.g., nucleosides or deoxynucleosides for RNA or DNA synthesis, amino acids for protein synthesis, acetate for lipids). Obviously this is a crude readout. A possible alternative might be to determine how compounds of interest affect the global pattern of gene expression in the cell; not necessarily with the hope of identifying the particular target or even the target pathway affected by the compound but rather with the expectation of inferring from the broad fingerprint of transcriptional changes observed whether the compound is likely to be acting by a target-specific mechanism. In the more distant future, we might imagine that the specific signature fingerprint associated with inhibition of a specific target protein will be known in advance. Such a signature might be obtained by placing expression of a specific essential gene under the control of a regulatable promoter. Downregulation of expression would then afford the opportunity to record the transcriptional profile associated with inhibition of that particular gene product. A comprehensive database of such gene product-depletion fingerprints might represent an extremely useful tool for establishing MOA.

What about the objection that cell-based screens for antibacterial activity can only detect compounds at or near their MIC? With the knowledge that a particular compound can inhibit a particular cellular target or target pathway, one might use DNA microarray technology to ask the questions: Does inhibition of the target lead to the induction or repression of a specific set of genes? And, more importantly, does that response occur at a concentration significantly below the MIC? In practice, the answer to that question is frequently yes (C. Murphy and P. Youngman, unpublished). When the answer is yes, then the opportunity exists to exploit the promoters induced by the compound to construct reporter gene-based screens for other inhibitors of the same pathway, screens that may be much more sensitive and specific than a screen designed simply to detect inhibition of bacterial growth.

5 CONCLUSIONS

Following a period of complacency in the 1970s when it appeared likely that derivatives of known antibiotic classes might be sufficient to remain one step ahead of nature's ability to evolve new mechanisms of resistance, the research, pharmaceutical, and clinical communities alike are now alarmed by the possibility that nature is winning the race. Everyone now agrees on the need for novel drugs with novel MOAs. Does this translate strictly into a need for novel targets, and does it mean that the most useful way in which genomics information and technologies can service the need for new drugs is to provide as many new targets as possible? Not necessarily.

That the novelty of the target itself is not necessarily a recipe for success is illustrated by the example of peptide deformylase (PDF), a novel target developed by researchers at Versicor Inc. (27). PDF is an attractive target for the development of antibacterial therapeutics, because it carries out a function (removal of the formyl group from the N-terminal methionine residue of newly synthesized proteins) that is unique to bacteria. Target-based screens against the PDF protein successfully yielded novel and potent inhibitors that might otherwise be regarded as promising candidates for clinical studies. Indeed, if they were to advance to the clinics, they would represent the first successful graduates from the target-based school of antibacterial drug discovery. Unfortunately, the frequency of acquiring complete resistance to PDF inhibitors through spontaneous mutation is approximately 10^{-6} in *Staphylococcus aureus* (28). Such a frequency is generally considered to be far too high for a clinical candidate. Although mutation to resistance in other bacteria may occur at a somewhat lower frequency, the future of PDF inhibitors does not appear to be promising. In contrast, as referred to previously, the oxazolidinones represent a novel structural class of inhibitor, with a novel MOA, that act against a target (the ribosome) that is far from novel. It might even be argued that the existence of known inhibitors, particularly of natural product origin, that affect the ribosome tends to validate it as a particularly vulnerable structure for small-molecule inhibitors. Thus, in the absence of evidence that the possibilities have been exhausted for discovery of new agents that interfere with translation by a novel mechanism, lack of novelty in this case might be viewed as a positive attribute rather than as a negative one.

I would not want to leave the impression that I believe that target-based screening is a bad way to discover novel antibacterial leads or that genomics information and technologies will not significantly enhance the effectiveness of this discovery paradigm. What I am trying to emphasize in this chapter is that to view the contribution of genomics technologies strictly in that context would miss the larger and more important point: knowledge of the DNA sequence of bacterial genomes (already in hand), knowledge of which genes are broadly essential for bacterial viability (already known), a complete understanding of gene expression networks and functional roles of bacterial genes (we are only scratching the surface), dramatic advancements in screening technologies (we are at a very early stage), dramatic improvements in the diversity and quality of synthetic screening libraries (we are also at a very early stage), and full integration of structural biology and computation chemistry into the target-selection and lead-optimization processes (in its infancy) will together revolutionize how we think about the discovery of the next generation of novel antibiotics in ways that we are only just beginning to comprehend. These exciting new technologies should not be viewed simply as ways to facilitate existing discovery paradigms but rather as opportunities to change the paradigms themselves.

REFERENCES

1. Slee AM, Wuanola MA, Mcripley RJ, Zajac I, Zawada MJ, Bartholomew PT, Gregory WA, Forbes M. Oxazolidinones, a new class of synthetic antibacterial agents: in vitro and in vivo activities of *Dup105* and *Dup721*. Antimicrob Agents Chemother 1987; 31:1791–1797.
2. Eustice DC, Feldman PA, Zajac I, Slee AM. Mechanism of action of Dup721: inhibition of an early event during initiation of protein synthesis. Antimicrob Agents Chemother 1988; 32:1218–1222.
3. Swaney SM, Shinabarger DL, Schaadt RD, Bock JH, Slightom JL, Zurenko GE. Oxazolidinone resistance is associated with a mutation in the peptidyl transferase region of 23S rRNA. 38th Interscience Conference on Antimicrobial Agents and Chemotherapy, 1998; Abstr. No. C-104.
4. Brickner SJ, Hutchinson DK, Barbachyn MR, Garmon SA, Grega KC, Hendges SK, Manninen PR, Toops DS, Ulanowicz DA, Kilburn JO, Glickman S, Zurenko GE, Ford CW. Synthesis of U-100592 and U-100766, two new oxazolidinone antibacterial agents in clinical trials for treatment of multiple-resistant gram-positive infections. 35th Interscience Conference on Antimicrobial Agents and Chemotherapy, 1995; Abstr. No. F221.
5. Fleischmann RD, Adams MD, White O, Clayton RA, Kirkness EF, Kerlavage AR, Bult CJ, Tomb JF, Dougherty BA, Merrick JM, McKenney K, Sutton G, FitzHugh W, Fields C, Gocayne JD, Scott J, Shirley R, Liu L, Glodek A, Kelley JM. Whole-genome random sequencing and assembly of *Haemophilus influenzae* Rd. Science 1995; 269:496–512.
6. Consortium. Genome sequence of the nematode *C. elegans*: a platform for investigating biology. Science 1998; 282:2012–2018.
7. DeRisi JL, Iyer VR, Brown PO. Exploring the metabolic and genetic control of gene expression on a genomic scale. Science 1997; 278:680–686.
8. Humphery-Smith I, Blackstock W. Proteome analysis: genomics via the output rather than the input code. J Protein Chem 1997; 16:537–544.
9. Smith TF. Functional genomics—bioinformatics is ready for the challenge. Trends Genet 1998; 14:291–293.
10. Gaasterland T. Structural genomics: bioinformatics in the driver's seat. Nat Biotechnol 1998; 16:625–627.
11. Arigoni F, Talabot F, Peitsch M, Edgerton MD, Meldrum E, Allet E, Fish R, Jamotte T, Curchod M-L, Loferer H. A genome-based approach for the identification of essential bacterial genes. Nat Biotechnol 1998; 16:851–856.
12. Woese CR, Maniloff J, Zablen LB. Phylogenetic analysis of the mycoplasmas. Proc Natl Acad Sci USA 1980; 77:494–498.
13. Fraser CM, Gocayne JD, White O, Adams MD, Clayton RA, Fleischmann RD. The minimal gene complement of *Mycoplasma genitalium*. Science 1995; 270: 397–403.
14. Blattner FR, Plunkett G III, Bloch CA, Perna NT, Burland V, Riley M, Collado-Vides J, Glasner JD, Rode CK, Mayhew GF, Gregor J, Davis NW, Kirkpatrick HA, Goeden MA, Rose DJ, Mau B, Shao Y. The complete genome sequence of *Escherichia coli* K-12. Science 1997; 277:1453–1462.

15. Link AJ, Phillips D, Church GM. Methods for generating precise deletions and insertions in the genome of wild-type *Escherichia coli*: application to open reading frame characterization. J Bacteriol 1997; 179:6228–6237.
16. Schmid MB, Kapur N, Isaacson DR, Lindroos P, Sharpe C. Genetic analysis of temperature-sensitive lethal mutants of *Salmonella typhimurium*. Genetics 1989; 123:625–633.
17. Schmid M. Novel approaches to the discovery of antimicrobial agents. Curr Opin Chem Biol 1998; 2:529–534.
18. Smith V, Botstein D, Brown P. Genetic fingerprinting: a genomic strategy for determining a gene's function given its sequence. Proc Natl Acad Sci USA 1995; 92:6479–6483.
19. Akerley BJ, Rubin EJ, Camilli A, Lampe DJ, Robertson HM, Mekalanos JJ. Systematic identification of essential genes by in vitro mariner mutagenesis. Proc Natl Acad Sci USA 1998; 95:8927–8932.
20. Lampe DJ, Churchill ME, Robertson HM. A purified mariner transposase is sufficient to mediate transposition in vitro. EMBO J 1997; 15:5470–5479.
21. Reich KA, Chovan L, Hessler P. Genome scanning in *Haemophilus influenzae* for identification of essential genes. J Bacteriol 1999; 181:4961–4968.
22. Vagner V, Dervyn E, Ehrlich SD. A vector for systematic gene disruption in *B. subtilis*. Microbiology 1998; 144:3097–3104.
23. Wach A. PCR-synthesis of marker cassettes with long flanking homology regions for gene disruptions in *S. cerevisiae*. Yeast 1996; 12:259–265.
24. Mahan MJ, Slauch JM, Mekalanos JJ. Selection of bacterial virulence genes that are specifically induced in host tissues. Science 1993; 259:686–688.
25. Valdivia RH, Falkow S. Fluorescence-based isolation of bacterial genes expressed within host cells. Science 1977; 277:2007–2011.
26. Hensel M, Shea JE, Gleeson C, Jones MD, Dalton E, Holden DW. Simultaneous identification of bacterial genes by negative selection. Science 1995; 269: 400–403.
27. Margolis P, Young D, Yuan Z, Wang W, Trias J. Peptide deformylase as a target for discovery of novel antibacterial agents. 39th Interscience Conference on Antimicrobial Agents and Chemotherapy, 1999; Abstr. No. 1793.
28. Yuan Z. Bacterial deforylase and its suitability as a novel antibacterial target. 5th International Anti-Bacterial Drug Discovery and Development Summit, 2000.
29. Simon R, Priefer U, Pühler A. A broad host range mobilization system for in vivo engineering: transposon mutagenesis in gram-negative bacteria. Biotechnology 1988; 1:784–791.

Index